Student's Solutions Manual

James Lapp
Rogue Community College

Mathematics All Around
Third Edition

Thomas L. Pirnot
Kutztown University of Pennsylvania

Boston   San Francisco   New York
London   Toronto   Sydney   Tokyo   Singapore   Madrid
Mexico City   Munich   Paris   Cape Town   Hong Kong   Montreal

Reproduced by Pearson Addison-Wesley from electronic files supplied by the author.

Copyright © 2007 Pearson Education, Inc.
Publishing as Pearson Addison-Wesley, 75 Arlington Street, Boston, MA 02116.

All rights reserved. No part of this publication may be reproduced, stored in a retrieval system, or transmitted, in any form or by any means, electronic, mechanical, photocopying, recording, or otherwise, without the prior written permission of the publisher. Printed in the United States of America.

ISBN   0-321-36859-2

   5 6 BRR 08 07

CONTENTS

Chapter 1 Set Theory: Using Mathematics to Classify Objects..................1
1.1 Problem Solving..1
1.2 Estimation...5
1.3 The Language Of Sets...6
1.4 Comparing Sets...7
1.5 Set Operations...10
1.6 Survey Problems...12
Chapter Review Exercises..14
Chapter Test..17
Of Further Interest: Infinite Sets..18

Chapter 2 Logic: The Study of What's True or False or Somewhere In Between..23
2.1 Inductive and Deductive Reasoning...23
2.2 Statements, Connectives, and Quantifiers Section..........................26
2.3 Truth Tables..28
2.4 The Conditional and Biconditional..32
2.5 Verifying Arguments..35
2.6 Using Euler Diagrams to Verify Syllogisms...................................40
Chapter Review Exercises..41
Chapter Test..45
Of Further Interest: Fuzzy Logic..48

Chapter 3 Graph Theory: The Mathematics of Relationships.....................51
3.1 Graphs, Puzzles And Map Coloring...51
3.2 The Traveling Salesman Problem..54
3.3 Directed Graphs..58
Chapter Review Exercises..60
Chapter Test..62
Of Further Interest: Scheduling Projects Using Pert.................................64

Chapter 4 Numeration Systems: Does It Matter How We Name Numbers?...........75
4.1 The Evolution of Numeration Systems..75
4.2 Place Value Systems...78
4.3 Calculating in Other Bases...84
Chapter Review Exercises..87
Chapter Test..89
Of Further Interest: Modular Systems...91

Chapter 5 Number Theory and the Real Number System: Understanding the
Numbers All Around Us.. 101
5.1   Number Theory.. 101
5.2   The Integers... 103
5.3   The Rational Numbers... 105
5.4   The Real Number System.. 111
5.5   Exponents and Scientific Notation... 114
Chapter Review Exercises... 117
Chapter Test.. 119
Of Further Interest: Sequences... 121

Chapter 6 Algebraic Models: How Do We Approximate Reality?..................... 129
6.1   Linear Equations.. 129
6.2   Modeling with Linear Equations.. 134
6.3   Modeling with Quadratic Equations.. 137
6.4   Exponential Equations and Growth... 144
6.5   Proportions and Variation.. 148
6.6   Functions... 152
Chapter Review Exercises... 154
Chapter Test.. 158
Of Further Interest: Dynamical Systems... 163

Chapter 7 Modeling with Systems of Linear Equations and Inequalities:
What's the Best Way to Do It?.. 167
7.1   Systems of Linear Equations.. 167
7.2   Systems of Linear Inequalities... 172
Chapter Review Exercises... 182
Chapter Test.. 184
Of Further Interest: Linear Programming... 186

Chapter 8 Geometry: Ancient and Modern Mathematics Embrace..................... 209
8.1   Lines, Angles, and Circles... 209
8.2   Polygons.. 211
8.3   Perimeter and Area.. 214
8.4   Volume and Surface Area.. 221
8.5   The Metric System and Dimensional Analysis............................ 225
8.6   Geometric Symmetry and Tessellations...................................... 228
Chapter Review Exercises... 232
Chapter Test.. 234
Of Further Interest: Fractals.. 237

Chapter 9 Legislative Apportionment: How Do We Measure Fairness?............... 241
9.1 Understanding Apportionment..241
9.2 The Huntington–Hill Apportionment Principle.....................................244
9.3 Applications of the Apportionment Principle......................................248
9.4 Other Paradoxes and Apportionment Methods...................................251
Chapter Review Exercises...264
Chapter Test...267
Of Further Interest: Fair Division...271

Chapter 10 Voting: Using Mathematics to Make Choices....................................275
10.1 Voting Methods...275
10.2 Defects in Voting Methods..278
10.3 Weighted Voting Systems...281
Chapter Review Exercises...289
Chapter Test...293
Of Further Interest: The Shapley–Shubik Index...297

Chapter 11 Consumer Mathematics: The Mathematics of Everyday Life.............309
11.1 Percent..309
11.2 Interest..312
11.3 Consumer Loans...315
11.4 Annuities..319
11.5 Amortization...323
Chapter Review Exercises...329
Chapter Test...333
Of Further Interest: The Annual Percentage Rate..336

Chapter 12 Counting: Just How Many Are There?...341
12.1 Introduction to Counting Methods..341
12.2 The Fundamental Counting Principle..348
12.3 Permutations and Combinations..349
Chapter Review Exercises...353
Chapter Test...354
Of Further Interest: Counting and Gambling...355

Chapter 13 Probability: What Are the Chances?...359
13.1 The Basics of Probability Theory..359
13.2 Complements and Unions of Events...362
13.3 Conditional Probability...365
13.4 Expected Value...371
Chapter Review Exercises...373
Chapter Test...375
Of Further Interest: Binomial Experiments...377

Chapter 14 Descriptive Statistics: What a Data Set Tells Us...................................383
14.1 Organizing and Visualizing Data..383
14.2 Measures of Central Tendency..386
14.3 Measures of Dispersion...390
14.4 The Normal Distribution...398
Chapter Review Exercises..401
Chapter Test...403
Of Further Interest: Linear Correlation..405

Chapter 1
SET THEORY: Using Mathematics To Classify Objects

Section 1.1 Problem Solving

1. Drawing pictures, choosing good names for unknowns, being systematic, looking for patterns, trying simpler versions of the problem, converting a new problem to an older one, and remembering that guessing is OK.

3. The Always Principle, the Counterexample Principle, the Order Principle, the Splitting Hairs Principle, the Analogies Principle, and the Three-way Principle. Examples may vary.

5. Good names make it easier to remember what the unknowns represent.

7. Examining simpler examples may help you see a pattern that can be applied to the original problem.

9. Drawings may vary.

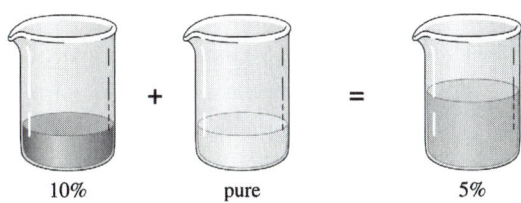

11. Drawings may vary.

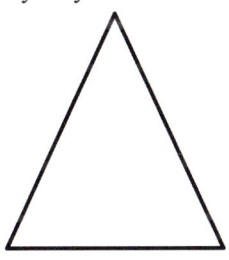

13. Answers may vary.
    Let w be the width of the bridge and l or $8w$ be the length.

15. Answers may vary.
    Let M be Monica, C be Chandler, R be Rachel, J be Joey, and S be Ross.

    Note: Since Rachel and Ross both start with "R", you cannot use R for both.

17. Answers may vary.
    Let c be the dollar amount invested in collectibles and p be the dollar amount invested in precious metals.

19. Answers (order) may vary.

Penny	Nickel
Heads	Heads
Heads	Tails
Tails	Heads
Tails	Tails

    Combinations would be *HH*, *HT*, *TH*, and *TT*.

21. Answers (order) may vary.
    There are six distinct segments:

    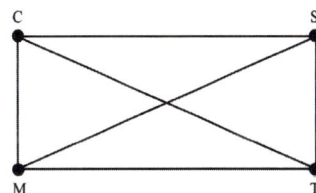

    Pairs would be *CS*, *CM*, *CT*, *SM*, *ST*, and *MT*.

23. r_3 is the set of people who are good singers and appeared on "American Idol". r_4 is the set of people who have appeared on "American Idol" and are not good singers.

25. $2 \times 2 \times 2 \times 2 = 16$

27. 7, 14, 21, 28, **35, 42, 49, 56, 63,...**

29. Answers may vary.
    *ab, ac, ad, ae, bc, bd, be, **bf, cd, ce, cf, cg,**...*

31. Answers may vary.
Three people are being honored for their work in reducing pollution. In how many ways can we line up these people for a picture?

Let the people be labeled A, B, and C.

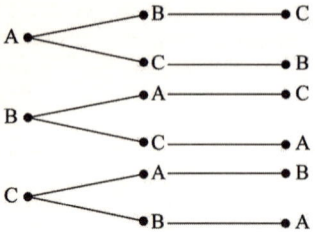

The possible orders are ABC, ACB, BAC, BCA, CAB, and CBA.
6 different ways

Four people are being honored for their work in reducing pollution. In how many ways can we line up these people for a picture?

Let the people be labeled A, B, C, and D.

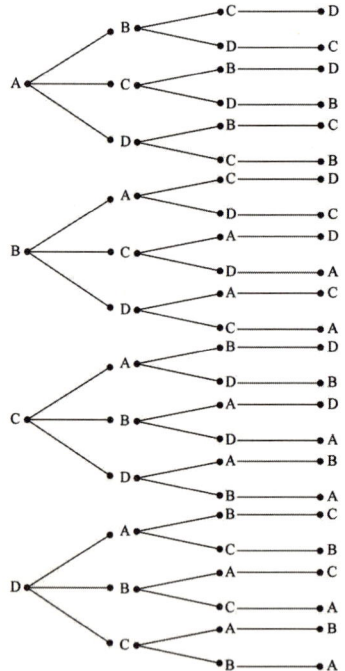

The possible orders are ABCD, ABDC, ACBD, ACDB, ADBC, ADCB, BACD, BADC, BCAD, BCDA, BDAC, BDCA, CABD, CADB, CBAD, CBDA, CDAB, CDBA, DABC, DACB, DBAC, DBCA, DCAB, and DCBA.
24 different ways

33. Answers may vary.
Using the first three letters of the alphabet, how many two-letter codes can we form if we are allowed to use the same letter twice?

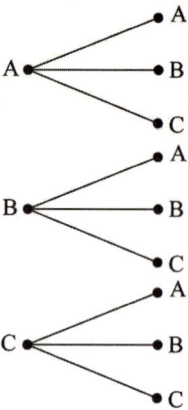

The possible codes are AA, AB, AC, BA, BB, BC, CA, CB, and CC.
9 different codes

Using the first five letters of the alphabet, how many two-letter codes can we form if we are allowed to use the same letter twice?

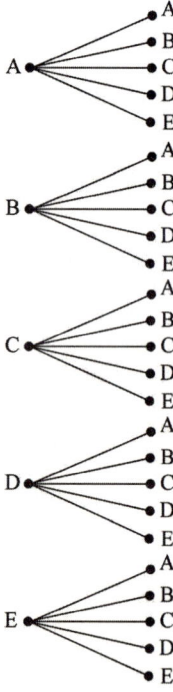

The possible codes are AA, AB, AC, AD, AE, BA, BB, BC, BD, BE, CA, CB, CC, CD, CE, DA, DB, DC, DD, DE, EA, EB, EC, ED, and EE.
25 different codes

35. Answers may vary.
An electric blue sports car comes with two options: air conditioning and a CD player. You may buy the car with any combination of the options (including none). How many different choices do you have?

Let A be air conditioning and C be CD player. If a feature is not included, it is indicated by a "0". If it is included, it is indicated by a "1".

A	C
0	0
0	1
1	0
1	1

4 different choices

An electric blue sports car comes with three options: air conditioning, CD player, and air bags. You may buy the car with any combination of the options (including none). How many different choices do you have?

Let A be air conditioning, C be CD player, and B be air bags. If a feature is not included, it is indicated by a "0". If it is included, it is indicated by a "1".

A	C	B
0	0	0
0	0	1
0	1	0
0	1	1
1	0	0
1	0	1
1	1	0
1	1	1

8 different choices

37. False, counterexamples may vary.
September has only 30 days.

39. False, counterexamples may vary.
$$\frac{1}{2}+\frac{3}{4}=\frac{2}{4}+\frac{3}{4}=\frac{5}{4}$$
$$\frac{1+3}{2+4}=\frac{4}{6}=\frac{2}{3}, \frac{5}{4}\neq\frac{2}{3}$$

41. False, A is the grandfather of C.

43. False, counterexamples may vary.
$-9 < -5$ but $(-9)^2 > (-5)^2$ since $81 > 25$.

45. False, counterexamples may vary.
If a square has a side of length 3 units, its area is 9 square units. If the length is doubled to 6 units, the area is 36 square units, not 18 square units.

47. Explanations may vary.
These two sequences do not give the same results. The question here is equivalent to asking if the algebraic statement, $x^2 + 5 = (x+5)^2$, is true.
If we let $x = 1$, we have a counterexample.
$$1^2 + 5 \stackrel{?}{=} (1+5)^2$$
$$1+5 \stackrel{?}{=} 6^2$$
$$6 \neq 36$$
Hence, the statement is false.

49. Explanations may vary.
These two sequences do give the same results. The question here is equivalent to asking if the algebraic statement, $\frac{x+y}{3} = \frac{x}{3} + \frac{y}{3}$, is true.
If you think of dividing by 3 equivalent to multiplying by $\frac{1}{3}$, then you can use the distributive property to prove this statement.
$$\frac{x+y}{3} = \frac{1}{3}(x+y) = \frac{1}{3}x + \frac{1}{3}y = \frac{x}{3} + \frac{y}{3}$$

51. Answers may vary.
5 is a number; {5} is a number with braces around it. Moreover, 5 is a singly listed element, while {5} is a set that contains the single element 5.

53. Answers may vary.
One is uppercase and the other is lowercase. Moreover, U usually denotes the universal set, and the lower case letters are usually elements that appear in some set.

55. Answers may vary.
{ } are different from ( ). Moreover, {1,2} is the set that contains only the elements 1 and 2. (1,2) is the interval that contains all real numbers between 1 and 2 (not including 1 and 2 themselves).

Note: Another interpretation of (1,2) is being a point in a plane. However, with {1,2} in this same problem, the interpretation should be relative to sets.

57. and 59. No solutions provided

61. Answers may vary.
Let O be the age of the older building and Y be the age of the younger building.

Guesses for O and Y	Good points	Weak Points
100, 221	Sum is 321.	The older is more than twice the younger.
110, 211	Sum is 321.	The older is less than twice the older.
107, 214	All conditions satisfied. We have the solution.	

63. Answers may vary.
Let S be the number of hours Janine worked in the sporting store and P be the number of hours Janine gave piano lessons.

Guesses for S and P	Good points	Weak Points
11 and 4	Sum is 15.	Amount earned is less than $119.25.
8 and 7	Sum is 15.	Amount earned is more than $119.25.
9 and 6	All conditions satisfied. We have the solution	

65. Answers may vary.
Let E be the number of times Emmitt carried the football, B be the number of times Barry carried the football, and R be the number of times Ricky carried the football.

Guesses for E, B and R	Good points	Weak Points
200, 171, and 251	E is 29 more than B and 51 less than R.	Sum is less than 742.
250, 221, and 301	E is 29 more than B and 51 less than R.	Sum is more than 742.
240, 211, and 291	All conditions satisfied. We have the solution.	

67. Answers may vary.
Let A be the amount invested at 8% and B be the amount invested at 6%.

Guesses for A and B	Good points	Weak Points
3000 and 5000	Sum is $8000.	Return is less than $550.
4000 and 4000	Sum is $8000.	Return is more than $550.
3500 and 4500	All conditions satisfied. We have the solution	

69. Answers may vary.
Let A be the number of administrators, S be the number of students, and F be the number of faculty members.

Guesses for A, S and F	Good points	Weak Points
10, 2, and 7	A is 5 times S and F is 5 more than S.	There are less than 26 people.
20, 4, and 9	A is 5 times S and F is 5 more than S.	There are more than 26 people.
15, 3, and 8	All conditions satisfied. We have the solution.	

There are 3 students.

71. – 75. No solutions provided

Section 1.2 Estimation

1. Answers may vary Possible answers include 3279 rounding up to 3300.

3. No solution provided

5. If the digit to the right of the place you are rounding is 5 or greater, we round up; otherwise, you round down.

7. Determining whether your estimate is too large or too small.

9. Answers may vary. Possible answers include not always having a calculator available.

11. Skill with mental manipulations will help you make quick, reliable estimates.

13. 37,300

15. 8,300

17. Answers may vary.
    $20 + 40 + 190 + 40 = 290$

19. Answers may vary.
    $35 - 15 = 20$

21. Answers may vary.
    $5 \times 16 = 80$

23. Answers may vary.
    $18 \div 3 = 6$

25. Answers may vary.
    $0.1 \times 800 = 80$

27. Answers may vary.
    $9\% \times 1,000 = 0.09 \times 1,000 = 90$

29. Answers may vary.
    $4 \times 5 \times 6 = 120$ miles
    The estimate is larger than the exact answer.
    Exact answer is 111 miles.

31. Answers may vary.
    $325 \div 50 = 6.5$ more hours, 7:30PM
    The estimate is earlier than the actual time.
    Actual answer is 7:50PM

33. Answers may vary.
    $\$80.00 \times 15\% = \$80.00 \times 0.15 = \$12.00$
    The estimate is less than the exact answer.

35. Answers may vary.
    $(3 \times \$3.00) + (4 \times \$1.50) + \$3.00 =$
    $\$9.00 + \$6.00 + \$3.00 = \$18.00$
    The estimate is larger than the exact answer.
    Exact answer is $17.20.

37. Answers may vary.
    It seems safe. Alicia probably weighs less than 200 pounds, so that leaves $2,300 - 200 = 2,100$ pounds for the 21 students. They probably each weigh less than 100 pounds.

39. Answers may vary.
    $\$40,000 \times 4\% = \$40,000 \times 0.04 = \$1,600$
    The estimate is larger than the exact answer.
    Exact answer is $1324.40.

41. Answers may vary.
    Her total expenses are about $100 per month; $100 \div 7 \approx 14$ $12 \times \$14 = \$168$
    The estimate is larger than the actual answer.
    Actual answer is $163 (you round to nearest dollar on deductions).

43. Estimated answers may vary.
    Exact answer is $80,000.

45. Estimated answers may vary.
    Exact answer is $24,500.

47. Estimated answers may vary.
    Exact answer is 36%.

49. Estimated answers may vary.
    Exact answer is DVD players; 65%.

51. $1{,}798 \times 0.41 = 737.18$
    $737.18 billion

53. $1{,}798 \times 0.09 = 161.82$
    $161.82 billion

55. $705{,}361 \times 0.302 \approx 213{,}019$
    The actual number of immigrants was 213,019.

57. $130{,}661 \div 705{,}361 \approx 0.185$
    18.5%

59. Answers may vary.
    The amount of lawn that needs to be fertilized is represented by the size of the lot, less the non-grassy areas such as the garden, driveway and house.
    $96 \cdot 169 - 96 \cdot 30 - 65 \cdot 28 - 18 \cdot 65 =$
    $16{,}224 - 2{,}880 - 1{,}820 - 1{,}170 = 10{,}354$
    If you divide the grassy area into rectangles, you get 10,354 square feet. They need slightly over two bags of fertilizer.

61. No solution provided

Section 1.3 The Language Of Sets

1. Answers may vary. Possible answers include $\{x : x \text{ is an even counting number}\}$ and $\{2, 4, 6, 8, \ldots\}$

3. Answers may vary. Possible answers include $3 \in \{1,3,5,7,8\}$ and $4 \notin \{1,3,5,7,8\}$.

5. No solution provided

7. when the set is too large to list all the elements

9. $\varnothing$ is the empty set, it contains no elements. $\{\varnothing\}$ is not empty, it contains 1 element, $\varnothing$.

11. $\{10, 11, 12, 13, 14, 15\}$

13. $\{17, 18, 19, 20, 21, 22, 23, 24, 25\}$

15. $\{4, 8, 12, 16, 20, 24, 28\}$

17. {Sunday, Monday, Tuesday, Wednesday, Thursday, Friday, Saturday}

19. $\varnothing$

21. {Ronald Reagan, George H.W. Bush, Bill Clinton}

23. Answers may vary. Possible answers include $\{x : x \text{ is a multiple of 3 between 3 and 12 inclusive}\}$.

25. Answers may vary. Possible answers include $\{y : y \text{ is a letter in the word } music\}$.

27. $\{-1, -2, -3, \ldots\}$

29. $\varnothing$

31. Answers may vary. Possible answers include $\{101, 102, 103, \ldots\}$.

33. Answers may vary. Possible answers include $\{x : x \text{ is an even natural number between 1 and 101}\}$.

35. well defined

37. not well defined

39. not well defined

41. well defined

43. $\notin$

45. $\in$

47. $\in$

49. $\in$

51. $\notin$

53. $\in$

55. 6

57. 0

59. 4

61. 2 elements; {1, 2}, {1, 2, 3}

63. 1 element; {{∅}}

65. finite

67. infinite

69. finite

71. finite

73. Answers may vary. Possible answers include 4.5.

75. Answers may vary. Possible answers include Sony.

77. Answers may vary. Possible answers include George W. Bush.

79. Answers may vary. Possible answers include Sunday.

81. $\{x : x \text{ is a humanities elective}\}$

83. {History012, History223, Geography115, Anthropology111}

85. No solution provided

87. If the barber shaves himself, then he (the barber) does not shave himself. If the barber does not shave himself, then he (the barber) does shave himself.
Conclusion: This is a paradox.

89. If $S \in S$, then $S \notin S$. If $S \notin S$, then $S \in S$.
Conclusion: This is a paradox.

Section 1.4 Comparing Sets

1. The sets contain exactly the same elements.

3. Every element added to a set doubles the number of subsets.

5. the order and repetition of the elements

7. Notations that looks similar may have very different meanings.

9. No solution provided

11. These two sets are equal. They have the same elements arranged in a different order.

13. These two sets are not equal. The second set contains (infinitely many) elements that don't appear in the first set.

15. These two sets are equal. They are both {1, 3, 5,..., 99}.

17. These two sets are equal. It is understood that the second set contains all twelve months of the year.

19. These two sets are equal. Common sense dictates that nobody born before 1800 should be living.

21. True. All the elements of the first set are understood to be elements of the second set.

23. False. The first set is the same as the second set, so the first set cannot be a proper subset.

25. False. The letter "y" is an element of the first set and not an element of the second set.

27. True. The null set is a subset of all sets.

8   Chapter 1: Set Theory

29. False. The set that contains the null set (which is not an empty set) is not a subset of the set {0}. Each of the two sets (left and right) contains one element, which are not the same.

31. The first set is equivalent to the second set because they both have the same number of elements.

33. The first set is equivalent to the second set because they both have that same number of elements, namely 4.

35. The first set is not equivalent to the second set. The first set has 0 elements while the second set has 1 element.

37. The first set is equivalent to the second set. They both have 8 elements.

39. The first set is not equivalent to the second set. The first set has 365 elements while the second set has 366 elements.

41. $\{1,2\}, \{1,3\}, \{2,3\}$ 43. $\{1,2,3\}, \{1,2,4\}, \{1,3,4\}, \{2,3,4\}$

45. There are $2^5 = 32$ subsets and $2^5 - 1 = 31$ proper subsets.

47. {Carmen, Frank, Ivana}= V

49. The set of lowerclassmen = L; Note: the **bolded** value indicates the set with the largest cardinality.
$n(U)= 4$, $\boldsymbol{n(L)= 6}$, $n(S)= 2$, $n(V)= 3$, $n(A)= 2$, $n(T)= 3$, and $n(D)=2$

51. 25 is not a power of 2.

53. Answers may vary.
Since Pete probably confused 5^2 with 2^5, then the set most likely has 5 elements.

55. There are 2^5 different subsets. The ones that contain zero, one, or two elements can be excluded because none of these would result in a sum of 9 or more. There should be 10 sets that have 3 elements, 5 with 4 elements and 1 set with five elements.

Set	Sum		Set	Sum	
{a,b,c}	7		{b,d,e}	8	
{a,b,d}	9	X	{c,d,e}	7	
{a,b,e}	6		{a,b,c,d}	11	X
{a,c,d}	8		{a,b,c,e}	8	
{a,c,e}	5		{a,b,d,e}	10	X
{a,d,e}	7		{a,c,d,e}	9	X
{b,c,d}	9	X	{b,c,d,e}	10	X
{b,c,e}	6		{a,b,c,d,e,}	12	X

7 subsets

57. Answers may vary.
At the first branching, the "yes" or "no" indicates that in forming a subset of $\{1, 2\}$, we will either take the 1 or omit it. The second branching in the tree indicate whether we are going to take the 2 as a member of the subset that we are forming. The tree shows all possible ways that we can decide to take 1 and 2 in forming a subset of $\{1, 2\}$. The top branch corresponds to the subset $\{1, 2\}$, the second branch corresponds to the subset $\{1\}$, the third branch corresponds to the subset $\{2\}$, and the bottom branch corresponds to the subset $\varnothing$.

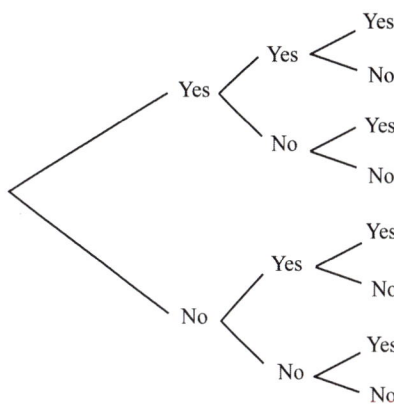

59. This line represents the counts of the number of subsets of a five-element set of sizes 0, 1, 2, 3, 4, and 5 elements.

61. The sixth, seventh, eighth, and ninth lines are:

    1 6 15 20 15 6 1
    1 7 21 35 35 21 7 1
    1 8 28 56 70 56 28 8 1
    1 9 36 **84** 126 126 84 36 9 1

So the newspaper editor can send three of them in 84 different ways.

63.
```
1                              → 1
1 1                            → 2
1 2 1                          → 4
1 3 3 1                        → 8
1 4 6 4 1                      → 16
1 5 10 10 5 1                  → 32
1 6 15 20 15 6 1               → 64
1 7 21 35 35 21 7 1            → 128
1 8 28 56 70 56 28 8 1         → 256
1 9 36 84 126 126 84 36 9 1    → 512
1 10 45 120 210 252 210 120 45 10 1 → 1024
```

The sum across the rows is always a power of 2, specifically 2^n, where n is the number of the row that is being summed. Note: Recall we start counting these lines with 0, not

65. Corresponding property: If $A \subseteq B$ and $B \subseteq C$, then $A \subseteq C$. If $x \in A$, then $x \in B$. Since $x \in B$, then $x \in C$. Therefore, if we have $x \in A$, we must also have $x \in C$. Note: We use capital letters for sets and lower case letters for elements.

67. No solution provided

Section 1.5 Set Operations

1. $A \cup B = \{x : x \in A \text{ or } x \in B\}$

3. The complement of A is the set of all elements in the universal set that are not elements of A.

5. $(A \cup B)' = A' \cap B'$; We cannot reverse the order of taking the union of the sets and taking the intersection of the complements of the sets.
   $(A \cap B)' = A' \cup B'$; We cannot reverse the order of taking the intersection of the sets and taking the union of the complements of the sets

7. "Union" implies joining together. "Intersection" implies overlapping.

9. Answers may vary. Possible answers include confusing DeMorgan's laws with the distributive property.

11. $A \cap B = \{1, 3, 5\}$

15. $A \cup \varnothing = \{1, 3, 5, 7, 9\} = A$

13. $B \cup C = \{1, 2, 3, 4, 5, 6, 7, 8\}$

17. $A \cup U = \{1, 2, 3, 4, 5, 6, 7, 8, 9, 10\} = U$

19. $A \cap (B \cup C) = A \cap \{1, 2, 3, 4, 5, 6, 7, 8\} = \{1, 3, 5, 7\}$

21. $(A - B) \cap (A - C) = \{7, 9\} \cap \{1, 3, 5, 9\} = \{9\}$

23. The set of all that will use the computer for education and business.

25. The set of all that will use the computer for all three.

27. The set of all that will use the computer for education and business but not home management.

29. $M \cap E = \{\text{potato chip, bread, pizza}\}$

31. $E - M = \{\text{apple, fish, banana}\}$

33. $M' \cap G' = \{\text{fish}\}$

35. $A - (B \cup C)$

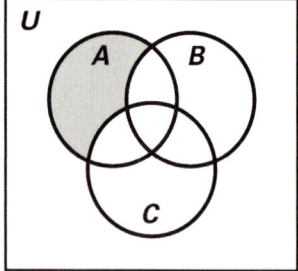

37. $(A \cap B) - C$

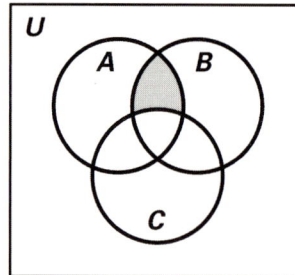

39. $A \cup (B - C)$

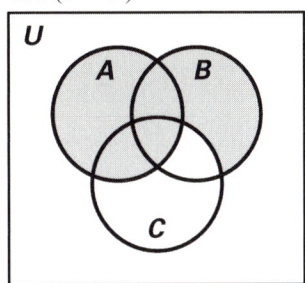

41. $(A \cup (B \cup C))'$

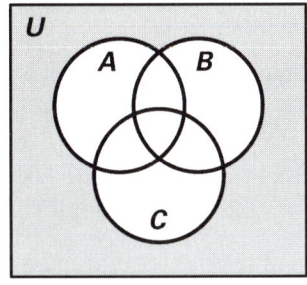

43. $B - A$

45. $(A \cup B)'$

47. $A \cap B \cap C$

49. $(A \cup C) - B$

51. 30

53. 28

55. 20

57. 27

59. A

61. B

63. False; It is possible that $A = B$ and hence $n(A) = n(B)$. Counterexamples may vary.

65. True

67. True

69. False; If $A \subset B$ then $n(A \cap B) = n(A) \geq 0$. However, if $A \subset B$ then $n(B) > n(A)$ thus $n(A) - n(B) < 0$. Counterexamples may vary.

71. True

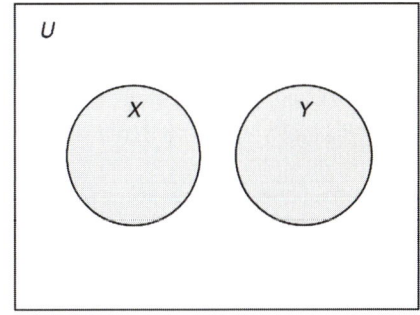

$X \cup Y$ is shaded and for all x such that $x \in X$, $x \in X \cup Y$

73. True

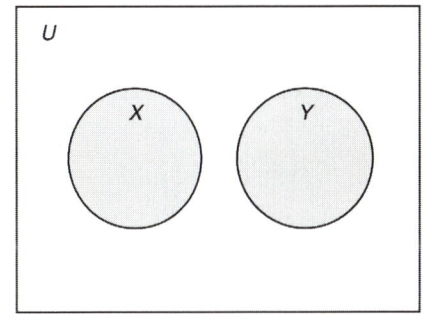

Since all $x \in X$ then all $x \in X \cup Y$. Therefore $n(x) \leq n(X \cup Y)$.

75. True

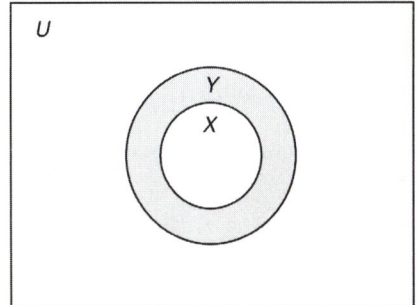

$Y - X$ is shaded. If the elements of X are removed from Y then those remaining are in $Y - X$.

77. $P \cap C$ = the set of cars whose price is above $12,000 and is compact = $\{d, g\}$

79. $W \cap G'$ = the set of cars that have a warranty of at least three years and doesn't have a good safety rating = $\{c, d, f, g\}$

81. $P \cap (G \cup W)$ = the set of cars that have a price above $12,000, and a good safety rating or a warrantee of at least three years = $\{b, d, g, h\} = P$

83. $P - (G \cup A)$ = the set of cars that have a price above $12,000 and don't have a good safety rating and don't have an antitheft package = $\{d\}$

85. $P \cap (B \cup A) = P \cap \{m, mc, hc\} = \{m, mc, hc\}$

87. $P \cup C \cup B = \{m, mc, bc, c, hc\}$

89. $A \cap B = B \cap A$; This is a true statement.

91. $(A \cap B) \cap C = A \cap (B \cap C)$; This is a true statement.

93. Sample Counterexample:
$$2 + (3 \cdot 4) \stackrel{?}{=} (2+3) \cdot (2+4)$$
$$2 + 12 \stackrel{?}{=} 5 \cdot 6$$
$$14 \neq 30$$

95. "And" is intersection $(\cap)$, and "Or" is union $(\cup)$.

96. These two requests are not the same. Additional answers may vary.

Section 1.6 Survey Problems

1. $(A - B) \cup (A \cap B) \cup (B - A)$

3. The 14 in $T \cap N \cap S$ was subtracted from the 22 in $T \cap N$.

5. They forget that r_2 excludes the elements in $A \cap B$.

7. r_2, r_3

9. r_3

11. r_2

13. r_2, r_3, r_5, r_6

15. r_5, r_6

17. r_4, r_7

19. r_2; $\{r_2, r_3, r_5, r_6\} - \{r_3, r_4, r_5, r_6, r_7, r_8\} = \{r_2\}$

21. r_7; $\{r_6, r_7\} \cap \{r_1, r_4, r_7, r_8\} = \{r_7\}$

23. r_6;
$\{r_3, r_4, r_6, r_7\} \cap \{r_5, r_6, r_7, r_8\} \cap \{r_2, r_3, r_5, r_6\} =$
$\{r_6, r_7\} \cap \{r_2, r_3, r_5, r_6\} = \{r_6\}$

25. $n(A) = 18, n(B) = 15,$ and $n(C) = 14$

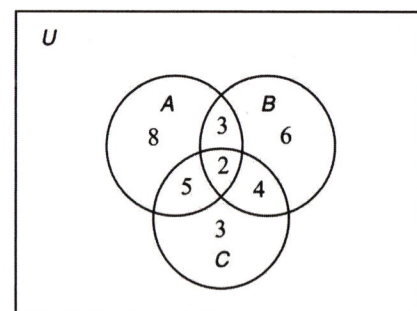

27. $n(A) = 5, n(B) = 14$, and $n(C) = 9$

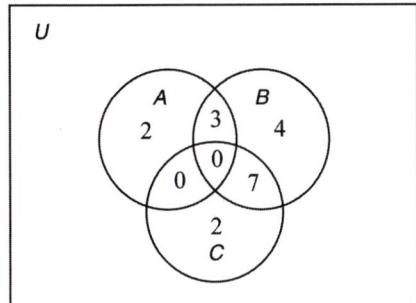

29. 59

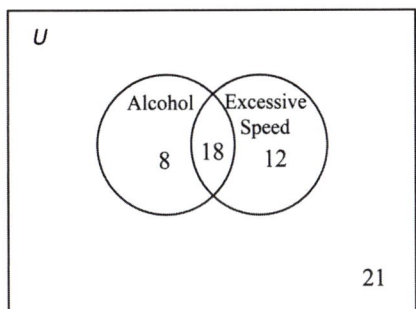

31. 35

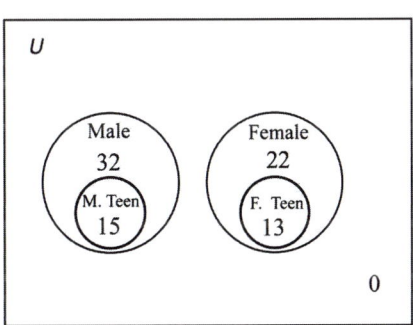

33. 30 attend the barbecue and 49 purchased a tour guide.

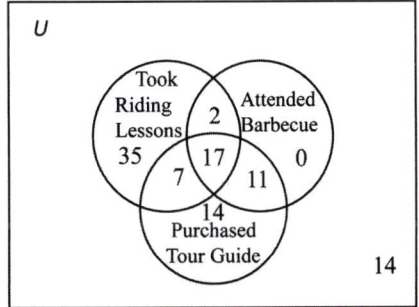

35. 37

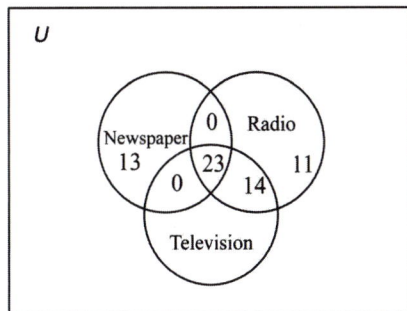

37.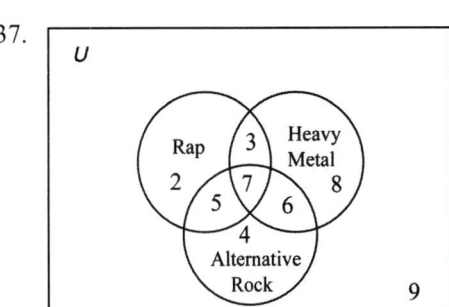

a) 44
b) 27
c) 8

39.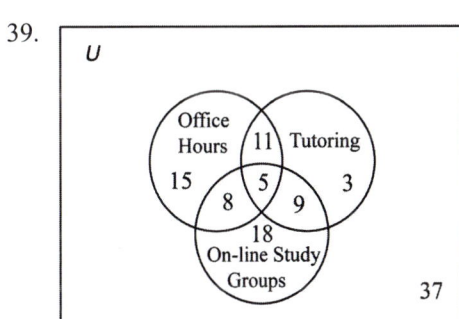

a) 106
b) 15
c) 40

41. $n(V \cup D) = 95 + 22 = 117$

43. $n(V - (M \cup S)) = 95 - (44 + 31)$
$= 95 - 75 = 20$

45. No solution provided

14 Chapter 1: Set Theory

47.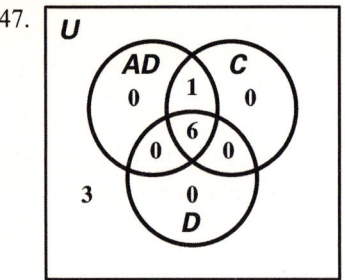

49. and 51. No solutions provided

Chapter Review Exercises

1. Answers (order) may vary.

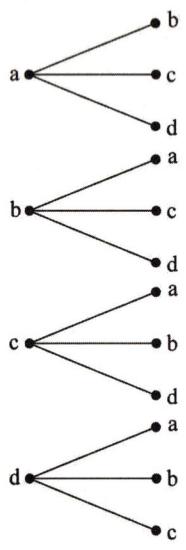

The pairs are *ab, ac, ad, ba, bc, bd, ca, cb, cd, da, db,* and *dc*.

2. Answers may vary.
   At a restaurant, you have 2 appetizers, 3 entrees, and 2 desserts. How many different meals can you choose if you select one appetizer, one entrée, and one desert?

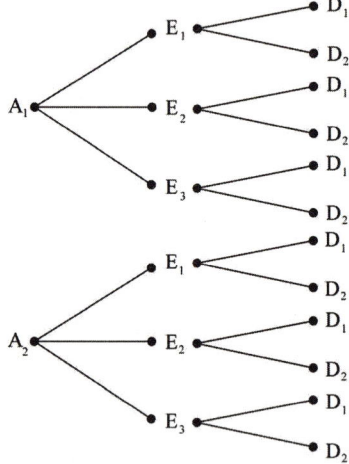

There are twelve different meals possible.

3. Answers may vary.
   Let S be the number of hours Hector worked as a stock person and I be the number of hours Hector worked as a ski instructor.

Guesses for S and I	Good points	Weak Points
9 and 11	Sum is 20.	Amount earned is less than $141.20.
7 and 13	Sum is 20.	Amount earned is more than $141.20.
8 and 12	All conditions satisfied. We have the solution.	

4. False
Counterexamples may vary.
$$\frac{1}{2}+\frac{3}{4}=\frac{2}{4}+\frac{3}{4}=\frac{5}{4}$$
$$\frac{1+3}{2+4}=\frac{4}{6}=\frac{2}{3}, \frac{5}{4} \neq \frac{2}{3}$$

5. No solution provided

6. a) 46,000
   b) 28,000

7. a) $210 - 60 = 150$
   b) $6 \times 15 = 90$

8. Estimated time left to travel would be $150 \div 50 = 3$ hours, arriving at 7:00 PM.

9. Estimated answers may vary.
   a) The exact value is $51,700.
   b) The exact value is $10,500.

10. a) Answers may vary. Possible answers include: $\{x : x$ is an odd counting number less than $10\}$.
    b) $\{h,a,p,i,n,e,s\}$
    c) $\{y : y$ is a letter of the English alphabet$\}$
    Note: If the case of the letters is of concern such as often times in computer passwords, then this set could be represented as $\{y : y$ is a lower case letter of the English alphabet$\}$.
    d) $\varnothing$

11. $\varnothing$ is the empty set; it contains no elements. $\{\varnothing\}$ is not empty; is contains $\varnothing$. 0 is not a set; it is a number.

12.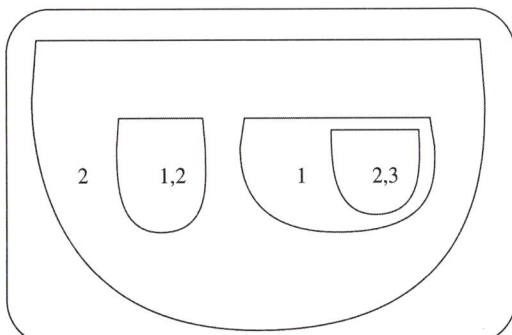

13. a) 5; b) 12

14. If $A \subseteq B$, all the elements of A are also elements of B, and the sets *may* be equal. If $A \subset B$, all the elements of A are also elements of B; however, the sets cannot be equal. The extra line in the notation $A \subseteq B$ suggests that the sets *may* me equal.

15. No solution provided

16. a) Yes, these two sets are equal. They have the same elements arranged in a different order.
    b) No, the two sets are not equal. The element 4, for example, appears in the second set but not in the first.
    c) Yes, these two sets are equal. Duplicate elements do not count as distinct (different) elements.
    d) No, these two sets are not the same. The first set contains no elements while the second set contains 1 element.

17. a) True. It is understood that all of the elements of the first set are also elements of the second set.
    b) False. In order to be a proper subset, the first set cannot be the same as the second set.
    c) False. The first set contains an element, namely e, that the second set doesn't have.
    d) True. The null set is a subset of all sets. We cannot find an element of the empty set that fails to be in $\{1,3,5\}$.
    e) False. In order to be a proper subset, the first set cannot be the same as the second set.

18. a) equivalent; b) equivalent; c) equivalent

19. a) There are $2^3 = 8$ subsets. They are $\varnothing$, $\{a\}$, $\{b\}$, $\{c\}$, $\{a,b\}$, $\{a, c\}$, $\{b, c\}$, and $\{a, b, c\}$.
    b) $2^{10} = 1,024$

20. a) $A \cap B = \{1, 3, 7\}$
    b) $B \cup C = \{1, 2, 3, 4, 5, 6, 7, 8\}$
    c) $C' = \{1, 8\}$
    d) $C - A = \{2, 4, 6\}$

16 Chapter 1: Set Theory

21. a) $(B' \cup C) \cap A = (\{2,5\} \cup C) \cap A =$
    $\{2,3,4,5,6,7\} \cap A = \{3,5,7\}$

    b) $(A-B) \cup (A-C) = \{5\} \ \{1\} = \{1,5\}$

22. a) $A \cap B$

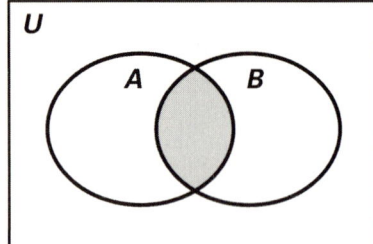

b) $B \cup C$

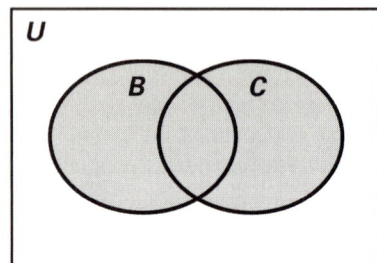

c) $A' \cap B'$

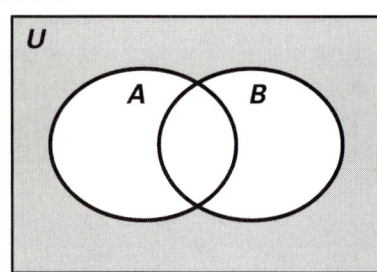

d) $A - (B \cup C)$

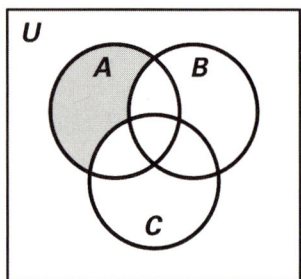

23. $A' \cup B'$

24. a) Any three of the following four are valid responses.
    1) Closure
    2) Commutativity; $A \cap B = B \cap A$
    3) Associativity;
       $(A \cap B) \cap C = A \cap (B \cap C)$
    4) Identity; $A \cap U = A$

    b) Any three of the following four are valid responses.
    1) Closure
    2) Commutativity; $A \cup B = B \cup A$
    3) Associativity;
       $(A \cup B) \cup C = A \cup (B \cup C)$
    4) Identity; $A \cup \emptyset = A$

    c) Union distributes over intersection
       $A \cup (B \cap C) = (A \cup B) \cap (A \cup C)$ and
       intersection distributes over union;
       $A \cap (B \cup C) = (A \cap B) \cup (A \cap C)$

25.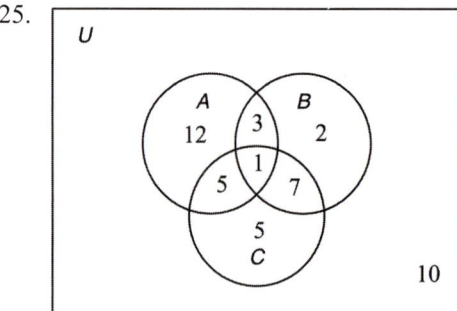

    a) $C - B = 5 + 5 = 10$
    b) $10 + 2 + 7 + 5 = 24$

26. 57 use the pool. 14 use the weight room only.

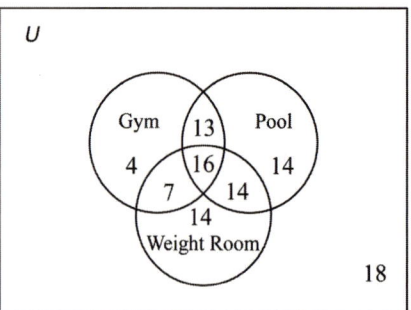

Chapter Test

1. No solution provided

2. a) {x: x is a counting number greater than 100}
   b) {Sunday, Monday, Tuesday, Wednesday, Thursday, Friday, Saturday}

3. Looking at mathematical ideas graphically, verbally, and using examples will help you better understand them.

4. Only the two sets in (a). They both equal {1, 2, 3, 5, 7, 9}

5. $\varnothing$, {a}, {b}, {c}, {a,b}, {a c}, {b c}, {a, b, c}

6. (3, 2), (3, 3), (3, 4), (4, 1), (4, 2)

7. Answer may vary. Possible answers include {x : x is a large number}. Different people may consider different numbers to be large.

8. 18,000

9. A = {2, 4, 6, 8, 10, 12, 14, 16, 18, 20, 22, 24}, so n(A) = 12.

10. Answers may vary. Possible answer is ($650 + $200 + $200 + $150)/3 = $400

11. The sets contain the exact same elements.

12. a) False; b) True; c) False; d) True

13. Answers may vary. One possible answer is {1, 3, 5, 7} and {2, 4, 6, 8} are equivalent.

14. a) $2^7 = 128$

15.
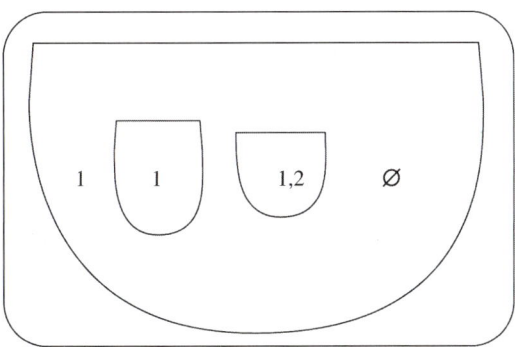

16. $(A \cap B)' = A' \cup B'$ or $(A \cup B)' = A' \cap B'$

17. a) $A \cap (B \cup C) =$
    $A \cap (\{2, 3, 5, 6, 7, 10\} \cup \{1, 3, 4, 5, 7, 8\}) =$
    $\{3, 5, 7, 8, 9\} \cap \{1, 2, 3, 4, 5, 6, 7, 8\} =$
    $\{3, 5, 7, 8\}$
    b) $A' \cap B' =$
    $\{1, 2, 4, 6, 10\} \cap \{1, 4, 8, 9\} =$
    $\{1, 4\}$
    c) $B - (A' \cup C) =$
    $B - (\{1, 2, 4, 6, 10\} \cup \{1, 3, 4, 5, 7, 8\}) =$
    $B - \{1, 2, 3, 4, 5, 6, 7, 8, 10\} =$
    $\varnothing$

18. 9

    $n(A) + n(B) - n(A \cap B)$
    $31 = 17 + 23 - n(A \cap B)$
    $n(A \cap B) = 17 + 23 - 31 = 9$

19. $(A \cup C) - B$

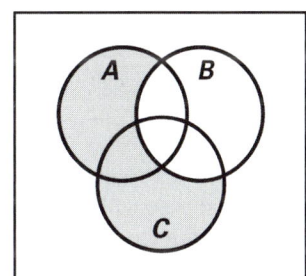

20.
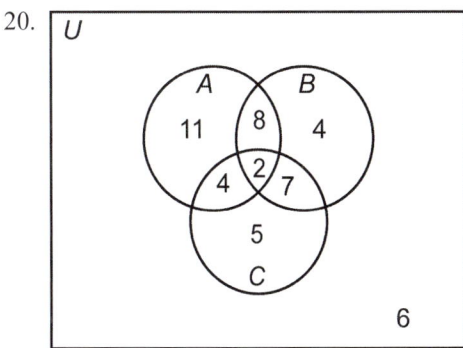

a) 11
b) 26

21.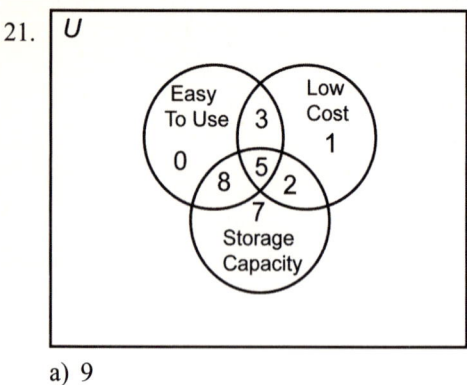

a) 9
b) 1

22. a) False; Let $A = \{1,2,3,4,5,6,7,8,9\}$, $B = \{1,3,5,7,9\}$, $C = \{3,4,5,6,7\}$.

$A - (B - C) = A - \{1,9\} = \{2,3,4,5,6,7,8\}$

$(A - B) - C = \{2,4,6,8\} - \{3,4,5,6,7\} = \{2,8\}$

Since $\{2,3,4,5,6,7,8\} \neq \{2,8\}$, $A - (B - C) \neq (A - B) - C$.

Of Further Interest: Infinite Sets

1. Answers may vary. Possible answers include matching each person with his predecessor, Reagan — Bush, Bush — Clinton, and Chaney — Gore

2. It was shown that the natural numbers has a one-to-one correspondence with a proper subset of itself. The subset was $\{2,4,6,\ldots\}$

3. It was shown that the set of positive rational numbers has a one-to-one correspondence with the natural numbers.

4. It was shown that if it was assumed that it were possible to put the real numbers between 0 and 1 in a one-to-one correspondence with the natural numbers, there would always be a number between 0 and 1 that we had omitted from the one-to-one correspondence.

5. $\{1,2,3,4,5\}$ does not have a one-to-one correspondence with any of its 31 proper subsets.

6. $\{2,4,6,8,\ldots\}$ has a one-to-one correspondence with the proper subset $\{2,6,10,14,\ldots\}$

7. 1 2 3 4 5 ... n ...
    ↕ ↕ ↕ ↕ ↕   ↕
    4 8 12 16 20... $4n$ ...

8. 1 2 3 4 5 ... n ...
    ↕ ↕ ↕ ↕ ↕   ↕
    5 10 15 20 25... $5n$ ...

9. Each term is 3 more than the previous term. The general term is represented by 8 plus 3, added $n-1$ times. So the general term would be $8 + (n-1) \cdot 3 = 8 + 3n - 3 = 3n + 5$.

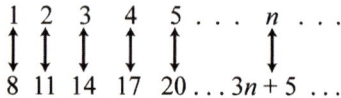

10. Each term is 4 more than the previous term. The general term is represented by 7 plus 4, added $n-1$ times. So the general term would be $7+(n-1)\cdot 4 = 7+4n-4 = 4n+3$.

$$\begin{array}{cccccc} 1 & 2 & 3 & 4 & 5 & \ldots & n & \ldots \\ \updownarrow & \updownarrow & \updownarrow & \updownarrow & \updownarrow & & \updownarrow \\ 7 & 11 & 15 & 19 & 23 & \ldots & 4n+3 & \ldots \end{array}$$

11. $$\begin{array}{cccccc} 1 & 2 & 3 & 4 & 5 & \ldots & n & \ldots \\ \updownarrow & \updownarrow & \updownarrow & \updownarrow & \updownarrow & & \updownarrow \\ 2 & 4 & 8 & 16 & 32 & \ldots & 2^n & \ldots \end{array}$$

12. $$\begin{array}{cccccc} 1 & 2 & 3 & 4 & 5 & \ldots & n & \ldots \\ \updownarrow & \updownarrow & \updownarrow & \updownarrow & \updownarrow & & \updownarrow \\ 3 & 9 & 27 & 81 & 243 & \ldots & 3^n & \ldots \end{array}$$

13. $$\begin{array}{cccccc} 1 & 2 & 3 & 4 & 5 & \ldots & n & \ldots \\ \updownarrow & \updownarrow & \updownarrow & \updownarrow & \updownarrow & & \updownarrow \\ 1 & 1/2 & 1/3 & 1/4 & 1/5 & \ldots & 1/n & \ldots \end{array}$$

14. $$\begin{array}{cccccc} 1 & 2 & 3 & 4 & 5 & \ldots & n & \ldots \\ \updownarrow & \updownarrow & \updownarrow & \updownarrow & \updownarrow & & \updownarrow \\ 1/2 & 2/3 & 3/4 & 4/5 & 5/6 & \ldots & n/(n+1) & \ldots \end{array}$$

15. $3\cdot 1 = 3$; $3\cdot 2 = 6$; $3\cdot 3 = 9$; $3\cdot 4 = 12$; $3\cdot 5 = 15$

$$\begin{array}{cccccc} 1 & 2 & 3 & 4 & 5 & \ldots & n & \ldots \\ \updownarrow & \updownarrow & \updownarrow & \updownarrow & \updownarrow & & \updownarrow \\ 3 & 6 & 9 & 12 & 15 & \ldots & 3n & \ldots \end{array}$$

16. $2\cdot 1+3 = 5$; $2\cdot 2+3 = 7$; $2\cdot 3+3 = 9$; $2\cdot 4+3 = 11$; $2\cdot 5+3 = 13$

$$\begin{array}{cccccc} 1 & 2 & 3 & 4 & 5 & \ldots & n & \ldots \\ \updownarrow & \updownarrow & \updownarrow & \updownarrow & \updownarrow & & \updownarrow \\ 5 & 7 & 9 & 11 & 13 & \ldots & 2n+3 & \ldots \end{array}$$

17. $3\cdot 1-2 = 1$; $3\cdot 2-2 = 4$; $3\cdot 3-2 = 7$; $3\cdot 4-2 = 10$; $3\cdot 5-2 = 13$

$$\begin{array}{cccccc} 1 & 2 & 3 & 4 & 5 & \ldots & n & \ldots \\ \updownarrow & \updownarrow & \updownarrow & \updownarrow & \updownarrow & & \updownarrow \\ 1 & 4 & 7 & 10 & 13 & \ldots & 3n-2 & \ldots \end{array}$$

18. $4\cdot 1+5 = 9$; $4\cdot 2+5 = 13$; $4\cdot 3+5 = 17$; $4\cdot 4+5 = 21$; $4\cdot 5+5 = 25$

$$\begin{array}{cccccc} 1 & 2 & 3 & 4 & 5 & \ldots & n & \ldots \\ \updownarrow & \updownarrow & \updownarrow & \updownarrow & \updownarrow & & \updownarrow \\ 9 & 13 & 17 & 21 & 25 & \ldots & 4n+5 & \ldots \end{array}$$

19. Match $\{2, 4, 6, 8, 10, \ldots\}$ with $\{4, 6, 8, 10, 12, \ldots\}$; in general, match $2n$ with $2(n+1) = 2n+2$.

20. Match $\{5, 10, 15, 20, 25, \ldots\}$ with $\{10, 15, 20, 25, 30, \ldots\}$; in general, match $5n$ with $5(n+1) = 5n+5$.

21. Since each term of $\{7, 10, 13, 16, 19, \ldots\}$ is 3 more than the previous, the general term is $7+3(n-1) = 7+3n-3 = 3n+4$. Match $\{7, 10, 13, 16, 19, \ldots\}$ with $\{10, 13, 16, 19, 22, \ldots\}$; in general, match $3n+4$ with $3(n+1)+4 = 3n+3+4 = 3n+7$.

22. Since each term of $\{6, 9, 12, 15, 18, \ldots\}$ is 3 more than the previous, the general term is $6+3(n-1) = 6+3n-3 = 3n+3$. Match $\{6, 9, 12, 15, 18, \ldots\}$ with $\{9, 12, 15, 18, 21, \ldots\}$; in general, match $3n+3$ with $3(n+1)+3 = 3n+3+3 = 3n+6$.

23. Match $\{2, 4, 8, 16, 32, ...\}$ with $\{4, 8, 16, 32, 64, ...\}$; in general, match 2^n with 2^{n+1}.

24. Match $\{3, 9, 27, 81, 243, ...\}$ with $\{9, 27, 81, 243, 729, ...\}$; in general, match 3^n with 3^{n+1}.

25. Match $\{1, \frac{1}{2}, \frac{1}{3}, \frac{1}{4}, \frac{1}{5}, ...\}$ with $\{\frac{1}{2}, \frac{1}{3}, \frac{1}{4}, \frac{1}{5}, \frac{1}{6}, ...\}$; in general, match $\frac{1}{n}$ with $\frac{1}{n+1}$.

26. Match $\{\frac{1}{2}, \frac{2}{3}, \frac{3}{4}, \frac{4}{5}, \frac{5}{6}, ...\}$ with $\{\frac{2}{3}, \frac{3}{4}, \frac{4}{5}, \frac{5}{6}, \frac{6}{7}, ...\}$; in general, match $\frac{n}{n+1}$ with $\frac{n+1}{(n+1)+1} = \frac{n+1}{n+2}$.

27. Match $\{\frac{1}{2}, \frac{1}{4}, \frac{1}{6}, \frac{1}{8}, \frac{1}{10}, ...\}$ with $\{\frac{1}{4}, \frac{1}{6}, \frac{1}{8}, \frac{1}{10}, \frac{1}{12} ...\}$; in general, match $\frac{1}{2n}$ with $\frac{1}{2(n+1)} = \frac{1}{2n+2}$.

28. Match $\{\frac{1}{2}, \frac{1}{4}, \frac{1}{8}, \frac{1}{16}, \frac{1}{32}, ...\}$ with $\{\frac{1}{4}, \frac{1}{8}, \frac{1}{16}, \frac{1}{32}, \frac{1}{64}, ...\}$; in general, match $\frac{1}{2^n}$ with $\frac{1}{2^{n+1}}$.

For Exercises 29 and 31 refer to the following diagram.

```
1/1  2/1  3/1  4/1  5/1  6/1  7/1  8/1  9/1  10/1 11/1 ...
1/2  2/2  3/2  4/2  5/2  6/2  7/2  8/2  9/2  10/2 11/2 ...
1/3  2/3  3/3  4/3  5/3  6/3  7/3  8/3  9/3  10/3 11/3 ...
1/4  2/4  3/4  4/4  5/4  6/4  7/4  8/4  9/4  10/4 11/4 ...
1/5  2/5  3/5  4/5  5/5  6/5  7/5  8/5  9/5  10/5 11/5 ..
1/6  2/6  3/6  4/6  5/6  6/6  7/6  8/6  9/6  10/6 11/6 ...
1/7  2/7  3/7  4/7  5/7  6/7  7/7  8/7  9/7  10/7 11/7 ...
1/8  2/8  3/8  4/8  5/8  6/8  7/8  8/8  9/8  10/8 11/8 ...
1/9  2/9  3/9  4/9  5/9  6/9  7/9  8/9  9/9  10/9 11/9 ...
1/10 2/10 3/10 4/10 5/10 6/10 7/10 8/10 9/10 10/10 11/10 ...
1/11 2/11 3/11 4/11 5/11 6/11 7/11 8/11 9/11 10/11 11/11 ...
```

29. 6; We skip 2/2, 2/4, 3/3, and 4/2.

30. 3/4; We skip 2/2, 2/4, 3/3, and 4/2.

31. 25; We skip 2/2, 2/4, 3/3, 4/2, 2/6, 4/4, 6/2, and 6/3.

32. 17; We skip 2/2, 2/4, 3/3, and 4/2.

33. No solution provided

34. 6

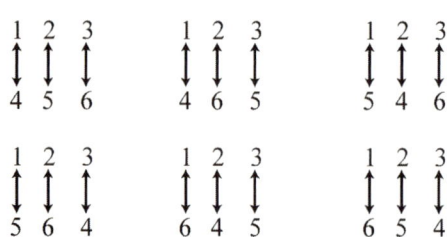

35. 24; Suppose the two sets were $\{1, 2, 3, 4\}$ and $\{a, b, c, d\}$. The 24 correspondences would be

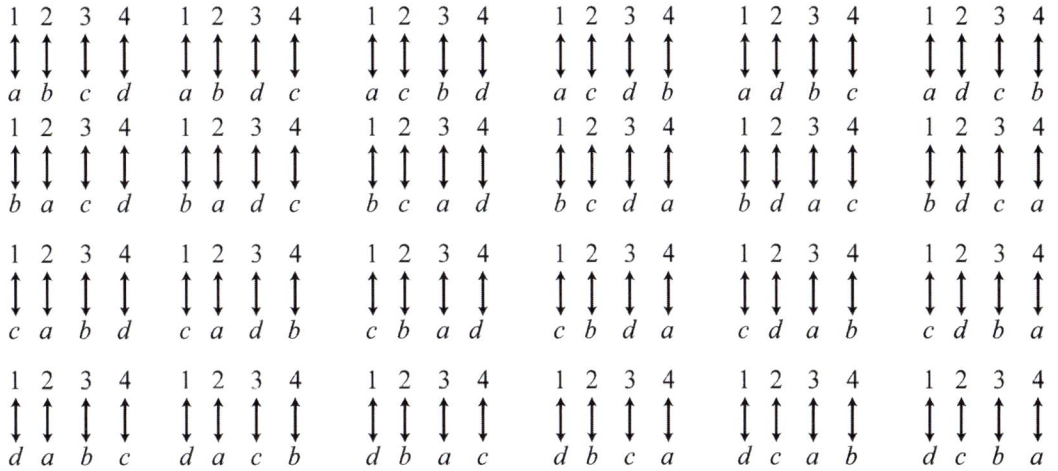

36. Answers may vary.

    If we take the union of $\{1\}$, which has cardinal number 1, and $\{2, 3, 4, ...\}$, which has cardinal number $\aleph_0$, we get $\{1, 2, 3, 4, ...\}$, which has cardinal number $\aleph_0$.

37. Answers may vary.

    Take the union of $\{1, 3, 5, ...\}$ and $\{2, 4, 6, ...\}$, both of which have cardinal number $\aleph_0$ to get $\{1, 2, 3, 4, 5, 6, ...\}$, which has cardinal number $\aleph_0$.

Chapter 2
LOGIC: The Study Of What's True Or False Or Somewhere In Between

Section 2.1 Inductive And Deductive Reasoning

1. Example 2; the test for divisibility by nine.

3. The reasoning uses accepted facts and general principles.

5. Answers may vary. Possible answers include: Inductive reasoning is the process of drawing a general conclusion by observing a pattern in specific instances. Deductive reasoning uses accepted facts and general principles to arrive at a specific conclusion.

7. inductive

9. deductive

11. inductive

13. deductive

15. inductive

17. 16

19. 96

21. $\frac{1}{64}$

23. 21

25.

27.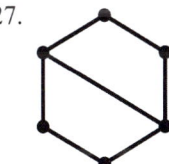

29. Hint: Think of prime numbers.

X	X		X		X				X		X				
2	3		5		7				11		13				

31. 4

		Units Digit
1	$2^1 = 2$	2
2	$2^2 = 4$	4
3	$2^3 = 8$	8
4	$2^4 = 16$	6
5	$2^5 = 32$	2
6	$2^6 = 64$	4
7	$2^7 = 128$	8
8	$2^8 = 256$	6
9	$2^9 = 512$	2
10	$2^{10} = 1024$	4

Since $50 = 4 \cdot 12 + 2$, 2^{50} has the same units digit as 2^2.

33. 4

		Units Digit
1	$4^1 = 4$	4
2	$4^2 = 16$	6
3	$4^3 = 64$	4
4	$4^4 = 256$	6
5	$4^5 = 1024$	4
6	$4^6 = 4096$	6
7	$4^7 = 16384$	4
8	$4^8 = 65536$	6
9	$4^9 = 262144$	4
10	$4^{10} = 1048576$	6

Since $95 = 2 \cdot 47 + 1$, 4^{47} has the same units digit as 4^1.

35. $3 + 13$ or $5 + 11$

37. $3 + 17$ or $7 + 13$

In Exercise 39, the generalized pattern is that for an $n \times n$ square, there are a total of $1^2 + 2^2 + \ldots + n^2$ squares of sizes $n \times n$, $(n-1) \times (n-1)$, ..., 2×2, 1×1.

39. There are a total of $1^2 + 2^2 + 3^2 + 4^2 + 5^2 = 1 + 4 + 9 + 16 + 25 = 55$ squares.
    1 5×5 square; 4 4×4 squares; 9 3×3 squares; 16 2×2 squares; 25 1×1 squares

In Exercise 41, the generalized pattern is that for an $n \times n$ triangle, there are a total of n^2 smaller 1×1 triangles.

41. There are a total of $10^2 = 100$ smaller 1×1 triangles.
    $1 + 3 + 5 + 7 + 9 + 11 + 13 + 15 + 17 + 19 = 100$ triangles

43. $1 + 4 + 9 + 16 = 30$

45. a) The total of all the numbers in the square is $1 + 2 + 3 + \ldots + 16 = 136$.

    b) The total of the numbers for each row, column, and diagonal would be $\frac{136}{4} = 34$.

    c) One can deduce the missing numbers to yield the following.

7	6	12	9
10	11	5	8
13	16	2	3
4	1	15	14

In Exercises 47 and 49, only the answers to Part b, the expected conjecture, and Part c, the algebraic justification, are provided.

47. In this trick, you will always get the result three.

    a) Call the number n.
    b) $3n$
    c) $3n+9$
    d) $\dfrac{3n+9}{3} = \dfrac{3(n+3)}{3} = n+3$
    e) $(n+3) - n = n+3-n = 3$

49. In this trick, you will always get a result that is twice the number that you started with.
    a) Call the number n.
    b) $8n$
    c) $8n+12$
    d) $\dfrac{8n+12}{4} = \dfrac{4(2n+3)}{4} = 2n+3$
    e) $(2n+3) - 3 = 2n+3-3 = 2n$

51.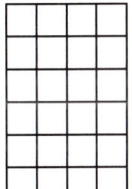
    (d) step 3

53. There are a total of 20 squares.
    2 3×3 squares; 6 2×2 squares; 12 1×1 squares

55. a) There are a total of 60 rectangles.

1×1	12
1×2	9
1×3	6
1×4	3
2×1	8
2×2	6
2×3	4
2×4	2
3×1	4
3×2	3
3×3	2
3×4	1

b) There are a total of 210 rectangles.

1×1	24		4×1	12
1×2	18		4×2	9
1×3	12		4×3	6
1×4	6		4×4	3
2×1	20		5×1	8
2×2	15		5×2	6
2×3	10		5×3	4
2×4	5		5×4	2
3×1	16		6×1	4
3×2	12		6×2	3
3×3	8		6×3	2
3×4	4		6×4	1

57. The base is a 6×4 rectangle. If you consider the diagram below as the base, we have $6 \times 4 = 24$ baseballs. In order to build the next level, we are looking for the number of places in which four baseballs (squares) meet. There are $5 \times 3 = 15$ such places.

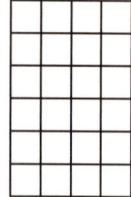

For the next level we would have $4 \times 2 = 8$ meeting places.

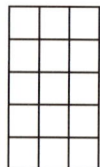

For the last level we would have $3 \times 1 = 3$ meeting places.

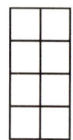

This yields a total of $6 \times 4 + 5 \times 3 + 4 \times 2 + 3 \times 1 = 24 + 15 + 8 + 3 = 50$ baseballs.

59. No solution provided

61. Answers may vary. Possible responses include that each new point needs to be connected to all of the previous points. So, if we have already connected four points, we need four new line segments to join a new point to the existing four.

63. If you expand the expression as follows,

$(2n+5) \cdot 50 + 1755 - 1987 = 100n + 250 + 1755 - 1987 = 100n + 2005 - 1987 = 100n + 18$,

you see that the $1755 + 250$ gives a multiple of 100 plus the current year (2005). If you then subtract the year that you were born, you get your age. If you have already had your birthday, you need to add the extra year, which is why we would then add 1756.

Section 2.2 Statements, Connectives, And Quantifiers

1. not (negation), and (conjunction), or (disjunction), if ... then (conditional), if and only if (biconditional)

3. Universal quantifiers state that all objects have a certain property.

5. "At least one is not."

7. In the first expression, the conjunction is taken before the negation; in the second expression, negations are taken before the conjunction.

9. To show that "not all objects have a property," it must be shown that "at least one does not have the property."

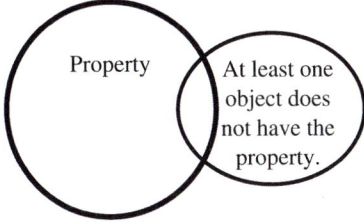

11. statement

13. not a statement

15. not a statement

17. statement

19. not a statement

21. compound; conjunction, negation, disjunction, conditional

23. compound; disjunction

25. simple

27. compound; conditional, disjunction

29. compound; conditional

31. $r \lor \sim s$

33. $(\sim r \land t) \to \sim s$

35. $\sim s \leftrightarrow t$

37. The radial tires are included or the sunroof is not extra.

39. It is not true that the sunroof is extra and the radial tires are included.

41. If the radial tires are included, then the sunroof is extra or power windows are not optional.

43. There exists at least one snake that is not poisonous.

45. All personal items are covered by this insurance policy.

47. All scientists believe that it is not true that an asteroid collision led to the extinction of the dinosaurs.

49. Some modern art is not difficult to understand.

51.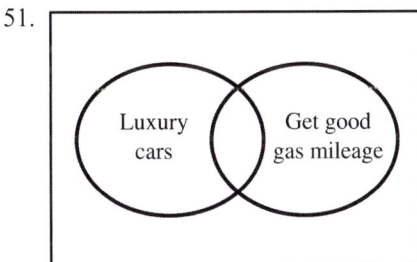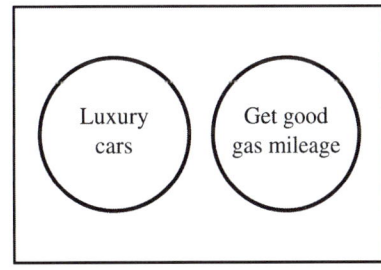

Some luxury cars do not get good gas mileage.

It is not true that some luxury cars get good gas mileage.

53. – 63. Answers for these exercises have not been provided. These statements are complex and sophisticated. You will probably find that you, your classmates, and your instructor do not always agree as to which connectives are present. Nevertheless, it is interesting to try to determine the form of these statements.

65. True

67. False; There exists a whole number that is not a natural number.

69. True

71. – 77. No solutions provided

Section 2.3 Truth Tables

1. The "and" is only true when both the components are true; the "or" is false only when both the components are false.

3. There is a difference between computing the "or" first before negating and negating before computing the "or". The situation is similar for "and."

5. line one

7. Answers may vary. One possibility is "paper or plastic?"

9. False

p	q	$p \wedge q$
T	F	T **F** F

11. True

		2 1
p	q	$p \wedge (\sim q)$
T	F	T **T** T

13. False

		2 1
p	q	$\sim (p \vee q)$
T	F	**F** T T F

15. True

		3 1 2
p	q	$\sim (\sim p \wedge q)$
T	F	**T** F F F

17.

		3 2 1
p	q	$\sim (p \vee \sim q)$

19.

		4 3 2 1
p	q	$p \wedge \sim (p \vee \sim q)$

21. T

23. T

25. F

27. T

29. $2^5 = 32$

31. Yes; $16 = 2^4$.

33. No; 9 cannot be represented as a power of 2.

35. Answers may vary. Possible responses include that we want the truth table to cover every possibility for the truth values for the three variables. There are eight different ways to assign truth values to the three variables.

37.

		2 1
p	q	$p \wedge (\sim q)$
T	T	T **F** F
T	F	T **T** T
F	T	F **F** F
F	F	F **F** T

39.

		3 1 2
p	q	$\sim (\sim p \wedge q)$
T	T	**T** F F T
T	F	**T** F F F
F	T	**F** T T T
F	F	**T** T F F

41.

		2	1	5	4	3
p	q	~	(p ∨ q)	∧	~	(p ∧ q)
T	T	F	T T T	**F**	F	T T T
T	F	F	T T F	**F**	T	T F F
F	T	F	F T T	**F**	T	F F T
F	F	T	F F F	**T**	T	F F F

43.

				1	4	3	2
p	q	r		(p ∨ r)	∧	(p ∧ ~q)
T	T	T		T T T	**F**	T F F
T	T	F		T T F	**F**	T F F
T	F	T		T T T	**T**	T T T
T	F	F		T T F	**T**	T T T
F	T	T		F T T	**F**	F F F
F	T	F		F F F	**F**	F F F
F	F	T		F T T	**F**	F F T
F	F	F		F F F	**F**	F F T

45.

				1	3	2
p	q	r		(p ∨ q)	∧	(p ∨ r)
T	T	T		T T T	**T**	T T T
T	T	F		T T T	**T**	T T F
T	F	T		T T F	**T**	T T T
T	F	F		T T F	**T**	T T F
F	T	T		F T T	**T**	F T T
F	T	F		F T T	**F**	F F F
F	F	T		F F F	**F**	F T T
F	F	F		F F F	**F**	F F F

47.

			3	2	1	4
p	q	r	~	(p ∨ ~q)	∧	r
T	T	T	F	T T F	**F**	T
T	T	F	F	T T F	**F**	F
T	F	T	F	T T T	**F**	T
T	F	F	F	T T T	**F**	F
F	T	T	T	F F F	**T**	T
F	T	F	T	F F F	**F**	F
F	F	T	F	F T T	**F**	T
F	F	F	F	F T T	**F**	F

49. exclusive or

51. inclusive or

53. *p*: Bill is tall.
    q: Bill is thin
    It is not true that Bill is tall and thin.
    $\sim(p \land q) \Leftrightarrow \sim p \lor \sim q$

    Negation: Bill is not tall or Bill is not thin.

55. *p*: Christian will apply for a loan.
    q: Christian will apply for work study.
    It is not true that Christian will apply for either a loan or work study.
    $\sim(p \lor q) \Leftrightarrow \sim p \land \sim q$

    Negation: Christian will not apply for a loan and he will not apply for work study.

57. *p*: Ken qualifies for a rebate.
    q: Ken qualifies for a reduced interest rate.
    It is not true that Ken qualifies for a rebate or a reduced interest rate.
    $\sim(p \lor q) \Leftrightarrow \sim p \land \sim q$

    Negation: Ken does not qualify for a rebate and he does not qualify for a reduced interest rate.

59. *p*: The number *x* is equal to five.
    q: The number *s* is odd.
    It is not true that The number *x* is not equal to five and *s* is not odd.
    $\sim(\sim p \land \sim q) \Leftrightarrow p \lor q$

    Negation: The number *x* is equal to five or *s* is odd.

61. Yes; $\sim(p \land \sim q)$ is logically equivalent to $\sim p \lor q$ by DeMorgan's laws.

63. No; $\sim(p\vee\sim q)\wedge\sim(p\vee q)$ is not logically equivalent to $p\vee(p\wedge q)$.

		3 2 1 6 5 4
p	q	$\sim(p\vee\sim q)\wedge\sim(p\vee q)$
T	T	F T T F **F** F TTT
T	F	F T T T **F** F TTF
F	T	T F F F **F** F FTT
F	F	F F T T **F** T FFF

		2 1
p	q	$p\vee(p\wedge q)$
T	T	T **T** TTT
T	F	T **T** TFF
F	T	F **F** FFT
F	F	F **F** FFF

65. Yes; $\sim(p\vee\sim q)\wedge\sim(p\vee q)$ is logically equivalent to $(\sim p\wedge q)\wedge(\sim p\wedge\sim q)$ by DeMorgan's laws.

67. Yes; $p\vee(\sim q\wedge r)$ is logically equivalent to $(p\vee(\sim q))\wedge(p\vee r)$.

			3 1 2
p	q	r	$p\vee(\sim q\wedge r)$
T	T	T	T **T** F FT
T	T	F	T **T** F FF
T	F	T	T **T** T TT
T	F	F	T **T** T FF
F	T	T	F **F** F FT
F	T	F	F **F** F FF
F	F	T	F **T** T TT
F	F	F	F **F** T FF

			2 1 4 3
p	q	r	$(p\vee(\sim q))\wedge(p\vee r)$
T	T	T	T T F **T** TTT
T	T	F	T T F **T** TTF
T	F	T	T T T **T** TTT
T	F	F	T T T **T** TTF
F	T	T	F F F **F** FTT
F	T	F	F F F **F** FFF
F	F	T	F T T **T** FTT
F	F	F	F T T **F** FFF

69. p: The earned income did reduce the tax you owe.
    q: The earned income gave you a refund.
    $\sim p\vee\sim q\Leftrightarrow\sim(p\wedge q)$
    It is not true that the earned income did reduce the tax you owe and gave you a refund.

71. p: You are single.
    q: You are the head of a household.
    $\sim p\wedge\sim q\Leftrightarrow\sim(p\vee q)$
    It is not true that you are single or the head of a household.

73. $(p\vee q)\vee(p\vee r)$ is logically equivalent to $p\vee(q\vee r)$.

			1 3 2
p	q	r	$(p\vee q)\vee(p\vee r)$
T	T	T	TTT **T** TTT
T	T	F	TTT **T** TTF
T	F	T	TTF **T** TTT
T	F	F	TTF **T** TTF
F	T	T	FTT **T** FTT
F	T	F	FTT **T** FFF
F	F	T	FFF **T** FTT
F	F	F	FFF **F** FFF

			2 1
p	q	r	$p\vee(q\vee r)$
T	T	T	T **T** TTT
T	T	F	T **T** TTF
T	F	T	T **T** FTT
T	F	F	T **T** FFF
F	T	T	F **T** TTT
F	T	F	F **T** TTF
F	F	T	F **T** FTT
F	F	F	F **F** FFF

73. (continued)

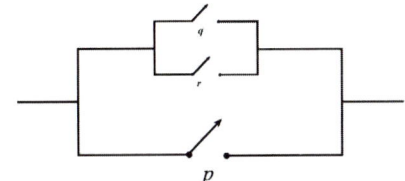

75. $(p \wedge q) \vee \sim q$ is logically equivalent to $p \vee \sim q$.

			1 3 2		
p	q	($p \wedge q$)$\vee \sim q$			
T	T	T	T	T	F
T	F	T	F	F	T
F	T	F	F	T	F
F	F	F	F	F	T

			2 1		
p	q	$p \vee \sim q$			
T	T	T	T		F
T	F	T	T		T
F	T	F	F		F
F	F	F	T		T

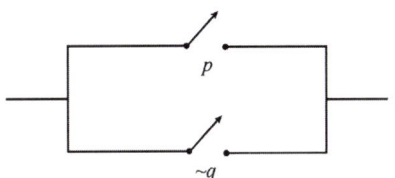

77. $(p \vee q) \wedge \sim q$ is logically equivalent to $p \wedge \sim q$ (One possible answer)

			1 3 2		
p	q	($p \vee q$)$\wedge \sim q$			
T	T	T	T	F	F
T	F	T	T	T	T
F	T	F	T	F	F
F	F	F	F	F	T

			1 3 2		
p	q	$p \wedge \sim q$			
T	T	T	F		F
T	F	T	T		T
F	T	F	F		F
F	F	F	F		T

79. $p \wedge (q \wedge r) \Leftrightarrow (p \wedge q) \wedge r$

81. $p \vee (q \wedge r) \Leftrightarrow (p \vee q) \wedge (p \vee r)$

83.

```
        q           r
              T ← (TTT)
          T <
              F ← (TTF)
     p
      T <
              T ← (TFT)
          F <
              F ← (TFF)

              T ← (FTT)
          T <
              F ← (FTF)
      F <
              T ← (FFT)
          F <
              F ← (FFF)
```

85. A "T" would imply that the element is in the subset, and an "F" would imply that it is not an element of the subset.

p	q	r	Set of Variables That Have True Values
T	T	T	{p, q, r}
T	T	F	{p, q}
T	F	T	{p, r}
T	F	F	{p}
F	T	T	{q, r}
F	T	F	{q}
F	F	T	{r}
F	F	F	∅

87. $p \wedge q$ is logically equivalent to $\sim(\sim p \vee \sim q)$ by DeMorgan's laws.

89. $p \mid q$ is logically equivalent to $\sim(p \wedge q)$.

		2	1	
p	q	$\sim(p \wedge q)$		$p \mid q$
T	T	F	T T T	F
T	F	T	T F F	T
F	T	T	F F T	T
F	F	T	F F F	T

91.

		1	3	2	
p	q	$(p \mid p)$	$\mid$	$(q \mid q)$	$p \vee q$
T	T	F	T	F	T
T	F	F	T	T	T
F	T	T	T	F	T
F	F	T	F	T	F

Section 2.4 The Conditional And Biconditional

1. The only way a statement of the form $p \rightarrow q$ can be false is the p is true and q is false.

3. Write the original statement symbolically in the form $p \rightarrow q$, then write the contrapositive in the form $\sim q \rightarrow \sim p$, and finally write the contrapositive in words.

5. Converse: switch clauses; inverse: negate both clauses; contrapositive: switch and negate both clauses.

7. $F \to T$; Inverse is $\sim F \to \sim T$ or $T \to F$.

9. True

		2	1	4	3
p	q	\multicolumn{4}{c}{$\sim(p \vee q) \to \sim p$}
T	F	F	TTF	**T**	F

11. True

			1	3	2
p	q	r	\multicolumn{3}{c}{$(p \wedge q) \to (q \vee r)$}
T	F	T	TFF	**T**	FTT

13. True

			1	3	2	4
p	q	r	\multicolumn{4}{c}{$(\sim p \vee \sim q) \to r$}
T	F	T	F T T	**T**	T

15. False

			3	1	2	5	4
p	q	r	\multicolumn{5}{c}{$\sim(\sim p \wedge q) \to \sim r$}
T	F	T	T	F	FF	**F**	F

17.

			2	1
p	q	\multicolumn{3}{c}{$p \to \sim q$}
T	T	T	**F**	F
T	F	T	**T**	T
F	T	F	**T**	F
F	F	F	**T**	T

19.

		2	1
p	q	\multicolumn{2}{c}{$\sim(p \to q)$}
T	T	**F**	TTT
T	F	**T**	TFF
F	T	**F**	FTT
F	F	**F**	FTF

21.

			1	4	3	2
p	q	r	\multicolumn{4}{c}{$(p \vee r) \to (p \wedge \sim q)$}
T	T	T	TTT	**F**	TFF
T	T	F	TTF	**F**	TFF
T	F	T	TTT	**T**	TTT
T	F	F	TTF	**T**	TTT
F	T	T	FTT	**F**	FFF
F	T	F	FFF	**T**	FFF
F	F	T	FTT	**F**	FFT
F	F	F	FFF	**T**	FFT

23.

			2	1	5	4	3
p	q	r	\multicolumn{5}{c}{$\sim(p \vee r) \to \sim(p \wedge q)$}
T	T	T	F TTT	**T**	F TTT
T	T	F	F TTF	**T**	F TTT
T	F	T	F TTT	**T**	T TFF
T	F	F	F TTF	**T**	T TFF
F	T	T	F FTT	**T**	T FFT
F	T	F	T FFF	**T**	T FFT
F	F	T	F FTT	**T**	T FFF
F	F	F	T FFF	**T**	T FFF

25.

			1	3	2
p	q	r	\multicolumn{3}{c}{$(p \vee q) \leftrightarrow (p \vee r)$}
T	T	T	TTT	**T**	TTT
T	T	F	TTT	**T**	TTF
T	F	T	TTF	**T**	TTT
T	F	F	TTF	**T**	TTF
F	T	T	FTT	**T**	FTT
F	T	F	FTT	**F**	FFF
F	F	T	FFF	**F**	FTT
F	F	F	FFF	**T**	FFF

27.

		1	2	5	3	4
p	q	\multicolumn{5}{c}{$(\sim p \to q) \leftrightarrow (\sim q \to p)$}
T	T	F T T	**T**	F T T
T	F	F T F	**T**	T T T
F	T	T T T	**T**	F T F
F	F	T F F	**T**	T F F

29. If it pours, then it rains.

31. If you do not buy all-weather radial tires, then they will not last for 80,000 miles.

33. If the sides of a geometric figure are not all equal in length, then it is not an equilateral triangle.

35. If x does not evenly divide 6, then x does not evenly divide 9.

37. Converse: $q \to \sim p$
    Inverse: $p \to \sim q$
    Contrapositive: $\sim q \to p$

39. Converse: $\sim(q \wedge r) \to (\sim p)$
    Inverse: $p \to (q \wedge r)$
    Contrapositive: $(q \wedge r) \to p$

41. The second statement is the contrapositive of the first. They are logically equivalent.

43. The second statement is the converse of the first. They are not logically equivalent.

45. If I finish my workout, then I'll take a break.

47. If you qualify for this deduction, then you complete Form 3093.

49. If you receive a free cell phone, then you sign up before March 1.

51. If you remain accident free for three years, then you get a reduction on your auto insurance.

53. $(p \leftrightarrow q) \Leftrightarrow (p \to q) \wedge (q \to p)$

			1			3	2	
p	q		$p \leftrightarrow q$			$(p \to q) \wedge (q \to p)$		
T	T	T	**T**	T	TTT	**T**	TTT
T	F	T	**F**	F	TFF	**F**	FTT
F	T	F	**F**	T	FTT	**F**	TFF
F	F	F	**T**	F	FTF	**T**	FTF

55. $(p \to q) \Leftrightarrow (\sim p \vee q)$

		1	4	3 2
p	q	$(p \to q) \leftrightarrow (\sim p \vee q)$		
T	T	TTT	**T**	FTT
T	F	TFF	**T**	FFF
F	T	FTT	**T**	TTT
F	F	FTF	**T**	TTF

57. $\sim(p \to q) \Leftrightarrow \sim(\sim p \vee q) \Leftrightarrow p \wedge \sim q$

59. If you can be claimed by someone else as a dependent, then your gross income is not more than $2,250.

61. If you decrease the amount being withheld from your pay, then the amount you overpaid is large.

63. True; $F \to T$

65. True; $F \to F$

67. Jamie has not been a member for at least ten years.

69. Using the contrapositive and DeMorgan's laws, $r \to (p \wedge q)$.

71. $p \vee (r \wedge (p \wedge q))$ is logically equivalent to p.

p	q	r	$p \vee$		$(r \wedge (p \wedge q))$
				3 2 1	
T	T	T	T	**T**	TTTTT
T	T	F	T	**T**	FFTTT
T	F	T	T	**T**	TFTFF
T	F	F	T	**T**	FFTFF
F	T	T	F	**F**	TFFFT
F	T	F	F	**F**	FFFFT
F	F	T	F	**F**	TFFFF
F	F	F	F	**F**	FFFFF

73. $(p \rightarrow q) \Leftrightarrow (\sim p \vee q)$

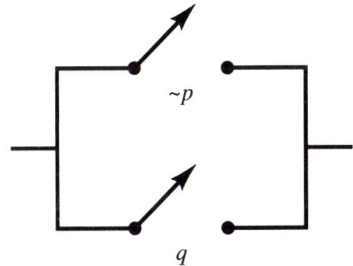

75. $(\sim p) \rightarrow \sim (q \wedge r) \Leftrightarrow p \vee \sim (q \wedge r) \Leftrightarrow p \vee (\sim q \vee \sim r)$

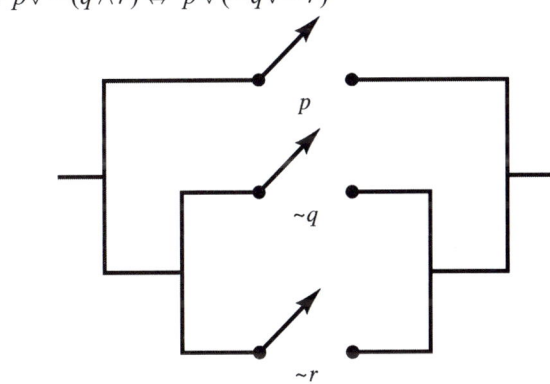

Section 2.5 Verifying Arguments

1. $(p_1 \wedge p_2) \rightarrow c$

3. The final column of the truth table had all trues.

5. The statement p allows us to detach q from $p \rightarrow q$.

7. In the disjunction $p \vee q$, if we don't have p, then we must have q.

9. If we have the premise $\sim p$, then we conclude $\sim q$, which is the inverse of $p \rightarrow q$.

11. *p:* The car has air bags.
    q: The car is safe.
    $p \rightarrow q$
    p
    _ _ _ _ _
    $\therefore q$
    valid argument
    Law of Detachment

13. *p:* The movie is exciting.
    q: The movie will gross a lot of money.
    $p \rightarrow q$
    q
    _ _ _ _ _
    $\therefore p$
    invalid argument
    Fallacy of the Converse

15. *p:* The computer has a DVD drive.
    q: The computer has a large monitor.
    $p \vee q$
    $\sim p$
    _ _ _ _ _
    $\therefore q$
    valid argument
    Disjunctive Syllogism

17. *p:* You pay your tuition late.
    q: You will pay a late payment fee.
    $p \rightarrow q$
    $\sim p$
    _ _ _ _ _
    $\therefore \sim q$
    invalid argument
    Fallacy of the Inverse

19. *p:* You watch *The Apprentice.*.
    q: You will succeed in business.
    r: A skyscraper will be named after you.
    $p \rightarrow q$
    $q \rightarrow r$
    _ _ _ _ _
    $\therefore p \rightarrow r$
    valid argument
    Law of Syllogism

21. *p:* You buy the sports package.
    q: You get the leather seats.
    $\sim p \rightarrow \sim q$
    q
    _ _ _ _ _
    $\therefore p$
    valid argument
    Law of Contraposition

23. *p:* Phillipe joins the basketball team.
    q: Phillipe will be able to work part time.
    $p \rightarrow \sim q$
    $\sim p$
    _ _ _ _ _
    $\therefore q$
    invalid argument
    Fallacy of the Inverse

25. *p:* Milik buys a satellite dish.
    q: Milik will get 128 TV channels.
    $p \rightarrow q$
    $\sim q$
    _ _ _ _ _
    $\therefore \sim p$
    valid argument
    Law of Contraposition

27. *p:* Minxia took the job in Boston.
    q: Minxia took the job in San Diego.
    $p \vee q$
    $\sim p$
    _ _ _ _ _
    $\therefore q$
    valid argument
    Disjunctive Syllogism

29. valid argument

| | | 3 2 1 5 4 | |
p	q	$[p \wedge (q \rightarrow \sim p)] \rightarrow \sim q$	
T	T	T F T F F	**T** F
T	F	T T F T F	**T** T
F	T	F F T T T	**T** F
F	F	F F F T T	**T** T

31. invalid argument

q	r									
		1	3	2		6	4	5		
		[~r	∧	(r	→	q)]	→	(~q	∧	r)
T	T	F	F	T	T	T	**T**	F	F	T
T	F	T	T	F	T	T	**F**	F	F	F
F	T	F	F	T	F	F	**T**	T	T	T
F	F	T	T	F	T	F	**F**	T	F	F

33. invalid argument

p	q	r											
			1	2	5	4	3	8	6	7			
			[(~q	→	p)	∧	(r	→~	q)]	→	(~p	→	r)
T	T	T	F	T	T	F	T	F	F	**T**	F	T	T
T	T	F	F	T	T	T	F	T	F	**T**	F	T	F
T	F	T	T	T	T	T	T	T	T	**T**	F	T	T
T	F	F	T	T	T	T	F	T	T	**T**	F	T	F
F	T	T	F	T	F	F	T	F	F	**T**	T	T	T
F	T	F	F	T	F	T	F	T	F	**F**	T	F	F
F	F	T	T	F	F	F	T	T	T	**T**	T	T	T
F	F	F	T	F	F	F	F	T	T	**T**	T	F	F

35. invalid argument

p	q	r										
			3	2	1	7	4	6	5		8	
			{p	∧	(q	→~	p)	∧	[~q	→	(r∨p)]}	→ r
T	T	T	T	F	T	F	F	F	F	T	TTT	**T**	T
T	T	F	T	F	T	F	F	F	F	T	FTT	**T**	F
T	F	T	T	T	F	T	F	T	T	T	TTT	**T**	T
T	F	F	T	T	F	T	F	T	T	T	FTT	**F**	F
F	T	T	F	F	T	T	T	F	F	T	TTF	**T**	T
F	T	F	F	F	T	T	T	F	F	T	FFF	**T**	F
F	F	T	F	F	F	T	T	F	T	T	TTF	**T**	T
F	F	F	F	F	F	T	T	F	T	F	FFF	**T**	F

37. valid argument

p	q	r									
			3	2	1	5	4	6			
			[r	∧	(r	→~	q)	∧	(p∨q)]	→	p
T	T	T	T	F	T	F	F	F	TTT	**T**	T
T	T	F	F	F	F	T	F	F	TTT	**T**	T
T	F	T	T	T	T	T	T	T	TTF	**T**	T
T	F	F	F	F	F	T	T	F	TTF	**T**	T
F	T	T	T	F	T	F	F	F	FTT	**T**	F
F	T	F	F	F	F	T	F	F	FTT	**T**	F
F	F	T	T	T	T	T	T	F	FFF	**T**	F
F	F	F	F	F	F	T	T	F	FTF	**T**	F

39. invalid argument

			2 1 5	4 3	8	6	7
p	q	r	$[(q \to\sim p) \wedge (r \to\sim q)]$		$\to$	$(\sim p \to r)$	
T	T	T	T F F	F T F F	**T**	F T T	
T	T	F	T F F	F F T F	**T**	F T F	
T	F	T	F T F	T T T T	**T**	F T T	
T	F	F	F T F	T F T T	**T**	F T F	
F	T	T	T T T	F T F F	**T**	T T T	
F	T	F	T T T	T F T F	**F**	T F F	
F	F	T	F T T	T T T T	**T**	T T T	
F	F	F	F T T	T F T T	**F**	T F F	

41. invalid argument
    p: The product has a lower price.
    q: The product has quality.
    r: The product is less reliable.

			2 1 8	3 5 4	7 6 9	10
p	q	r	$\{(p \to\sim q) \wedge [(\sim p \vee \sim q) \to\sim r] \wedge p\}$		$\to r$	
T	T	T	T F F	F F F F	T F F T	**T** T
T	T	F	T F F	F F F F	T T F T	**T** F
T	F	T	T T T	F F T T	F F F T	**T** T
T	F	F	T T T	T F T T	T T T T	**F** F
F	T	T	F T F	F T T F	F F F F	**T** T
F	T	F	F T F	T T T F	T T F F	**T** F
F	F	T	F T T	F T T T	F F F F	**T** T
F	F	F	F T T	T T T T	T T F F	**T** F

43. valid argument
    p: You buy from a reputable breeder.
    q: Your puppy requires shots.
    r: It will cost you extra.

			2 1 5	3 4	7 6	9 8
p	q	r	$\{(p \to\sim q) \wedge (\sim p \to r) \wedge \sim r\}$			$\to\sim q$
T	T	T	T F F	F F T T	F F	**T** F
T	T	F	T F F	F F T F	F T	**T** F
T	F	T	T T T	T F T T	F F	**T** T
T	F	F	T T T	T F T F	T T	**T** T
F	T	T	F T F	T T T T	F F	**T** F
F	T	F	F T F	F T F F	F T	**T** F
F	F	T	F T T	T T T T	F F	**T** T
F	F	F	F T T	F T F F	F T	**T** T

Section 2.5: Verifying Arguments

45. valid argument
    p: Dave is alone tonight.
    q: Dave will come to the party.
    r: Dave will work on his term paper tonight.

			2 1 7 3 5 4 6 9 8 11 10
p	q	r	$\{(p \to \sim q) \wedge [(\sim p \vee \sim q) \to r] \wedge \sim r\} \to \sim p$
T	T	T	T F F F F F F T T F F **T** F
T	T	F	T F F F F F F T F F T **T** F
T	F	T	T T T T F T T T T F F **T** F
T	F	F	T T T F F T T F F F T **T** F
F	T	T	F T F T T T F T T F F **T** T
F	T	F	F T F F T T F F F F T **T** T
F	F	T	F T T T T T T T T F F **T** T
F	F	F	F T T F T T T F F F T **T** T

Note: In the last conjunction before the conditional — all F's appear. This implies that the conclusion is irrelevant. The conclusion could have been "Therefore Nadia and Adam will also go to the party." and the argument would have been valid.

47. invalid argument
    p: Jaimie is fluent in Spanish.
    q: Jaimie will work in Madrid.
    r: Jaimie will visit Mexico.

			2 1 6 3 5 4 8
p	q	r	$[p \wedge (p \to q) \wedge (\sim r \vee \sim q)] \to r$
T	T	T	T T T T T F F F F **T** T
T	T	F	T T T T T T T T F **F** F
T	F	T	T F T F F F F T T **T** T
T	F	F	T F T F F T T T T **T** F
F	T	T	F F F T T F F F F **T** T
F	T	F	F F F T T F T T F **T** F
F	F	T	F F F T F F F T T **T** T
F	F	F	F F F T F F T T T **T** F

49. and 51. No solutions provided

53. $a \wedge b$
    $b \to c$
    $d \to \sim c$
    - - - - -
    $\therefore \sim d$

   (1) If we assume that $a \wedge b$ is true, then b must be true.
   (2) By the law of detachment if b and $b \to c$ are true, then c is true.
   (3) Because $d \to \sim c$ is equivalent to its contrapositive, we know $c \to \sim d$ is true.
   (4) By the law of detachment again, with c true and $c \to \sim d$ true we conclude $\sim d$ is true.

Section 2.6 Using Euler Diagrams To Verify Syllogisms

1. A syllogism is valid if whenever all its premises are true, then the conclusion is also true.

3. Whenever you draw an Euler diagram, you will always draw some extra conditions that were not stated as premises.

5. and 7. Many diagrams are possible.

9. valid

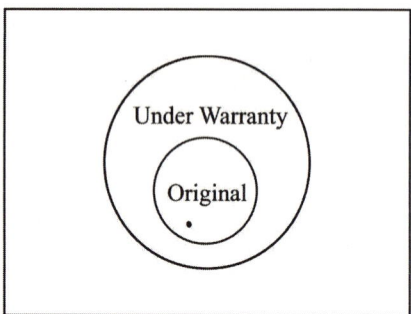

15. invalid

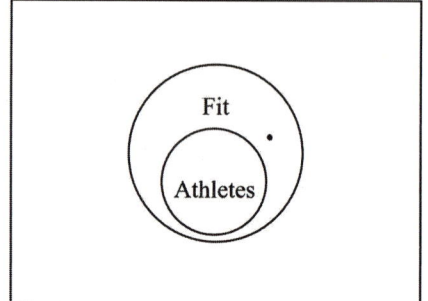

11. invalid

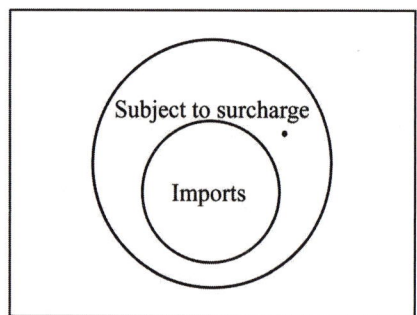

17. valid

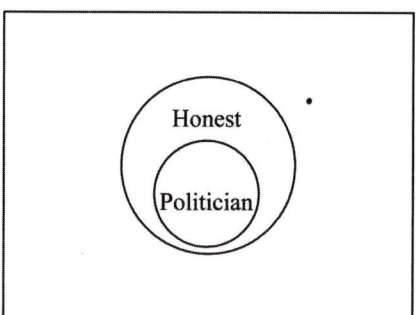

13. invalid

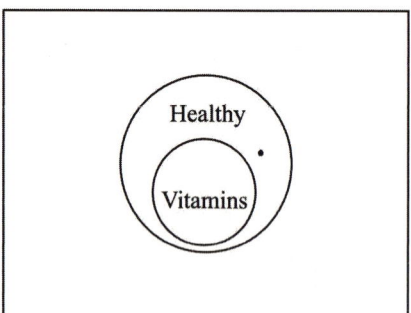

19. invalid

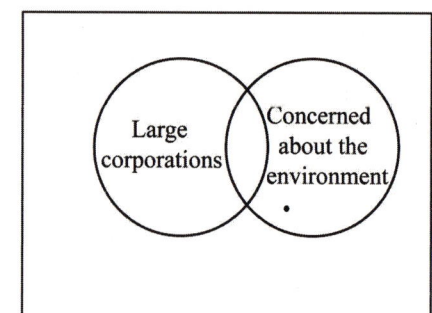

21. invalid

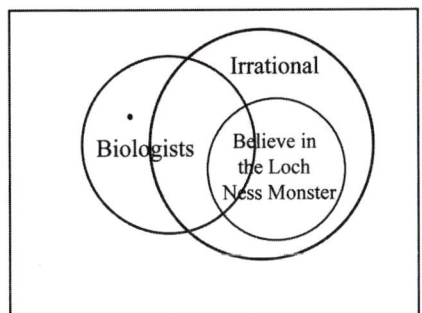

23. valid

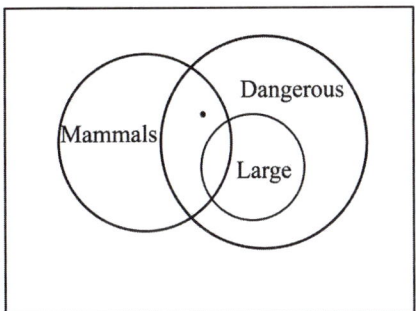

25. Some taxes should be abolished.

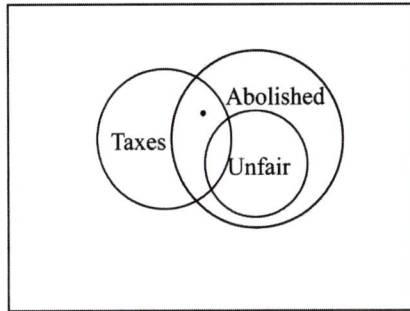

27. Some teams that wear red uniforms do not play in a domed stadium.

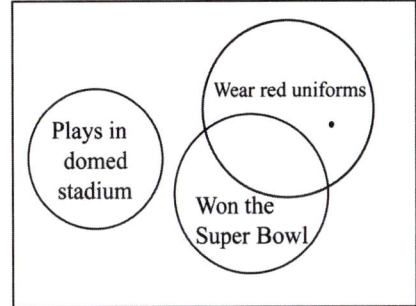

Note: The "Plays in domed stadium" and "Wear red uniforms" could be overlapping (like logic would suggest), but the conclusion would be the same.

29. – 33. No solutions provided

Chapter Review Exercises

1. a) inductive
   b) deductive

2. Answers may vary. Possible answers include: Inductive reasoning is the process of drawing a general conclusion by observing a pattern in specific instances. In deductive reasoning, we use accepted facts and general principles to arrive at a specific conclusion.

3. a) 27
   b) 47

4.
		X
		X

5. $5+43$, $7+41$, $11+37$, $17+31$, or $19+29$

6. In this exercise, only the answer to Part b, the expected conjecture, and Part c, the algebraic justification, are provided.

   In this trick, you will always get twice the number you started with.
   a) Call the number n.
   b) $8n$
   c) $8n+12$
   d) $\dfrac{8n+12}{4} = \dfrac{4(2n+3)}{4} = 2n+3$
   e) $(2n+3)-3 = 2n+3-3 = 2n$

7. a) Not a statement
   b) Statement
   c) Not a statement

8. a) $\sim b \vee s$
   b) $\sim(b \wedge \sim s)$

9. a) It is not true that Antonio is fluent in Spanish and he has not lived in Spain for a semester.
   b) Antonio is not fluent in Spanish or he has not lived in Spain for a semester.

10. a) It is not true that all writers are passionate.
       There is at least one writer who is not passionate.
    b) It is not true that some graduates received several job offers.
       No graduates received several job offers. –or– All graduates do not receive several job offers.

11. a) True

		2 1
p	q	$p \wedge (\sim q)$
T	F	T **T** T

    b) False

		2 1
p	q	$\sim(p \vee q)$
T	F	**F** TTF

    c) True

		2 1
p	q	$\sim(p \wedge q)$
T	F	**T** TFF

12. a) $2^3 = 8$
    b) $2^5 = 32$

13. a)

		3 2 1
p	q	$\sim(p \vee \sim q)$
T	T	**F** TTF
T	F	**F** TTT
F	T	**T** FFF
F	F	**F** FTT

    b)

			3 2 1 5 4
p	q	r	$\sim(p \vee \sim q) \wedge \sim r$
T	T	T	F TTF **F** F
T	T	F	F TTF **F** T
T	F	T	F TTT **F** F
T	F	F	F TTT **F** T
F	T	T	T FFF **F** F
F	T	F	T FFF **T** T
F	F	T	F FTT **F** F
F	F	F	F FTT **F** T

14. a) p: You qualify for extended warranty.
    q: You qualify for the maintenance contract.
    $\sim (p \vee q) \Leftrightarrow \sim p \wedge \sim q$
    You do not qualify for the extended warranty and you do not qualify for the free maintenance contract.

   b) p: I will sign the lease.
    q: I will accept the housing agreement.
    $\sim (\sim p \vee \sim q) \Leftrightarrow p \wedge q$
    I will sign the lease and I will accept the housing agreement.

15. a) $\sim (p \wedge \sim q)$ is logically equivalent to $\sim p \vee q$ by DeMorgan's laws.

   b) $\sim (p \vee \sim q) \wedge \sim (p \vee q)$ is not logically equivalent to $p \vee (p \wedge q)$.

		3 2 1 6 5 4					2 1
p	q	$\sim (p \vee \sim q) \wedge \sim (p \vee q)$			p	q	$p \vee (p \wedge q)$
T	T	F T T F **F** F TTT			T	T	T **T** TTT
T	F	F T T T **F** F TTF			T	F	T **T** TFF
F	T	T F F F **F** F FTT			F	T	F **F** FFT
F	F	F F T T **F** T FFF			F	F	F **F** FFF

16. a) True

		2 1 4 3
p	q	$\sim (p \vee q) \to \sim p$
T	F	F TTF **T** F

  b) False

			1 3 2
p	q	r	$(p \wedge q) \leftrightarrow (q \vee r)$
T	F	T	TFF **F** FTT

  c) True

			1 3 2 4
p	q	r	$(\sim p \vee \sim q) \to r$
T	F	T	F T T **T** T

17. a)

		1 2
p	q	$\sim p \to q$
T	T	F **T** T
T	F	F **T** F
F	T	T **T** T
F	F	T **F** F

   b)

			2 1 5 4 3
p	q	r	$\sim (p \wedge r) \leftrightarrow \sim (p \vee q)$
T	T	T	F TTT **T** F TTT
T	T	F	T TFF **F** F TTT
T	F	T	F TTT **T** F TTF
T	F	F	T TFF **F** F TTF
F	T	T	T FFT **F** F FTT
F	T	F	T FFF **F** F FTT
F	F	T	T FFT **T** T FFF
F	F	F	T FFF **T** T FFF

18. Converse: If we cannot recover the data, then the disk drive has been damaged.
    Inverse: If the disk drive has not been damaged, then we can recover the data.
    Contrapositive: If we can recover the data, then the disk drive has not been damaged.

19. a) If the Knicks get to the finals, then they beat the Lakers.
    b) If you are an astronaut, then you have a pilot's license.

20. a) Fallacy of the Inverse (invalid argument)
    p: You invest in stock.
    q You risk your savings.

    $p \to q$

    $\sim p$

    $\therefore \sim q$

    b) Law of Contraposition (valid argument)
    p: Felisha enjoys spicy food.
    q: Felisha enjoys Cajun chicken.

    $p \to q$

    $\sim q$

    $\therefore \sim p$

21. invalid argument

			1 3	2	6	4	5	7	
p	q	r	$\{\sim p \wedge (q \to p) \wedge$		$[(p \vee q)$	$\to r]\}$		$\to r$	
T	T	T	F F	TTT	F	TTT	T T	**T**	T
T	T	F	F F	TTT	F	TTT	F F	**T**	F
T	F	T	F F	FTT	F	TTF	T T	**T**	T
T	F	F	F F	FTT	F	TTF	F F	**T**	F
F	T	T	T F	TFF	F	FTT	T T	**T**	T
F	T	F	T F	TFF	F	FTT	F F	**T**	F
F	F	T	T T	FTF	T	FFF	T T	**T**	T
F	F	F	T T	FTF	T	FFF	T F	**F**	F

22. valid argument
    p: The product has a higher price.
    q: The product has quality.
    r: The product is reliable.

			1	4	2	3	5	6	
p	q	r	$\{(p \to q) \wedge$	$[(p \vee q)$	$\to r]$	$\wedge p\}$		$\to r$	
T	T	T	TTT	T TTT	TT	TT	**T**	T
T	T	F	TTT	F TTT	FF	FT	**T**	F
T	F	T	TFF	F TTF	TT	FT	**T**	T
T	F	F	TFF	F TTF	FF	FT	**T**	F
F	T	T	FTT	T FTT	TT	FF	**T**	T
F	T	F	FTT	F FTT	FF	FF	**T**	F
F	F	T	FTF	T FFF	TT	FF	**T**	T
F	F	F	FTF	T FFF	TF	FF	**T**	F

23. invalid argument

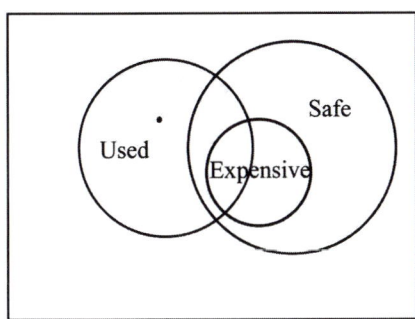

Chapter Test

1. Answers may vary. Possible answers include: Inductive reasoning is the process of drawing a general conclusion by observing a pattern in specific instances. Deductive reasoning uses accepted facts and general principles to arrive at a specific conclusion.

2. a) 23
   b) 49

3. $3+47$, $7+43$, $13+37$, $19+31$, or $29+31$

4. a) Statement b) Not a statement

5. a) It is not true that all rock starts are fine musicians.
      There is at least one rock star who is not a fine musician.
   b) It is not true that some dogs are aggressive.
      No dogs are aggressive. –or– All dogs are not aggressive.

6. a) $p \vee \sim f$
   b) $\sim(\sim p \wedge f)$

7. a) It is not true that The Mets will win the series or Pedro will not win the Cy Young Award..
   b) The Mets will not win the series and Pedro will not win the Cy Young Award..

8. $2^5 = 32$

9. a) True b) False

		3	2 1
p	q	$\sim(p \vee \sim q)$	
T	F	**F**	T T T

		1 3 2
p	q	$\sim p \wedge \sim q$
T	F	F **F** T

10. a)

		3 2 1
p	q	~(p∧~q)
T	T	**T** T F F
T	F	**F** T T T
F	T	**T** F F F
F	F	**T** F F T

b)

			1 2 3 5 4
p	q	r	(~p ∨ ~q) ∧ r
T	T	T	F F F **F** T
T	T	F	F F F **F** F
T	F	T	F T T **T** T
T	F	F	F T T **F** F
F	T	T	T T F **T** T
F	T	F	T T F **F** F
F	F	T	T T T **T** T
F	F	F	T T T **F** F

11. a) If one surfs the Internet, then one is well informed.
    b) If I hold the apartment, then you will give me a deposit.

12. a) *p*: You can take the final.
    q: You can write a term paper.
    ~(p ∨ q) ⇔ ~p ∧ ~q
    You cannot take the final exam and you cannot write a term paper.

    b) *p*: I will finish the painting..
    q: I will show it at the gallery.
    ~(~p ∨ ~q) ⇔ p ∧ q
    I will finish the painting and I will show it at the gallery.

13. a) ~(p ∨ ~q) is logically equivalent to ~p ∧ q by DeMorgan's laws.
    b) (~p ∨ ~q) ∧ (~p ∨ q) is not logically equivalent to ~p ∧ q.

		1 3 2 6 4 5
p	q	(~p ∨ ~q) ∧ (~p ∨ q)
T	T	F F F **F** F T T
T	F	F T T **F** F F F
F	T	T T F **T** T T T
F	F	T T T **T** T T F

		1 2
p	q	~p ∧ q
T	T	F **F** T
T	F	F **F** F
F	T	T **T** T
F	F	T **F** F

14. Converse: If it is gold, then it glitters.
    Inverse: If it does not glitter, then it is not gold.
    Contrapositive: If it is not gold, then it does not glitter.

15. a) True

		2 1 4 3
p	q	(p ∨ ~q) → ~q
T	F	T T T **T** T

c) True

		2 1 5 4 3
p	q	(p ∨ ~q) ↔ ~(p ∧ q)
T	F	T T T **T** T T F F

b) True

			1 2 4 3
p	q	r	(~p ∧ q) → ~r
T	F	T	F F F **T** F

16. a)

		1	2	5	4	3	
p	q	(~p	∨ q)	→	~(p	∧ q)	
T	T	F	T T	**F**	F	T T T	
T	F	F	F F	**T**	T	T F F	
F	T	T	T T	**T**	T	F F T	
F	F	T	T F	**T**	T	F F F	

b)

			1	3	2	5	4	
p	q	r	(~p	∨	~q)	↔	r	
T	T	T	F	F	F	**F**	T	
T	T	F	F	F	F	**T**	F	
T	F	T	F	T	T	**T**	T	
T	F	F	F	T	T	**F**	F	
F	T	T	T	T	F	**T**	T	
F	T	F	T	T	F	**F**	F	
F	F	T	T	T	T	**T**	T	
F	F	F	T	T	T	**F**	F	

17. Valid argument

			4	1	3	2	8	5	7	6	9	
p	q	r	{p	∧	[~q	→~r]	∧	[(q ∨ r)	→~	p]}	→	~r
T	T	T	T	T	F T F	F	TTT	F	F	**T**	F
T	T	F	T	T	F T T	F	TTF	F	F	**T**	T
T	F	T	T	F	T F F	F	FTT	F	F	**T**	F
T	F	F	T	T	T T T	T	FFF	T	F	**T**	T
F	T	T	F	F	F T F	F	TTT	T	T	**T**	F
F	T	F	F	F	F T T	F	TTF	F	T	**T**	T
F	F	T	F	F	T F F	F	FTT	T	T	**T**	F
F	F	F	F	F	T T T	F	FFF	T	T	**T**	T

18. a) Fallacy of the Inverse (invalid argument)
   p: It ain't broke.
   q don't fix it.

   p → q
   ~p
   ─────
   ∴ ~q

   b) Disjunctive Syllogism (valid argument)
   p: I'll major in music.
   q: I'll major in art history.

   p ∨ q
   ~p
   ─────
   ∴ q

19. invalid argument
   p: The concert is cancelled.
   q: You will get a refund.
   r: You will get a replacement ticket.

			2	1	8	3	4	9	5	7	6	10	
p	q	r	{[p	→ (q ∨ r)]	∧	[~r	→ q]	∧	[~p	∨ ~q]}	→	r	
T	T	T	T	T TTT	T	F TT	F	F	F F	**T**	T	
T	T	F	T	T TTF	T	T TT	F	F	F F	**T**	F	
T	F	T	T	F FTT	F	F TF	F	F	T T	**T**	T	
T	F	F	T	F FFF	F	T FF	F	F	T T	**T**	F	
F	T	T	F	T TTT	T	F TT	T	T	T F	**T**	T	
F	T	F	F	T TTF	T	T TT	T	T	T F	**F**	F	
F	F	T	F	T FTT	F	F TF	T	T	T T	**T**	T	
F	F	F	F	T FFF	F	T FF	T	T	T T	**T**	F	

20. invalid argument

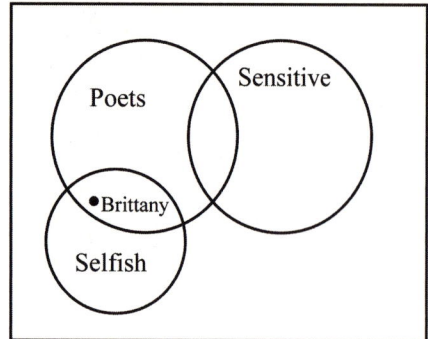

Of Further Interest: Fuzzy Logic

1. There are an infinite number of values between 0 and 1, inclusive.

2. 1 − the truth value of p

3. the minimum of the truth values for p and q

4. the maximum of the truth values for p and q

5. If we only allowed the values 1 (for true) and 0 (for false), then the rules for computing truth values in fuzzy logic are exactly the same as the rules for computing truth tables.

6. – 14. No solutions provided

15. $1 - 0.95 = 0.05$

16. $1 - 0.15 = 0.85$

17. $1 - 0.75 = 0.25$

18. $1 - 0.40 = 0.60$

19. $0.75 \vee 0.85 = 0.85$

20. $0.75 \wedge 0.85 = 0.75$

21. $0.75 \wedge \sim 0.85 = 0.75 \wedge 0.15 = 0.15$

22. $\sim(0.75 \vee 0.85) = \sim 0.85 = 0.15$

23. $0.27 \wedge \sim(0.64) = 0.27 \wedge 0.36 = 0.27$

24. $\sim(0.27 \vee 0.64) = \sim 0.64 = 0.36$

25. $\sim(0.27 \vee \sim 0.64) = \sim(0.27 \vee 0.36) = \sim 0.36 = 0.64$

26. $(0.27 \vee 0.71) \wedge (0.27 \wedge \sim 0.64) = 0.71 \wedge (0.27 \wedge 0.36) = 0.71 \wedge 0.27 = 0.27$

27. $\sim(0.27 \vee \sim 0.64) \wedge \sim 0.71 = \sim(0.27 \vee 0.36) \wedge 0.29 = \sim 0.36 \wedge 0.29 = 0.64 \wedge 0.29 = 0.29$

28. $0.27 \rightarrow 0.71 = \sim 0.27 \vee 0.71 = 0.73 \vee 0.71 = 0.73$

29. $0.64 \rightarrow (0.27 \vee 0.71) = \sim 0.64 \vee (0.27 \vee 0.71) = 0.36 \vee 0.71 = 0.71$

30. $\sim(0.27 \vee 0.64) \rightarrow \sim(0.64 \wedge 0.71) = (0.27 \vee 0.64) \vee \sim(0.64 \wedge 0.71) = 0.64 \vee \sim 0.64 = 0.64 \vee 0.36 = 0.64$

31.

	i	$\sim i$	Marketing Trainee c	Marketing Trainee $\sim i \vee c$	Data Analyst c	Data Analyst $\sim i \vee c$
Salary	0.70	0.30	0.60	0.60	0.65	0.65
Interesting	0.80	0.20	0.50	0.50	0.80	0.80
Work with People	0.60	0.40	0.90	0.90	0.30	0.40
Flexible Hours	0.75	0.25	0.80	0.80	0.60	0.60
Conjunction of all $\sim i \vee c$				0.50		0.40

According to this rating, it is better for you to take the Marketing Trainee position.

32.

	i	$\sim i$	Home c	Home $\sim i \vee c$	Apartment c	Apartment $\sim i \vee c$
Cost	0.70	0.30	0.80	0.80	0.65	0.65
Close to Job	0.60	0.40	0.40	0.40	0.80	0.80
Adequate Space	0.55	0.45	0.90	0.90	0.75	0.75
Near City Attractions	0.65	0.35	0.55	0.55	0.90	0.90
Conjunction of all $\sim i \vee c$				0.40		0.65

According to this rating, it is better for you to continue to rent an apartment.

33.

	i	$\sim i$	LSU c	LSU $\sim i \vee c$	GOS c	GOS $\sim i \vee c$	MFU c	MFU $\sim i \vee c$
Size (not too large)	0.60	0.40	0.95	0.95	0.60	0.60	0.40	0.40
Cost	0.80	0.20	0.40	0.40	0.80	0.80	0.30	0.30
Academics	0.75	0.25	0.80	0.80	0.70	0.70	0.65	0.65
Social Life	0.90	0.10	0.55	0.55	0.75	0.75	0.70	0.70
Close to Home	0.30	0.70	0.70	0.70	0.65	0.70	0.95	0.95
Conjunction of all $\sim i \vee c$				0.40		0.60		0.30

According to this rating, it is better for her to attend Good Old State (GOS).

34.

	i	$\sim i$	A c	A $\sim i \vee c$	B c	B $\sim i \vee c$	C c	C $\sim i \vee c$
Past Performance	0.85	0.15	0.80	0.80	0.65	0.65	0.90	0.90
Safety	0.60	0.40	0.45	0.45	0.60	0.60	0.80	0.80
Liquidity	0.75	0.25	0.60	0.60	0.75	0.75	0.75	0.75
Minimum Amount to Invest	0.35	0.65	0.85	0.85	0.60	0.65	0.65	0.65
Management Fee	0.55	0.45	0.45	0.45	0.80	0.80	0.75	0.75
Conjunction of all $\sim i \vee c$				0.45		0.60		0.65

According to this rating, it is better for you to choose investment Option C.

35. – 37. No solutions provided

Chapter 3
GRAPH THEORY: The Mathematics Of Relationships

Section 3.1 Graphs, Puzzles And Map Coloring

1. An odd vertex is connected to an odd number of vertices; an even vertex is connected to an even number of vertices.

3. If FC were traversed and removed, the graph would become disconnected and the remaining graph could not be traced.

5. Every time the vertex is entered by one edge, it must be left by another.

7. No solution provided

9. Connected; A and B are odd vertices; C and D are even vertices.

11. Connected; A, B, E, and F are odd vertices; C and D are even vertices.

13. Not connected; All vertices are even.

15. Connected; A, B, E, and F are odd vertices; C and D are even vertices.

17. Graphs may vary.

   A   B

   C   D

19. not possible

21. Graphs may vary.

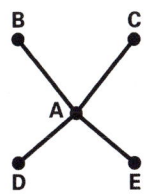

27. Yes, this graph can be traced.

29. No, this graph cannot be traced because it has four odd vertices.

23. Graphs may vary.

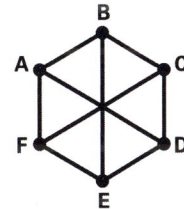

25. Examples and responses may vary.
    Each edge connects to two vertices, so it contributes two of the total of all the degrees. The total of all degrees of all of the edges will be twice the number of edges.

31. No, this graph cannot be traced because it is not connected.

33. No, this graph cannot be traced because it has four odd vertices.

35. This is a Eulerian graph because all vertices are even. One possible Euler circuit is:
    ABFHI GCBEF GECDA

37. This is not an Eulerian graph.
    Duplicate edges GH, GI, and FG.

    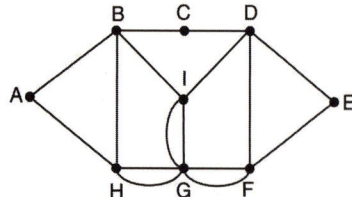

    A possible path is ABCDE FDIGF GIBHG HA.

39. This is not an Eulerian graph.
    Duplicate edges CF, DE, and EI.

    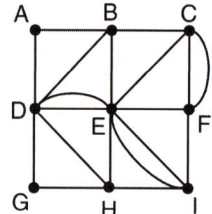

    A possible path is ABDEB CEFCF IEIHE DHGDA.

41. No. If you consider the intersection of two streets as a vertex, then there are four odd vertices. The graph can therefore not be traced, and the ice cream vendor must travel over parts of the route more than once.

43. If you label the intersections that represent odd vertices as follows you can duplicate edges AB, DE, CF, and GH. Follow the possible path indicated in numerical order.

    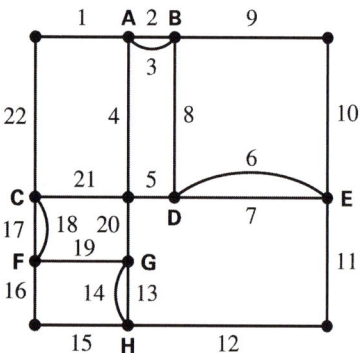

45.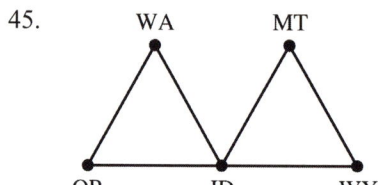

47.

49. The trip is possible because the graph in Exercise 45 has no odd vertices.

51. The trip is not possible because the graph in Exercise 47 has more than two odd vertices.

53. It is not possible. The room can be graphed as follows:

    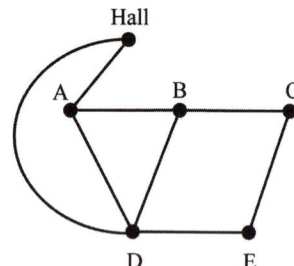

    Since there are two odd vertices, it is possible to trace this graph. The tracing (locking of doors) must begin at one odd vertex and terminate at the other. Here this would imply that the security guard starts in either Room B and ends in Room A or starts in Room A and ends in Room B. Either way, he would lock himself in the building because he cannot enter or exit through the hallway.

55. Answers may vary. The edges in the following graph represent family members that are unfriendly to each other.

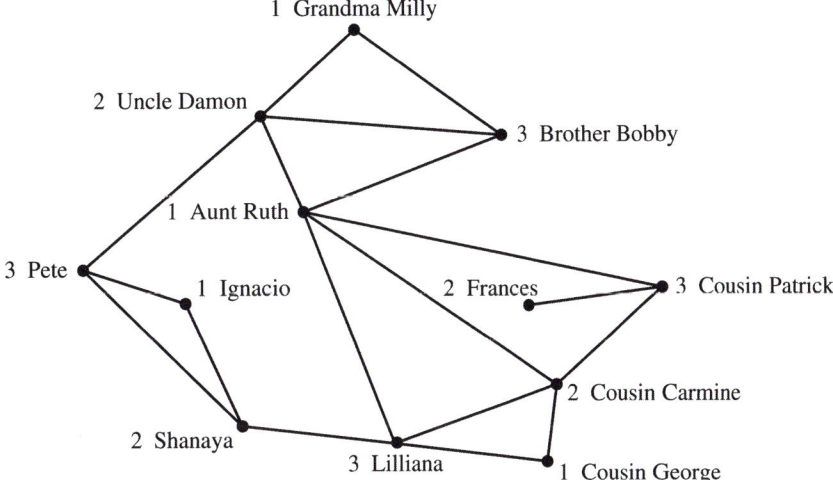

It is possible to color this graph with three colors (indicated by 1, 2, and 3 on the graph). This tells us that three tables will be satisfactory. A satisfactory seating arrangement would be:

Table 1) Grandma Milly, Ignacio, Aunt Ruth, and Cousin George,

Table 2) Uncle Damon, Shanaya, Cousin Carmine, and Frances,

Table 3) Pete, Lilliana, Cousin Patrick, and Brother Bobby.

57. Answers may vary.
The edges in the following graph represent committees that would have a conflict because they have common members.

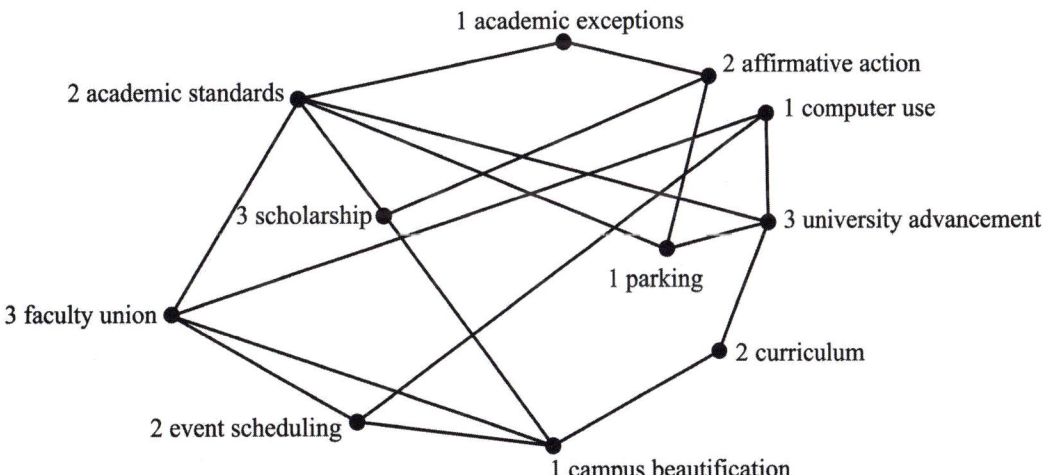

It is possible to color this graph with three colors (indicated by 1,2, and 3). This tells us that three times will be satisfactory to schedule meetings. A satisfactory schedule would be:

Time 1) academic exceptions, computer use, campus beautification, and parking,

Time 2) academic standards , affirmative action, event scheduling, and curriculum,

Time 3) university advancement, faculty union, and scholarship.

54 Chapter 3: Graph Theory

59. Answers may vary.
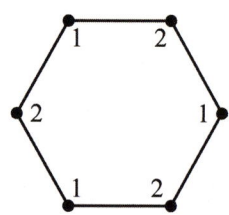

61. Answers may vary.
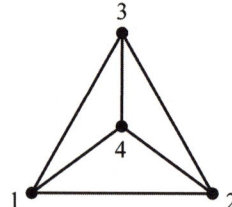

63. One possible sequence is C, F, E, C#, D, A, E^b, B.

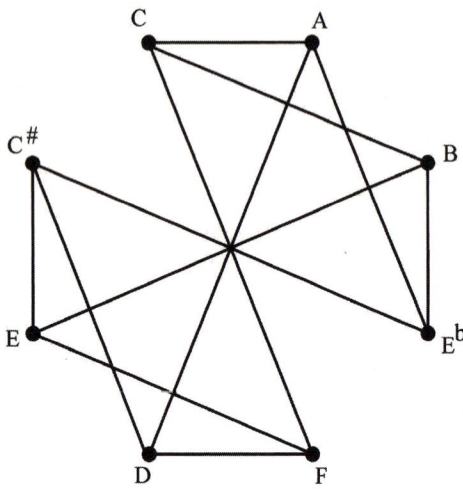

Section 3.2 The Traveling Salesman Problem

1. Two vertices remain.

3. Choose the vertex that is connected to the present vertex by the edge with the smallest weight.

5. A Hamilton circuit goes through every vertex but does not travel over every edge as an Euler circuit does.

7. They are faster than the brute force algorithm.

9. Answers may vary.
   a) Possible answers are ADBCEA, ADCBEA, ADCEBA, or ADECBA.
   b) Possible answers are EBCDAE or EBADCE.

11. Answers may vary.
    a) Possible answers are ADCBFEA, ADCEFBA, ADBFCEA, ADBCEFA, ADBCFEA, ADEFCBA, ADECBFA, or ADECFBA.
    b) Possible answers are EADCBFE, EABDCFE, EABFCDE, EADBFCE, EADBCFE, EAFBDCE, EAFCBDE, or EAFBCDE.

13. Possible circuits are ABCDEA, ABCEDA, ABDCEA, and ABECDA.

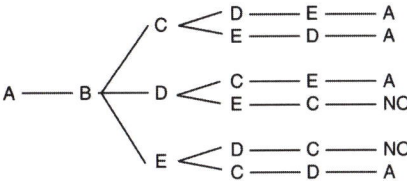

17. K_6

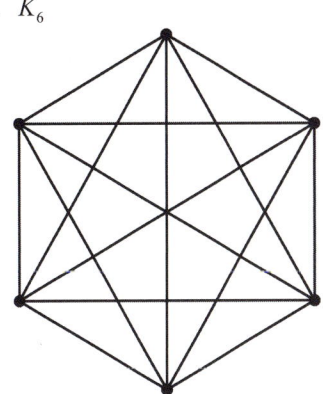

15. $(7-1)! = 6! = 720$

19. a) The weight of path AGEDCB would be $2 + 4 + 2 + 8 + 4 = 20$.
    b) The weight of path ABCDGEF would be $3 + 4 + 8 + 1 + 4 + 3 = 23$.

21. Since Hamilton circuits occur in pairs, we need only calculate the weight of three Hamilton circuits.

Hamilton Circuit	Weight
ABCDA	3+4+5+3=15
ABDCA	3+6+5+1=15
ACBDA	**1+4+6+3=14**

    Since 14 is the smallest number in the right hand column, the circuits ACBDA and ADBCA have minimal weight.

23. Since Hamilton circuits occur in pairs, we need only calculate the weight of three Hamilton circuits.

Hamilton Circuit	Weight
ABCDA	23+14+37+36=110
ABDCA	23+31+37+18=109
ACBDA	**18+14+31+36=99**

    Since 99 is the smallest number in the right hand column, the circuits ACBDA and ADBCA have minimal weight.

25. ABDECA

27. ADCFEBA

29. Start by selecting DE which has the smallest weight (weight 2). Next choose edges AB and CD (weight 3). Edge BD (weight 4) cannot next be chosen because that will create a vertex that joins three edges. We can next choose edge AC (weight 5). Edge AE (weight 6) cannot next be chosen because that will create a vertex that joins three edges. Next choose edge BE (weight 6) to complete the circuit of ABEDCA or ACDEBA.

31. Start by selecting CF which has the smallest weight (weight 4). Next choose edge AD (weight 5). Next choose edge BE (weight 6). Next choose edge CE (weight 7). Edge CD (weight 8) cannot next be chosen because that will create a vertex that joins three edges. Edge AC (weight 9) cannot next be chosen because that will create a vertex that joins three edges. Next choose edge BD (weight 9). Next choose edge AF (weight 13) to complete the circuit of ADBECFA or AFCEBDA.

33. Answers may vary.
    If you start at any vertex, you will see that you cannot follow edges to visit every vertex exactly once. For example, if you start at vertex A and go to B, you must then go to C. At C you have a choice to go to D or E. If you visit D, then you cannot go to E without visiting A again. If you instead go to E from C, then you cannot go to D without visiting A again.

56 Chapter 3: Graph Theory

35. Answers may vary.
If you start at any vertex, you will see that you cannot follow edges to visit every vertex exactly once. For example, if you start at vertex A you would have a choice to visit either B or E or F. If you choose to visit B then you go to C and then D. At D you have a choice to go to either F or E. If you go to F, you must go through A again in order to visit E. Also, if you go to E from D, you must visit A again prior to going to F. If back at A you decided to go to E instead of B initially, you would have similar problems. At E you would then go to D and here you would have the choice to go to F or C. If you go to F from D then you must visit A again prior to going to B. If at D you instead decided to go to C then B, you must visit A again prior to going to F. If back at A you decided to go to F instead of E or B initially, you would have again similar problems. At F you would then go to D and would be then forced to go to E then A would have to be revisited before reaching B and C.

37. a) This is a K_7 drawing. Since this graph has 7 vertices, there will be 6!=720 Hamilton circuits. We won't count reversals so $\dfrac{720}{2} = 360$ circuits need to be considered.

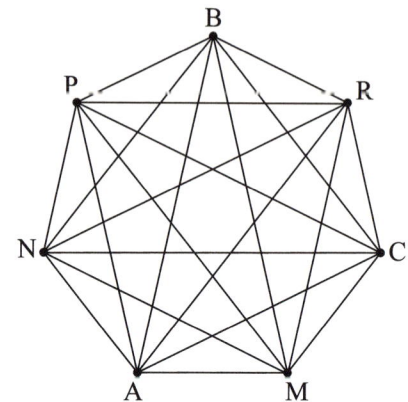

b) If you consider 360 circuits at one per minute, then it would take 360 minutes or 6 hours to do this problem by the brute force algorithm.

c) Using the nearest neighbor algorithm, the path Danielle should go is PNBRAMCP. The cost of this trip would be 210 + 180 + 290 + 90 + 170 + 240 + 420= $1,600.

d) Using the best edge algorithm, the edge with the smallest weight is AR with a weight of 90. The next edge of choice would be MR (weight 120). Edge AM (weight 170) cannot be next chosen because it will create a circuit. Edge BN (weight 180) would be chosen next. Edge NP (weight 210) would be chosen next followed by edge AP (weight 230). Edge PR (weight 240) cannot be chosen because it will create a vertex in which three edges meet. The next choice would be edge CM (weight 240). The remaining choice of BC (weight 380) would be made to complete the circuit. The path would therefore be PNBCMRAP or PARMCBNP. The cost would be 210 + 180 + 380 + 240 + 120 + 90 + 230 = $1,450.

39. Find a table to relate the distance between the stops and the store.

	A	B	D	E	H	F	M
A	0	2	3	6	4	9	3
B	2	0	3	4	4	7	1
D	3	3	0	5	3	8	2
E	6	4	5	0	4	3	3
H	4	4	3	4	0	5	3
F	9	7	8	3	5	0	6
M	3	1	2	3	3	6	0

39. (continued)

    a) ABMDHEFA

    b) Using the best edge algorithm, we can first choose edge BM because is has the lowest weight (weight 1). The next edges that can be chosen are AB and DM (weight 2). Of the edges of weight 3, you don't want to choose EM, AM, HM or BD because any of these will cause three edges to meet at one vertex. The next edge of weight 3 to choose would be EF. You now have to make a choice between edges AD and DH. They both have weight 3 and can't both be chosen because the combination of these two edges will cause three edges will meet at a vertex. AD cannot be chosen however because it will close a circuit that does not include all the stops. Therefore, if you next choose DH (weight 3), then the next to be chosen would be EH (weight 4). The next edge to choose would be AF (weight 9) to complete the circuit. The circuits that could be found would be ABMDHEFA or AFEHDMBA

41. a) Find a table to relate the distance between the stops and the company.

	X	A	B	C	D	E	F	P
X	0	9	8	9	8	5	2	7
A	9	0	3	2	3	4	7	6
B	8	3	0	5	4	3	6	5
C	9	2	5	0	1	6	7	4
D	8	3	4	1	0	5	6	3
E	5	4	3	6	5	0	3	4
F	2	7	6	7	6	3	0	5
P	7	6	5	4	3	4	5	0

    Using nearest neighbor algorithm the path followed should be XFEBACDPX

    b) Using the best edge algorithm, we can first choose edge CD because it has the lowest weight (weight 1). The next edges that can be chosen are AC and XF (weight 2). Of the edges of weight 3, you can choose EF, DP, AB and BE. If edge AD were chosen then it would cause three vertices to meet. The next edge to choose would be XP (weight 7) to complete the circuit. The circuits that could be found would be XFEBACDPX or XPDCABEFX.

43. A K_{10} figure would have 9!=362,880 circuits. We won't count reversals so $\frac{362,880}{2} = 181,440$ circuits need to be considered. If a computer can examine 1,000 Hamilton circuits per second, then it would take the computer 181.44 seconds. This is just over 3 minutes.

45. Using the nearest neighbor algorithm starting at A we have the circuit **ABDECA**. The weight of this circuit is $3 + 4 + 2 + 7 + 5 = $ **21**.
    Using the nearest neighbor algorithm starting at B we have the circuit **BACDEB**. The weight of this circuit is $3 + 5 + 3 + 2 + 6 =$**19**.
    Using the nearest neighbor algorithm starting at C we have the circuit **CDEBAC** or **CDEABC**. The weight of these circuits are $3 + 2 + 6 + 3 + 5 = $ **19** and $3 + 2 + 6 + 3 + 9 = $ **23**, respectively.
    Using the nearest neighbor algorithm starting at D we have the circuit **DEABCD** or **DEBACD**. The weight of these circuits are $2 + 6 + 3 + 9 + 3 = $ **23** and $2 + 6 + 3 + 5 + 3 = $ **19**, respectively.
    Using the nearest neighbor algorithm starting at E we have the circuit **EDCABE**. The weight of this circuit is $2 + 3 + 5 + 3 + 6 = $ **19**.
    The smallest weight among these is 19. There is essentially one circuit (not counting reversing the order) here that starts and stops at A. It is ABEDCA.

47. No solution provided

49. Since all vertices connect to each other in this type of figure, you will always be able to create a Hamilton circuit.

58 Chapter 3: Graph Theory

51. If one constructs a tree for a K_n similar to the one drawn in Example 3, then one can see that given a starting vertex, there are $n-1$ directions to go. For each of these branches, there are $n-2$ directions to go (because two of the vertices cannot be revisited). This continues until all of the vertices are visited and the circuit is completed.

53. and 55. No solutions provided

Section 3.3 Directed Graphs

1. because the rumors only travel in one direction

3. the total of the direct and two-stage influences of each committee member

5. when the relationship goes in only one direction

7. I and B are in a block of vertices that do not have any directed edges leaving them. (Meaning J, D, or A could not be infected.)

9. The graph would have to have a directed edge or path from F to I and from F to J. We cannot do this by changing the direction of only one edge.

11. a) ABCE or ABCDE
    b) ABC
    c) ABCE
    d) Not possible because you must first go from A to B. From B you cannot go directly to E.
    e) Answers may vary. Possible answers include ABCEDA or ABCEBA.

13. a) BCE
    b) Answers may vary. Possible answers included ABCFG or ACFG.
    c) ACFGCE
    d) Not possible because you cannot go to D. It is possible to go from D.
    e) ABCFGC

15.

		To				
		A	B	C	D	E
	A	1	1	1	0	0
	B	1	1	1	1	1
From	C	1	1	0	2	2
	D	1	2	0	1	1
	E	2	1	1	1	1

17.

		To						
		A	B	C	D	E	F	G
	A	0	1	2	0	1	1	0
	B	0	0	1	0	1	1	0
	C	0	0	0	0	1	1	1
From	D	0	0	1	0	2	1	0
	E	0	0	0	0	0	0	0
	F	0	0	1	0	0	0	1
	G	0	0	1	0	1	1	0

19. Any of the five neighbors. If Alicia starts, it could spread to Ted, then Mike, then Bob, then Kara. If Bob starts it could spread to Alicia, then Ted, then Mike, then Kara. If Kara starts it could spread to Ted, then Mike, then Bob, then Alicia. If Ted starts, it could spread to Mike, then Kara, then back to Ted, then back to Mike, then Bob, then Alicia. If Mike starts, it could spread to Bob, then Kara, then Ted, then back to Mike, then back to Bob, then Alicia.

21. The only possible sequence would be chief financial officer, then sales manager, then production manager, then company president, then marketing director. If the secretary starts with the sales manager, then he/she will not be able to obtain the signature of the chief financial officer. If the secretary starts with the production manager, then he/she will not be able to get the signatures of the sales manager and the chief financial officer. If the secretary starts with the company president, then he/she will not be able to get the signatures of the sales manager, chief financial officer, and the production manager. If the secretary starts with the marketing director, then he/she will not be able to get the signatures of the chief financial officer, sales manager, production manager, and the company president.

23.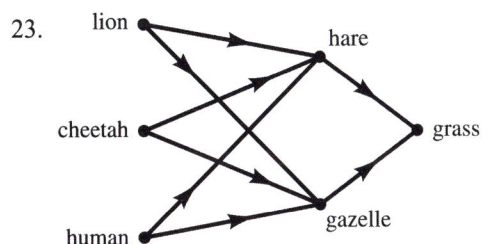

25.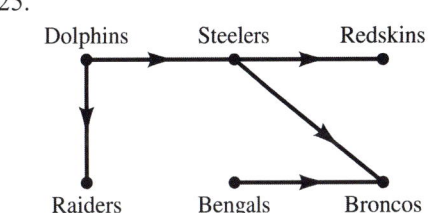

27.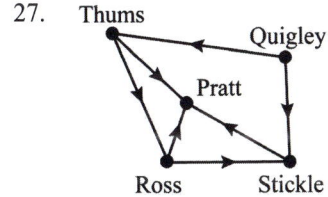

Row	Sum of Entries
Thums	0 + 2 + 0 + 1 + 1 = 4
Pratt	0 + 0 + 0 + 0 + 0 = 0
Quigley	1 + 2 + 0 + 1 + 1 = 5
Ross	0 + 2 + 0 + 0 + 1 = 3
Stickle	0 + 1 + 0 + 0 + 0 = 1

Quigley (5), Thums (4), Ross (3), Stickle (1), and Pratt (0)

		To				
		Thums	Pratt	Quigley	Ross	Stickle
From	Thums	0	2	0	1	1
	Pratt	0	0	0	0	0
	Quigley	1	2	0	1	1
	Ross	0	2	0	0	1
	Stickle	0	1	0	0	0

60   Chapter 3: Graph Theory

29.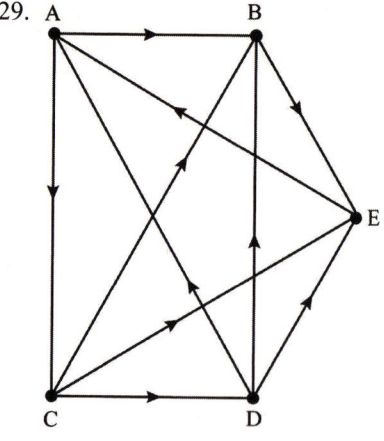

Row	Sum of Entries
A	0 + 2 + 1 + 1 + 2 = 6
B	1 + 0 + 0 + 0 + 1 = 2
C	2 + 2 + 0 + 1 + 3 = 8
D	2 + 2 + 1 + 0 + 2 = 7
E	1 + 1 + 1 + 0 + 0 = 3

C (8), D (7), A (6), E (3), and B (2)

		To				
		A	B	C	D	E
	A	0	2	1	1	2
	B	1	0	0	0	1
From	C	2	2	0	1	3
	D	2	2	1	0	2
	E	1	1	1	0	0

31. and 33. No solutions provided

Chapter Review Exercises

1. a) 12
   b) A, B, C, D, and E are even. F, G, H, and I are odd.
   c) Yes. The graph is connected.
   d) Yes. The bridges are FH, GH, and HI.

2. Answers may vary.
   The objects are represented by vertices. Any two objects that are related are joined by an edge in the graph.

3. Graph (a) can be traced because it has only two odd vertices. Graph (b) cannot be traced because it has more than two odd vertices, namely vertices A, B, D, and E.

4. Answers may vary.
   One possible circuit is ABCDB EDFGH FKHIJ KCA.

5. This is not an Eulerian graph. Duplicate edges BC, EI, HL, and NO so that all vertices are even. One possible route would be ABCDH GCBFG KLHLP ONJKO NMIEF JIEA.

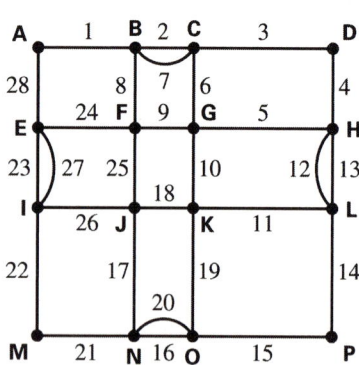

6.

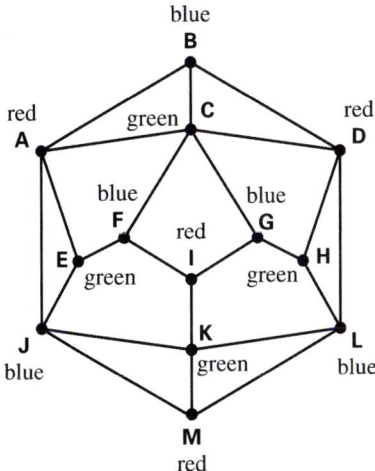

7. ABCDEA, ABCEDA, ABDCEA, ABDECA, ABECDA, and ABEDCA.

8.

Hamilton Circuit	Weight
ABCEDA	4+8+1+2+3=18
ABDECA	**4+4+2+1+2=13**
ACEBDA	2+1+5+4+3=15
ACBEDA	2+8+5+2+3=20
ACEDBA	**2+1+2+4+4=13**
ADBECA	3+4+5+1+2=15
ADEBCA	3+2+5+8+2=20
ADECBA	3+2+1+8+4=18

Since 13 is the smallest number in the right hand column, the circuits ABDECA and ACEDBA have minimal weight.

9. Using the nearest neighbor algorithm, the circuit would be ACEDBA for a weight of 2+1+2+4+4=13.

10. Using the best edge algorithm, the edge with the smallest weight is EC with a weight of 1. The next edges of choice would be DE and AC(weight 2). Edge AD (weight 3) cannot be next chosen because it will create a circuit. Edges AB and BD (weight 4) would be chosen next to complete the circuit. The path would therefore be ABDECA or ACEDBA. The weight would be 4+4+2+1+2=13.

11. a) CDEB
    b) No. There is only one path from C to B. It is CDEB.
    c) HGEBF
    d) FGEB and FHGEB

12. Answers may vary
    Directed graphs are used when the relationship modeled may apply in one direction. For example, object X is related to Y, but Y may not be related to X.

13.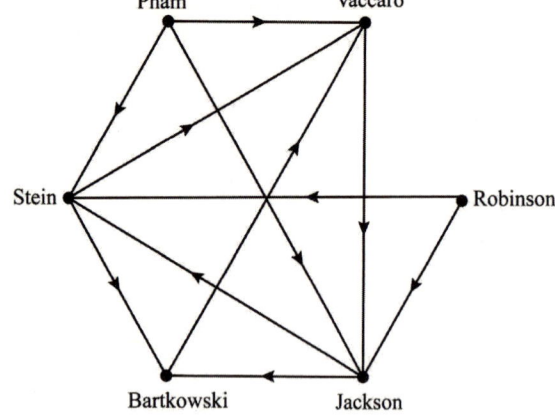

	To					
		Pham	Vaccaro	Robinson	Jackson	Bartkowski	Stein
From	Pham	0	2	0	2	2	2
	Vaccaro	0	0	0	1	1	1
	Robinson	0	1	0	1	2	2
	Jackson	0	2	0	0	2	1
	Bartkowski	0	1	0	1	0	0
	Stein	0	2	0	1	1	0

Committee Member	Sum of Entries
Pham	0 + 2 + 0 + 2 + 2 + 2 = 8
Vaccaro	0 + 0 + 0 + 1 + 1 + 1 = 3
Robinson	0 + 1 + 0 + 1 + 2 + 2 = 6
Jackson	0 + 2 + 0 + 0 + 2 + 1 = 5
Bartkowski	0 + 1 + 0 + 1 + 0 + 0 = 2
Stein	0 + 2 + 0 + 1 + 1 + 0 = 4

Thus, Pham is most influential.

Chapter Test

1. a) 9
   b) E and F are odd; A, B, C, D, G, and H are even.
   c) Yes, the graph is connected.
   d) Yes, EF is a bridge.

2. a) This graph cannot be traced because it has more than two odd vertices, namely A, B, C, E, G, and H.
   b) can be traced because it has two odd vertices.

3. Answers may vary. One possible circuit is ABDACEGIHGFEDFA.

4. ADBCEA, ADBECA, ADCBEA, ADCEBA, ADEBCA, ADECBA.

5. This is not an Eulerian graph. Duplicate the edges as shown so that all vertices are even. One possible route is numbered on the diagram.

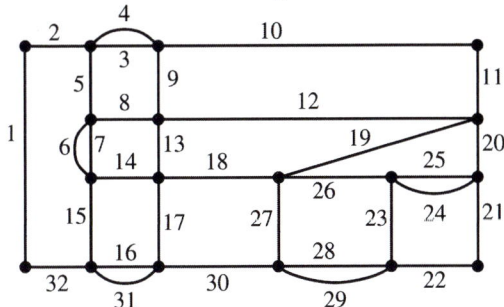

6. Two colors are required.

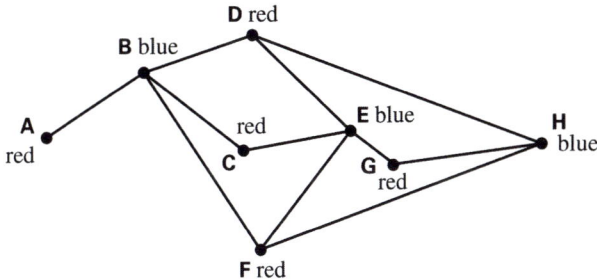

7.

Hamilton Circuit	Weight
ABCEDA	5+5+3+7+8=28
ABDECA	**5+2+7+3+2=19**
ACEBDA	**2+3+4+2+8=19**
ACBEDA	2+5+4+7+8=26
ACEDBA	2+3+7+2+5=19
ADBECA	8+2+4+3+2=19
ADEBCA	8+7+4+5+2=26
ADECBA	8+7+3+5+5=28

Since 19 is the smallest number in the right hand column, the circuits ABDECA and ACEBDA have minimal weight.

8. Using the nearest neighbor algorithm, the circuit would be ACEBDA for a weight of 2+3+4+2+8=19.

9. Using the best edge algorithm, the edges with the smallest weight are AC and BD (weight 2). The next edge of choice would be CE (weight 3). Edge BE (weight 4) cannot be next chosen because it will create a circuit. Edges AB and AD (weight 5) would be chosen next to complete the circuit. The path would therefore be ADBECA or ACEBDA. The weight would be 2+3+4+2+8=19.

10. a) BAEF
    b) There is no path from C to G since no edges go towards G.

11.

	To					
From	A	B	C	D	E	F
A	0	0	1	0	0	0
B	1	0	3	1	0	1
C	0	0	0	0	0	0
D	1	1	2	0	1	2
E	0	0	1	1	0	1
F	0	1	2	1	1	0

Person	Sum of Entries
A	0 + 0 + 1 + 0 + 0 + 0 = 1
B	1 + 0 + 3 + 1 + 0 + 1 = 6
C	0 + 0 + 0 + 0 + 0 + 0 = 0
D	1 + 1 + 2 + 0 + 1 + 2 = 7
E	0 + 0 + 1 + 1 + 0 + 1 = 3
F	0 + 1 + 2 + 1 + 1 + 0 = 5

Thus, D is most influential.

Of Further Interest: Scheduling Projects Using PERT

1. the time required, in months, to build the shell

2. The directed path "Build shell, Assemble shell, Install life-support systems" already implies that "Build shell" precedes "Install life-support systems."

3. No, because it does not lie on a critical path for the project.

4. First, find a critical path for T. Then schedule T to begin after the sum of the times along its critical path.

5. Add the times along a critical path for "End."

6. No, because "Assemble shell" would no longer be on a critical path.

7. a) Begin,A,B,E,I

Directed Path	Time to Complete
Begin,A,B,D,I	3+6+3+2=14 days
Begin,A,B,E,I	**3+6+5+2=16 days**
Begin,A,C,E,I	3+4+5+2=14 days
Begin,A,C,F,I	3+4+6+2=15 days

b) Begin,A,B,E

Directed Path	Time to Complete
Begin,A,B,E	**3+6+5=14 days**
Begin,A,C,E	3+4+5=12 days

c) On day 14

Directed Path	Time to Complete
Begin,A,C,F,H	**3+4+6=13 days**
Begin,A,C,G,H	3+4+3=10 days

d) After 16 days

Directed Path	Time to Complete
Begin,A,B,D,I	3+6+3+2=14 days
Begin,A,B,E,I	**3+6+5+2=16 days**
Begin,A,C,E,I	3+4+5+2=14 days
Begin,A,C,F,I	3+4+6+2=15 days

7. (continued)

   e) 17 days

Directed Path	Time to Complete
Begin,A,B,D,I,J,End	3+6+3+2+1=15 days
Begin,A,B,E,I,J,End	**3+6+5+2+1=17 days**
Begin,A,C,E,I,J,End	3+4+5+2+1=15 days
Begin,A,C,F,I,J,End	3+4+6+2+1=16 days
Begin,A,C,F,H,J,End	**3+4+6+3+1=17 days**
Begin,A,C,G,H,J,End	3+4+3+3+1=14 days

   f) Begin,A,B,E,I,J,End or Begin,A,C,F,H,J,End

8. a) Begin,A,B,E,H

Directed Path	Time to Complete
Begin,A,B,D,H	5+6+3+4=18 days
Begin,A,B,E,H	**5+6+5+4=20 days**
Begin,A,C,E,H	5+4+5+4=18 days

   b) Begin,A,B,E,G

Directed Path	Time to Complete
Begin,A,B,E,G	**5+6+5+5=21days**
Begin,A,C,E,G	5+4+5+5=19 days
Begin,A,C,F,G	5+4+6+5=20 days

   c) Since it will take 5+6+5=16 days to complete the tasks that precede task G, task G should begin on day 17.

   d) After 20 days

Directed Path	Time to Complete
Begin,A,B,D,H	5+6+3+4=18days
Begin,A,B,E,H	**5+6+5+4=20 days**
Begin,A,C,E,H	5+4+5+4=18 days

   e) 22 days

Directed Path	Time to Complete
Begin,A,B,D,H,I,End	5+6+3+4+1=19 days
Begin,A,B,E,H,I,End	5+6+5+4+1=21 days
Begin,A,B,E,G,I,End	**5+6+5+5+1=22 days**
Begin,A,C,E,H,I,End	5+4+5+4+1=19 days
Begin,A,C,E,G,I,End	5+4+5+5+1=20 days
Begin,A,C,F,G,I,End	5+4+6+5+1=21 days

   f) Begin,A,B,E,G,I,End

9. a) Begin,B,E,F,G

Directed Path	Time to Complete
Begin,A,D,G	2+4+2=8 days
Begin,A,E,G	2+2+2=6 days
Begin,B,E,G	3+2+2=7 days
Begin,B,E,F,G	**3+2+3+2=10 days**
Begin,C,F,G	3+3+2=8 days

   b) Begin,B,E,F,H

Directed Path	Time to Complete
Begin,A,E,H	2+2+1=5 days
Begin,A,E,F,H	2+2+3+1=8 days
Begin,B,E,H	3+2+1=6 days
Begin,B,E,F,H	**3+2+3+1=9 days**
Begin,C,F,H	3+3+1=7 days

9. (continued)

c) Begin,B,E,F,G,I,End

Directed Path	Time to Complete
Begin,A,D,G,I,End	2+4+2+4=12 days
Begin,A,E,G,I,End	2+2+2+4=10 days
Begin,A,E,H,I,End	2+2+1+4=9 days
Begin,A,E,F,G,I,End	2+2+3+2+4=13 days
Begin,A,E,F,H,I,End	2+2+3+1+4=12 days
Begin,B,E,G,I,,End	3+2+2+4=11 days
Begin,B,E,H,I,End	3+2+1+4=10 days
Begin,B,E,F,G,I,End	**3+2+3+2+4=14 days**
Begin,B,E,F,H,I,End	3+2+3+1+4=13 days
Begin,C,F,G,I,End	3+3+2+4=12 days
Begin,C,F,H,I,End	3+3+1+4=11 days

d) Since it will take 3+2=5 days to complete the tasks that precede task F, task F should begin on day 6.

Directed Path	Time to Complete
Begin,A,E,F	2+2+3=7 days
Begin,B,E,F	**3+2+3=8 days**
Begin,C,F	3+3=6 days

e) From part a) we see that a critical path for G is Begin, B, E, F, G. Since it will take 3+2+3=8 days to complete the tasks that precede task G, task G should begin on day 9.

f) From part c) we determined that the critical path for "End" was Begin,B,E,F,G,I,End. This implies that it should take 14 days to complete the project.

10. a) Begin,C,E,G

Directed Path	Time to Complete
Begin,A,D,G	2+3+4=9 days
Begin,B,D,G	3+3+4=10 days
Begin,C,E,G	**3+5+4=12 days**
Begin,C,F,G	3+4+4=11 days

b) Begin,C,E,H

Directed Path	Time to Complete
Begin,C,E,H	**3+5+6=14 days**
Begin,C,F,H	3+4+6=13 days

c) Begin,C,E,H,I,End

Directed Path	Time to Complete
Begin,A,D,G,I,End	2+3+4+3=12 days
Begin,B,D,G,I,End	3+3+4+3=13 days
Begin,C,E,G,I,End	3+5+4+3=15 days
Begin,C,E,H,I,End	**3+5+6+3=17 days**
Begin,C,F,G,I,End	3+4+4+3=14 days
Begin,C,F,H,I,End	3+4+6+3=16 days

d) From part b) we determined that the critical path for task H was Begin,C,E,H. Since it will take 3+5=8 days to complete the tasks that precede task H, task H should begin on day 9.

e) From part a) we determined that the critical path for task G was Begin,C,E,G. Since it will take 3+5=8 days to complete the tasks that precede task G, task G should begin on day 9.

f) From part c) we determined that the critical path for "End" was Begin,C,E,H,I,End. This implies that it should take 17 days to complete the project.

11. Both tasks A and B can begin on day 1. We must however find the critical paths for tasks C,D,E,F,G,and H.

   C: Since it will take 2 days to complete the tasks that precede task C, task C should begin on day 3.

Directed Path	Time to Complete
Begin,A,C	1+3=4 days
Begin,B,C	**2+3=5 days**

   D: Since it will take 2 days to complete the tasks that precede task D, task D should begin on day 3.

Directed Path	Time to Complete
Begin,A,D	1+5=6 days
Begin,B,D	**2+5=7 days**

   E: Since it will take 2+3=5 days to complete the tasks that precede task E, task E should begin on day 6.

Directed Path	Time to Complete
Begin,A,C,E	1+3+5=9 days
Begin,B,C,E	**2+3+5=10 days**

   F: Since it will take 2+5=7 days to complete the tasks that precede task F, task F should begin on day 8.

Directed Path	Time to Complete
Begin,A,C,F	1+3+4=8 days
Begin,A,D,F	1+5+4=10 days
Begin,B,C,F	2+3+4=9 days
Begin,B,D,F	**2+5+4=11 days**

   G: Since it will take 2+5=7 days to complete the tasks that precede task G, task G should begin on day 8.

Directed Path	Time to Complete
Begin,A,D,G	1+5+2=8 days
Begin,B,D,G	**2+5+2=9 days**

   H: Since it will take 2+5+4=11 days to complete the tasks that precede task H, task H should begin on day 12.

Directed Path	Time to Complete
Begin,A,C,E,H	1+3+5+6=15 days
Begin,A,C,F,H	1+3+4+6=14 days
Begin,A,D,F,H	1+5+4+6=16 days
Begin,A,D,G,H	1+5+2+6=14 days
Begin,B,C,E,H	2+3+5+6=16 days
Begin,B,C,F,H	2+3+4+6=15 days
Begin,B,D,F,H	**2+5+4+6=17 days**
Begin,B,D,G,H	2+5+2+6=15 days

68   Chapter 3: Graph Theory

12. Both tasks A and B can begin on day 1. Task C can begin on day 2 and Task D can begin on day 3. We must however find the critical paths for tasks E,F,G,and H.

   E: Since it will take 1+5=6 days to complete the tasks that precede task E, task E should begin on day 7.

Directed Path	Time to Complete
Begin,A,C,E	**1+5+5=11 days**
Begin,B,E	2+5=7 days

   F: Since it will take 1+5=6 days to complete the tasks that precede task F, task F should begin on day 7.

Directed Path	Time to Complete
Begin,A,C,F	**1+5+2=8 days**
Begin,B,D,F	2+3+2=7 days

   G: Since it will take 2+3=5 days to complete the tasks that precede task G, task G should begin on day 6.

Directed Path	Time to Complete
Begin,A,G	1+8=9 days
Begin,B,D,G	**2+3+8=13 days**

   H: Since it will take 2+3+8=13 days to complete the tasks that precede task H, task H should begin on day 14.

Directed Path	Time to Complete
Begin,A,C,E,H	1+5+5+4=15 days
Begin,A,C,F,H	1+5+2+4=12 days
Begin,A,G,H	1+8+4=13 days
Begin,B,E,H	2+5+4=11 days
Begin,B,D,F,H	2+3+2+4=11 days
Begin,B,D,G,H	**2+3+8+4=17 days**

13. Tasks A, B, and C can begin on day 1. We must however find the critical paths for tasks D,E,F,G,H, I, and J.

   D: Since it will take 2 days to complete the tasks that precede task D, task D should begin on day 3.

Directed Path	Time to Complete
Begin,A,D	1+2=3 days
Begin,B,D	**2+2=4 days**

   E: Since it will take 3 days to complete the tasks that precede task E, task E should begin on day 4.

Directed Path	Time to Complete
Begin,B,E	2+3=5 days
Begin,C,E	**3+3=6 days**

   F: Since it will take 2+2=4 days to complete the tasks that precede task F, task F should begin on day 5.

Directed Path	Time to Complete
Begin,A,D,F	1+2+5=8 days
Begin,B,D,F	**2+2+5=9 days**

   G: Since it will take 3+3=6 days to complete the tasks that precede task G, task G should begin on day 7.

Directed Path	Time to Complete
Begin,A,D,G	1+2+4=7 days
Begin,B,D,G	2+2+4=8 days
Begin,B,E,G	2+3+4=9 days
Begin,C,E,G	**3+3+4=10 days**

13. (continued)

H: Since it will take 3+3=6 days to complete the tasks that precede task H, task H should begin on day 7.

Directed Path	Time to Complete
Begin,B,E,H	2+3+2=7 days
Begin,C,E,H	**3+3+2=8 days**

I: Since it will take 3+3+4=10 days to complete the tasks that precede task I, task I should begin on day 11.

Directed Path	Time to Complete
Begin,A,D,F,I	1+2+5+1=9 days
Begin,A,D,G,I	1+2+4+1=8 days
Begin,B,D,F,I	2+2+5+1=10 days
Begin,B,E,G,I	2+3+4+1=10 days
Begin,C,E,G,I	**3+3+4+1=11 days**

J: Since it will take 3+3+4=10 days to complete the tasks that precede task J, task J should begin on day 11.

Directed Path	Time to Complete
Begin,A,D,G,J	1+2+4+3=10 days
Begin,B,D,G,J	2+2+4+3=11 days
Begin,B,E,G,J	2+3+4+3=12 days
Begin,B,E,H,J	2+3+2+3=10 days
Begin,C,E,G,J	**3+3+4+3=13 days**
Begin,C,E,H,J	3+3+2+3=11 days

14. Tasks A, B, and C can begin on day 1. Task D can begin on day 6. Task G can begin on day 4. We must however find the critical paths for tasks E,F,H,I and J.

E: Since it will take 6 days to complete the tasks that precede task E, task E should begin on day 7.

Directed Path	Time to Complete
Begin,B,E	3+4=7 days
Begin,C,E	**6+4=10 days**

F: Since it will take 6+4=10 days to complete the tasks that precede task F, task F should begin on day 11.

Directed Path	Time to Complete
Begin,A,D,F	5+3+3=11 days
Begin,B,E,F	3+4+3=10 days
Begin,C,E,F	**6+4+3=13 days**

H: Since it will take 6+4=10 days to complete the tasks that precede task H, task H should begin on day 11.

Directed Path	Time to Complete
Begin,A,D,H	5+3+6=14 days
Begin,B,E,H	3+4+6=13 days
Begin,C,E,H	**6+4+6=16 days**

I: Since it will take 6+4+3=13 days to complete the tasks that precede task I, task I should begin on day 14.

Directed Path	Time to Complete
Begin,A,D,F,I	5+3+3+2=13 days
Begin,B,E,F,I	3+4+3+2=12 days
Begin,B,G,I	3+7+2=12 days
Begin,C,E,F,I	**6+4+3+2=15 days**

14. (continued)

J: Since it will take 6+4+6=16 days to complete the tasks that precede task J, task J should begin on day 17.

Directed Path	Time to Complete
Begin,A,D,H,J	5+3+6+3=17 days
Begin,B,E,H,J	3+4+6+3=16 days
Begin,C,E,H,J	**6+4+6+3=19 days**

15.

Funding and Permits can begin in week 1. Programs can begin in week 3. We must determine the critical paths for the rest of the tasks.

Rent tents: Since it will take 2+2=4 weeks to complete the tasks that precede renting tents, renting tents should begin on week 5.

Directed Path	Time to Complete
Begin,1,3,4	**2+2+2=6 weeks**
Begin,2,4	1+2=3 weeks

Arrange for speakers, etc.: Since it will take 2+2+2=6 weeks to complete the tasks that precede arranging for speakers, arranging for speakers should begin on week 7.

Directed Path	Time to Complete
Begin,1,3,4,5	**2+2+2+4=10 weeks**
Begin,2,4,5	1+2+4=7 weeks

Advertise: Since it will take 2+2+2+4=10 weeks to complete the tasks that precede arranging for advertising, arranging for advertising should begin on week 11.

Directed Path	Time to Complete
Begin,1,3,4,5,6	**2+2+2+4+2=12 weeks**
Begin,2,4,5,6	1+2+4+2=9 weeks

Set up tents, etc.: Since it will take 2+2+2=6 weeks to complete the tasks that precede setting up tents, setting up tents should begin on week 7.

Directed Path	Time to Complete
Begin,1,3,4,7	**2+2+2+1=7 weeks**
Begin,2,4,7	1+2+1=4 weeks

Set up festival: Since it will take 2+2+2+4+2=12 weeks to complete the tasks that precede setting up the festival, setting up the festival should begin on week 13.

Directed Path	Time to Complete
Begin,1,3,4,5,6,8	**2+2+2+4+2+1=13 weeks**
Begin,2,4,5,6,8	1+2+4+2+1=10 weeks
Begin,1,3,4,7,8	2+2+2+1+1=8 weeks
Begin,2,4,7,8	1+2+1+1=5 weeks

16.

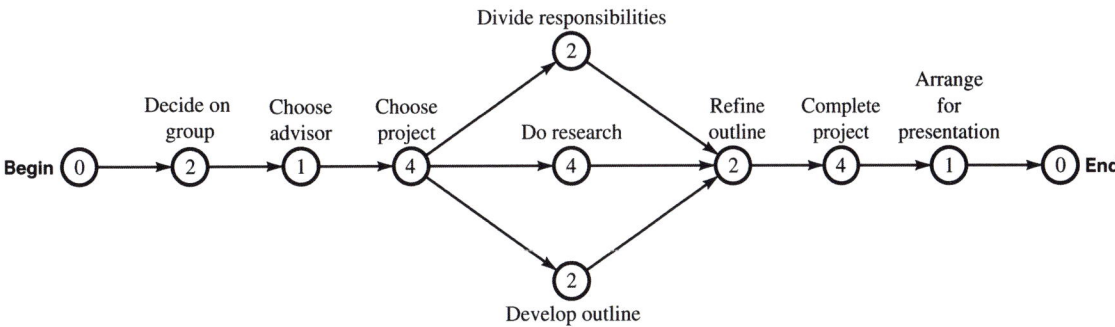

Deciding on a group can begin on week 1. Choosing an advisor can begin on week 3. Choosing a project can begin on week 4. Dividing responsibilities, doing research, and developing a rough outline can all begin on week 8. We must however determine the critical paths for the rest of the tasks.

Refine outline: Since it will take 2+1+4+4=11 weeks to complete the tasks that precede refining the outline, refining the outline should begin on week 12.

Directed Path	Time to Complete
Begin,1,2,3,4,7	2+1+4+2+2=11 weeks
Begin,1,2,3,5,7	**2+1+4+4+2=13 weeks**
Begin,1,2,3,6,7	2+1+4+2+2=11 weeks

Complete project: Since it will take 2+1+4+4+2=13 weeks to complete the tasks that precede completing the project, completing the project should begin on week 14.

Directed Path	Time to Complete
Begin,1,2,3,4,7,8	2+1+4+2+2+4=15 weeks
Begin,1,2,3,5,7,8	**2+1+4+4+2+4=17 weeks**
Begin,1,2,3,6,7,8	2+1+4+2+2+4=15 weeks

Arrange for presentation: Since it will take 2+1+4+4+2+4=17 weeks to complete the tasks that precede arrange for the presentation, arrange for the presentation should begin on week 18.

Directed Path	Time to Complete
Begin,1,2,3,4,7,8,9	2+1+4+2+2+4+1=16 weeks
Begin,1,2,3,5,7,8,9	**2+1+4+4+2+4+1=18 weeks**
Begin,1,2,3,6,7,8,9	2+1+4+2+2+4+1=16 weeks

17.

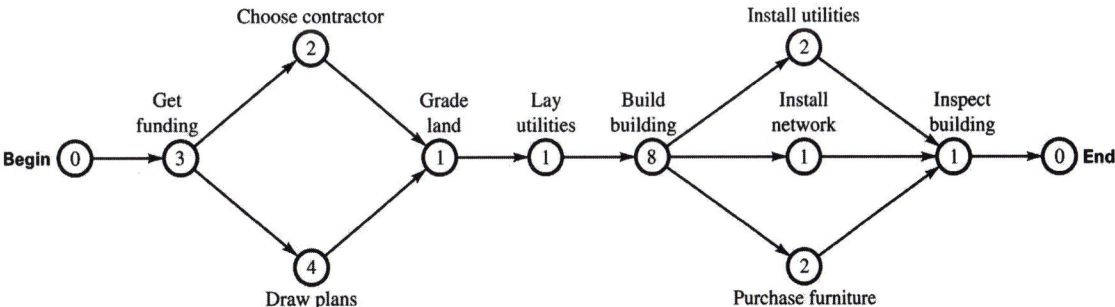

Getting funding can begin on month 1. Choosing a contractor and drawing the plans can begin on month 4. We must however determine the critical paths for the rest of the tasks.

17. (continued)

Grade land: Since it will take 3+4=7 months to complete the tasks that precede grading the land, grading the land should begin on month 8.

Directed Path	Time to Complete
Begin,1,2,4	3+2+1=6 months
Begin,1,3,4	**3+4+1=8 months**

Lay underground utilities: Since it will take 3+4+1=8 months to complete the tasks that precede laying underground utilities, laying underground utilities should begin on month 9.

Directed Path	Time to Complete
Begin,1,2,4,5	3+2+1+1=7 months
Begin,1,3,4,5	**3+4+1+1=9 months**

Build buildings: Since it will take 3+4+1+1=9 months to complete the tasks that precede building buildings, building buildings should begin on month 10.

Directed Path	Time to Complete
Begin,1,2,4,5,6	3+2+1+1+8=15 months
Begin,1,3,4,5,6	**3+4+1+1+8=17 months**

Install utilities in building: Since it will take 3+4+1+1+8=17 months to complete the tasks that precede installing utilities in buildings, installing utilities in buildings should begin on month 18.

Directed Path	Time to Complete
Begin,1,2,4,5,6,7	3+2+1+1+8+2=17 months
Begin,1,3,4,5,6,7	**3+4+1+1+8+2=19 months**

Install computers: Since it will take 3+4+1+1+8=17 months to complete the tasks that precede installing computers, installing computers should begin on month 18.

Directed Path	Time to Complete
Begin,1,2,4,5,6,8	3+2+1+1+8+1=16 months
Begin,1,3,4,5,6,8	**3+4+1+1+8+1=18 months**

Purchase furniture: Since it will take 3+4+1+1+8=17 months to complete the tasks that precede purchase furniture, purchase furniture should begin on month 18.

Directed Path	Time to Complete
Begin,1,2,4,5,6,9	3+2+1+1+8+2=17 months
Begin,1,3,4,5,6,9	**3+4+1+1+8+2=19 months**

Inspect Building: Since it will take 3+4+1+1+8+2=19 months to complete the tasks that precede inspecting the buildings, inspecting the buildings should begin on month 20.

Directed Path	Time to Complete
Begin,1,2,4,5,6,7,10	3+2+1+1+8+2+1=18 months
Begin,1,2,4,5,6,8,10	3+2+1+1+8+1+1=17 months
Begin,1,2,4,5,6,9,10	3+2+1+1+8+2+1=18 months
Begin,1,3,4,5,6,7,10	**3+4+1+1+8+2+1=20 months**
Begin,1,3,4,5,6,8,10	3+4+1+1+8+1+1=19 months
Begin,1,3,4,5,6,9,10	**3+4+1+1+8+2+1=20 months**

18.

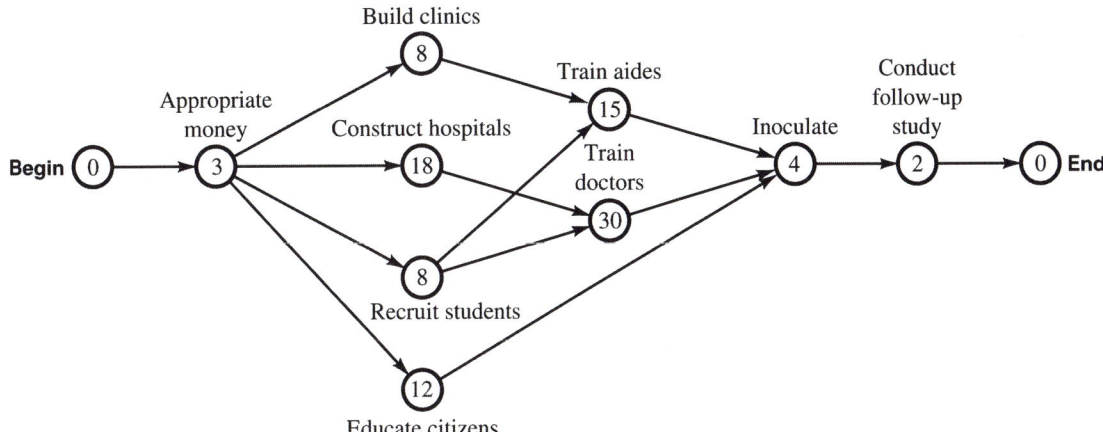

Appropriating money can begin on month 1. Building health clinics, constructing hospitals, recruiting students, and educating citizens in health practices can all begin on month 4. We must however determine the critical paths for the rest of the tasks.

Train aides: Since it will take 3+8=11 months to complete the tasks that precede training aides, training aids should begin on month 12.

Directed Path	Time to Complete
Begin,1,2,7	**3+8+15=26 months**
Begin,1,5,7	**3+8+15=26 months**

Train doctors: Since it will take 3+18=21 months to complete the tasks that precede training doctors, training doctors should begin on month 22.

Directed Path	Time to Complete
Begin,1,3,6	**3+18+30=51 months**
Begin,1,5,6	3+8+30=41 months

Inoculate: Since it will take 3+18+30=51 months to complete the tasks that precede inoculations, inoculations should begin on month 52.

Directed Path	Time to Complete
Begin,1,2,7,8	3+8+15+4=30 months
Begin,1,5,7,8	3+8+15+4=30 months
Begin,1,3,6,8	**3+18+30+4=55 months**
Begin,1,5,6,8	3+8+30+4=45 months
Begin,1,4,8	3+12+4=19 months

Follow-up study: Since it will take 3+18+30+4=55 months to complete the tasks that precede conducting a follow-up study, conducting a follow-up study should begin on month 56.

Directed Path	Time to Complete
Begin,1,2,7,8,9	3+8+15+4+2=32 months
Begin,1,5,7,8,9	3+8+15+4+2=32 months
Begin,1,3,6,8,9	**3+18+30+4+2=57 months**
Begin,1,5,6,8,9	3+8+30+4+2=47 months
Begin,1,4,8,9	3+12+4+2=21 months

19.

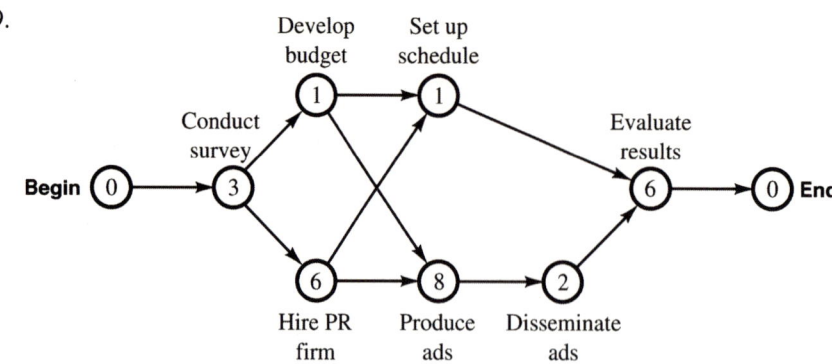

Conducting a survey can begin on month 1. Developing a budget and hiring a PR firm can begin on month 4. We must however determine the critical paths for the rest of the tasks.

Set up production schedule: Since it will take 3+6=9 months to complete the tasks that precede setting up the production schedule, setting up the production schedule should begin on month 10.

Directed Path	Time to Complete
Begin,1,2,4	3+1+1=5 months
Begin,1,3,4	**3+6+1=10 months**

Produce ads: Since it will take 3+6=9 months to complete the tasks that precede producing ads, producing ads should begin on month 10.

Directed Path	Time to Complete
Begin,1,2,5	3+1+8=12 months
Begin,1,3,5	**3+6+8=17 months**

Disseminate ads: Since it will take 3+6+8=17 months to complete the tasks that precede disseminating ads, disseminating ads should begin on month 18.

Directed Path	Time to Complete
Begin,1,2,5,6	3+1+8+2=14 months
Begin,1,3,5,6	**3+6+8+2=19 months**

Evaluate results: Since it will take 3+6+8+2=19 months to complete the tasks that precede evaluating results, evaluating results should begin on month 20.

Directed Path	Time to Complete
Begin,1,2,4,7	3+1+1+6=11 months
Begin,1,3,4,7	3+6+1+6=16 months
Begin,1,2,5,6,7	3+1+8+2+6=20 months
Begin,1,3,5,6,7	**3+6+8+2+6=25 months**

20. – 22. No solutions provided

Chapter 4
NUMERATION SYSTEMS: Does It Matter How We Name Numbers?

Section 4.1 The Evolution Of Numeration Systems

1. X, ✝, ∩, ⟨

3. One scroll was rewritten as ten heel bones.

5. A number tells "how many." A numeral is a symbol for writing a number.

7. To omit some power of ten, simply don't use the symbol for that power.

9. $300 + 20 + 4 = 324$

11. $20,000 + 1,000 + 300 + 20 + 4 = 21,324$

13. $2,000,000 + 100,000 + 20,000 + 1,000 + 200 + 10 = 2,121,210$

15. 99∩∩∩|||||

17. 𓆼𓆼𓆼 99∩∩∩|||||

19. 𓆛𓍢𓏽𓏽𓏽𓏽 𓆼𓆼𓆼𓆼 999∩

21.
```
              999∩∩|||||
   +99999999∩∩ |||||||
   9999999999∩∩∩∩ |||||||||||||
   ten scrolls equals   ten strokes equals
   1 lotus flower       1 heel bone
              𓆼9∩∩∩∩∩|||
```

23.
```
   𓆼𓏽𓏽𓏽𓏽𓏽𓆼99∩
   +𓆼𓍢𓏽𓏽𓏽𓏽𓏽𓏽9
   𓆼𓆼𓍢𓏽𓏽𓏽𓏽𓏽𓏽𓏽𓏽𓏽𓆼999∩
   ten pointing fingers
   equals 1 fish
        𓆼𓆼𓍢𓏽𓆼999∩
```

25.
```
                   999∩∩∩∩∩∩ ||
          +  999999∩∩∩∩|||||
          99999999∩∩∩∩∩∩∩∩∩∩ |||||
          ten heel bones
          equals 1 scroll
          9999999999∩∩ |||||||
          ten scrolls equals
          1 lotus flower
                       𓆼∩∩ ||||||
```

27.
```
          one scroll equals
          ten heel bones
                ↓
          999999∩∩ |||||||
        −  999∩∩∩∩∩|||||
        99999∩∩∩∩∩∩∩∩∩∩|||||||
        −  999∩∩∩∩∩|||
        99∩∩∩∩∩∩∩|||||
```

29.

                         one pointing finger   one scroll equals
                      equals ten lotus flowers   ten heel bones
                                ↓   ↓
                            𓂀𓂀𓂀 𓆓𓆓𓆓𓆓𓆓
            − 𓂀𓂀𓆼𓆼𓆼𓆼𓆼 𓆓𓆓𓆓 ∩∩∩∩∩ ||||

                                         one heel bone
                                     equals ten staffs
                                              ↓
     𓂀𓂀𓆼𓆼𓆼𓆼𓆼𓆼𓆼𓆼𓆼𓆼 𓆓𓆓𓆓𓆓 ∩∩∩∩∩∩∩∩∩∩
        − 𓂀𓂀𓆼𓆼𓆼𓆼𓆼 𓆓𓆓𓆓 ∩∩∩∩∩ ||||

     𓂀𓂀𓆼𓆼𓆼𓆼𓆼𓆼𓆼𓆼𓆼𓆼 𓆓𓆓𓆓𓆓 ∩∩∩∩∩∩∩∩∩ ||||||||||
        − 𓂀𓂀𓆼𓆼𓆼𓆼𓆼 𓆓𓆓𓆓 ∩∩∩∩∩ ||||
                 𓆼𓆼𓆼𓆼𓆼 𓆓 ∩∩∩∩ ||||||

31.

        one scroll equals   one heel bone
         ten heel bones   equals ten staffs
                ↓   ↓
              𓆓𓆓 ∩∩∩ ||
        − 𓆓 ∩∩∩∩ |||||

     𓆓 ∩∩∩∩∩∩∩∩∩∩∩∩ ||||||||||||
        − 𓆓 ∩∩∩∩ |||||
          ∩∩∩∩∩∩∩ |||||||

33. $14 = 2 + 4 + 8$

Powers of 2	Times 43
1	43
2	43 + 43 = 86
4	86 + 86 = 172
8	172 + 172 = 344

$86 + 172 + 344 = 602$

35. $12 = 4 + 8$

Powers of 2	Times 107
1	107
2	107 + 107 = 214
4	214 + 214 = 428
8	428 + 428 = 856

$428 + 856 = 1,284$

37. $35 = 1 + 2 + 32$

Powers of 2	Times 121
1	121
2	121 + 121 = 242
4	242 + 242 = 484
8	484 + 484 = 968
16	968 + 968 = 1,936
32	1,936 + 1,936 = 3,872

$121 + 242 + 3,872 = 4,235$

39. $50+(10-1)=50+9=59$

41. $500+50+10+(5-1)=$
    $500+60+4=564$

43. $1,000+(1,000-100)+50+10+3=1,000+900+60+3=1,963$

45. $5\times1,000+2,000+500+(50-10)+(5-1)=5,000+2,000+500+40+4=$
    $7,000+500+40+4=7,544$

47. $500\times100+200+50+10+2=50,000+200+60+2=50,262$

49. $5\times1,000\times100+1,000+(500-100)+20=500,000+1,000+400+20=501,420$

51. XLIV

53. CCLXXVIII

55. CDXLIV

57. $\overline{\text{IV}}$DCCXCV

59. $\overline{\text{LXXXIX}}$CDXXIII

61. $4\times100+3\times10+6=400+30+6=436$

63. $5\times1,000+6\times10+7=5,000+60+7=5,067$

65. $9\times1,000+9\times100+9\times10+9=9,000+900+90+9=9,999$

67. 六十八

69. [Chinese numeral]

71. [Chinese numeral]

73. [Chinese numeral]

75. [Egyptian numeral addition]

$$\begin{array}{r}\overset{1}{234,540}\\+102,080\\\hline 336,620\end{array}$$

ten heel bones equals 1 scroll

77. MCMXXXIX

79. MCMXCIV

81. [Chinese numerals]

83. 9,999,999

78   Chapter 4: Numeration Systems

85. In the solution to this exercise, the symbol $\sqrt{.}$ is being used instead of $\sqrt{}$.

   a) $\alpha\alpha\alpha \nabla \otimes\otimes\otimes\otimes \sqrt{.}\sqrt{.}$; $\alpha\alpha\alpha\alpha\alpha\alpha\alpha\alpha \nabla\nabla\sqrt{.}\sqrt{.}\sqrt{.}$

   b)

$$\alpha\alpha\alpha \nabla \otimes\otimes\otimes\otimes \sqrt{.}\sqrt{.}$$
$$+ \alpha\alpha\alpha\alpha\alpha\alpha\alpha\alpha \nabla\nabla\sqrt{.}\sqrt{.}\sqrt{.}$$
$$\overline{\alpha\alpha\alpha\alpha\alpha\alpha\alpha\alpha\alpha\alpha\alpha \nabla\nabla \otimes\otimes\otimes\otimes \sqrt{.}\sqrt{.}\sqrt{.}\sqrt{.}}$$

   regroup $\approx \nabla\nabla\nabla \otimes\otimes\otimes\otimes \sqrt{.}\sqrt{.}\sqrt{.}\sqrt{.}$

   c)

$$\alpha\alpha\alpha\alpha\alpha\alpha\alpha\alpha \nabla\nabla\sqrt{.}\sqrt{.}\sqrt{.}$$
$$- \alpha\alpha\alpha \nabla \otimes\otimes\otimes\otimes \sqrt{.}\sqrt{.}$$

   regroup

$$\alpha\alpha\alpha\alpha\alpha\alpha\alpha\alpha \nabla \otimes\otimes\otimes\otimes\otimes\otimes\otimes\otimes \sqrt{.}\sqrt{.}\sqrt{.}$$
$$- \alpha\alpha\alpha \nabla \otimes\otimes\otimes\otimes \sqrt{.}\sqrt{.}$$
$$\overline{\alpha\alpha\alpha\alpha\alpha \otimes\otimes\otimes\otimes\otimes \sqrt{.}}$$

87. and 89.   No solutions provided

Section 4.2   Place Value Systems

1. Babylonian, Chinese, Mayan, Hindu–Arabic.

3. The Mayan system had a zero.

5. Numerals can be written more efficiently and there is no need for different symbols for different powers of the base.

7. In the galley method, all partial products are computed separately before being added to get the final product. In our method, several smaller partial products are combined into a single partial product before adding.

9. 32

11. $30 - 2 = 28$

13. $(10+2) \times 60 + (10+1) = 12 \times 60 + 11 = 720 + 11 = 731$

15. $2 \times 60^2 + (20-4) \times 60 + (20+7) = 2 \times 3{,}600 + 16 \times 60 + 27 = 7{,}200 + 960 + 27 = 8{,}187$

17. $8{,}235 = 2 \times 60^2 + 17 \times 60 + 15$

$8{,}235 = 2 \times 60^2 + (20-3) \times 60 + 15$

19. $18{,}397 = 5 \times 60^2 + 6 \times 60 + 37$

$18{,}397 = 5 \times 60^2 + 6 \times 60 + (40-3)$

21. $123{,}485 = 34 \times 60^2 + 18 \times 60 + 5$

$123{,}485 = 34 \times 60^2 + (20-2) \times 60 + 5$

23. $188,289 = 52 \times 60^2 + 18 \times 60 + 9$ ❰❰❰❰❰❰❙❙ ❰❙❙❙❙❙❙❙❙ ❙❙❙❙❙❙❙❙❙ or

$188,289 = 52 \times 60^2 + (20-2) \times 60 + 9$ ❰❰❰❰❰❰❙❙ ❰❰❙❙ ❰❙❙

25.
•• $2 \times 20 \times 18 \times 20 = 14,400$
⋮• $12 \times 20 \times 18 = 4,320$
⊜ $0 \times 20 = 0$
⋮•• $13 = 13$
                          $18,733$

27.
•• $2 \times 20 \times 18 \times 20^2 = 288,000$
⊜ $0 \times 20 \times 18 \times 20 = 0$
• $6 \times 20 \times 18 = 2,160$
⊜ $0 \times 20 = 0$
⋮•• $13 = 13$
                         $290,173$

29. five hundred

31. five thousand

33. $2 \times 10^4 + 5 \times 10^3 + 3 \times 10^2 + 8 \times 10^1 + 9 \times 10^0$

35. $2 \times 10^5 + 7 \times 10^4 + 8 \times 10^3 + 0 \times 10^2 + 6 \times 10^1 + 3 \times 10^0$

37. $1 \times 10^6 + 2 \times 10^5 + 0 \times 10^4 + 0 \times 10^3 + 0 \times 10^2 + 4 \times 10^1 + 5 \times 10^0$

39. 5,368

41. 370,082

43. 3,070,502

45. $2,863 = 2 \times 10^3 + 8 \times 10^2 + 6 \times 10^1 + 3 \times 10^0$
$\underline{+425 = + 4 \times 10^2 + 2 \times 10^1 + 5 \times 10^0}$
$2 \times 10^3 + 12 \times 10^2 + 8 \times 10^1 + 8 \times 10^0$

$2 \times 10^3 + 12 \times 10^2 + 8 \times 10^1 + 8 \times 10^0 = 2 \times 10^3 + (10+2) \times 10^2 + 8 \times 10^1 + 8 \times 10^0$
$= 2 \times 10^3 + 10 \times 10^2 + 2 \times 10^2 + 8 \times 10^1 + 8 \times 10^0$
$= 2 \times 10^3 + 10^3 + 2 \times 10^2 + 8 \times 10^1 + 8 \times 10^0$
$= (2+1) \times 10^3 + 2 \times 10^2 + 8 \times 10^1 + 8 \times 10^0$
$= 3 \times 10^3 + 2 \times 10^2 + 8 \times 10^1 + 8 \times 10^0$
$= 3,288$

47. $3,482 = 3 \times 10^3 + 4 \times 10^2 + 8 \times 10^1 + 2 \times 10^0$
$\underline{+2,756 = +\ 2 \times 10^3 + 7 \times 10^2 + 5 \times 10^1 + 6 \times 10^0}$
$5 \times 10^3 + 11 \times 10^2 + 13 \times 10^1 + 8 \times 10^0$

$5 \times 10^3 + 11 \times 10^2 + 13 \times 10^1 + 8 \times 10^0 = 5 \times 10^3 + (10+1) \times 10^2 + (10+3) \times 10^1 + 8 \times 10^0$
$= 5 \times 10^3 + 10 \times 10^2 + 1 \times 10^2 + 10 \times 10^1 + 3 \times 10^1 + 8 \times 10^0$
$= 5 \times 10^3 + 10^3 + 1 \times 10^2 + 10^2 + 3 \times 10^1 + 8 \times 10^0$
$= (5+1) \times 10^3 + (1+1) \times 10^2 + 3 \times 10^1 + 8 \times 10^0$
$= 6 \times 10^3 + 2 \times 10^2 + 3 \times 10^1 + 8 \times 10^0$
$= 6,238$

80   Chapter 4: Numeration Systems

49. $926 = 9 \times 10^2 + 2 \times 10^1 + 6 \times 10^0$
     $\underline{-784 = -7 \times 10^2 + 8 \times 10^1 + 4 \times 10^0}$

$$9 \times 10^2 + 2 \times 10^1 + 6 \times 10^0 = (8+1) \times 10^2 + 2 \times 10^1 + 6 \times 10^0$$
$$= 8 \times 10^2 + 1 \times 10^2 + 2 \times 10^1 + 6 \times 10^0$$
$$= 8 \times 10^2 + 10 \times 10^1 + 2 \times 10^1 + 6 \times 10^0$$
$$= 8 \times 10^2 + (10+2) \times 10^1 + 6 \times 10^0$$
$$= 8 \times 10^2 + 12 \times 10^1 + 6 \times 10^0$$

$926 = 8 \times 10^2 + 12 \times 10^1 + 6 \times 10^0$
$\underline{-784 = -7 \times 10^2 + 8 \times 10^1 + 4 \times 10^0}$
$1 \times 10^2 + 4 \times 10^1 + 2 \times 10^0$

or 142

51. $5{,}238 = 5 \times 10^3 + 2 \times 10^2 + 3 \times 10^1 + 8 \times 10^0$
     $\underline{-1{,}583 = -1 \times 10^3 + 5 \times 10^2 + 8 \times 10^1 + 3 \times 10^0}$

$$5 \times 10^3 + 2 \times 10^2 + 3 \times 10^1 + 8 \times 10^0 = 5 \times 10^3 + (1+1) \times 10^2 + 3 \times 10^1 + 8 \times 10^0$$
$$= 5 \times 10^3 + 1 \times 10^2 + 1 \times 10^2 + 3 \times 10^1 + 8 \times 10^0$$
$$= 5 \times 10^3 + 1 \times 10^2 + 10 \times 10^1 + 3 \times 10^1 + 8 \times 10^0$$
$$= 5 \times 10^3 + 1 \times 10^2 + (10+3) \times 10^1 + 8 \times 10^0$$
$$= 5 \times 10^3 + 1 \times 10^2 + 13 \times 10^1 + 8 \times 10^0$$

$$5 \times 10^3 + 1 \times 10^2 + 13 \times 10^1 + 8 \times 10^0 = (4+1) \times 10^3 + 1 \times 10^2 + 13 \times 10^1 + 8 \times 10^0$$
$$= 4 \times 10^3 + 1 \times 10^3 + 1 \times 10^2 + 13 \times 10^1 + 8 \times 10^0$$
$$= 4 \times 10^3 + 10 \times 10^2 + 1 \times 10^2 + 13 \times 10^1 + 8 \times 10^0$$
$$= 4 \times 10^3 + (10+1) \times 10^2 + 13 \times 10^1 + 8 \times 10^0$$
$$= 4 \times 10^3 + 11 \times 10^2 + 13 \times 10^1 + 8 \times 10^0$$

$5{,}238 = 4 \times 10^3 + 11 \times 10^2 + 13 \times 10^1 + 8 \times 10^0$
$\underline{-1{,}583 = -1 \times 10^3 + 5 \times 10^2 + 8 \times 10^1 + 3 \times 10^0}$
$\phantom{-1{,}583 = -}3 \times 10^3 + 6 \times 10^2 + 5 \times 10^1 + 5 \times 10^0$

or 3,655

Note: In part b of the solutions to Exercises 53 – 57, the smaller number is placed below the larger in the conventional method.

53. a) [Lattice multiplication diagram with top factor 4, side factors 0, 9, 4; cells containing 0/0, 0/8, 1/2, 2/0 giving partial products, with result reading 0, 9, 4, 0]

b) 235
    $\underline{\times 4}$
    20
    120
    $\underline{800}$
    940

55. a)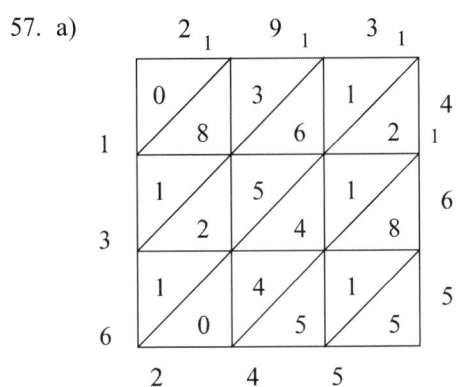

b)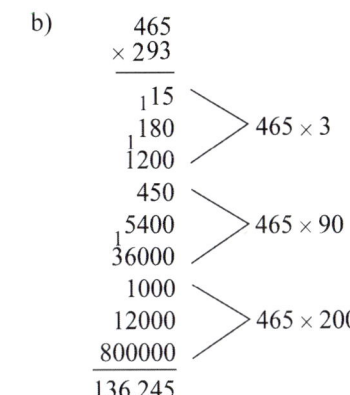
```
    876
  ×  23
  -----
     18  ⎫
    210  ⎬ 876 × 3
   2400  ⎭
    120  ⎫
   1400  ⎬ 876 × 20
  16000  ⎭
  ------
  20,148
```

57. a)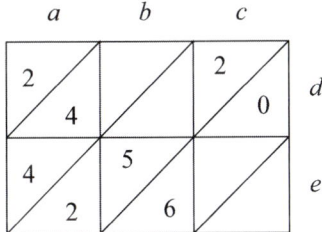

b)
```
     465
   × 293
   -----
      15  ⎫
     180  ⎬ 465 × 3
    1200  ⎭
     450  ⎫
    5400  ⎬ 465 × 90
   36000  ⎭
    1000  ⎫
   12000  ⎬ 465 × 200
  800000  ⎭
  -------
  136,245
```

59. $ad = 24$, $cd = 20$, $ae = 42$, and $be = 56 \Rightarrow ad = 6 \cdot 4$, $cd = 5 \cdot 4$, $ae = 6 \cdot 7$, and $be = 8 \cdot 7$

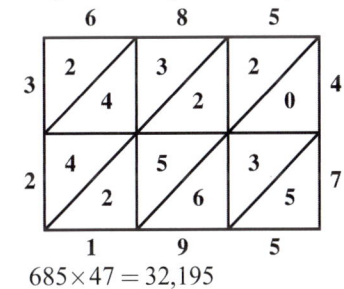

$685 \times 47 = 32{,}195$

61. 1,884

63. 3,936

65. 5,544

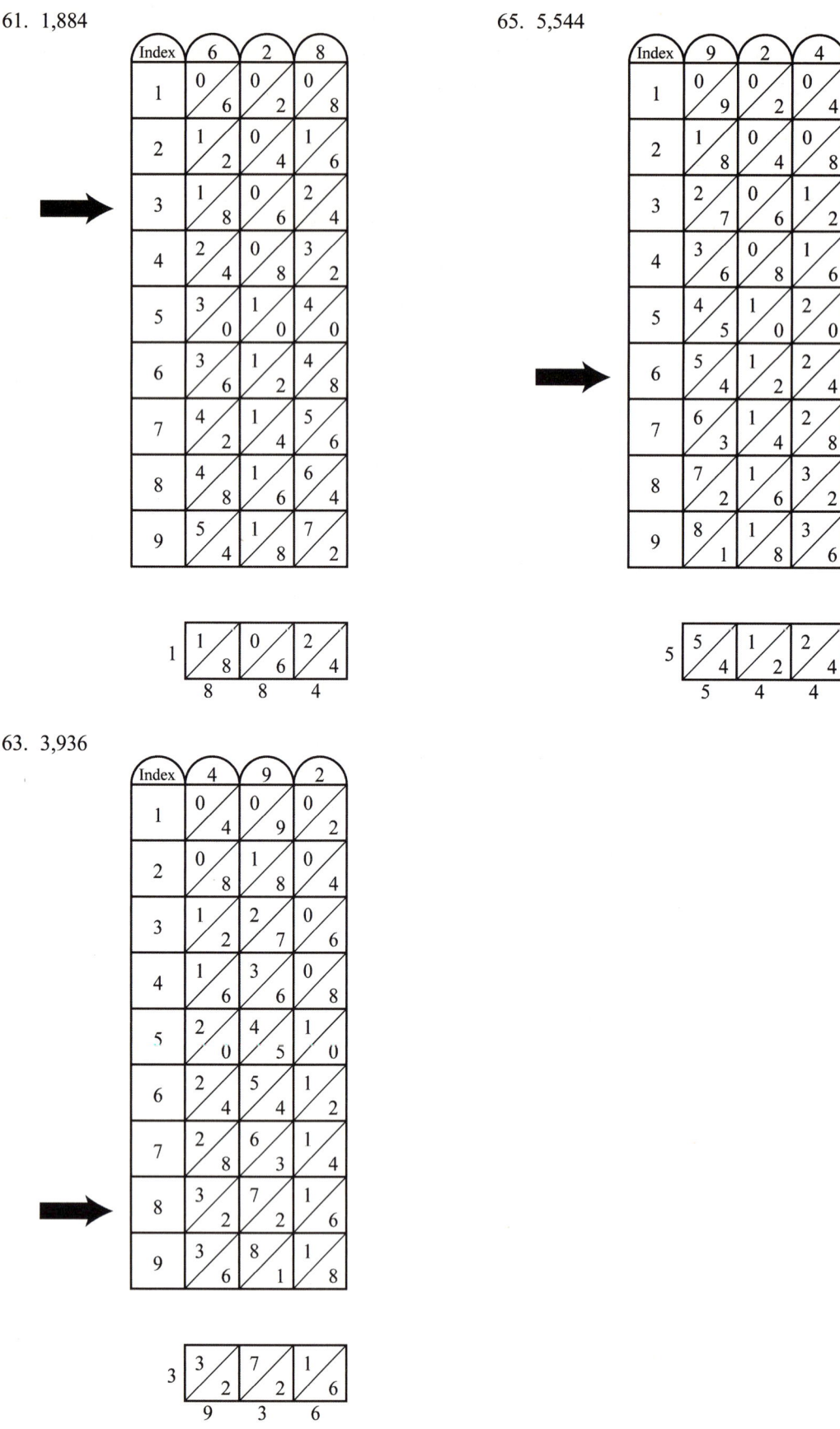

67. Answers may vary.

If you think of 324 as $300 + 20 + 4 = 3 \cdot 100 + 2 \cdot 10 + 4$ then you can use indices of 3, 2, and 4 for 615. You would position the rectangles to consider the place values of the 3, 2, and 4. Dashed boxes are included for ease in adding. The result is 199,260.

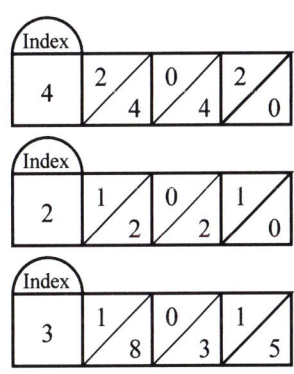

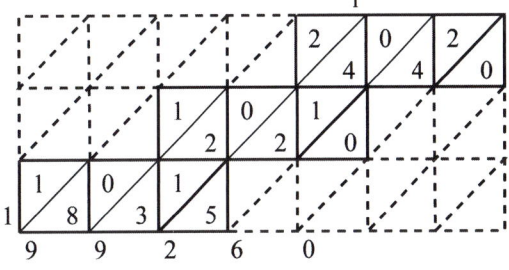

69. Answers may vary.

In writing ▏ ▏▏▏▏▏, it could be difficult to tell whether there is one space or two spaces between the two groups of symbols. If there is one space, the number is $1 \times 60 + 5 = 60 + 5 = 65$. If there are two spaces, the number is $1 \times 60^2 + 5 = 3,600 + 5 = 3,605$.

71. 日三大

73. 五〇六七

75. ḥOOḥ

77. Answers may vary.

The ancient Chinese system requires a separate symbol for each power of ten, e.g. one billion, ten billion, and so on. The Hindu–Arabic system uses different powers of 10, 10^9, 10^{10}, and so on. The positioning of symbols in the numeral indicates these powers of ten.

79. Examples and explanations may vary.

c) tally marks (early primitive systems); d) simple grouping (Egyptian); b) multiplicative grouping (Chinese); a) place value system (Hindu–Arabic)

Section 4.3 Calculating In Other Bases

1. 4

3. five

5. Base 8: 64 facts; The eight symbols 0, 1, ... 7 are combined with themselves.
   Base 16: 256 facts; The sixteen symbols 0, 1, ...9, A, B, ... F are combined with themselves.

7. The number 10 always represents the base, which is 16 in this case and a single symbol was needed to represent 10.

9. 23_5 and 30_5

11. 455_6 and 501_6

13. 1010_2 and 1100_2

15. 66_8 and 70_8

17. FD_{16} and FF_{16}

19. $432_5 = 4 \times 5^2 + 3 \times 5^1 + 2 \times 5^0$
    $= 4 \times 25 + 3 \times 5 + 2 \times 1$
    $= 100 + 15 + 2$
    $= 117$

21. $504_6 = 5 \times 6^2 + 0 \times 6^1 + 4 \times 6^0$
    $= 5 \times 36 + 0 \times 6 + 4 \times 1$
    $= 180 + 0 + 4$
    $= 184$

23. $100111_2 = 1 \times 2^5 + 0 \times 2^4 + 0 \times 2^3 + 1 \times 2^2 + 1 \times 2^1 + 1 \times 2^0$
    $= 1 \times 32 + 0 \times 16 + 0 \times 8 + 1 \times 4 + 1 \times 2 + 1 \times 1$
    $= 32 + 0 + 0 + 4 + 2 + 1$
    $= 39$

25. $1110101_2 = 1 \times 2^6 + 1 \times 2^5 + 1 \times 2^4 + 0 \times 2^3 + 1 \times 2^2 + 0 \times 2^1 + 1 \times 2^0$
    $= 1 \times 64 + 1 \times 32 + 1 \times 16 + 0 \times 8 + 1 \times 4 + 0 \times 2 + 1 \times 1$
    $= 64 + 32 + 16 + 0 + 4 + 0 + 1$
    $= 117$

27. $267_8 = 2 \times 8^2 + 6 \times 8^1 + 7 \times 8^0$
    $= 2 \times 64 + 6 \times 8 + 7 \times 1$
    $= 128 + 48 + 7$
    $= 183$

29. $704_8 = 7 \times 8^2 + 0 \times 8^1 + 4 \times 8^0$
    $= 7 \times 64 + 0 \times 8 + 4 \times 1$
    $= 448 + 0 + 4$
    $= 452$

31. $2F4_{16} = 2 \times 16^2 + 15 \times 16^1 + 4 \times 16^0$
    $= 2 \times 256 + 15 \times 16 + 4 \times 1$
    $= 512 + 240 + 4$
    $= 756$

33. $D08_{16} = 13 \times 16^2 + 0 \times 16^1 + 8 \times 16^0$
    $= 13 \times 256 + 0 \times 16 + 8 \times 1$
    $= 3328 + 0 + 8$
    $= 3,336$

35. 2314_5

    5 | 334   4
    5 | 66   1
    5 | 13   3
    5 | 2   2
        0

37. 12302_6

```
6 | 1838    2
6 | 306     0
6 | 51      3
6 | 8       2
6 | 1       1
    0
```

39. 1100111_2

```
2 | 103    1
2 | 51     1
2 | 25     1
2 | 12     0
2 | 6      0
2 | 3      1
2 | 1      1
    0
```

41. 1011110_2

```
2 | 94     0
2 | 47     1
2 | 23     1
2 | 11     1
2 | 5      1
2 | 2      0
2 | 1      1
    0
```

43. 6513_8

```
8 | 3403    3
8 | 425     1
8 | 53      5
8 | 6       6
    0
```

45. 6115_8

```
8 | 3149    5
8 | 393     1
8 | 49      1
8 | 6       6
    0
```

47. $AE8_{16}$

```
16 | 2792    8
16 | 174     E (14)
16 | 10      A (10)
     0
```

49. DEA_{16}

```
16 | 3562    A (10)
16 | 222     E (14)
16 | 13      D (13)
     0
```

51. $\quad \overset{1}{}3412_5$
    $+\ \ 231_5$
    $\overline{4143_5}$

53. $\quad \overset{1\ \ 11}{4013_5}$
    $+\ 1242_5$
    $\overline{10310_5}$

55. $\quad \overset{1}{}2A18_{16}$
    $+\ \ 43B_{16}$
    $\overline{2E53_{16}}$

57. $\quad \overset{1\,1\,1\,1\,1}{11011_2}$
    $+\ 10101_2$
    $\overline{110000_2}$

59. $\quad \overset{3}{2\overset{1}{4}12_5}$
    $-\ \ 321_5$
    $\overline{2041_5}$

61. $\quad A\overset{7}{\overset{1}{8}}3_{16}$
    $-\ 43B_{16}$
    $\overline{648_{16}}$

63. $\quad 11\overset{0}{\overset{1}{1}}011_2$
    $-\ 10101_2$
    $\overline{100110_2}$

65. $\quad\quad 41_5$
    $\times\ 23_5$
    $\overline{\quad {}_1 223_5}$
    132
    $\overline{\quad 2043_5}$

67.
$$\begin{array}{r}\overset{1}{}\overset{1}{302}_5\\ \times\ 43_5\\ \hline \overset{1}{1411}_5\\ 2213\\ \hline 24041_5\end{array}$$

69. $114_5\ R\ 11_5$

$$\begin{array}{r}114\\ 24_5\overline{)3412_5}\\ \underline{24}\\ 101\\ \underline{24}\\ 222\\ \underline{211}\\ 11\end{array}$$

71. $44_5\ R\ 24_5$

$$\begin{array}{r}44\\ 42_5\overline{)4132_5}\\ \underline{323}\\ 402\\ \underline{323}\\ 24\end{array}$$

73. 1355_8 $2ED_{16}$

1 011 101 101   10 1110 1101
1 3   5   5   2   E   D

75. 1075_8 $23D_{16}$

1 000 111 101   10 0011 1101
1 0   7   5   2   3   D

77. 1751_8 $3E9_{16}$

1 111 101 001   11 1110 1001
1 7   5   1   3   E   9

79. 10100110_2

 2   4   6
010 100 110

81. 101000111110_2

 A   3   E
1010 0011 1110

83. 754_{16}

 3   5   2   4
011 101 010 100
0111 0101 0100
 7   5   4

85. 1654_8

 3   A   C
0011 1010 1100
001 110 101 100
 1   6   5   4

For Exercises 87 and 89 refer to the following table.

65	66	67	68	69	70	71	72	73	74	75	76	77
A	B	C	D	E	F	G	H	I	J	K	L	M

78	79	80	81	82	83	84	85	86	87	88	89	90
N	O	P	Q	R	S	T	U	V	W	X	Y	Z

For Exercises 87 and 89 refer to the table on the previous page.

87. CANDY

1000011	1000001	1001110	1000100	1011001
$2^6+2^1+2^0=$	$2^6+2^0=$	$2^6+2^3+2^2+2^1=$	$2^6+2^2=$	$2^6+2^4+2^3+2^0=$
$64+2+1=67$	$64+1=65$	$64+8+4+2=78$	$64+4=68$	$64+16+8+1=89$

89. LOVE

1001100	1001111	1010110	1000101
$2^6+2^3+2^2=$	$2^6+2^3+2^2+2^1+2^0=$	$2^6+2^4+2^2+2^1=$	$2^6+2^2+2^0=$
$64+8+4=76$	$64+8+4+2+1=79$	$64+16+4+2=86$	$64+4+1=69$

91. 60; 5

93. If you choose A and B to represent 10 and 11, respectively, in base twelve then the first twenty-five counting numbers would be

$$1, 2, 3, 4, 5, 6, 7, 8, 9, A, B, 10_{12}, 11_{12}, 12_{12}, 13_{12},$$
$$14_{12}, 15_{12}, 16_{12}, 17_{12}, 18_{12}, 19_{12}, 1A_{12}, 1B_{12}, 20_{12}, 21_{12}.$$

Another choice that is commonly used is T and E to represent 10 and 11, respectively, in base twelve.

95. ⊕, ≠, $, @, %, ⊕⊤, ⊕⊕, ⊕≠, ⊕$, ⊕@, ⊕%, ≠⊤, ≠⊕, ≠≠, ≠$, ≠@, ≠%, $⊤, $⊕, $≠

97. %⊤

```
6 | 30      0  (⊤)
6 | 5       5  (%)
    0
```

99. No solution provided

Chapter Review Exercises

1. $1,000,000 + 200,000 + 30,000 + 2,000 + 200 + 10 = 1,232,210$

2. one scroll equals one heel bone
   ten heel bones equals ten staffs

3. $37 = 1 + 4 + 32$

Powers of 2	Times 53
1	53
2	53 + 53 = 106
4	106 + 106 = 212
8	212 + 212 = 424
16	424 + 424 = 848
32	848 + 848 = 1,696

$1{,}696 + 212 + 53 = 1{,}961$

4. MMMMDCCXCV

5. $7 \times 1{,}000 + 5 \times 100 + 9 \times 10 + 3 = 7{,}000 + 500 + 90 + 3 = 7{,}593$

6. Answers may vary.
No. A number tells us "how many," whereas a numeral is a symbol or symbols that we use to represent the number.

7. Answers may vary.
The Roman system had the subtraction principle, which allowed them to write certain numbers more efficiently. They also had a multiplication principle.

8. $11{,}292 = 3 \times 60^2 + 8 \times 60 + 12$ ▼▼▼ ▼▼▼▼▼▼▼▼ ⟨▼▼ or

$11{,}292 = 3 \times 60^2 + (10-2) \times 60 + 12$ ▼▼▼ ⟨⃔▼▼ ⟨▼▼

9. $4{,}237 = 4 \times 10^3 + 2 \times 10^2 + 3 \times 10^1 + 7 \times 10^0$
   $\underline{-2{,}673 = - 2 \times 10^3 + 6 \times 10^2 + 7 \times 10^1 + 3 \times 10^0}$

$$\begin{aligned}
4 \times 10^3 + 2 \times 10^2 + 3 \times 10^1 + 7 \times 10^0 &= 4 \times 10^3 + (1+1) \times 10^2 + 3 \times 10^1 + 7 \times 10^0 \\
&= 4 \times 10^3 + 1 \times 10^2 + 1 \times 10^2 + 3 \times 10^1 + 7 \times 10^0 \\
&= 4 \times 10^3 + 1 \times 10^2 + 10 \times 10^1 + 3 \times 10^1 + 7 \times 10^0 \\
&= 4 \times 10^3 + 1 \times 10^2 + (10+3) \times 10^1 + 7 \times 10^0 \\
&= 4 \times 10^3 + 1 \times 10^2 + 13 \times 10^1 + 7 \times 10^0
\end{aligned}$$

$$\begin{aligned}
4 \times 10^3 + 1 \times 10^2 + 13 \times 10^1 + 7 \times 10^0 &= (3+1) \times 10^3 + 1 \times 10^2 + 13 \times 10^1 + 7 \times 10^0 \\
&= 3 \times 10^3 + 1 \times 10^3 + 1 \times 10^2 + 13 \times 10^1 + 7 \times 10^0 \\
&= 3 \times 10^3 + 10 \times 10^2 + 1 \times 10^2 + 13 \times 10^1 + 7 \times 10^0 \\
&= 3 \times 10^3 + (10+1) \times 10^2 + 13 \times 10^1 + 7 \times 10^0 \\
&= 3 \times 10^3 + 11 \times 10^2 + 13 \times 10^1 + 7 \times 10^0
\end{aligned}$$

$4{,}237 = 3 \times 10^3 + 11 \times 10^2 + 13 \times 10^1 + 7 \times 10^0$
$\underline{-2{,}673 = - 2 \times 10^3 + 6 \times 10^2 + 7 \times 10^1 + 3 \times 10^0}$
$\phantom{-2{,}673 = -} 1 \times 10^3 + 5 \times 10^2 + 6 \times 10^1 + 4 \times 10^0$

or 1,564

10. 4,738

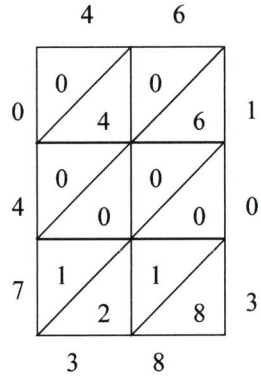

11. Answers may vary.
    We need to be able to indicate which powers of the base (say ten) that are missing from the numeral.

12. $342_5 = 3 \times 5^2 + 4 \times 5^1 + 2 \times 5^0$ $B3D_{16} = 11 \times 16^2 + 3 \times 16^1 + 13 \times 16^0$
    $ = 3 \times 25 + 4 \times 5 + 2 \times 1$ and $\phantom{B3D_{16}} = 11 \times 256 + 3 \times 16 + 13 \times 1$
    $ = 75 + 20 + 2$ $\phantom{B3D_{16}} = 2816 + 48 + 13$
    $ = 97$ $\phantom{B3D_{16}} = 2{,}877$

13. 6513_8

    ```
    8 | 3403    3
    8 | 425     1
    8 | 53      5
    8 | 6       6
        0
    ```

15. $134_5 \; R \; 20_5$

    ```
            134
    23₅ ) 4312₅
          23
          201
          124
          222
          202
           20
    ```

14. $\overset{1\,1\,1\,1\,1}{10111_2}$
    $\underline{+\,11001_2}$
    110000_2

16. 1342_8 $2E2_{16}$

    1 011 100 010   10 1110 0010
    1 3   4   2   2   E   2

Chapter Test

1. MMMDCLXXXV

2. $2 \times 1{,}000{,}000 + 1 \times 100{,}000 + 2 \times 10{,}000 + 3 \times 1{,}000 + 4 \times 100 + 2 \times 10 =$
   $2{,}000{,}000 + 100{,}000 + 20{,}000 + 3{,}000 + 400 + 20 = 2{,}123{,}420$

90   Chapter 4: Numeration Systems

3.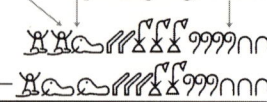

4. $5 \times 1{,}000 + 3 \times 100 + 4 \times 10 + 8 = 5{,}000 + 300 + 40 + 8 = 5{,}348$

5. Hindu–Arabic numeration has zero and place value.

6. $59 = 1 + 2 + 8 + 16 + 32$

Powers of 2	Times 26
1	26
2	26 + 26 = 52
4	52 + 52 = 104
8	104 + 104 = 208
16	208 + 208 = 416
32	416 + 416 = 832

$832 + 416 + 208 + 52 + 26 = 1{,}534$

7. $264_7 = 2 \times 7^2 + 6 \times 7^1 + 4 \times 7^0$
$= 2 \times 49 + 6 \times 7 + 4 \times 1$
$= 98 + 42 + 4$
$= 144$

and

$A3E_{16} = 10 \times 16^2 + 3 \times 16^1 + 14 \times 16^0$
$= 10 \times 256 + 3 \times 16 + 14 \times 1$
$= 2560 + 48 + 14$
$= 2{,}622$

8. 10, X, +

9. $10{,}937 = 3 \times 60^2 + 2 \times 60 + 17$ ▼▼▼ ▼▼ ⟨▼▼▼▼▼▼▼ or

$10{,}937 = 3 \times 60^2 + 2 \times 60 + (20 - 3)$ ▼▼▼ ▼▼ ⟨⟨▼▼▼▼

10. 15,946

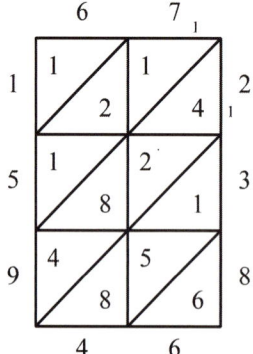

11. 4401_8

$\begin{array}{r|rl} 8 & 2305 & 1 \\ 8 & 288 & 0 \\ 8 & 36 & 4 \\ 8 & 4 & 4 \\ & 0 & \end{array}$

12. $1,738 = 1 \times 10^3 + 7 \times 10^2 + 3 \times 10^1 + 8 \times 10^0$

$ 526 = + 5 \times 10^2 + 2 \times 10^1 + 6 \times 10^0$

$ 1 \times 10^3 + 12 \times 10^2 + 5 \times 10^1 + 14 \times 10^0$

$\begin{aligned} 1 \times 10^3 + 12 \times 10^2 + 5 \times 10^1 + 14 \times 10^0 &= 1 \times 10^3 + (10+2) \times 10^2 + 5 \times 10^1 + (10+4) \times 10^0 \\ &= 1 \times 10^3 + 10 \times 10^2 + 2 \times 10^2 + 5 \times 10^1 + 10 \times 10^0 + 4 \times 10^0 \\ &= 1 \times 10^3 + 1 \times 10^3 + 2 \times 10^2 + 5 \times 10^1 + 1 \times 10^1 + 4 \times 10^0 \\ &= (1+1) \times 10^3 + 2 \times 10^2 + (5+1) \times 10^1 + 4 \times 10^0 \\ &= 2 \times 10^3 + 2 \times 10^2 + 6 \times 10^1 + 4 \times 10^0 \\ &= 2,264 \end{aligned}$

13. Answers may vary. Possible answers include: More symbols were required to write their numerals.

14. 110_5 R2_5

    110_5
    $24_5 \overline{)3142_5}$
    $\underline{24}$
    24
    $\underline{24}$
    2
    2

15. $10\overset{1}{1}\overset{1}{1}\overset{1}{0}\overset{1}{1}_2$
    $\underline{+110101_2}$
    1100010_2

16. 3113_5

    $\overset{1}{4}2_5$
    $\underline{\times 34_5}$
    $\overset{1}{3}23$
    $\underline{231}$
    3133_5

17. 6536_8 $\qquad$ $D5E_{16}$

    110 101 011 110
     6   5   3   6

    1101 0101 1110
     D   5   E

Of Further Interest: Modular Systems

1. In doing modulo 12 computations, all multiples of 12 are ignored.

2. Beginning at 0, count to 53 on the clock to determine where to stop. Divide by twelve and keep the remainder.

3. Answers may vary. One example is –3, which means moving counterclockwise 3 positions. This is the same as moving clockwise 9 positions so –3 and 9 are the same on a 12-hour clock.

4. To check the validity in the UPC code.

5. Answers may vary. One possible answer is to perform the operations as usual and then count off the result on an m-hour clock.

6. The set of all integers that correspond to the same position on the clock. Examples may vary. The congruence class containing 3 in a modulo 8 system would be $\{\ldots, -13, -5, 3, 11, 19, \ldots\}$.

7. Try the numbers 0, 1, 2, …, 11 to see which ones satisfy the congruences.

8. Solve both congruences separately as in Exercise 7 and then find which solutions are common to both congruences.

9. 7 since $43 = 3 \times 12 + 7$

10. 3 since $21 = 3 \times 6 + 3$

11. 4 since $39 = 5 \cdot 7 + 4$

12. 5 since $69 = 8 \cdot 8 + 5$

13. True since $59 - 35 = 24$ is evenly divisible by 12.

14. False since $31 - 14 = 17$ is not evenly divisible by 7.

15. True since $43 - 11 = 32$ is evenly divisible by 8.

16. False since $29 - 18 = 11$ is not evenly divisible by 6.

17. False since $78 - 46 = 32$ is not evenly divisible by 10.

18. True since $75 - 53 = 22$ is evenly divisible by 11.

19. 8

20. 10

21. 5, 12, 19, 26, 33,…

22. 8, 20, 32, 44, 56,…

23. 6, 16, 26, 36, 46,….

24. 3, 12, 21, 30, 39,…

25. 5 since $8 + 9 \equiv 5 \pmod{12}$

26. 3 since $4 + 7 \equiv 3 \pmod{8}$

27. 2 since $3 + 4 \equiv 2 \pmod{5}$

28. 3 since $6 + 8 \equiv 3 \pmod{11}$

29. 7 since $4 - 9 \equiv 7 \pmod{12}$

30. 7 since $5 - 6 \equiv 7 \pmod{8}$

31. 4 since $3 - 4 \equiv 4 \pmod{5}$

32. 6 since $3 - 8 \equiv 6 \pmod{11}$

33. 0 since $4 \times 9 \equiv 0 \pmod{12}$

34. 6 since $5 \times 6 \equiv 6 \pmod{8}$

35. 3 since $2 \times 5 \equiv 3 \pmod{7}$

36. 13 since $9 \times 3 \equiv 13 \pmod{14}$

37. 10

$3 + 0 \equiv 1 \pmod{12}$; false
$2 - 1 \equiv 8 \pmod 9$; false
$3 + 2 \equiv 1 \pmod{12}$; false
$3 + 3 \equiv 1 \pmod{12}$; false
$3 + 4 \equiv 1 \pmod{12}$; false
$3 + 5 \equiv 1 \pmod{12}$; false
$3 + 6 \equiv 1 \pmod{12}$; false
$3 + 7 \equiv 1 \pmod{12}$; false
$3 + 8 \equiv 1 \pmod{12}$; false
$3 + 9 \equiv 1 \pmod{12}$; false
$3 + 10 \equiv 1 \pmod{12}$; **true**
$3 + 11 \equiv 1 \pmod{12}$; false

38. 9

$9 + 0 \equiv 6 \pmod{12}$; false
$9 + 1 \equiv 6 \pmod{12}$; false
$9 + 2 \equiv 6 \pmod{12}$; false
$9 + 3 \equiv 6 \pmod{12}$; false
$9 + 4 \equiv 6 \pmod{12}$; false
$9 + 5 \equiv 6 \pmod{12}$; false
$9 + 6 \equiv 6 \pmod{12}$; false
$9 + 7 \equiv 6 \pmod{12}$; false
$9 + 8 \equiv 6 \pmod{12}$; false
$9 + 9 \equiv 6 \pmod{12}$; **true**
$9 + 10 \equiv 6 \pmod{12}$; false
$9 + 11 \equiv 6 \pmod{12}$; false

39. 6

$6 + 0 \equiv 4 \pmod 8$; false
$6 + 1 \equiv 4 \pmod 8$; false
$6 + 2 \equiv 4 \pmod 8$; false
$6 + 3 \equiv 4 \pmod 8$; false
$6 + 4 \equiv 4 \pmod 8$; false
$6 + 5 \equiv 4 \pmod 8$; false
$6 + 6 \equiv 4 \pmod 8$; **true**
$6 + 7 \equiv 4 \pmod 8$; false

40. 3

$5 + 0 \equiv 2 \pmod 6$; false
$5 + 1 \equiv 2 \pmod 6$; false
$5 + 2 \equiv 2 \pmod 6$; false
$5 + 3 \equiv 2 \pmod 6$; **true**
$5 + 4 \equiv 2 \pmod 6$; false
$5 + 5 \equiv 2 \pmod 6$; false

41. 6

$3 - 0 \equiv 4 \pmod 7$; false
$3 - 1 \equiv 4 \pmod 7$; false
$3 - 2 \equiv 4 \pmod 7$; false
$3 - 3 \equiv 4 \pmod 7$; false
$3 - 4 \equiv 4 \pmod 7$; false
$3 - 5 \equiv 4 \pmod 7$; false
$3 - 6 \equiv 4 \pmod 7$; **true**

42. 8

$5 - 0 \equiv 7 \pmod{10}$; false
$5 - 1 \equiv 7 \pmod{10}$; false
$5 - 2 \equiv 7 \pmod{10}$; false
$5 - 3 \equiv 7 \pmod{10}$; false
$5 - 4 \equiv 7 \pmod{10}$; false
$5 - 5 \equiv 7 \pmod{10}$; false
$5 - 6 \equiv 7 \pmod{10}$; false
$5 - 7 \equiv 7 \pmod{10}$; false
$5 - 8 \equiv 7 \pmod{10}$; **true**
$5 - 9 \equiv 7 \pmod{10}$; false

43. 3

$2 - 0 \equiv 8 \pmod 9$; false
$2 - 1 \equiv 8 \pmod 9$; false
$2 - 2 \equiv 8 \pmod 9$; false
$2 - 3 \equiv 8 \pmod 9$; **true**
$2 - 4 \equiv 8 \pmod 9$; false
$2 - 5 \equiv 8 \pmod 9$; false
$2 - 6 \equiv 8 \pmod 9$; false
$2 - 7 \equiv 8 \pmod 9$; false
$2 - 8 \equiv 8 \pmod 9$; false

44. 4

$3 - 0 \equiv 4 \pmod 5$; false
$3 - 1 \equiv 4 \pmod 5$; false
$3 - 2 \equiv 4 \pmod 5$; false
$3 - 3 \equiv 4 \pmod 5$; false
$3 - 4 \equiv 4 \pmod 5$; **true**

45. 4

$3 \times 0 \equiv 5 \pmod 7$; false
$3 \times 1 \equiv 5 \pmod 7$; false
$3 \times 2 \equiv 5 \pmod 7$; false
$3 \times 3 \equiv 5 \pmod 7$; false
$3 \times 4 \equiv 5 \pmod 7$; **true**
$3 \times 5 \equiv 5 \pmod 7$; false
$3 \times 6 \equiv 5 \pmod 7$; false

46. 7

$5 \times 0 \equiv 3 \pmod 8$; false
$5 \times 1 \equiv 3 \pmod 8$; false
$5 \times 2 \equiv 3 \pmod 8$; false
$5 \times 3 \equiv 3 \pmod 8$; false
$5 \times 4 \equiv 3 \pmod 8$; false
$5 \times 5 \equiv 3 \pmod 8$; false
$5 \times 6 \equiv 3 \pmod 8$; false
$5 \times 7 \equiv 3 \pmod 8$; **true**

47. 3, 8

$4 \times 0 \equiv 2 \pmod{10}$; false
$4 \times 1 \equiv 2 \pmod{10}$; false
$4 \times 2 \equiv 2 \pmod{10}$; false
$4 \times 3 \equiv 2 \pmod{10}$; **true**
$4 \times 4 \equiv 2 \pmod{10}$; false
$4 \times 5 \equiv 2 \pmod{10}$; false
$4 \times 6 \equiv 2 \pmod{10}$; false
$4 \times 7 \equiv 2 \pmod{10}$; false
$4 \times 8 \equiv 2 \pmod{10}$; **true**
$4 \times 9 \equiv 2 \pmod{10}$; false

48. 2, 6

$6 \times 0 \equiv 4 \pmod 8$; false
$6 \times 1 \equiv 4 \pmod 8$; false
$6 \times 2 \equiv 4 \pmod 8$; **true**
$6 \times 3 \equiv 4 \pmod 8$; false
$6 \times 4 \equiv 4 \pmod 8$; false
$6 \times 5 \equiv 4 \pmod 8$; false
$6 \times 6 \equiv 4 \pmod 8$; **true**
$6 \times 7 \equiv 4 \pmod 8$; false

49. For $x \equiv 3 \pmod 7$, x is in the congruence class containing 3 modulo 7 which is

$$3, 10, 17, \mathbf{24}, 31, 38, \ldots$$

For $x \equiv 4 \pmod 5$, x is in the congruence class containing 4 modulo 5 which is

$$4, 9, 14, 19, \mathbf{24}, 29, \ldots$$

The smallest number that is found in both of these classes is 24.

50. For $x \equiv 1 \pmod 6$, x is in the congruence class containing 1 modulo 6 which is

$$1, \mathbf{7}, 13, 19, 25, 31, \ldots$$

For $x \equiv 7 \pmod 8$, x is in the congruence class containing 7 modulo 8 which is

$$\mathbf{7}, 15, 23, 31, 39, 47, \ldots$$

The smallest number that is found in both of these classes is 7.

51. For $x \equiv 2 \pmod{10}$, x is in the congruence class containing 2 modulo 10 which is
$$2, 12, 22, 32, \mathbf{42}, 52, 62, 72\ldots$$
For $x \equiv 2 \pmod{8}$, x is in the congruence class containing 2 modulo 8 which is
$$2, 10, 18, 26, 34, \mathbf{42}, 50, 58 \ldots$$

For $x \equiv 0 \pmod{6}$, x is in the congruence class containing 0 modulo 6 which is
$$0, 6, 12, 18, 24, 30, 36, \mathbf{42}, \ldots$$

The smallest number that is found in these three classes is 42.

52. For $x \equiv 5 \pmod{9}$, x is in the congruence class containing 5 modulo 9 which is
$$5, 14, \mathbf{23}, 32, 41, 50, \ldots$$
For $x \equiv 1 \pmod{11}$, x is in the congruence class containing 1 modulo 11 which is
$$1, 12, \mathbf{23}, 34, 45, 56, \ldots$$

For $x \equiv 7 \pmod{8}$, x is in the congruence class containing 7 modulo 8 which is
$$7, 15, \mathbf{23}, 31, 39, 47, \ldots$$

The smallest number that is found in these three classes is 23.

53. 3 since $80 = 11 \cdot 7 + 3$

54. 1 since $71 = 10 \cdot 7 + 1$

55. This problem is the same as finding the smallest positive integer that solves both $x \equiv 4 \pmod{6}$ and $x \equiv 3 \pmod{5}$.

    For $x \equiv 4 \pmod{6}$, x is in the congruence class containing 4 modulo 6 which is
    $$4, 10, 16, 22, \mathbf{28}, 34, \ldots$$
    For $x \equiv 3 \pmod{5}$, x is in the congruence class containing 3 modulo 5 which is
    $$3, 8, 13, 18, 23, \mathbf{28}, \ldots$$

    The smallest number that is found in both of these classes is 28.

96 Chapter 4: Numeration Systems

56. This problem is the same as finding the smallest positive integer that solves $x \equiv 2 \pmod 8$, $x \equiv 2 \pmod{10}$, and $x \equiv 6 \pmod{12}$.

    For $x \equiv 2 \pmod 8$, x is in the congruence class containing 2 modulo 8 which is
    $$2, 10, 18, 26, 34, \mathbf{42}, \ldots$$
    For $x \equiv 2 \pmod{10}$, x is in the congruence class containing 2 modulo 10 which is
    $$2, 12, 22, 32, \mathbf{42}, 52, \ldots$$
    For $x \equiv 6 \pmod{12}$, x is in the congruence class containing 6 modulo 12 which is
    $$6, 18, 30, \mathbf{42}, 54, 66, \ldots$$

    The smallest number that is found in these three classes is 42.

57. We are looking for years that are congruent to 2 modulo 12. Since this book was written in 2003 and Chris appears to be less than 30, we will start in the year $2003 - 29 = 1974$.

    $1974 \equiv 2 \pmod{12}$; false $1980 \equiv 2 \pmod{12}$; false
    $1975 \equiv 2 \pmod{12}$; false $1981 \equiv 2 \pmod{12}$; false
    $1976 \equiv 2 \pmod{12}$; false $1982 \equiv 2 \pmod{12}$; **true**
    $1977 \equiv 2 \pmod{12}$; false $1983 \equiv 2 \pmod{12}$; false
    $1978 \equiv 2 \pmod{12}$; false $1984 \equiv 2 \pmod{12}$; false
    $1979 \equiv 2 \pmod{12}$; false $1985 \equiv 2 \pmod{12}$; false

    Chris was born in 1982.

58. We are looking for years that are congruent to 3 modulo 12. Since the person is between 40 and 50 years old, we will start in the year $2003 - 49 = 1954$ and end in the year $2003 - 41 = 1962$.

    $1954 \equiv 3 \pmod{12}$; false $1959 \equiv 3 \pmod{12}$; **true**
    $1955 \equiv 3 \pmod{12}$; false $1960 \equiv 3 \pmod{12}$; false
    $1956 \equiv 3 \pmod{12}$; false $1961 \equiv 3 \pmod{12}$; false
    $1957 \equiv 3 \pmod{12}$; false $1962 \equiv 3 \pmod{12}$; false
    $1958 \equiv 3 \pmod{12}$; false

    The person was born in 1959.

59. 2

10×0	0
9×0	0
8×6	48
7×6	42
$6 \times d$	$6d$
5×0	0
4×9	36
3×4	12
2×4	8
Total	$296 + 4d$

Since 7 is the check digit, we need to find the smallest integer that satisfies
$11 - 7 = 4 \equiv (146 + 6d)(\bmod 11)$.

60. 4

10×0	0
9×8	72
8×0	0
7×5	35
6×0	0
5×7	35
4×1	4
$3 \times d$	$3d$
2×5	10
Total	$156 + 3d$

Since 8 is the check digit, we need to find the smallest integer that satisfies
$11 - 8 = 3 \equiv (156 + 3d)(\bmod 11)$.

61. 5

10×0	0
9×8	72
8×1	8
7×9	63
$6 \times d$	$6d$
5×6	30
4×7	28
3×1	3
2×2	4
Total	$208 + 6d$

Since 4 is the check digit, we need to find the smallest integer that satisfies
$11 - 4 = 7 \equiv (208 + 6d)(\bmod 11)$.

$4 \equiv (146 + 6 \times 0)(\bmod 11)$; false
$4 \equiv (146 + 6 \times 1)(\bmod 11)$; false
$4 \equiv (146 + 4 \times 6)(\bmod 11)$; **true**
$4 \equiv (146 + 6 \times 3)(\bmod 11)$; false
$4 \equiv (146 + 6 \times 4)(\bmod 11)$; false
$4 \equiv (146 + 6 \times 5)(\bmod 11)$; false
$4 \equiv (146 + 6 \times 6)(\bmod 11)$; false
$4 \equiv (146 + 6 \times 7)(\bmod 11)$; false
$4 \equiv (146 + 6 \times 8)(\bmod 11)$; false
$4 \equiv (146 + 6 \times 9)(\bmod 11)$; false
$4 \equiv (146 + 6 \times 10)(\bmod 11)$; false

$3 \equiv (156 + 3 \times 0)(\bmod 11)$; false
$3 \equiv (156 + 3 \times 1)(\bmod 11)$; false
$3 \equiv (156 + 3 \times 2)(\bmod 11)$; false
$3 \equiv (156 + 3 \times 3)(\bmod 11)$; false
$3 \equiv (156 + 3 \times 4)(\bmod 11)$; **true**
$3 \equiv (156 + 3 \times 5)(\bmod 11)$; false
$3 \equiv (156 + 3 \times 6)(\bmod 11)$; false
$3 \equiv (156 + 3 \times 7)(\bmod 11)$; false
$3 \equiv (156 + 3 \times 8)(\bmod 11)$; false
$3 \equiv (156 + 3 \times 9)(\bmod 11)$; false
$3 \equiv (156 + 3 \times 10)(\bmod 11)$; false

$7 \equiv (208 + 6 \times 0)(\bmod 11)$; false
$7 \equiv (208 + 6 \times 1)(\bmod 11)$; false
$7 \equiv (208 + 6 \times 2)(\bmod 11)$; false
$7 \equiv (208 + 6 \times 3)(\bmod 11)$; false
$7 \equiv (208 + 6 \times 4)(\bmod 11)$; false
$7 \equiv (208 + 6 \times 5)(\bmod 11)$; **true**
$7 \equiv (208 + 6 \times 6)(\bmod 11)$; false
$7 \equiv (208 + 6 \times 7)(\bmod 11)$; false
$7 \equiv (208 + 6 \times 8)(\bmod 11)$; false
$7 \equiv (208 + 6 \times 9)(\bmod 11)$; false
$7 \equiv (208 + 6 \times 10)(\bmod 11)$; false

62. 2

10×0	0
9×6	54
8×8	64
7×9	63
6×8	48
5×6	30
$4 \times d$	$4d$
3×7	21
2×8	0
Total	$296 + 4d$

Since 4 is the check digit, we need to find the smallest integer that satisfies
$11 - 4 = 7 \equiv (296 + 4d)(\mod 11)$.

$7 \equiv (296 + 4 \times 0)(\mod 11)$; false
$7 \equiv (296 + 4 \times 1)(\mod 11)$; false
$7 \equiv (296 + 4 \times 2)(\mod 11)$; **true**
$7 \equiv (296 + 4 \times 3)(\mod 11)$; false
$7 \equiv (296 + 4 \times 4)(\mod 11)$; false
$7 \equiv (296 + 4 \times 5)(\mod 11)$; false
$7 \equiv (296 + 4 \times 6)(\mod 11)$; false
$7 \equiv (296 + 4 \times 7)(\mod 11)$; false
$7 \equiv (296 + 4 \times 8)(\mod 11)$; false
$7 \equiv (296 + 4 \times 9)(\mod 11)$; false
$7 \equiv (296 + 4 \times 10)(\mod 11)$; false

63. $12^5 (\mod 35) = 248832 (\mod 35) = 17$

64. $15^5 (\mod 35) = 759375 (\mod 35) = 15$

65. $22^5 (\mod 35) = 5153632 (\mod 35) = 22$

66. $5^5 (\mod 35) = 3125 (\mod 35) = 10$

67. Explanations may vary.
In this modulo 8 system, find 6 times what number (or numbers) is (are) congruent to 4.
If $4/6 \equiv x (\mod 8)$ then $4 \equiv 6x (\mod 8)$

$4 \equiv (6 \times 0)(\mod 8)$; false
$4 \equiv (6 \times 1)(\mod 8)$; false
$4 \equiv (6 \times 2)(\mod 8)$; **true**
$4 \equiv (6 \times 3)(\mod 8)$; false
$4 \equiv (6 \times 4)(\mod 8)$; false
$4 \equiv (6 \times 5)(\mod 8)$; false
$4 \equiv (6 \times 6)(\mod 8)$; **true**
$4 \equiv (6 \times 7)(\mod 8)$; false   Answer: 4, 6

a) If $5/3 (\mod 8) \equiv x$ then $5 \equiv 3x (\mod 8)$

$5 \equiv (3 \times 0)(\mod 8)$; false
$5 \equiv (3 \times 1)(\mod 8)$; false
$5 \equiv (3 \times 2)(\mod 8)$; false
$5 \equiv (3 \times 3)(\mod 8)$; false
$5 \equiv (3 \times 4)(\mod 8)$; false
$5 \equiv (3 \times 5)(\mod 8)$; false
$5 \equiv (3 \times 6)(\mod 8)$; false
$5 \equiv (3 \times 7)(\mod 8)$; **true**   Answer: 7

b) If $2/6 (\mod 8) \equiv x$ then $2 \equiv 6x (\mod 8)$

$2 \equiv (6 \times 0)(\mod 8)$; false
$2 \equiv (6 \times 1)(\mod 8)$; false
$2 \equiv (6 \times 2)(\mod 8)$; false
$2 \equiv (6 \times 3)(\mod 8)$; **true**
$2 \equiv (6 \times 4)(\mod 8)$; false
$2 \equiv (6 \times 5)(\mod 8)$; false
$2 \equiv (6 \times 6)(\mod 8)$; false
$2 \equiv (6 \times 7)(\mod 8)$; **true**   Answer: 3, 7

68. If $2/x \equiv 5 \pmod 6$ then $2 \equiv 5x \pmod 6$

   $2 \equiv (5 \times 0) \pmod 6$; false
   $2 \equiv (5 \times 1) \pmod 6$; false
   $2 \equiv (5 \times 2) \pmod 6$; false
   $2 \equiv (5 \times 3) \pmod 6$; false
   $2 \equiv (5 \times 4) \pmod 6$; **true**
   $2 \equiv (5 \times 5) \pmod 6$; false   Answer: 4

69. – 70. No solutions provided

Chapter 5
NUMBER THEORY AND THE REAL NUMBER SYSTEM: Understanding The Numbers All Around Us

Section 5.1 Number Theory

1. The square root of 83 is between 9 and 10. Since 7 is the largest prime less than 10, there will be no prime factors.

3. When each branch ended in a prime number.

5. 17

7. The largest powers of any primes that occur in either number are multiplied.

9. True

11. False

13. False

15. True

17. 4 is a factor of 48; 48 is a multiple of 4

19. 7 is a factor of 28; 7 divides 28

21. 9 divides 72; 72 is a multiple of 9

23. 53, 59, 61, 67, 71, 73, 79, 83, 89, 97

25. Since $81 < 95 < 100$, $9 < \sqrt{95} < 10$. Thus $a = 9$.

27. Since $144 < 153 < 169$, $12 < \sqrt{153} < 13$. Thus $a = 12$.

29. 231 is composite. Possible factorizations into the product of two natural numbers include 1×231, 3×77, 7×33, and 11×21.

31. 113 is prime.

For Exercise 23 refer to the following.

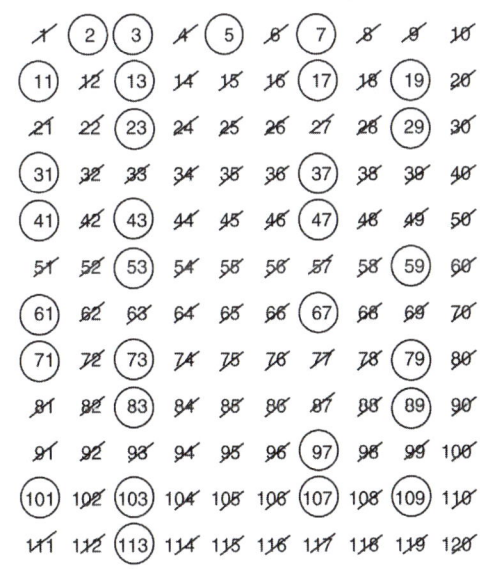

33. 197 is prime.

35. 119 is composite. Possible factorizations into the product of two natural numbers include 1×119 and 7×17.

101

37. 141,270 is divisible by **2** because the last digit, 0, is divisible by 2.
    141,270 is divisible by **3** because the sum of the digits, 15, is divisible by 3.
    141,270 is not divisible by 4 because the number formed by the last two digits, 70, is not divisible by 4.
    141,270 is divisible by **5** because the last digit, 0, is a 0 or a 5.
    141,270 is divisible by **6** because the number is divisible by both 2 and 3.
    141,270 is not divisible by 8 because the number formed by the last three digits, 270, is not divisible by 8.
    141,270 is not divisible by 9 because the sum of the digits, 15, is not divisible by 9.
    141,270 is divisible by **10** because the last digit, 0, is 0.

39. 47,385 is not divisible by 2 because the last digit, 5, is not divisible by 2.
    47,385 is divisible by **3** because the sum of the digits, 27, is divisible by 3.
    47,385 is not divisible by 4 because the number formed by the last two digits, 85, is not divisible by 4.
    47,385 is divisible by **5** because the last digit, 5, is a 0 or a 5.
    47,385 is not divisible by 6 because the number is not divisible by both 2 and 3.
    47,385 is not divisible by 8 because the number formed by the last three digits, 385, is not divisible by 8.
    47,385 is divisible by **9** because the sum of the digits, 27, is divisible by 9.
    47,385 is not divisible by 10 because the last digit, 5, is not 0.

41. Since any number divisible by both 2 and 3 is also divisible by 6, the smallest possible number would be $2 \times 3 \times 5 = 30$.

43. Since 10 is the product of 2 and 5, any number divisible by 4 (2×2) and 5 would be divisible by 10. The smallest possible number would be $4 \times 5 = 20$.

45. Answers may vary. Possible values for *a* are 4, 12, 20, 28, 36, etc.

47. Answers may vary. Possible values for *a* are 20, 60, 100, 140, 180, etc.

49. $980 = 2^2 \times 5 \times 7^2$

51. $9900 = 2^2 \times 3^2 \times 5^2 \times 11$

53. $621 = 3^3 \times 23$

55. $319 = 11 \times 29$

57. $20 = 2^2 \times 5$ and $24 = 2^3 \times 3$
    $GCD(20, 24) = 2^2 = 4$
    $LCM(20, 24) = 2^3 \times 3 \times 5 = 120$

59. $56 = 2^3 \times 7$ and $70 = 2 \times 5 \times 7$
    $GCD(56, 70) = 2 \times 7 = 14$
    $LCM(56, 70) = 2^3 \times 5 \times 7 = 280$

61. $216 = 2^3 \times 3^3$ and $288 = 2^5 \times 3^2$
    $GCD(216, 288) = 2^3 \times 3^2 = 72$
    $LCM(216, 288) = 2^5 \times 3^3 = 864$

63. $147 = 3 \times 7^2$ and $567 = 3^4 \times 7$
    $GCD(147, 567) = 3 \times 7 = 21$
    $LCM(147, 567) = 3^4 \times 7^2 = 3969$
    $LCM(275, 363) = 3 \times 5^2 \times 11^2 = 9075$

65. 28

67. 45

69. $GCD(12,27) = 3$

$$12 \overline{)\underset{\underline{24}}{27}}^{2} \quad 3\overline{)\underset{\underline{12}}{12}}^{4}$$
$$\;\;\boxed{3} \quad\quad\quad\; 0$$

$LCM(12,27) = \dfrac{12 \cdot 27}{3} = 4 \cdot 27 = 108$

71. $GCD(90,120) = 30$

$$90 \overline{)\underset{\underline{90}}{120}}^{1} \quad 30\overline{)\underset{\underline{90}}{90}}^{3}$$
$$\;\boxed{30} \quad\quad\quad\; 0$$

$LCM(90,120) = \dfrac{90 \cdot 120}{30} = 3 \cdot 120 = 360$

73. In this exercise we need to find $LCM(20,35)$ which is 140.; 140 minutes

75. In this exercise we need to find $GCD(30,36)$ which is 6.; 6 packages

77. In this exercise we need to find $GCD(21,33)$ which is 3.; 3 feet by 3 feet

79. In this exercise we need to find $LCM(48,54)$ which is 432.; 432 hours

81. Three will always divide a multiple of nine, that is, $5 \cdot 999 + 7 \cdot 99 + 1 \cdot 9$, so all we have to check is to see if three divides the sum of the digits, $5 + 7 + 1 + 2 = 15$, which it does.

83. a) $80 = 7 + 73 = 13 + 67 = 19 + 61 = 37 + 43$
    b) $100 = 3 + 97 = 11 + 89 = 17 + 83 = 27 + 73 = 29 + 71 = 41 + 59 = 47 + 53$
    c) $200 = 3 + 197 = 7 + 193 = 19 + 181 = 37 + 163 = 43 + 157 = 61 + 139 = 73 + 127 = 97 + 103$

85. In each case, the two numbers have common factors greater than one.

87. Answers may vary.
    The numbers 2 and 3 have no factors in common greater than one; however, 2 and 4 do have a common factor greater than one.

89. Answers may vary.
    It means that if we express a natural number as a product of prime numbers, we cannot do it in any other way, which is what the Fundamental Theorem of Arithmetic tells us.

Section 5.2 The Integers

1. The definition $a - b = a + (-b)$ was used and negative numbers were interpreted as movement to the left on the number line and positive numbers were interpreted as movement to the right.

3. The number –2 was the only number that solved the equation $+12 = (\;\;)\times(-6)$.

5. Such memory devices only work in certain situations, for example (–3) + (–5) is not a positive number.

7. $(+8) + (-5) = +3$

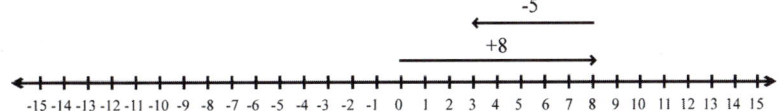

9. $(+4)+(+6)=+10$

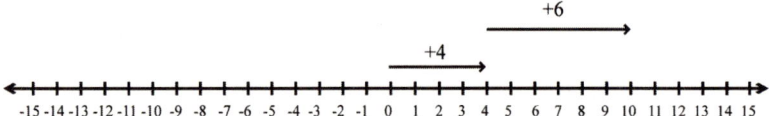

11. It is not practical to draw a number line when calculating with such large numbers; however, you can still imagine movements along the number line. If you visualize moving 128 units to the left and then 137 units to the right, you can see that the net effect is that you moved $137-128=9$ units to the right, so $(-128)+(+137)=+9$.

13. It is not practical to draw a number line when calculating with such large numbers; however, you can still imagine movements along the number line. If you visualize moving 57 units to the right and then 38 units to the left, you can see that the net effect is that you moved $57-38=19$ units to the right, so $(+57)+(-38)=+19$.

15. $(+18)-(-5)=(+18)+(+5)=+23$

19. $(-28)-(+37)=(-28)+(-37)=-65$

17. $(+6)-(+19)=(+6)+(-19)=-13$

21. $(+32)-(-18)=(+32)+(+18)=+50$

23. For the next four days you will spend $6, so in four days, your net change in finances will be $-\$24$. Thus, $(+4)(-6)=-24$.

25. For the past three days you received $7, so three days ago, you had $21 less. Thus $(-3)(+7)=-21$.

27. $(-5)(-7)=+35$

49. $+15$

29. $(-7)(+8)=-56$

51. $+5$

53. -6

31. $(+8)(+6)=+48$

55. $+6$

33. $(+19)(-2)=-38$

57. -4

35. $(-9)(+7)=-63$

59. $(-2)(-3)+(+4)(-8)=$
$(+6)+(-32)=-26$

37. $(-8)(-9)=+72$

61. $(-4)(-2)+(+4)(-5)=$
$(+8)+(-20)=-12$

39. $+16=(-2)\cdot c$; $c=-8$

41. $-14=(-2)\cdot c$; $c=+7$

63. $(-5)(-6)+(+2)(-7)=$
$(+30)+(-14)=+16$

43. $-40=(+8)\cdot c$; $c=-5$

45. $+12=(+3)\cdot c$; $c=+4$

65. $(-3)(+6)+(-4)(+8)=$
$(-18)+(-32)=-50$

47. -3

67. There are an infinite number of integers that are whole numbers. You can choose from 0, 1, 2, 3, …

69. not possible; the whole numbers are 0, 1, 2, 3, … There are no negative integers in this list.

71. There are an infinite number of integers that are not natural numbers. You can choose from 0, –1, –2, –3,…

73. not possible; the integers are …, –3, –2, –1, 0, 1, 2, 3, … All natural numbers appear in this list.

75. True

77. False; Counterexamples may vary; $(-5)+(+8)=+3$

79. False; counterexamples may vary; $\dfrac{-6}{-3}=+2$

81. True

83. $29,028-(-28,232)=57,260$ feet

85. $(-28,232)-(-35,433)=7,201$ feet

87. $136-(-129)=265$ degrees

89. $59-(-7)=66$ floors

91. $1368-(-551)=1,919$ years

93. $65+4(-7)=65+(-28)=37$ degrees

95. No; The sums in row 3 and column 4 are not the same as the other sums.

−8	−1	4	3	−2
5	2	−7	−2	−2
−5	−4	7	1	−1
6	1	−6	−3	−2
−2	−2	−2	−2	−1	−2

97. If $\dfrac{8}{0}=x$, then $8=0\cdot x$, which is not possible.

99. The sum of the entries, 333, is nine times the center number.

101. Put a point, c, in the middle of the 4-by-4 square and notice that any two numbers that are symmetric with respect to c have the same total. There are eight such pairs in a 4-by-4 box; therefore add two numbers in the box that are symmetric with respect to the center, c, and multiply this total by eight to get the total of all sixteen numbers.

103. Answers may vary.
 $5+x=3$

Section 5.3 The Rational Numbers

1. Cross multiply and see if the same result are the same.

3. denominators

5. Cancel common factors from the numerators and denominators.

7. Divide the denominator, d, into the numerator to get a quotient q and a remainder r. Then, write the mixed number as $q\dfrac{r}{d}$.

9. equal; $\dfrac{2}{3} = \dfrac{8}{12}$ since $2 \cdot 12 = 3 \cdot 8 \ (24 = 24)$

11. not equal; $\dfrac{12}{14} \neq \dfrac{14}{16}$ since $12 \cdot 16 \neq 14 \cdot 14 \ (192 \neq 196)$

13. not equal; $\dfrac{22}{14} \neq \dfrac{30}{21}$ since $22 \cdot 21 \neq 14 \cdot 30 \ (462 \neq 420)$

15. equal; $\dfrac{5}{14} = \dfrac{15}{42}$ since $5 \cdot 42 = 14 \cdot 15 \ (210 = 210)$

17. not equal; $\dfrac{32}{14} \neq \dfrac{18}{12}$ since $32 \cdot 12 \neq 14 \cdot 18 \ (384 \neq 252)$

19. $\dfrac{15}{35} = \dfrac{\cancel{5} \cdot 3}{\cancel{5} \cdot 7} = \dfrac{3}{7}$

21. $-\dfrac{24}{72} = -\dfrac{\cancel{24} \cdot 1}{\cancel{24} \cdot 3} = -\dfrac{1}{3}$

23. $-\dfrac{77}{126} = -\dfrac{\cancel{7} \cdot 11}{\cancel{7} \cdot 18} = -\dfrac{11}{18}$

25. $\dfrac{225}{350} = \dfrac{\cancel{25} \cdot 9}{\cancel{25} \cdot 14} = \dfrac{9}{14}$

27. $\dfrac{143}{154} = \dfrac{\cancel{11} \cdot 13}{\cancel{11} \cdot 14} = \dfrac{13}{14}$

29. $\dfrac{2}{3} + \dfrac{1}{2} = \dfrac{2 \cdot 2 + 3 \cdot 1}{3 \cdot 2} = \dfrac{4+3}{6} = \dfrac{7}{6}$

31. $\dfrac{1}{6} - \dfrac{1}{2} = \dfrac{1}{6} - \dfrac{1 \cdot 3}{2 \cdot 3} = \dfrac{1}{6} - \dfrac{3}{6} = -\dfrac{2}{6} = -\dfrac{1}{3}$

33. $\dfrac{7}{16} - \dfrac{1}{3} = \dfrac{7 \cdot 3 - 16 \cdot 1}{16 \cdot 3} = \dfrac{21-16}{48} = \dfrac{5}{48}$

35. $\dfrac{5}{24} + \dfrac{7}{18} = \dfrac{5 \cdot 3}{24 \cdot 3} + \dfrac{7 \cdot 4}{18 \cdot 4} = \dfrac{15}{72} + \dfrac{28}{72} = \dfrac{43}{72}$

37. $\dfrac{3}{4} + \dfrac{5}{6} + \dfrac{7}{8} = \dfrac{3 \cdot 3}{4 \cdot 3} + \dfrac{5 \cdot 2}{6 \cdot 2} + \dfrac{7}{8} = \dfrac{9}{12} + \dfrac{10}{12} + \dfrac{7}{8} = \dfrac{19}{12} + \dfrac{7}{8} = \dfrac{19 \cdot 2}{12 \cdot 2} + \dfrac{7 \cdot 3}{8 \cdot 3} = \dfrac{38}{24} + \dfrac{21}{24} = \dfrac{59}{24}$

39. $\dfrac{1}{8} - \dfrac{2}{3} + \dfrac{1}{2} = \dfrac{1 \cdot 3}{8 \cdot 3} - \dfrac{2 \cdot 8}{3 \cdot 8} + \dfrac{1}{2} = \dfrac{3}{24} - \dfrac{16}{24} + \dfrac{1}{2} = \dfrac{-13}{24} + \dfrac{1}{2} = \dfrac{-13}{24} + \dfrac{1 \cdot 12}{2 \cdot 12} = \dfrac{-13}{24} + \dfrac{12}{24} = \dfrac{-1}{24} = -\dfrac{1}{24}$

41. $\dfrac{\cancel{2}^{1}}{3} \cdot \dfrac{1}{\cancel{2}_{1}} = \dfrac{1}{3}$

43. $\dfrac{1}{6} \div \dfrac{1}{2} = \dfrac{1}{\cancel{6}_{3}} \cdot \dfrac{\cancel{2}^{1}}{1} = \dfrac{1}{3}$

45. $\dfrac{7}{8} \div \left(-\dfrac{5}{24}\right) = \dfrac{7}{\cancel{8}_{1}} \cdot \left(-\dfrac{\cancel{24}^{3}}{5}\right) = -\dfrac{21}{5}$

47. $\dfrac{7}{\cancel{18}_{6}} \cdot \left(-\dfrac{\cancel{3}^{1}}{4}\right) = -\dfrac{7}{24}$

49. $\left(\dfrac{7}{\underset{6}{\cancel{18}}}\cdot\left(-\dfrac{\overset{1}{\cancel{3}}}{4}\right)\right)\div\left(\dfrac{7}{9}\right)=-\dfrac{7}{24}\div\dfrac{7}{9}=-\dfrac{\overset{1}{\cancel{7}}}{\underset{8}{\cancel{24}}}\cdot\dfrac{\overset{3}{\cancel{9}}}{\underset{1}{\cancel{7}}}=-\dfrac{3}{8}$

51. $\left(\dfrac{11}{30}\div\left(-\dfrac{1}{6}\right)\right)\cdot\left(\dfrac{15}{4}\right)=\left(\dfrac{11}{\underset{5}{\cancel{30}}}\cdot\left(-\dfrac{\overset{1}{\cancel{6}}}{1}\right)\right)\cdot\left(\dfrac{15}{4}\right)=-\dfrac{11}{\underset{1}{\cancel{5}}}\cdot\dfrac{\overset{3}{\cancel{15}}}{4}=-\dfrac{33}{4}$

53. $\dfrac{5}{3}\cdot\left(\dfrac{8}{15}+\dfrac{2}{3}\right)=\dfrac{5}{3}\cdot\left(\dfrac{8}{15}+\dfrac{2\cdot 5}{3\cdot 5}\right)=\dfrac{5}{3}\cdot\left(\dfrac{8}{15}+\dfrac{10}{15}\right)=\dfrac{\overset{1}{\cancel{5}}}{\underset{1}{\cancel{3}}}\cdot\left(\dfrac{\overset{\overset{2}{\cancel{6}}}{\cancel{18}}}{\underset{\underset{1}{\cancel{3}}}{\cancel{15}}}\right)=\dfrac{2}{1}=2$

55. $\dfrac{22}{27}\cdot\left(\dfrac{3}{11}+\dfrac{2}{3}\right)=\dfrac{22}{27}\cdot\left(\dfrac{3\cdot 3+11\cdot 2}{11\cdot 3}\right)=\dfrac{22}{27}\cdot\left(\dfrac{9+22}{33}\right)=\dfrac{\overset{2}{\cancel{22}}}{27}\cdot\left(\dfrac{31}{\underset{3}{\cancel{33}}}\right)=\dfrac{62}{81}$

57. $\dfrac{7}{30}\div\left(\dfrac{1}{6}-\dfrac{3}{14}\right)=\dfrac{7}{30}\div\left(\dfrac{1\cdot 7}{6\cdot 7}-\dfrac{3\cdot 3}{14\cdot 3}\right)=\dfrac{7}{30}\div\left(\dfrac{7}{42}-\dfrac{9}{42}\right)=\dfrac{7}{30}\div\left(-\dfrac{2}{42}\right)=\dfrac{7}{\underset{5}{\cancel{30}}}\cdot\left(-\dfrac{\overset{7}{\cancel{42}}}{2}\right)=-\dfrac{49}{10}$

59. $\dfrac{79}{5}=15\dfrac{4}{5}$

61. $\dfrac{29}{3}=9\dfrac{2}{3}$

63. $\dfrac{95}{12}=7\dfrac{11}{12}$

65. $\begin{array}{r}6\\4\overline{)27}\\\underline{24}\\3\end{array}$; $6\dfrac{3}{4}$

67. $\begin{array}{r}8\\15\overline{)121}\\\underline{120}\\1\end{array}$; $8\dfrac{1}{15}$

69. $\begin{array}{r}60\\17\overline{)1036}\\\underline{1020}\\16\end{array}$; $60\dfrac{16}{17}$

71. $\dfrac{2\cdot 4+3}{4}=\dfrac{8+3}{4}=\dfrac{11}{4}$

73. $\dfrac{9\cdot 6+1}{6}=\dfrac{54+1}{6}=\dfrac{55}{6}$

75. $\dfrac{11\cdot 3+2}{3}=\dfrac{33+2}{3}=\dfrac{35}{3}$

77. 0.75

$\begin{array}{r}0.75\\4\overline{)3.00}\\\underline{28}\\20\\\underline{20}\\0\end{array}$

79. 0.1875

$\begin{array}{r}0.1875\\16\overline{)3.0000}\\\underline{16}\\140\\\underline{128}\\120\\\underline{112}\\80\\\underline{80}\\0\end{array}$

81. $5.\overline{3}$

```
    5.3
3)16.0
   15
   ‾
   10
    9
    ‾
   10
```

83. $0.\overline{81}$

```
    0.81
11)9.00
   88
   ‾
   20
   11
   ‾
    90
```

85. $0.\overline{307692}$

```
     0.307692
13)4.000000
   39
   ‾
   10
    0
    ‾
   100
    91
    ‾
    90
    78
    ‾
    120
    117
    ‾
     30
     26
     ‾
     40
```

87. $0.64 = \dfrac{64}{100} = \dfrac{16 \cdot 4}{25 \cdot 4} = \dfrac{16}{25}$

89. $0.836 = \dfrac{836}{1000} = \dfrac{209 \cdot 4}{250 \cdot 4} = \dfrac{209}{250}$

91. $2.345 = \dfrac{2345}{1000} = \dfrac{469 \cdot 5}{200 \cdot 5} = \dfrac{469}{200}$

93. $12.2 = \dfrac{122}{10} = \dfrac{61 \cdot 2}{5 \cdot 2} = \dfrac{61}{5}$

95. $10 \cdot x = 4.444444...$
    $- x = -0.444444...$
    $\overline{}$
    $9 \cdot x = 4 \Rightarrow x = \dfrac{4}{9}$

97. $1000 \cdot x = 189.189189...$
    $- x = - 0.189189...$
    $\overline{}$
    $999 \cdot x = 189 \Rightarrow x = \dfrac{189}{999} = \dfrac{7 \cdot 27}{37 \cdot 27} = \dfrac{7}{37}$

99. $100 \cdot x = 31.81818...$
    $- x = - 0.31818...$
    $\overline{}$
    $99 \cdot x = 31.5 \Rightarrow x = \dfrac{31.5}{99} = \dfrac{31.5 \cdot 10}{99 \cdot 10} = \dfrac{315}{990} = \dfrac{7 \cdot 45}{22 \cdot 45} = \dfrac{7}{22}$

    or

    $1000 \cdot x = 318.1818...$
    $-\ 10 \cdot x = -\ 3.1818...$
    $\overline{}$
    $990 \cdot x = 315 \Rightarrow x = \dfrac{315}{990} = \dfrac{7 \cdot 45}{22 \cdot 45} = \dfrac{7}{22}$

101. $1{,}000{,}000 \cdot x = 384{,}615.384615384615...$
     $\phantom{1{,}000{,}000 \cdot}- x = - 0.384615384615...$
     $\overline{}$
     $999{,}999 \cdot x = 384{,}615 \Rightarrow x = \dfrac{384{,}615}{999{,}999} = \dfrac{5 \cdot 76{,}923}{13 \cdot 76{,}923} = \dfrac{5}{13}$

103. Andre spends $\frac{1}{3}+\frac{1}{4}+\frac{1}{6}=\frac{1\cdot 4+3\cdot 1}{3\cdot 4}+\frac{1}{6}=\frac{4+3}{12}+\frac{1}{6}=\frac{7}{12}+\frac{1}{6}=$

$\frac{7}{12}+\frac{1\cdot 2}{6\cdot 2}=\frac{7}{12}+\frac{2}{12}=\frac{9}{12}=\frac{3\cdot 3}{4\cdot 3}=\frac{3}{4}$ of his paycheck on rent, food, and utilities.

That leaves $1-\frac{3}{4}=\frac{4}{4}-\frac{3}{4}=\frac{1}{4}$ of his paycheck for other expenses.

105. $\frac{1}{3}+\frac{1}{4}=\frac{1\cdot 4+3\cdot 1}{3\cdot 4}=\frac{4+3}{12}=\frac{7}{12}$ of the purse goes to the winner and the person in second place. That leaves $1-\frac{7}{12}=\frac{12}{12}-\frac{7}{12}=\frac{5}{12}$ for the last four drivers. Each would get $\frac{1}{4}\cdot\frac{5}{12}=\frac{5}{48}$ of the purse.

107. The height of the room is $17\cdot 12=204$ inches. We need to find $204\div 8\frac{1}{2}$ and determine if it represents a natural number. You could perform long division or simplify $\frac{204}{8\frac{1}{2}}$. Using the second approach we have $\frac{204}{8\frac{1}{2}}=\frac{204}{\frac{17}{2}}=\frac{\overset{12}{\cancel{204}}}{1}\cdot\frac{2}{\underset{1}{\cancel{17}}}=\frac{24}{1}=24$. Thus we will have 24 rows and the tiles do not need to be cut.

109. In order to increase the recipe that serves 4 to a recipe that serves 10, you must increase the ingredients by or $2\frac{1}{2}$ times. We should use $2\frac{1}{2}\cdot 1\frac{1}{2}=\frac{5}{2}\cdot\frac{3}{2}=\frac{15}{4}=3\frac{3}{4}$ tablespoons of lemon juice.

111. The stock dropped by
$37\frac{1}{8}-35\frac{3}{4}=\left(35+2+\frac{1}{8}\right)-\left(35+\frac{3}{4}\right)=2\frac{1}{8}-\frac{3}{4}=\frac{17}{8}-\frac{3}{4}=\frac{17}{8}-\frac{3\cdot 2}{4\cdot 2}=\frac{17}{8}-\frac{6}{8}=\frac{11}{8}=1\frac{3}{8}$.
or
The stock dropped by $37\frac{1}{8}-35\frac{3}{4}=\frac{297}{8}-\frac{143}{4}=\frac{297}{8}-\frac{143\cdot 2}{4\cdot 2}=\frac{297}{8}-\frac{286}{8}=\frac{11}{8}=1\frac{3}{8}$.

113. $5\frac{1}{4}=5.25$ and $7\frac{1}{8}=7.125$

The unit cost of the $5\frac{1}{4}$-ounce tube would be about 49.5 cents per ounce (rounded).

```
           0.4952
    5.25 )2.600000
              2100
              5000
              4725
              2750
              2625
              1250
              1050
               200
```

The unit cost of the $7\frac{1}{8}$-ounce tube would be about 50.5 cents per ounce (rounded).

```
            0.5052
    7.125 )3.6000000
              35625
               3750
                  0
              37500
              35625
              18750
              14250
               4500
```

The smaller tube is a better buy.

110   Chapter 5: Number Theory and the Real Number System

115. He can get $12 \div 3\dfrac{3}{8} = 12 \div \dfrac{27}{8} = \dfrac{\cancel{12}^{\,4}}{1} \cdot \dfrac{8}{\cancel{27}_{\,9}} = \dfrac{32}{9} = 3\dfrac{5}{9}$ strips from a twelve-foot strip and

$15 \div 3\dfrac{3}{8} = 15 \div \dfrac{27}{8} = \dfrac{\cancel{15}^{\,5}}{1} \cdot \dfrac{8}{\cancel{27}_{\,9}} = \dfrac{40}{9} = 4\dfrac{4}{9}$ strips from a fifteen-foot strip. Since $\dfrac{5}{9} > \dfrac{4}{9}$, there is more

waste from the twelve-foot strips.

117. Each piece would be $10\dfrac{7}{8} \div 4 = \dfrac{87}{8} \div \dfrac{4}{1} = \dfrac{87}{8} \cdot \dfrac{1}{4} = \dfrac{87}{32} = 2\dfrac{23}{32}$ feet

119. There are many approaches to this exercise. One approach is to first determine the distance from the edge of the mirror to the edge of the wall with the mirror centered on the wall. We can do this by finding the difference between the length of the wall and the mirror and then dividing this difference by 2. We can then add to this measurement one-third of the length of the mirror. This will yield the desired measurement.

Distance from edge of mirror to edge of wall:

$\left(72 - 40\tfrac{1}{2}\right) \div 2 = \left(\tfrac{72}{1} - \tfrac{81}{2}\right) \div 2 = \left(\tfrac{72 \cdot 2}{1 \cdot 2} - \tfrac{81}{2}\right) \div 2 = \left(\tfrac{144}{2} - \tfrac{81}{2}\right) \div 2 = \left(\tfrac{63}{2}\right) \div 2 = \tfrac{63}{2} \cdot \tfrac{1}{2} = \tfrac{63}{4} = 15\tfrac{3}{4}$ inches

One third of the length the mirror: $\dfrac{1}{3} \cdot \left(40\dfrac{1}{2}\right) = \dfrac{1}{\cancel{3}_{\,1}} \cdot \dfrac{\cancel{81}^{\,27}}{2} = \dfrac{27}{2} = 13\dfrac{1}{2}$ inches

Distance from the edge of the wall to a nail would therefore be:

$15\dfrac{3}{4} + 13\dfrac{1}{2} = 15 + \dfrac{3}{4} + 13 + \dfrac{1}{2} = (15 + 13) + \left(\dfrac{3}{4} + \dfrac{1}{2}\right) = 28 + \left(\dfrac{3}{4} + \dfrac{1 \cdot 2}{2 \cdot 2}\right) =$

$= 28 + \left(\dfrac{3}{4} + \dfrac{2}{4}\right) = 28 + \dfrac{5}{4} = 28 + 1\dfrac{1}{4} = 29\dfrac{1}{4}$ inches

121. Responses may vary. If we assume that $\dfrac{x+5}{y+5} = \dfrac{x}{y}$, then we can cross multiply to get $(x+5) \cdot y = (y+5) \cdot x$. Thus, $xy + 5y = xy + 5x$. This would imply that $5y = 5x$ or $y = x$.

123. $\dfrac{1}{2} \cdot \dfrac{4}{5} = \dfrac{4}{10} = \dfrac{2}{5}$

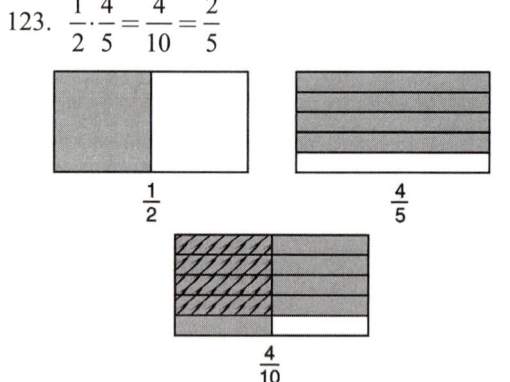

125. $\dfrac{2}{3} \cdot \dfrac{2}{5} = \dfrac{4}{15}$

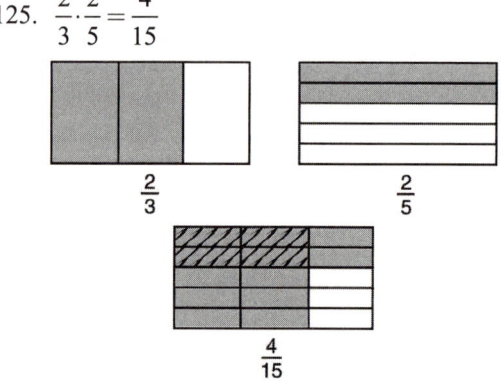

127. $\dfrac{\frac{3}{2}}{\frac{1}{4}} = \dfrac{\frac{3}{2}\cdot\frac{4}{1}}{\frac{1}{4}\cdot\frac{4}{1}} = \dfrac{\frac{3}{2}\cdot\frac{4}{1}}{1} = \dfrac{3}{\underset{1}{\cancel{2}}}\cdot\dfrac{\overset{2}{\cancel{4}}}{1} = \dfrac{6}{1} = 6$

129. $\dfrac{\frac{5}{12}}{\frac{3}{8}} = \dfrac{\frac{5}{12}\cdot\frac{8}{3}}{\frac{3}{8}\cdot\frac{8}{3}} = \dfrac{\frac{5}{12}\cdot\frac{8}{3}}{1} = \dfrac{5}{\underset{3}{\cancel{12}}}\cdot\dfrac{\overset{2}{\cancel{8}}}{3} = \dfrac{10}{9}$

Section 5.4 The Real Number System

1. Irrational numbers have nonrepeating expansions; rational numbers have repeating or terminating expansions.

3. $\sqrt{ab} = \sqrt{a}\sqrt{b}$, $\sqrt{\dfrac{a}{b}} = \dfrac{\sqrt{a}}{\sqrt{b}}$

5. perfect squares

7. The Always Principle

9. rational

11. irrational

13. rational

15. irrational

17. Examples may vary; $\sqrt{9} = 3$

19. $\sqrt{18} = \sqrt{9\cdot 2} = \sqrt{9}\sqrt{2} = 3\sqrt{2}$

21. $\sqrt{12} = \sqrt{4\cdot 3} = \sqrt{4}\sqrt{3} = 2\sqrt{3}$

23. $\sqrt{75} = \sqrt{25\cdot 3} = \sqrt{25}\sqrt{3} = 5\sqrt{3}$

25. $\sqrt{48} = \sqrt{16\cdot 3} = \sqrt{16}\sqrt{3} = 4\sqrt{3}$

27. $\sqrt{189} = \sqrt{9\cdot 21} = \sqrt{9}\sqrt{21} = 3\sqrt{21}$

29. $3\sqrt{5} + 8\sqrt{5} = 11\sqrt{5}$

31. not possible

33. $\sqrt{20} + 6\sqrt{5} = \sqrt{4\cdot 5} + 6\sqrt{5} = \sqrt{4}\sqrt{5} + 6\sqrt{5} = 2\sqrt{5} + 6\sqrt{5} = 8\sqrt{5}$

35. not possible
$5\sqrt{12} + 13\sqrt{18} = 5\sqrt{4\cdot 3} + 13\sqrt{9\cdot 2} = 5\sqrt{4}\sqrt{3} + 13\sqrt{9}\sqrt{2} = 5\cdot 2\sqrt{3} + 13\cdot 3\sqrt{2} = 10\sqrt{3} + 39\sqrt{2}$

37. not possible
$\sqrt{50} + 2\sqrt{75} = \sqrt{25\cdot 2} + 2\sqrt{25\cdot 3} = \sqrt{25}\sqrt{2} + 2\sqrt{25}\sqrt{3} = 5\sqrt{2} + 2\cdot 5\sqrt{3} = 5\sqrt{2} + 10\sqrt{3}$

39. $\sqrt{18}\sqrt{2} = \sqrt{18\cdot 2} = \sqrt{9\cdot 2\cdot 2} = \sqrt{9}\sqrt{2\cdot 2} = 3\sqrt{4} = 3\cdot 2 = 6$

41. $\sqrt{12}\sqrt{15} = \sqrt{12\cdot 15} = \sqrt{4\cdot 3\cdot 3\cdot 5} = \sqrt{4}\sqrt{3\cdot 3\cdot 5} = 2\sqrt{9\cdot 5} = 2\sqrt{9}\sqrt{5} = 2\cdot 3\sqrt{5} = 6\sqrt{5}$

43. $\sqrt{28}\sqrt{21} = \sqrt{28\cdot 21} = \sqrt{4\cdot 7\cdot 7\cdot 3} = \sqrt{4}\sqrt{7\cdot 7\cdot 3} = 2\sqrt{49\cdot 3} = 2\sqrt{49}\sqrt{3} = 2\cdot 7\sqrt{3} = 14\sqrt{3}$

45. $\dfrac{\sqrt{24}}{\sqrt{6}} = \sqrt{\dfrac{24}{6}} = \sqrt{4} = 2$

47. $\dfrac{\sqrt{32}}{\sqrt{18}} = \sqrt{\dfrac{32}{18}} = \sqrt{\dfrac{16}{9}} = \dfrac{\sqrt{16}}{\sqrt{9}} = \dfrac{4}{3}$

49. $\dfrac{\sqrt{96}}{\sqrt{72}} = \dfrac{\sqrt{16\cdot 6}}{\sqrt{36\cdot 2}} = \dfrac{4\sqrt{6}}{6\sqrt{2}} = \dfrac{4}{6}\cdot\dfrac{\sqrt{6}}{\sqrt{2}} = \dfrac{2}{3}\sqrt{\dfrac{6}{2}} = \dfrac{2}{3}\sqrt{3} = \dfrac{2\sqrt{3}}{3}$

51. $\dfrac{3}{\sqrt{5}} = \dfrac{3 \cdot \sqrt{5}}{\sqrt{5} \cdot \sqrt{5}} = \dfrac{3\sqrt{5}}{\sqrt{5 \cdot 5}} = \dfrac{3\sqrt{5}}{\sqrt{25}} = \dfrac{3\sqrt{5}}{5}$

53. $\dfrac{12}{\sqrt{6}} = \dfrac{12 \cdot \sqrt{6}}{\sqrt{6} \cdot \sqrt{6}} = \dfrac{12\sqrt{6}}{\sqrt{6 \cdot 6}} = \dfrac{12\sqrt{6}}{\sqrt{36}} = \dfrac{12\sqrt{6}}{6} = 2\sqrt{6}$

55. $\dfrac{10}{\sqrt{22}} = \dfrac{10\sqrt{22}}{\sqrt{22} \cdot \sqrt{22}} = \dfrac{10\sqrt{22}}{\sqrt{22 \cdot 22}} = \dfrac{10\sqrt{22}}{\sqrt{484}} = \dfrac{10\sqrt{22}}{22} = \dfrac{5\sqrt{22}}{11}$

57. $\dfrac{4\sqrt{6}}{\sqrt{18}} = \dfrac{4\sqrt{6} \cdot \sqrt{2}}{\sqrt{9 \cdot 2} \cdot \sqrt{2}} = \dfrac{4 \cdot \sqrt{12}}{\sqrt{36}} = \dfrac{4\sqrt{4 \cdot 3}}{6} = \dfrac{4\sqrt{4}\sqrt{3}}{6} = \dfrac{4 \cdot 2\sqrt{3}}{6} = \dfrac{8\sqrt{3}}{6} = \dfrac{4\sqrt{3}}{3}$

An easier approach for this exercise however is to simplify as much as possible before rationalizing the denominator.

$\dfrac{4\sqrt{6}}{\sqrt{18}} = 4\sqrt{\dfrac{6}{18}} = 4\sqrt{\dfrac{1}{3}} = 4\dfrac{\sqrt{1}}{\sqrt{3}} = \dfrac{4 \cdot 1\sqrt{3}}{\sqrt{3} \cdot \sqrt{3}} = \dfrac{4\sqrt{3}}{\sqrt{9}} = \dfrac{4\sqrt{3}}{3}$

59. $\dfrac{3\sqrt{10}}{\sqrt{72}} = \dfrac{3 \cdot \sqrt{10}\sqrt{2}}{\sqrt{36 \cdot 2} \cdot \sqrt{2}} = \dfrac{3\sqrt{20}}{\sqrt{144}} = \dfrac{3\sqrt{4 \cdot 5}}{12} = \dfrac{3\sqrt{4} \cdot \sqrt{5}}{12} = \dfrac{3 \cdot 2\sqrt{5}}{12} = \dfrac{6\sqrt{5}}{12} = \dfrac{\sqrt{5}}{2}$

An easier approach for this exercise however is to simplify as much as possible before rationalizing the denominator.

$\dfrac{3\sqrt{10}}{\sqrt{72}} = 3\sqrt{\dfrac{10}{72}} = 3\sqrt{\dfrac{5}{36}} = 3\dfrac{\sqrt{5}}{\sqrt{36}} = 3\dfrac{\sqrt{5}}{6} = \dfrac{\sqrt{5}}{2}$

61. False. $\sqrt{2}$ is irrational; 1.414212362 is a rational, decimal approximation of $\sqrt{2}$.

63. False. Counterexample: $\sqrt{25} = 5$, which is rational.

65. True. $\sqrt{3}\sqrt{5} = \sqrt{15}$, and 15 is not a perfect square.

67. True. If a is a rational number and b is an irrational number and $a + b = c$, then if c is rational then we could say that $b = c - a$. But $c - a$ is the difference of two rational numbers and therefore rational. This would contradict the fact that b is an irrational number.

69. Answers may vary.
    0.435; 0.435121121112111112…

71. Answers may vary.
    $0.\overline{4578} = 0.457845784578\ldots$
    $0.45\overline{78} = 0.457878787878\ldots$
    0.45785; 0.45785121121112111112…

73. Answers may vary.
    $\dfrac{4}{7} = 0.\overline{571428}$
    $\dfrac{5}{7} = 0.\overline{714285}$
    0.6; 0.612112111211112…

75. Answers may vary.
    0.1211̲2111211112…
    $0.\overline{12} = 0.1212̲121212\ldots$
    0.12113; 0.12113121121112111112…

77. a, d, b, c
   a) 0.345345000000
   b) 0.345345345345...
   c) 0.345454545454...
   d) 0.345345343434...

79. a, c, d, b
   a) 0.4444444...
   b) 0.5555555...
   c) 0.4545454...
   d) 0.5545454...

81. distributive

83. commutative (addition)

85. associative (addition)

87. identity element for addition

89. commutative (addition)

91. $x^2 = 12^2 + 5^2 = 144 + 25 = 169 \Rightarrow x = \sqrt{169} = 13$ pounds

93. $x^2 = 120^2 + 80^2 = 14{,}400 + 6{,}400 = 20{,}800 \Rightarrow x = \sqrt{20{,}800} \approx 144.22$ pounds

95. $D = \sqrt{2(6+42)} = \sqrt{2 \cdot 48} = \sqrt{2 \cdot 3 \cdot 16} = \sqrt{6 \cdot 16} = \sqrt{16} \cdot \sqrt{6} = 4\sqrt{6} \approx 9.8$ miles.

97. $v = 2\sqrt{5 \cdot 120} = 2\sqrt{600} = 2\sqrt{100 \cdot 6} = 2\sqrt{100} \cdot \sqrt{6} = 2 \cdot 10 \cdot \sqrt{6} = 20\sqrt{6} \approx 49.0$ mph.

99. $t = \dfrac{\sqrt{200}}{4} = \dfrac{\sqrt{100 \cdot 2}}{4} = \dfrac{\sqrt{100}\sqrt{2}}{4} = \dfrac{10\sqrt{2}}{4} = \dfrac{5\sqrt{2}}{2} \approx 3.5$ sec.

101.
   a) we can cancel common factors from the numerator and denominator
   b) square both sides of a)
   c) multiply through equation by b^2
   d) 2 divides the left side of c), so 2 divides the right side of c)
   e) if a were odd then a^2 would be odd, but it is not
   f) a is even
   g) substitute $2k$ for a in c)
   h) simplify $(2k)^2 = (2k)(2k) = 4k^2$
   i) divide both sides of h) by 2
   j) 2 divides b^2, since 2 divides $2k^2$
   k) if b were odd, then b^2 would be odd
   l) We assumed that $\dfrac{a}{b}$ was reduced, so a and b cannot both be even.

103. $\sqrt{71}$ is greater than 8 and less than 9 because $8^2 = 64$ and $9^2 = 81$. Since $8.1^2 = 65.61$, $8.2^2 = 67.24$, $8.3^2 = 68.89$, $8.4^2 = 70.56$, and $8.5^2 = 72.25$ we know that $8.4 < \sqrt{71} < 8.5$. Now $8.41^2 = 70.7281$, $8.42^2 = 70.8964$, and $8.43^2 = 71.0649$. We have that $\sqrt{71}$ is approximately 8.42.

114 Chapter 5: Number Theory and the Real Number System

105. $\sqrt{106}$ is greater than 10 and less than 11 because $10^2 = 100$ and $11^2 = 121$. Since $10.1^2 = 102.01$, $10.2^2 = 104.04$, and $10.3^2 = 106.09$ we know that $10.2 < \sqrt{106} < 10.3$. Since $\sqrt{106}$ must be closer to 10.3 than 10.2, we'll start with $10.25^2 = 105.0625$, and continue with $10.26^2 = 105.26676$, $10.27^2 = 105.4729$, $10.28^2 = 105.6784$, and $10.29^2 = 105.8841$. We have that $\sqrt{106}$ is approximately 10.29.

107. 2

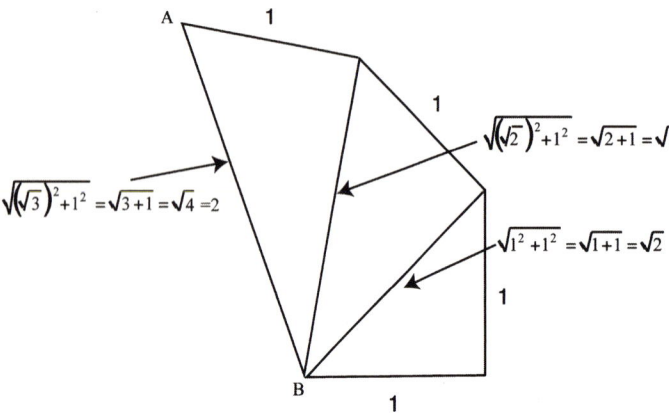

109. Answers may vary.

These numbers are possible lengths of the three sides of a right triangle. If a right triangle has legs of lengths 3 and 4 and hypotenuse of length 5, then $3^2 + 4^2 = 5^2$.

Section 5.5 Exponents And Scientific Notation

1. $x^2 \cdot x^3 = xx \cdot xxx = xxxxx = x^5$; $x^m \cdot x^n = x^{m+n}$

3. $\dfrac{x^8}{x^3} = \dfrac{\cancel{xxx}xxxxx}{\cancel{xxx}} = xxxxx = x^5$; $\dfrac{x^m}{x^n} = x^{m-n}$

5. Examples may vary. Refer to the solution to Exercise 1.

7. Examples may vary. Refer to the solution to Exercise 3.

9. 32

11. $-2^4 = -(2^4) = -(16) = -16$

13. $-3^2 = -(3^2) = -(9) = -9$

15. 9

17. 1

19. 0

21. $3^{2+4} = 3^6 = 729$

23. $(7^2)^3 = 7^{2 \cdot 3} = 7^6 = 117{,}649$

25. $5^4 \cdot 5^{-6} = 5^{4+(-6)} = 5^{-2} = \dfrac{1}{5^2} = \dfrac{1}{25}$

27. $(3^2)^{-3} = 3^{2 \cdot (-3)} = 3^{-6} = \dfrac{1}{3^6} = \dfrac{1}{729}$

29. $(-3)^{-2}(-3)^3 = (-3)^{-2+3} = (-3)^1 = -3$

31. $\dfrac{5^9}{5^7} = 5^{9-7} = 5^2 = 25$

33. $\dfrac{6^{-2}}{6^{-4}} = 6^{-2-(-4)} = 6^2 = 36$

39. $\dfrac{11^{-2}}{11^{-3}} = 11^{-2-(-3)} = 11^1 = 11$

35. $\dfrac{-3^6}{-3^9} = \dfrac{3^6}{3^9} = 3^{6-9} = 3^{-3} = \dfrac{1}{3^3} = \dfrac{1}{27}$

41. $\left(-2^2\right)^{-3} = (-4)^{-3} = \dfrac{1}{(-4)^3} = \dfrac{1}{-64} = -\dfrac{1}{64}$

37. $\left(3^{-4}\right)^0 = 3^{-4 \cdot 0} = 3^0 = 1$

43. $\dfrac{2^9}{4^3} = \dfrac{2^9}{\left(2^2\right)^3} = \dfrac{2^9}{2^6} = 2^{9-6} = 2^3 = 8$

45. In evaluating -2^4, we raise 2 to the fourth power to get 16 and then negate the result to get -16. In evaluating $(-2)^4$, we first negate 2 and then raise that negative number to the fourth power to get 16.

47. 4.356×10^6

61. 0.00178

49. 7.83×10^2

63. 63

51. 2.4×10^{-3}

65. 0.0627

53. 3.824×10^5

67. 0.00000045

55. 4.0×10^{-1}

69. $1{,}000{,}000$

57. 8.0×10^{-3}

71. 2.381×10^7

59. $32{,}500$

73. 8.4×10^2

75. $\left(3 \times 10^6\right)\left(2 \times 10^5\right) = (3 \cdot 2) \times \left(10^6 \cdot 10^5\right) = 6.0 \times 10^{6+5} = 6.0 \times 10^{11}$

77. $\left(1.2 \times 10^{-3}\right)\left(3 \times 10^5\right) = (1.2 \cdot 3) \times \left(10^{-3} \cdot 10^5\right) = 3.6 \times 10^{-3+5} = 3.6 \times 10^2$

79. $\left(7.24 \times 10^{-5}\right)\left(3.6 \times 10^8\right) = (7.24 \cdot 3.6) \times \left(10^{-5} \cdot 10^8\right) = 26.064 \times 10^{-5+8}$
$= 26.064 \times 10^3 = 2.6064 \times 10^4$

81. $\left(8 \times 10^{-2}\right) \div \left(2 \times 10^3\right) = \dfrac{8 \times 10^{-2}}{2 \times 10^3} = \dfrac{8}{2} \times \dfrac{10^{-2}}{10^3} = 4 \times 10^{-2-3} = 4.0 \times 10^{-5}$

83. $\dfrac{\left(5.44 \times 10^8\right)\left(2.1 \times 10^{-3}\right)}{\left(3.4 \times 10^6\right)} = \dfrac{5.44 \cdot 2.1}{3.4} \times \dfrac{10^8 \cdot 10^{-3}}{10^6} = \dfrac{11.424}{3.4} \times \dfrac{10^{8+(-3)}}{10^6} = 3.36 \times \dfrac{10^5}{10^6}$
$= 3.36 \times 10^{5-6} = 3.36 \times 10^{-1}$

85. $\dfrac{\left(9.6368 \times 10^3\right)\left(4.15 \times 10^{-6}\right)}{\left(1.52 \times 10^4\right)} = \dfrac{9.6368 \cdot 4.15}{1.52} \times \dfrac{10^3 \cdot 10^{-6}}{10^4} = \dfrac{39.99272}{1.52} \times \dfrac{10^{3+(-6)}}{10^4} = 26.311 \times \dfrac{10^{-3}}{10^4}$
$= 26.311 \times 10^{-3-4} = 26.311 \times 10^{-7} = 2.6311 \times 10^{-6}$

87. $67{,}300{,}000 \times 1{,}200 = \left(6.73 \times 10^7\right)\left(1.2 \times 10^3\right) = (6.73 \cdot 1.2) \times \left(10^7 \cdot 10^3\right)$
$= 8.076 \times 10^{7+3} = 8.076 \times 10^{10}$

89. $6{,}800{,}000 \times 2{,}300{,}000 = (6.8 \times 10^6)(2.3 \times 10^6) = (6.8 \cdot 2.3) \times (10^6 \cdot 10^6)$
$= 15.64 \times 10^{6+6} = 15.64 \times 10^{12} = 1.564 \times 10^{13}$

91. $0.00016 \times 0.0025 = (1.6 \times 10^{-4})(2.5 \times 10^{-3}) = (1.6 \cdot 2.5) \times (10^{-4} \cdot 10^{-3})$
$= 4.0 \times 10^{-4+(-3)} = 4.0 \times 10^{-7}$

93. $0.00165 \times 0.0004 = (1.65 \times 10^{-3})(4.0 \times 10^{-4}) = (1.65 \cdot 4.0) \times (10^{-3} \cdot 10^{-4})$
$= 6.6 \times 10^{-3+(-4)} = 6.6 \times 10^{-7}$

95. $\dfrac{256 \times 10^9}{151 \times 10^6} = \dfrac{2.56 \times 10^{11}}{1.51 \times 10^8} = \dfrac{2.56}{1.51} \times \dfrac{10^{11}}{10^8} \approx 1.70 \times 10^{11-8} = 1.7 \times 10^3 = \$1{,}700$ per person.

97. $\dfrac{6.379 \times 10^9}{296 \times 10^6} = \dfrac{6.379 \times 10^9}{2.96 \times 10^8} = \dfrac{6.379}{2.96} \times \dfrac{10^9}{10^8} \approx 2.16 \times 10^{9-8} = 2.16 \times 10^1 = 21.6$ times larger.

99. $\dfrac{375.2 \times 10^9}{296 \times 10^6} = \dfrac{3.752 \times 10^{11}}{2.96 \times 10^8} = \dfrac{3.752}{2.96} \times \dfrac{10^{11}}{10^8} \approx 1.27 \times 10^{11-8} = 1.27 \times 10^3 = \$1{,}270$ per person

101. Assuming there are 365 days in a year, we have $60 \times 60 \times 24 \times 365$ seconds in one year.
$\dfrac{1.0 \times 10^9}{60 \times 60 \times 24 \times 365} = \dfrac{1.0 \times 10^9}{31{,}536{,}000} = \dfrac{1.0 \times 10^9}{3.1536 \times 10^7} = \dfrac{1.0}{3.1536} \times \dfrac{10^9}{10^7} \approx 0.317 \times 10^{9-7}$
$= 0.317 \times 10^2 = 3.17 \times 10^1 = 31.7$ years

103. (a) $454 \cdot 130 = 59020 = 5.902 \times 10^4$

(b) $\dfrac{5.902 \times 10^4}{1 \times 10^{-3}} = \dfrac{5.902}{1} \times \dfrac{10^4}{10^{-3}} = 5.902 \times 10^{4-(-3)} = 5.902 \times 10^7 = 59.02 \times 10^6$

(c) The 130 pound person weighs as much as 59 million mosquitoes.

105. $\dfrac{296 \times 10^6}{3.537 \times 10^6} = \dfrac{2.96 \times 10^8}{3.537 \times 10^6} = \dfrac{2.96}{3.537} \times \dfrac{10^8}{10^6} \approx 0.837 \times 10^{8-6} = 0.837 \times 10^2$
$= 8.37 \times 10^1 = 83.7$ people per square mile.

107. $\dfrac{93{,}000{,}000}{25{,}000} = \dfrac{9.3 \times 10^7}{2.5 \times 10^4} = \dfrac{9.3}{2.5} \times \dfrac{10^7}{10^4} = 3.72 \times 10^{7-4} = 3.72 \times 10^3 = 3{,}720$ hours

109. $\dfrac{3.6 \times 10^9}{93 \times 10^6} = \dfrac{3.6 \times 10^9}{9.3 \times 10^7} = \dfrac{3.6}{9.3} \times \dfrac{10^9}{10^7} \approx 0.387 \times 10^{9-7} = 0.387 \times 10^2 = 3.87 \times 10^1 = 38.7$ times larger

111. $\dfrac{56.42 \times 10^9}{2.6 \times 10^6} = \dfrac{5.642 \times 10^{10}}{2.6 \times 10^6} = \dfrac{5.642}{2.6} \times \dfrac{10^{10}}{10^6} = 2.17 \times 10^{10-6} = 2.17 \times 10^4$ seconds

$\dfrac{2.17 \times 10^4}{60 \cdot 60} = \dfrac{2.17 \times 10^4}{3{,}600} = \dfrac{2.17 \times 10^4}{3.6 \times 10^3} = \dfrac{2.17}{3.6} \times \dfrac{10^4}{10^3} \approx 0.603 \times 10^{4-3} = 0.603 \times 10^1 \approx 6$ hours

113. Scientific notation allows us to compute easily with very large or very small numbers.

115. $\left(\dfrac{a}{b}\right)^n = \dfrac{a^n}{b^n}$

117. No solution provided

Chapter Review Exercises

1. Refer to the table for Exercise 23 in the solutions for Section 5.1.
   71, 73, 79, 83, 89

2. Since $169 < 180 < 196$, $13 < \sqrt{180} < 14$. Thus $a = 13$.

3. 191 is prime. You need to check for divisibility by 2, 3, 5, 7, 11, and 13.
   441 is composite. $441 = 3^2 \times 7^2$

4. 1,080,036 is divisible by **3** because the sum of the digits, 18, is divisible by 3.
   1,080,036 is divisible by **4** because the number formed by the last two digits, 36, is divisible by 4.
   1,080,036 is not divisible by 5 because the last digit, 6, is not a 0 or a 5.
   1,080,036 is divisible by **6** because the number is divisible by both 2 and 3. (You know the number is divisible by 2 because the last digit, 6, is divisible by 2.)
   1,080,036 is not divisible by 8 because the number formed by the last three digits, 36, is not divisible by 8.
   1,080,036 is divisible by **9** because the sum of the digits, 18, is divisible by 9.
   1,080,036 is not divisible by 10 because the last digit, 6, is not 0.

5. $396 = 2^2 \times 3^2 \times 11$ and $330 = 3^4 \times 7$; $GCD(396, 330) = 2 \times 3 \times 11 = 66$
   $LCM(396, 330) = 2^2 \times 3^2 \times 5 \times 11 = 1,980$

6. To calculate the GCD, multiply the *smallest* powers of any primes that are common to both numbers. To calculate the LCM, multiply the *largest* powers that occur in either number.

7. a) 9 c) 72
   b) 11 d) -16

8. For the past four days you received $3, so four days ago, you had $12 less. Thus $(-4)(+3) = -12$.

9. If $\dfrac{-24}{-8} = c$, then $-24 = (-8) \cdot c$. Thus $c = +3$.

10. $17 - (-3) = 20$ degrees

11. If $\dfrac{5}{0} = c$, then $5 = 0 \cdot c$, which is not possible.

12. not equal; $\dfrac{28}{65} \neq \dfrac{14}{35}$ since $28 \cdot 35 \neq 65 \cdot 14$ $(980 \neq 910)$

13. a) $\dfrac{4}{9}\cdot\left(\dfrac{3}{4}-\dfrac{1}{3}\right)=\dfrac{4}{9}\cdot\left(\dfrac{3\cdot 3-4\cdot 1}{4\cdot 3}\right)=\dfrac{4}{9}\cdot\left(\dfrac{9-4}{12}\right)=\dfrac{\overset{1}{\cancel{4}}}{9}\cdot\dfrac{5}{\underset{3}{\cancel{12}}}=\dfrac{5}{27}$

  b) $\dfrac{3}{7}\div\left(\dfrac{2}{3}+\dfrac{3}{14}\right)=\dfrac{3}{7}\div\left(\dfrac{2\cdot 14+3\cdot 3}{3\cdot 14}\right)=\dfrac{3}{7}\div\left(\dfrac{28+9}{42}\right)=\dfrac{3}{7}\div\left(\dfrac{37}{42}\right)=\dfrac{3}{\underset{1}{\cancel{7}}}\cdot\dfrac{\overset{6}{\cancel{42}}}{37}=\dfrac{18}{37}$

14. $8\overline{\smash{)}53}\,;\ 6\dfrac{5}{8}$
    $\underline{48}$
    5

15. $\left(3\dfrac{1}{2}\right)\left(4\dfrac{1}{7}\right)=\dfrac{7}{2}\cdot\dfrac{29}{7}=\dfrac{\overset{1}{\cancel{7}}}{2}\cdot\dfrac{29}{\underset{1}{\cancel{7}}}=\dfrac{29}{2}=14\dfrac{1}{2}$

16. a) $0.375=\dfrac{375}{1000}=\dfrac{3\cdot 125}{8\cdot 125}=\dfrac{3}{8}$

    b) $100\cdot x = 63.6363...$
       $\underline{x = -\ 0.6363...}$
       $99\cdot x = 63 \Rightarrow x=\dfrac{63}{99}=\dfrac{7\cdot 9}{11\cdot 9}=\dfrac{7}{11}$

17. In order to increase the recipe which serves 16 to a recipe that serves 24, you must increase the ingredients by $1\tfrac{1}{2}$ times. We should use $1\tfrac{1}{2}\cdot 2\tfrac{1}{2}=\tfrac{3}{2}\cdot\tfrac{5}{2}=\tfrac{15}{4}=3\tfrac{3}{4}$ cups of hot peppers and $1\tfrac{1}{2}\cdot 3\tfrac{1}{4}=\tfrac{3}{2}\cdot\tfrac{13}{4}=\tfrac{39}{8}=4\tfrac{7}{8}$ cups of tomatoes.

18. No solution provided

19. Rational numbers have repeating expansions, and irrational numbers have nonrepeating expansions.

20. a) $\sqrt{108}=\sqrt{36\cdot 3}=\sqrt{36}\sqrt{3}=6\sqrt{3}$

    b) $\dfrac{\sqrt{15}}{\sqrt{5}}=\sqrt{\dfrac{15}{5}}=\sqrt{3}$

21. Examples may vary.
    $8\div(4\div 2)\ne(8\div 4)\div 2$
    $8\div 2\ne 2\div 2$
    $4\ne 1$

22. a) $3^6\cdot 3^{-2}=3^{6+(-2)}=3^4=81$

    b) $\left(2^4\right)^{-2}=2^{4(-2)}=2^{-8}=\dfrac{1}{2^8}=\dfrac{1}{256}$

    c) $\dfrac{6^8}{6^5}=6^{8-5}=6^3=216$

    d) $\dfrac{8^{-6}}{8^{-4}}=8^{-6-(-4)}=8^{-2}=\dfrac{1}{8^2}=\dfrac{1}{64}$

23. In evaluating -2^4, we raise 2 to the fourth power to get 16 and then negate the result to get -16. In evaluating $(-2)^4$, we first negate 2 and then raise that negative number to the fourth power to get positive 16.

24. $0.000456=4.56\times 10^{-4}$ and $1{,}230{,}000=1.23\times 10^6$

25. $1.325\times 10^6=1{,}325{,}000$ and $8.63\times 10^{-5}=0.0000863$

26. $\dfrac{\left(3.6\times 10^3\right)\left(2.8\times 10^{-5}\right)}{\left(4.2\times 10^4\right)}=\dfrac{3.6\cdot 2.8}{4.2}\times\dfrac{10^3\cdot 10^{-5}}{10^4}=\dfrac{10.08}{4.2}\times\dfrac{10^{3+(-5)}}{10^4}=2.4\times\dfrac{10^{-2}}{10^4}$
    $=2.4\times 10^{-2-4}=2.4\times 10^{-6}$

27. $\dfrac{6.157\times 10^9}{102\times 10^6}=\dfrac{6.157\times 10^9}{1.02\times 10^8}=\dfrac{6.157}{1.02}\times\dfrac{10^9}{10^8}\approx 6.036\times 10^{9-8}=6.036\times 10^1=60.36$ times

Chapter Test

1. Refer to the table for Exercise 23 in the solutions for Section 5.1.
   101, 103, 107, 109, 113

2. Since $196 < 200 < 225$, $14 < \sqrt{200} < 15$. Thus $a = 14$.

3. 241 is prime. You need to check for divisibility by 2, 3, 5, 7, 11, and 13.
   539 is composite. $539 = 7^2 \times 11$

4. 2,542,128 is divisible by **3** because the sum of the digits, 24, is divisible by 3.
   2,542,128 is divisible by **4** because the number formed by the last two digits, 28, is divisible by 4.
   2,542,128 is not divisible by 5 because the last digit, 8, is not a 0 or a 5.
   2,542,128 is divisible by **6** because the number is divisible by both 2 and 3. (You know the number is divisible by 2 because the last digit, 8, is divisible by 2.)
   2,542,128 is divisible by 8 because the number formed by the last three digits, 128, is divisible by 8.
   2,542,128 is not divisible by **9** because the sum of the digits, 24, is not divisible by 9.
   2,542,128 is not divisible by 10 because the last digit, 8, is not 0.

5. $1716 = 2^2 \times 3 \times 11 \times 13$ and $936 = 2^3 \times 3^2 \times 13$; $GCD(1716, 936) = 2 \times 3 \times 11 \times 13 = 858$
   $LCM(1716, 936) = 2^3 \times 3^2 \times 11 \times 13 = 10,296$

6. To calculate the GCD, multiply the *smallest* powers of any primes that are common to both numbers.
   To calculate the LCM, multiply the *largest* powers that occur in either number.
   $GCD(a,b) = 2^3 \times 3^6 \times 5^2 \times 7^2$
   $LCM(a,b) = 2^8 \times 3^9 \times 5^3 \times 7^9$

7. a) –6   c) –54
   b) +3   d) +8

8. If you spend $5 for the next seven days, you will have $35 less. Thus $(+7)(-5) = -35$.

9. If $\dfrac{-21}{3} = c$, then $-21 = (3) \cdot c$. Thus $c = -7$.

10. $50 - (-19) = 69$ degrees.

11. If $\dfrac{8}{0} = c$, then $8 = 0 \cdot c$, which is not possible.

12. not equal; $\dfrac{198}{213} \neq \dfrac{68}{72}$ since $198 \cdot 72 \neq 68 \cdot 213$ $(14,256 \neq 14,484)$

13. a) $\dfrac{3}{5} \cdot \left(\dfrac{7}{9} - \dfrac{3}{4}\right) = \dfrac{3}{5} \cdot \left(\dfrac{7 \cdot 4 - 3 \cdot 9}{9 \cdot 4}\right) = \dfrac{3}{5} \cdot \left(\dfrac{28 - 27}{36}\right) = \dfrac{\overset{1}{3}}{5} \cdot \left(\dfrac{1}{\underset{12}{36}}\right) = \dfrac{1}{60}$

b) $\dfrac{4}{5} \div \left(\dfrac{2}{5} - \dfrac{1}{4}\right) = \dfrac{4}{5} \div \left(\dfrac{2 \cdot 4 - 1 \cdot 5}{5 \cdot 4}\right) = \dfrac{4}{5} \div \left(\dfrac{8 - 5}{20}\right) = \dfrac{4}{5} \div \left(\dfrac{3}{20}\right) = \dfrac{4}{\underset{1}{5}} \cdot \dfrac{\overset{4}{20}}{3} = \dfrac{16}{3}$

14. $\begin{array}{r}15\\3\overline{)47}\\30\\\overline{17}\\15\\\overline{2}\end{array}$; $15\dfrac{2}{3}$

16. a) $0.573 = \dfrac{573}{1000}$

    b) $100 \cdot x = 57.5757...$
       $\underline{-x = -0.5757...}$
       $99 \cdot x = 57 \Rightarrow x = \dfrac{57}{99} = \dfrac{19}{33}$

15. $4\dfrac{1}{2} \div 3\dfrac{3}{8} = \dfrac{9}{2} \div \dfrac{27}{8} = \dfrac{\overset{1}{9}}{\underset{1}{2}} \cdot \dfrac{\overset{4}{8}}{\underset{3}{27}} = \dfrac{4}{3}$

17. The photo is being enlarged by $5\tfrac{1}{4}$. The dimensions should be $23\tfrac{5}{8}$ inches, by $33\tfrac{1}{4}$ inches.
    $5\tfrac{1}{4} \cdot 4\tfrac{1}{2} = \tfrac{21}{4} \cdot \tfrac{9}{2} = \tfrac{189}{8} = 23\tfrac{5}{8}$ $\qquad$ $5\tfrac{1}{4} \cdot 6\tfrac{1}{3} = \tfrac{21}{4} \cdot \tfrac{19}{3} = \tfrac{399}{12} = 33\tfrac{1}{4}$

18. No solution provided

19. Rational numbers have repeating or terminating expansions and irrational numbers have nonrepeating expansions.

20. a) $\sqrt{180} = \sqrt{36 \cdot 5} = \sqrt{36}\sqrt{5} = 6\sqrt{5}$

    b) $\dfrac{3}{\sqrt{15}} = \dfrac{3 \cdot \sqrt{15}}{\sqrt{15} \cdot \sqrt{15}} = \dfrac{\overset{1}{3} \cdot \sqrt{15}}{\underset{5}{15}} = \dfrac{\sqrt{15}}{5}$

22. a) $2^7 \cdot 2^{-4} = 2^{7+(-4)} = 2^3 = 8$

    b) $(3^2)^{-2} = 3^{2(-2)} = 3^{-4} = \dfrac{1}{3^4} = \dfrac{1}{81}$

    c) $\dfrac{5^7}{5^4} = 5^{7-4} = 5^3 = 125$

    d) $\dfrac{3^{-2}}{3^{-5}} = 3^{-2-(-5)} = 3^3 = 27$

21. Answers may vary. Possible example is:

    $\dfrac{5}{2} \div \left(\dfrac{3}{2} \div \dfrac{1}{2}\right) \ne \left(\dfrac{5}{2} \div \dfrac{3}{2}\right) \div \dfrac{1}{2}$

    $\dfrac{5}{2} \div \left(\dfrac{3}{2} \cdot \dfrac{2}{1}\right) \ne \left(\dfrac{5}{2} \cdot \dfrac{2}{3}\right) \div \dfrac{1}{2}$

    $\dfrac{5}{2} \div \dfrac{3}{1} \ne \dfrac{5}{3} \div \dfrac{1}{2}$

    $\dfrac{5}{2} \cdot \dfrac{1}{3} \ne \dfrac{5}{3} \cdot \dfrac{2}{1}$

    $\dfrac{5}{6} \ne \dfrac{10}{3}$

23. In evaluating -3^2, 3 is raised to the second power to get 9 before being negated to get -9. In evaluating $(-3)^2$, 3 is negated first and then that negative number is raised to the second power to get 9.

24. $15{,}460{,}000 = 1.546 \times 10^7$ and $0.00000000623 = 6.23 \times 10^{-9}$

25. $\dfrac{1.3 \times 10^9}{2.9 \times 10^8} = \dfrac{1.3 \times 10^9}{2.9 \times 10^8} = \dfrac{1.3}{2.9} \times \dfrac{10^9}{10^8} \approx 0.44827 \times 10^{9-8} = 4.48 \times 10^1 = 44.8$ times larger

Of Further Interest: Sequences

1. We considered the terms of the sequence to all be of the form 3 plus some multiple of 4 to see a pattern.

2. We considered the terms of the sequence to all be of the form 2 times some power of 3 to see a pattern.

3. – 6. Answers will vary.

7. This is an arithmetic sequence. There is a common difference of 3 between consecutive terms. The next two terms are 17 and 20.

8. This is an arithmetic sequence. There is a common difference of -4 between consecutive terms. The next two terms are -5 and -9.

9. This is a geometric sequence. There is a common ratio of 3 between consecutive terms. The next two terms are 648 and 1,944.

10. This is an arithmetic sequence. There is a common difference of 8 between consecutive terms. The next two terms are 36 and 44.

11. This is a geometric sequence. There is a common ratio of $\frac{1}{2}$ between consecutive terms. The next two terms are $\frac{1}{16}$ and $\frac{1}{32}$.

12. This is a geometric sequence. There is a common ratio of 0.1 between consecutive terms. The next two terms are 0.00001 and 0.000001.

13. This is an arithmetic sequence. There is a common difference of -5 between consecutive terms. The next two terms are -10 and -15.

14. This is an arithmetic sequence. There is a common difference of 13 between consecutive terms. The next two terms are 56 and 69.

15. This is a geometric sequence. There is a common ratio of 2 between consecutive terms. The next two terms are 32 and 64.

16. This is a geometric sequence. There is a common ratio of $\frac{1}{3}$ between consecutive terms. The next two terms are $\frac{1}{81}$ and $\frac{1}{243}$.

17. This is an arithmetic sequence. There is a common difference of 0.5 between consecutive terms. The next two terms are 3.5 and 4.0.

18. This is a geometric sequence. There is a common ratio of -1 between consecutive terms. The next two terms are 1 and -1.

19. a) Since $a_1 = 5$, $d = 3$, and $n = 11$, we have $a_{11} = 5 + (11-1) \cdot 3 = 5 + 10 \cdot 3 = 5 + 30 = 35$.

    b) Since $a_1 = 5$, $a_{11} = 35$, and $n = 11$, the sum of the first 11 terms is $\dfrac{11(5+35)}{2} = \dfrac{11 \cdot 40}{2} = 11 \cdot 20 = 220$.

122   Chapter 5: Number Theory and the Real Number System

20. a) Since $a_1 = 11$, $d = 6$, and $n = 9$, we have $a_9 = 11 + (9-1) \cdot 6 = 11 + 8 \cdot 6 = 11 + 48 = 59$.

   b) Since $a_1 = 11$, $a_9 = 59$, and $n = 9$, the sum of the first 9 terms is $\dfrac{9(11+59)}{2} = \dfrac{9 \cdot 70}{2} = 9 \cdot 35 = 315$.

21. a) Since $a_1 = 2$, $d = 6$, and $n = 15$, we have $a_{15} = 2 + (15-1) \cdot 6 = 2 + 14 \cdot 6 = 2 + 84 = 86$.

   b) Since $a_1 = 2$, $a_{15} = 86$, and $n = 15$, the sum of the first 15 terms is $\dfrac{15(2+86)}{2} = \dfrac{15 \cdot 88}{2} = 15 \cdot 44 = 660$.

22. a) Since $a_1 = -6$, $d = 4$, and $n = 22$, we have $a_{22} = -6 + (22-1) \cdot 4 = -6 + 21 \cdot 4 = -6 + 84 = 78$.

   b) Since $a_1 = -6$, $a_{22} = 78$, and $n = 22$, the sum of the first 22 terms is $\dfrac{22(-6+78)}{2} = \dfrac{22 \cdot 72}{2} = 22 \cdot 36 = 792$.

23. a) Since $a_1 = 1$, $d = 0.5$, and $n = 20$, we have $a_{20} = 1 + (20-1) \cdot 0.5 = 1 + 19 \cdot 0.5 = 1 + 9.5 = 10.5$.

   b) Since $a_1 = 1$, $a_{20} = 10.5$, and $n = 20$, the sum of the first 20 terms is $\dfrac{20(1+10.5)}{2} = \dfrac{20 \cdot 11.5}{2} = 10 \cdot 11.5 = 115$.

24. a) Since $a_1 = 3$, $d = 0.25$, and $n = 11$, we have $a_{11} = 3 + (11-1) \cdot 0.25 = 3 + 10 \cdot 0.25 = 3 + 2.5 = 5.5$.

   b) Since $a_1 = 3$, $a_{11} = 5.5$, and $n = 11$, the sum of the first 11 terms is $\dfrac{11(3+5.5)}{2} = \dfrac{11 \cdot 8.5}{2} = \dfrac{93.5}{2} = 46.75$.

25. We need to find the sum of the first 21 terms of this sequence and subtract the sum of the first 13 terms. Since $a_1 = 5$, $d = 3$, and $n = 21$, we have $a_{21} = 5 + (21-1) \cdot 3 = 5 + 20 \cdot 3 = 5 + 60 = 65$. Thus, the sum of the first 21 terms is $\dfrac{21(5+65)}{2} = \dfrac{21 \cdot 70}{2} = 21 \cdot 35 = 735$. Since $a_1 = 5$, $d = 3$, and $n = 13$, we have $a_{13} = 5 + (13-1) \cdot 3 = 5 + 12 \cdot 3 = 5 + 36 = 41$. Thus the sum of the first 13 terms is $\dfrac{13(5+41)}{2} = \dfrac{13 \cdot 46}{2} = 13 \cdot 23 = 299$. The desired sum is therefore $735 - 299 = 436$.

26. We need to find the sum of the first 37 terms of this sequence and subtract the sum of the first 20 terms. Since $a_1 = 11$, $d = 6$, and $n = 37$, we have $a_{37} = 11 + (37-1) \cdot 6 = 11 + 36 \cdot 6 = 11 + 216 = 227$. Thus, the sum of the first 37 terms is $\dfrac{37(11+227)}{2} = \dfrac{37 \cdot 238}{2} = 37 \cdot 119 = 4{,}403$. Since $a_1 = 11$, $d = 6$, and $n = 20$, we have $a_{20} = 11 + (20-1) \cdot 6 = 11 + 19 \cdot 6 = 11 + 114 = 125$. Thus the sum of the first 20 terms is $\dfrac{20(11+125)}{2} = \dfrac{20 \cdot 136}{2} = 10 \cdot 136 = 1{,}360$. The desired sum is therefore $4{,}403 - 1{,}360 = 3{,}043$.

27. a) Since $a_1 = 1$, $r = 3$, and $n = 11$, we have $a_{11} = 1 \cdot 3^{11-1} = 1 \cdot 3^{10} = 1 \cdot 59{,}049 = 59{,}049$.

b) Since $a_1 = 1$, $r = 3$, and $n = 11$, the sum of the first 11 terms is $\dfrac{1 \cdot (3^{11} - 1)}{3 - 1} = \dfrac{1 \cdot (177{,}147 - 1)}{2} = \dfrac{1 \cdot 177{,}146}{2} = \dfrac{177{,}146}{2} = 88{,}573$.

28. a) Since $a_1 = 3$, $r = 2$, and $n = 9$, we have $a_9 = 3 \cdot 2^{9-1} = 3 \cdot 2^8 = 3 \cdot 256 = 768$.

b) Since $a_1 = 3$, $r = 2$, and $n = 9$, we the sum of the first 9 terms is $\dfrac{3 \cdot (2^9 - 1)}{2 - 1} = \dfrac{3 \cdot (512 - 1)}{1} = \dfrac{3 \cdot 511}{1} = \dfrac{1{,}533}{1} = 1{,}533$.

29. a) Since $a_1 = 1$, $r = \frac{1}{2}$, and $n = 7$, we have $a_7 = 1 \cdot \left(\frac{1}{2}\right)^{7-1} = 1 \cdot \left(\frac{1}{2}\right)^6 = 1 \cdot \frac{1}{64} = \frac{1}{64}$.

b) Since $a_1 = 1$, $r = \frac{1}{2}$, and $n = 7$, the sum of the first 7 terms is $\dfrac{1 \cdot \left(\left(\frac{1}{2}\right)^7 - 1\right)}{\frac{1}{2} - 1} = \dfrac{1 \cdot \left(\frac{1}{128} - 1\right)}{-\frac{1}{2}} = \dfrac{1 \cdot \left(-\frac{127}{128}\right)}{-\frac{1}{2}} = \dfrac{-\frac{127}{128}}{-\frac{1}{2}} = \left(-\frac{127}{128}\right) \cdot \left(-\frac{2}{1}\right) = \frac{127}{64}$.

30. a) Since $a_1 = 2$, $r = -2$, and $n = 10$, we have $a_{10} = 2 \cdot (-2)^{10-1} = 2 \cdot (-2)^9 = 2 \cdot (-512) = -1{,}024$.

b) Since $a_1 = 2$, $r = -2$, and $n = 10$, the sum of the first 10 terms is $\dfrac{2 \cdot \left((-2)^{10} - 1\right)}{-2 - 1} = \dfrac{2 \cdot (1{,}024 - 1)}{-3} = \dfrac{2 \cdot 1{,}023}{-3} = \dfrac{2{,}046}{-3} = -682$.

31. a) Since $a_1 = 2$, $r = 0.1$, and $n = 6$, we have $a_6 = 2 \cdot 0.1^{6-1} = 2 \cdot 0.1^5 = 2 \cdot 0.00001 = 0.00002$.

b) Since $a_1 = 2$, $r = 0.1$, and $n = 6$, the sum of the first 6 terms is $\dfrac{2 \cdot (0.1^6 - 1)}{0.1 - 1} = \dfrac{2 \cdot (0.000001 - 1)}{-0.9} = \dfrac{2 \cdot (-0.999999)}{-0.9} = 2(1.11111) = 2.22222$.

32. a) Since $a_1 = 5$, $r = 10$, and $n = 9$, we have $a_9 = 5 \cdot 10^{9-1} = 5 \cdot 10^8 = 5 \cdot 100{,}000{,}000 = 500{,}000{,}000$.

b) Since $a_1 = 5$, $r = 10$, and $n = 9$, the sum of the first 9 terms is $\dfrac{5 \cdot (10^9 - 1)}{10 - 1} = \dfrac{5 \cdot (1{,}000{,}000{,}000 - 1)}{9} = \dfrac{5 \cdot 999{,}999{,}999}{9} = 5 \cdot 111{,}111{,}111 = 555{,}555{,}555$.

33. 144

$F_{13} = F_{11} + F_{12}$
$233 = F_{12} + 89$
$144 = F_{12}$

34. 10,946

$F_{23} = F_{21} + F_{22}$
$28{,}657 = F_{21} + 17{,}711$
$10{,}946 = F_{21}$

35. 377

$F_{15} = F_{13} + F_{14}$
$610 = 233 + F_{14}$
$377 = F_{14}$

36. 75,025

$F_{25} = F_{23} + F_{24}$
$F_{25} = 28,657 + 46,368$
$F_{25} = 75,025$

37. $a_1 = 2,875$, $d = 35$, and $n = 12$, thus $a_{12} = 2,875 + (12-1) \cdot 35 = 2,875 + 11 \cdot 35 = 2,875 + 385 = 3,260$.

Therefore, $\dfrac{12(2,875 + 3,260)}{2} = \dfrac{12 \cdot 6,135}{2} = 6 \cdot 6,135 = \$36,810$ is the amount earned in one year.

38. You are finding the sum $11 + 10 + 9 + 8 + 7 + 6 + 5 + 4 + 3 + 2 + 1$. $a_1 = 11$, $a_{11} = 1$, and $n = 11$,

thus there are $\dfrac{11(11+1)}{2} = \dfrac{11 \cdot 12}{2} = 11 \cdot 6 = 66$ cans in the stack.

39. In this exercise, you need to find the seventh term of this geometric sequence. The reason for the seventh term is because you are initially making a deposit. The amount in the account in six years is the amount of money you have starting the seventh year. The common ratio would be 1.035 because you retain the amount plus receive a percentage of the amount from one year to the next. Thus the amount you will receive will be $1,200(1.035)^{7-1} = 1,200(1.035)^6 \approx \$1,475.11$.

40. $a_1 = 8$ and $r = 8$.

n	$a_n = 8 \cdot 8^{n-1} = 8^n$
1	$a_1 = 8^1 = 8$
2	$a_2 = 8^2 = 64$
3	$a_3 = 8^3 = 512$
4	$a_4 = 8^4 = 4,096$
5	$a_5 = 8^5 = 32,768$
6	$a_6 = 8^6 = 262,144$
7	$a_7 = 8^7 = 2,097,152$
8	$a_8 = 8^8 = 16,777,216$
9	$a_9 = 8^9 = 134,217,728$
10	$a_{10} = 8^{10} = 1,073,741,824$
11	$a_{11} = 8^{11} = 8,589,934,592$

At the 11th stage

41. In this geometric sequence, the height after the fifth bounce would be the sixth term where $a_1 = 8$ and

$r = \frac{7}{8}$. $a_6 = 8 \cdot \left(\frac{7}{8}\right)^{6-1} = 8 \cdot \left(\frac{7}{8}\right)^5 = 8 \cdot \dfrac{16,807}{32,768} = \dfrac{16,807}{4,096} \approx 4.10$ feet

42. $a_1 = 3$ and $r = 2$

n	Amount bet $a_n = 3 \cdot 2^{n-1}$	Total amount bet $\dfrac{3 \cdot (2^n - 1)}{2 - 1} = \dfrac{3 \cdot (2^n - 1)}{1} = 3(2^n - 1)$
1	$a_1 = 3 \cdot 2^{1-1} = 3 \cdot 2^0 = 3 \cdot 1 = 3$	$3(2^1 - 1) = 3(2 - 1) = 3 \cdot 1 = 3$
2	$a_2 = 3 \cdot 2^{2-1} = 3 \cdot 2^1 = 3 \cdot 2 = 6$	$3(2^2 - 1) = 3(4 - 1) = 3 \cdot 3 = 9$
3	$a_3 = 3 \cdot 2^{3-1} = 3 \cdot 2^2 = 3 \cdot 4 = 12$	$3(2^3 - 1) = 3(8 - 1) = 3 \cdot 7 = 21$
4	$a_4 = 3 \cdot 2^{4-1} = 3 \cdot 2^3 = 3 \cdot 8 = 24$	$3(2^4 - 1) = 3(16 - 1) = 3 \cdot 15 = 45$
5	$a_5 = 3 \cdot 2^{5-1} = 3 \cdot 2^4 = 3 \cdot 16 = 48$	$3(2^5 - 1) = 3(32 - 1) = 3 \cdot 31 = 93$
6	$a_6 = 3 \cdot 2^{6-1} = 3 \cdot 2^5 = 3 \cdot 32 = 96$	$3(2^6 - 1) = 3(64 - 1) = 3 \cdot 63 = 189$
7	$a_7 = 3 \cdot 2^{7-1} = 3 \cdot 2^6 = 3 \cdot 64 = 192$	$3(2^7 - 1) = 3(128 - 1) = 3 \cdot 127 = 381$
8	$a_8 = 3 \cdot 2^{8-1} = 3 \cdot 2^7 = 3 \cdot 128 = 384$	$3(2^8 - 1) = 3(256 - 1) = 3 \cdot 255 = 765$
9	$a_9 = 3 \cdot 2^{9-1} = 3 \cdot 2^8 = 3 \cdot 256 = 768$	$3(2^9 - 1) = 3(512 - 1) = 3 \cdot 511 = 1{,}533$
10	$a_{10} = 3 \cdot 2^{10-1} = 3 \cdot 2^9 = 3 \cdot 512 = 1{,}536$	$3(2^{10} - 1) = 3(1024 - 1) = 3 \cdot 1023 = 3{,}069$

After 9 bets, you have lost \$1,533 and can no longer continue doubling your bets.

43. Answers may vary.

Consider the sum $a_k + a_{k+1} + \ldots + a_{n-1} + a_n$ and the same sum written in reverse order as $a_n + a_{n-1} + \ldots + a_{k+1} + a_k$. If we place these sums under one another and add, we get

$$\begin{array}{ccccccccc} a_k & + & a_{k+1} & + \ldots + & a_{n-1} & + & a_n \\ a_n & + & a_{n-1} & + \ldots + & a_{k+1} & + & a_k \end{array}$$

$$(a_k + a_n) + (a_{k+1} + a_{n-1}) + \ldots + (a_{n-1} + a_{k+1}) + (a_n + a_k).$$

It can be shown that each term represents the same quantity, namely $a_k + a_n$. For example $a_{k+1} + a_{n-1} = (a_k + d) + (a_n - d) = a_k + a_n$, where d is the common difference. This term is being added $n - k + 1$ times, so $(n - k + 1)(a_k + a_n)$ is twice the desired sum, so the formula would be

$$\dfrac{(n - k + 1)(a_k + a_n)}{2}.$$

An alternate, lengthier approach would be to find the sum of the first n terms of the original sequence and subtract from that the sum of the first $k - 1$ terms.

$$\dfrac{n(a_1 + a_n)}{2} - \dfrac{(k - 1)(a_1 + a_{k-1})}{2}$$

Since $a_n = a_1 + (n - 1) \cdot d = a_1 + nd - d$ and

$a_{k-1} = a_1 + ((k - 1) - 1) \cdot d = a_1 + (k - 2) \cdot d = a_1 + kd - 2d$, we have

126   Chapter 5: Number Theory and the Real Number System

43. (continued)

$$\frac{n(a_1+a_n)}{2} - \frac{(k-1)(a_1+a_{k-1})}{2} = \frac{n(a_1+a_1+nd-d)}{2} - \frac{(k-1)(a_1+a_1+kd-2d)}{2}$$

$$= \frac{n(2a_1+nd-d)}{2} - \frac{(k-1)(2a_1+kd-2d)}{2}$$

$$= \frac{2na_1+n^2d-nd}{2} - \frac{2ka_1+k^2d-2kd-2a_1-kd+2d}{2}$$

$$= \frac{na_1+na_1+n^2d-nd-2ka_1-k^2d+2kd+2a_1+kd-2d}{2}$$

$$= \frac{n(a_1+nd-d)+na_1-ka_1+nkd-nkd-ka_1-k^2d+kd+kd+a_1+(a_1+kd-d)-d+nd-nd}{2}$$

$$= \frac{n(a_1+nd-d)-ka_1-ka_1+n(a_1+kd-d)-nkd-k^2d+kd+kd+(a_1+nd-d)+(a_1+kd-d)}{2}$$

$$= \frac{n(a_1+(n-1)d)-ka_1-ka_1+n(a_1+(k-1)d)-nkd-k^2d+kd+kd+(a_1+(n-1)d)+(a_1+(k-1)d)}{2}$$

$$= \frac{na_n-ka_1-ka_1+na_k-nkd-k^2d+kd+kd+a_n+a_k}{2}$$

$$= \frac{na_n+a_n-k(a_1+nd-d)-k(a_1+kd-d)+na_k+a_k}{2}$$

$$= \frac{na_n+a_n-k(a_1+(n-1)d)-k(a_1+(k-1)d)+na_k+a_k}{2}$$

$$= \frac{na_n+a_n-ka_n-ka_k+na_k+a_k}{2} = \frac{na_n-ka_n+a_n+na_k-ka_k+a_k}{2} = \frac{(n-k+1)a_n+(n-k+1)a_k}{2}$$

$$= \frac{(n-k+1)(a_n+a_k)}{2}$$

44. a) This is the sum of the terms a_1 to a_n in a geometric sequence.

    b) Use the distributive property to factor out a.

    c) If you multiply $(1-r)(1+r+r^2+\ldots+r^{n-1})$ you will get $1-r^n$. Divide both sides of the identity $(1-r)(1+r+r^2+\ldots+r^{n-1})=1-r^n$ by $(1-r)$, you get the result in c).

    d) Multiply numerator and denominator by -1 to get the final result.

45. You are finding the tenth term in a geometric sequence where $a_1=174$, $r=1.013$, and $n=10$. The common ratio is 1.013 because you retain the population from one year to the next and then add a percentage of the population. Thus the population in 2010 would be $174\cdot 1.013^{10-1}=174\cdot 1.013^9 \approx 195.45$ million.

46. You are finding the tenth term in a geometric sequence where $a_1=1273$, $r=1.01$, and $n=10$. The common ratio is 1.01 because you retain the population from one year to the next and then add a percentage of the population. Thus the population in 2010 would be $1273\cdot 1.01^{10-1}=1273\cdot 0.01^9 \approx 1{,}392.26$ million.

47. $F_1 + F_2 + \ldots + F_n = F_{n+2} - 1$

n	F_n	$F_1 + F_2 + \ldots + F_n$
1	1	$1 = 2 - 1 = F_3 - 1$
2	1	$2 = 3 - 1 = F_4 - 1$
3	2	$4 = 5 - 1 = F_5 - 1$
4	3	$7 = 8 - 1 = F_6 - 1$
5	5	$12 = 13 - 1 = F_7 - 1$
6	8	$20 = 21 - 1 = F_8 - 1$
7	13	$33 = 34 - 1 = F_9 - 1$
8	21	$54 = 55 - 1 = F_{10} - 1$
9	34	$88 = 89 - 1 = F_{11} - 1$
10	55	$143 = 144 - 1 = F_{12} - 1$
11	89	$232 = 233 - 1 = F_{13} - 1$
12	144	$376 = 377 - 1 = F_{14} - 1$
13	233	$609 = 610 - 1 = F_{15} - 1$
14	377	
15	610	

48. $F_n^2 + F_{n+1}^2 = F_{2n+1}$

n	F_n	$F_n^2 + F_{n+1}^2$
1	1	$1^2 + 1^2 = 1 + 1 = 2 = F_3 = F_{2+1}$
2	1	$1^2 + 2^2 = 1 + 4 = 5 = F_5 = F_{4+1}$
3	2	$2^2 + 3^2 = 4 + 9 = 13 = F_7 = F_{6+1}$
4	3	$3^2 + 5^2 = 9 + 25 = 34 = F_9 = F_{8+1}$
5	5	$5^2 + 8^2 = 25 + 64 = 89 = F_{11} = F_{10+1}$
6	8	$8^2 + 13^2 = 64 + 169 = 233 = F_{13} = F_{12+1}$
7	13	$13^2 + 21^2 = 169 + 441 = 610 = F_{15} = F_{14+1}$
8	21	
9	34	
10	55	
11	89	
12	144	
13	233	
14	377	
15	610	

49. $F_5 = \dfrac{\left(\dfrac{1+\sqrt{5}}{2}\right)^5 + \left(\dfrac{1-\sqrt{5}}{2}\right)^5}{\sqrt{5}} = \dfrac{\dfrac{80\sqrt{5}+176}{32} + \dfrac{176-80\sqrt{5}}{32}}{\sqrt{5}} = \dfrac{\dfrac{352}{32}}{\sqrt{5}} = \dfrac{11}{\sqrt{5}} \approx 4.9193$, rounded to the nearest whole number would be 5.

50. $F_7 = \dfrac{\left(\dfrac{1+\sqrt{5}}{2}\right)^7 + \left(\dfrac{1-\sqrt{5}}{2}\right)^7}{\sqrt{5}} = \dfrac{\dfrac{832\sqrt{5}+1{,}856}{128} + \dfrac{1{,}856-832\sqrt{5}}{128}}{\sqrt{5}} = \dfrac{\dfrac{3{,}712}{128}}{\sqrt{5}} = \dfrac{29}{\sqrt{5}} \approx 12.9692$, rounded to the nearest whole number would be 13.

51. $F_9 = \dfrac{\left(\dfrac{1+\sqrt{5}}{2}\right)^9 + \left(\dfrac{1-\sqrt{5}}{2}\right)^9}{\sqrt{5}} = \dfrac{\dfrac{8{,}704\sqrt{5}+19{,}456}{512} + \dfrac{19{,}456-8{,}704\sqrt{5}}{512}}{\sqrt{5}} = \dfrac{\dfrac{38{,}912}{512}}{\sqrt{5}} = \dfrac{76}{\sqrt{5}} \approx 33.9882$, rounded to the nearest whole number would be 34.

52. $F_{11} = \dfrac{\left(\dfrac{1+\sqrt{5}}{2}\right)^{11} + \left(\dfrac{1-\sqrt{5}}{2}\right)^{11}}{\sqrt{5}} = \dfrac{\dfrac{91{,}136\sqrt{5}+203{,}776}{2{,}048} + \dfrac{203{,}776-91{,}136\sqrt{5}}{2{,}048}}{\sqrt{5}} = \dfrac{\dfrac{407{,}552}{2{,}048}}{\sqrt{5}} = \dfrac{199}{\sqrt{5}} \approx 88.9955$, rounded to the nearest whole number would be 89.

Chapter 6
ALGEBRAIC MODELS: How Do We Approximate Reality?

Section 6.1 Linear Equations

1. Add, subtract, multiply, and divide both sides by the same quantity.

3. 12.50 and 15; 0.22.

5. the y-coordinate

7. the x- and y-intercepts

9. $3x+4=5x-6$
$-2x+4=-6$
$-2x=-10$
$x=5$

11. $4-2y=8+3y$
$4=8+5y$
$-4=5y$
$\dfrac{-4}{5}=y$
$y=-\dfrac{4}{5}$

13. $\dfrac{1}{2}x+4=\dfrac{3}{4}x-6$
$4\cdot\left[\dfrac{1}{2}x+4\right]=\left[\dfrac{3}{4}x-6\right]\cdot 4$
$2x+16=3x-24$
$-x+16=-24$
$-x=-40$
$x=40$

15. $\dfrac{1}{3}y+4=\dfrac{1}{4}y+3$
$12\cdot\left[\dfrac{1}{3}y+4\right]=\left[\dfrac{1}{4}y+3\right]\cdot 12$
$4y+48=3y+36$
$y+48=36$
$y=-12$

17. $0.2x+6=3x-0.4$
$10\cdot[0.2x+6]=[3x-0.4]\cdot 10$
$2x+60=30x-4$
$-28x+60=-4$
$-28x=-64$
$x=\dfrac{-64}{-28}=\dfrac{16}{7}$

19. $0.3y+2=0.5y-3$
$10\cdot[0.3y+2]=[0.5y-3]\cdot 10$
$3y+20=5y-30$
$20=2y-30$
$50=2y$
$25=y$
$y=25$

21. $P=2l+2w$
$P-2l=2w$
$\dfrac{P-2l}{2}=w$
$w=\dfrac{P-2l}{2}$

23.
$$z = \frac{x-\mu}{\sigma}$$
$$\sigma \cdot z = \left[\frac{x-\mu}{\sigma}\right] \cdot \sigma$$
$$\sigma z = x - \mu$$
$$\sigma z - x = -\mu$$
$$-(\sigma z - x) = \mu$$
$$x - \sigma z = \mu$$
$$\mu = x - \sigma z$$

25.
$$A = P(1+rt)$$
$$A = P + Prt$$
$$A - P = Prt$$
$$\frac{A-P}{Pt} = \frac{Prt}{Pt} = r$$
$$r = \frac{A-P}{Pt}$$

27.
$$2x + 3y = 6$$
$$2x = -3y + 6$$
$$x = \frac{-3y+6}{2}$$

29.
$$V = lwh$$
$$\frac{V}{wh} = \frac{lwh}{wh} = l$$
$$l = \frac{V}{wh}$$

31.
$$S = 2\pi rh + 2\pi r^2$$
$$S - 2\pi r^2 = 2\pi rh$$
$$\frac{S - 2\pi r^2}{2\pi r} = \frac{2\pi rh}{2\pi r} = h$$
$$h = \frac{S - 2\pi r^2}{2\pi r}$$

33. x-intercept (when $y = 0$)
$$3x + 2 \cdot 0 = 12$$
$$3x + 0 = 12$$
$$3x = 12$$
$$x = 4 \Rightarrow (4, 0)$$

y-intercept (when $x = 0$)
$$3 \cdot 0 + 2y = 12$$
$$0 + 2y = 12$$
$$2y = 12$$
$$y = 6 \Rightarrow (0, 6)$$

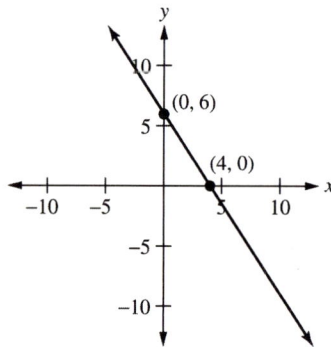

35. x-intercept (when $y = 0$)
$$4x - 3 \cdot 0 = 16$$
$$4x - 0 = 16$$
$$4x = 16$$
$$x = 4 \Rightarrow (4, 0)$$

y-intercept (when $x = 0$)
$$4 \cdot 0 - 3y = 16$$
$$0 - 3y = 16$$
$$-3y = 16$$
$$y = -\frac{16}{3} \Rightarrow \left(0, -\frac{16}{3}\right)$$

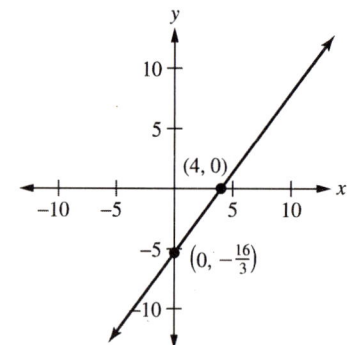

37. *x*-intercept (when $y = 0$)

$$\frac{1}{3}x + \frac{1}{2} \cdot 0 = 3$$

$$\frac{1}{3}x + 0 = 3$$

$$\frac{1}{3}x = 3$$

$$x = 9 \Rightarrow (9, 0)$$

y-intercept (when $x = 0$)

$$\frac{1}{3} \cdot 0 + \frac{1}{2}y = 3$$

$$0 + \frac{1}{2}y = 3$$

$$\frac{1}{2}y = 3$$

$$y = 6 \Rightarrow (0, 6)$$

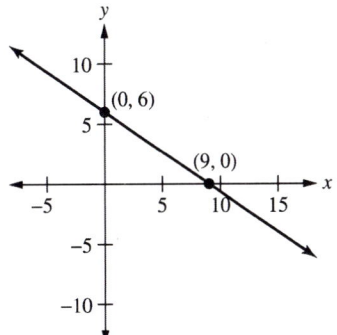

39. *x*-intercept (when $y = 0$)

$$\frac{1}{6}x - 2 \cdot 0 = \frac{3}{4} \Rightarrow \frac{1}{6}x - 0 = \frac{3}{4}$$

$$\frac{1}{6}x = \frac{3}{4} \Rightarrow 12 \cdot \left[\frac{1}{6}x\right] = \left[\frac{3}{4}\right] \cdot 12$$

$$2x = 9 \Rightarrow x = \frac{9}{2} \Rightarrow \left(\frac{9}{2}, 0\right)$$

y-intercept (when $x = 0$)

$$\frac{1}{6} \cdot 0 - 2y = \frac{3}{4} \Rightarrow \frac{1}{6}x - 0 = \frac{3}{4}$$

$$0 - 2y = \frac{3}{4}$$

$$-2y = \frac{3}{4} \Rightarrow y = -\frac{3}{8} \Rightarrow \left(0, -\frac{3}{8}\right)$$

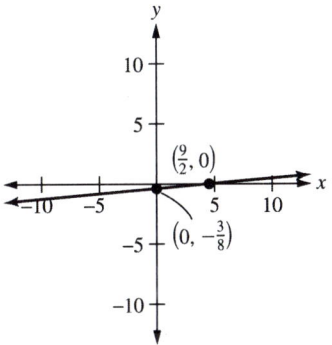

41. *x*-intercept (when $y = 0$)

$$0.2x = 4 \cdot 0 + 1.6$$

$$0.2x = 0 + 1.6$$

$$0.2x = 1.6$$

$$x = 8 \Rightarrow (8, 0)$$

y-intercept (when $x = 0$)

$$0.2 \cdot 0 = 4y + 1.6$$

$$0 = 4y + 1.6$$

$$-1.6 = 4y$$

$$-0.4 = y \Rightarrow (0, -0.4)$$

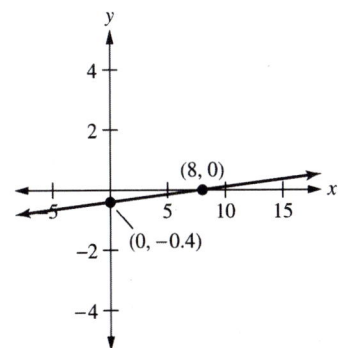

43. x-intercept (when $y=0$)
$$0.4x - 0.3\cdot 0 = 1.2$$
$$0.4x - 0 = 1.2$$
$$x = 3 \Rightarrow (3, 0)$$

y-intercept (when $x=0$)
$$0.4\cdot 0 - 0.3y = 1.2$$
$$0. - 0.3y = 1.2$$
$$y = -4 \Rightarrow (0, -4)$$

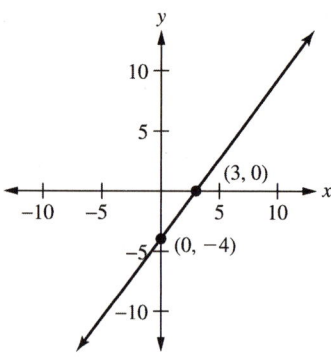

45. $m = \dfrac{y_2 - y_1}{x_2 - x_1} = \dfrac{8-5}{6-2} = \dfrac{3}{4}$

47. $m = \dfrac{y_2 - y_1}{x_2 - x_1} = \dfrac{2-6}{8-3} = \dfrac{-4}{5} = -\dfrac{4}{5}$

49. $m = \dfrac{y_2 - y_1}{x_2 - x_1} = \dfrac{1-(-4)}{5-3} = \dfrac{5}{2}$

51. $m = \dfrac{y_2 - y_1}{x_2 - x_1} = \dfrac{8-5}{6-6} = \dfrac{3}{0}$;

   slope does not exist

53. a and d

55. f

57. The rise is zero.

59. y-intercept is $(0, -3)$; slope is 4

61. y-intercept is $(0, -3)$; slope is -5

63. Let h be the number of hours Paola used the club last year.
$$95 + 2.50h = 515$$
$$2.50h = 420$$
$$h = 168$$
168 hours

65. Let p be the number of pages.
$$16 + 1.30(p - 10) = 44.60$$
$$16 + 1.30p - 13 = 44.60$$
$$3 + 1.30p = 44.60$$
$$1.30p = 41.60$$
$$p = 32 \text{ pages}$$

67. Let x be the number of computer systems Shaun sells.
$y = 45x + 225$ is the amount Shaun would receive from Best Deal Electronics.
$y = 20x + 400$ is the amount Shaun would receive from Circuit Town.
To find when both deals are the same, we solve $45x + 225 = 20x + 400$.
$$45x + 225 = 20x + 400$$
$$25x + 225 = 400$$
$$25x = 175$$
$$x = 7$$
If Shaun sells 7 computer systems, then the two deals are the same. Because he would receive more per computer system sold, if Shaun sells more than seven systems, then Best Deal is better for him.

69. Let x be the number of minutes Cassandra spends on the cell phone.
$y = 0.07x + 12.75$ is the amount Cassandra spends with BT&T.
$y = 0.05x + 14.15$ is the amount Cassandra spends with Cingleton.
To find when both deals are the same, we solve $0.07x + 12.75 = 0.05x + 14.15$.

$$0.07x + 12.75 = 0.05x + 14.15$$
$$100 \cdot [0.07x + 12.75] = [0.05x + 14.15] \cdot 100$$
$$7x + 1275 = 5x + 1415$$
$$2x + 1275 = 1415$$
$$2x = 140$$
$$x = 70$$

If Cassandra spends 70 minutes talking on her cell phone, then the two deals are the same. Because she would spend less per minute, if Cassandra talks more than 70 minutes, then Cingleton is better for her.

71. Let n be the net pay.
Let g be the gross pay.

$n = g - 10 - 0.0675g - 0.14g - 0.021g - 17$
$n = 0.7715g - 27$
$n = 0.7715 \cdot 400 - 27$
$n = 308.6 - 27$
$n = 281.60$

$281.60

73. Let c be the cost for the car after fees.
Let b be the base price.

$c = b + 35 + 0.06b + 0.012b + 45$
$c = 1.072b + 80$
$c = 1.072 \cdot 2200 + 80$
$c = 2{,}358.40 + 80$
$c = 2{,}438.40$

$2,438.40

75. Let l be the number of cups of lettuce.
Let t be the number of tomatoes.

$10l + 25t = 200$

77. $22l + 13t = 341$

79. Let t be the number of shares with Time-Warner.
Let d be the number of shares with Dell.

$35t + 55d = 14{,}500$

81. Let e be the number of end tables to be constructed.
Let c be the number of coffee tables to be constructed.

$9e + 15c = 342$

83. $y = -23.85x + 262.35$
The slope, -23.85, is the rate at which the DVD player is being paid off.
The y-intercept, 262.35, is the amount initially owed on the player.

134   Chapter 6: Algebraic Models

85. *x* is the number of days.
   y is the cost in dollars.

   For Save-U: $y = 183 + 32x$
   For Cheap Car: $y = 119 + 43x$

   a) After six days the Save-U is the cheaper plan.
   b) The two slopes (rates) are different.

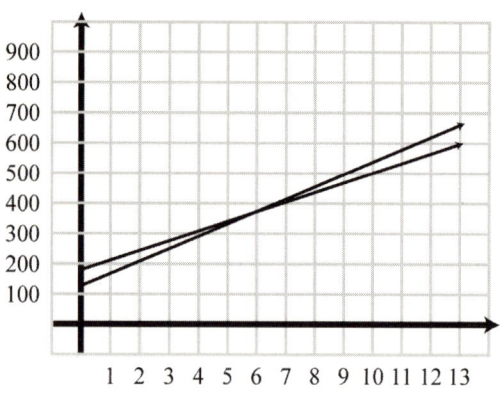

87. Let *x* be the number of tokens.
$$8 + 0.30x = 0.35x$$
$$8 = 0.05x$$
$$160 = x$$
after 160 tokens

89. amount of interest on the CD: $5000(0.038) = 190$
   amount of tax paid to the government: $190(0.14) = 26.60$
   amount realized on the CD: $190 - 26.60 = 163.40$
   Let *r* be the interest rate needed to be earned on the bonds.
$$163.40 = 5000 \cdot r$$
$$0.03268 = r$$
   The bonds would need to earn 3.268%.

91. a) $y = 115d$

   b) $d = \dfrac{y}{115}$

93. $e = 1.1d \Rightarrow \dfrac{e}{1.1} = d$

   $D = 264d \Rightarrow \dfrac{D}{264} = d$

   $\dfrac{e}{1.1} = \dfrac{D}{264} \Rightarrow D = \dfrac{264}{1.1} e = 240e$

95. $m = 0.08 = \dfrac{18}{x}$

   $x = \dfrac{18}{0.08} = 225$ inches

   $225 - 12 = 213$ inches

   $\dfrac{213}{12} = 17.75$ feet

97. No solution provided

Section 6.2   Modeling With Linear Equations

1. Write the equation of the line in standard form; state the *x*- and *y*-intercepts; specify the slope and *y*-intercept.

3. Two points were again used to find the equation of the line.

5. The rate of smoking dropped at a constant rate.

7. $y = mx + b$
$y = 3x + b$
$1 = 3 \cdot 2 + b \Rightarrow 1 = 6 + b \Rightarrow -5 = b$
$y = 3x - 5$

9. $y = mx + b$
$y = 4x + b$
$6 = 4 \cdot (-2) + b \Rightarrow 6 = -8 + b \Rightarrow 14 = b$
$y = 4x + 14$

11. $y = mx + b$
$y = -2x + b$
$3 = -2 \cdot 4 + b \Rightarrow 3 = -8 + b \Rightarrow 11 = b$
$y = -2x + 11$

13. $y = mx + b$
$y = -5x + b$
$-1 = -5 \cdot (-6) + b \Rightarrow -1 = 30 + b \Rightarrow$
$b = -31$
$y = -5x - 31$

15. $m = \dfrac{y_2 - y_1}{x_2 - x_1} = \dfrac{9 - 3}{5 - 2} = \dfrac{6}{3} = 2$
$y = mx + b$
$y = 2x + b$
$3 = 2 \cdot 2 + b \Rightarrow 3 = 4 + b \Rightarrow -1 = b$
$y = 2x - 1$

17. $m = \dfrac{y_2 - y_1}{x_2 - x_1} = \dfrac{10 - 12}{9 - 17} = \dfrac{-2}{-8} = \dfrac{1}{4}$
$y = mx + b$
$y = \dfrac{1}{4}x + b$
$10 = \dfrac{1}{4} \cdot 9 + b \Rightarrow 10 = \dfrac{9}{4} + b \Rightarrow \dfrac{31}{4} = b$
$y = \dfrac{1}{4}x + \dfrac{31}{4}$

19. $m = \dfrac{y_2 - y_1}{x_2 - x_1} = \dfrac{2 - (-4)}{-8 - 11} = \dfrac{6}{-19} = -\dfrac{6}{19}$
$y = mx + b$
$y = -\dfrac{6}{19}x + b$
$2 = -\dfrac{6}{19} \cdot (-8) + b \Rightarrow 2 = \dfrac{48}{19} + b \Rightarrow$
$b = -\dfrac{10}{19}$
$y = -\dfrac{6}{19}x - \dfrac{10}{19}$

21. $m = \dfrac{y_2 - y_1}{x_2 - x_1} = \dfrac{-1 - (-8)}{-4 - (-6)} = \dfrac{7}{2}$
$y = mx + b$
$y = \dfrac{7}{2}x + b$
$-1 = \dfrac{7}{2} \cdot (-4) + b$
$-1 = -14 + b$
$13 = b$
$y = \dfrac{7}{2}x + 13$

23. a) Let x be the number of years after 2001.
Let y be the life expectancy in years.
$y = 0.4x + 80.2$
   b) $y = 0.4 \cdot 9 + 80.2 = 3.6 + 80.2 = 83.8$ years

25. a) Let x be the number of years after 1999.
Let y be the number of male students enrolled in college (in thousands).
$y = 25x + 1{,}984$
   b) $y = 25 \cdot 11 + 1{,}984 = 275 + 1{,}984 = 2{,}259$
2,259,000 estimated male students enrolled

136 Chapter 6: Algebraic Models

27. a) Let x be the number of years after 2000.
Let y be the price of tuition.
$$y = 157x + 3{,}506$$

b) $y = 157 \cdot 10 + 3{,}506 = 1{,}570 + 3{,}506 = 5{,}076$
$\$5{,}076$ estimated price of tuition

29. a) $m = \dfrac{v_2 - v_1}{t_2 - t_1} = \dfrac{495 - 685}{7 - 0}$
$= \dfrac{-190}{7} = -\dfrac{190}{7}$

b) $v = mt + b$
$v = -\dfrac{190}{7}t + b$
$685 = -\dfrac{190}{3} \cdot 0 + b$
$685 = 0 + b$
$685 = b$
$v = -\dfrac{190}{3}t + 685$

c) $v = -\dfrac{190}{7} \cdot 15 + 685 = -\dfrac{2850}{7} + 685$
$= -\dfrac{2850}{7} + \dfrac{4795}{7} \approx 277.9$

31. Let x be the number of years after 1998.
Let y be the number of database administrators (in thousands).
The given information corresponds to the two ordered pairs $(0,87)$ and $(10,155)$.
$$m = \dfrac{y_2 - y_1}{x_2 - x_1} = \dfrac{155 - 87}{10 - 0} = \dfrac{68}{10} = 6.8$$
$y = mx + b$
$y = 6.8x + b$
$87 = 6.8 \cdot 0 + b$
$87 = 0 + b$
$87 = b$
$y = 6.8x + 87$
$y = 6.8 \cdot 12 + 87 = 81.6 + 87 = 168.6$

168,600 estimated database administrators in 2010

33. Let x be the number of years after 1900.
Let y be the number of hours of sleep.
The given information corresponds to the two ordered pairs $(0,8.5)$ and $(102, 6.9)$.
$$m = \dfrac{y_2 - y_1}{x_2 - x_1} = \dfrac{6.9 - 8.5}{104 - 0} = \dfrac{-1.6}{104} = -\dfrac{16}{1040} = -\dfrac{1}{65} \approx -0.0154$$
$y = mx + b$
$y = -0.0154x + b$
$8.5 = -0.0154 \cdot 0 + b$
$8.5 = 0 + b$
$8.5 = b$
$y = -0.0154x + 8.5$
$0 = -0.0154 \cdot x + 8.5$
$0.0154x = 8.5$
$x \approx 552$

In the year $1900 + 552 = 2452$, it is estimated that Americans won't be sleeping at all on a weeknight.

35. a) $m = \dfrac{l_2 - l_1}{t_2 - t_1} = \dfrac{146.5 - 142.6}{3 - 0} = \dfrac{3.9}{3} = 1.3$

   b) $l = mt + b$
   $l = 1.3t + b$
   $142.6 = 1.3 \cdot 0 + b$
   $142.6 = 0 + b$
   $142.6 = b$
   $l = 1.3t + 142.6$

   c) $l = 1.3 \cdot 10 + 142.6 = 13 + 142.6 = 155.6$ million
   A U.S. civilian labor force of 155,600,000 is estimated in the year 2010.

37.

	2000 (year 0)	2001 (year 1)	2002 (year 2)	2003 (year 3)
Actual data	142.6	143.7	144.9	146.5
Values predicted by your model	$l = 1.3 \cdot 0 + 142.6$ $= 0 + 142.6$ $= 142.6$	$l = 1.3 \cdot 1 + 142.6$ $= 1.3 + 142.6$ $= 143.9$	$l = 1.3 \cdot 2 + 142.6$ $= 2.6 + 142.6$ $= 145.2$	$l = 1.3 \cdot 3 + 142.6$ $= 3.9 + 142.6$ $= 146.5$
Values predicted by line of best fit	$y = 1.29 \cdot 0 + 142.49$ $= 0 + 142.49$ $= 142.49$	$y = 1.29 \cdot 1 + 142.49$ $= 1.29 + 142.49$ $= 143.78$	$y = 1.29 \cdot 2 + 142.49$ $= 2.58 + 142.49$ $= 145.07$	$y = 1.29 \cdot 3 + 142.49$ $= 3.87 + 142.49$ $= 146.36$

39. No solution provided

Section 6.3   Modeling With Quadratic Equations

1. To solve a quadratic equation.

3. $b^2 - 4ac$; If it is negative, there are no real roots for the equation; if it is zero, there is one real root; if it is positive, there are two distinct real roots.

5. By using the quadratic formula, $x = \dfrac{-b \pm \sqrt{b^2 - 4ac}}{2a}$

   $x = \dfrac{-(-10) \pm \sqrt{(-10)^2 - 4(1)(16)}}{2(1)} = \dfrac{10 \pm \sqrt{100 - 64}}{2} = \dfrac{10 \pm \sqrt{36}}{2} = \dfrac{10 \pm 6}{2}$

   $x = \dfrac{10 + 6}{2} = \dfrac{16}{2} = 8$ or $x = \dfrac{10 - 6}{2} = \dfrac{4}{2} = 2$

7. $x = \dfrac{-(-5) \pm \sqrt{(-5)^2 - 4(2)(3)}}{2(2)} = \dfrac{5 \pm \sqrt{25 - 24}}{4} = \dfrac{5 \pm \sqrt{1}}{4} = \dfrac{5 \pm 1}{4}$

   $x = \dfrac{5 + 1}{4} = \dfrac{6}{4} = \dfrac{3}{2}$ or $x = \dfrac{5 - 1}{4} = \dfrac{4}{4} = 1$

9. $x = \dfrac{-(7) \pm \sqrt{(7)^2 - 4(3)(-6)}}{2(3)} = \dfrac{-7 \pm \sqrt{49 - (-72)}}{6} = \dfrac{-7 \pm \sqrt{121}}{6} = \dfrac{-7 \pm 11}{6}$

$x = \dfrac{-7 + 11}{6} = \dfrac{4}{6} = \dfrac{2}{3}$ or $x = \dfrac{-7 - 11}{6} = \dfrac{-18}{6} = -3$

11. $x = \dfrac{-(-17) \pm \sqrt{(-17)^2 - 4(5)(-12)}}{2(5)} = \dfrac{17 \pm \sqrt{289 - (-240)}}{10} = \dfrac{17 \pm \sqrt{529}}{10} = \dfrac{17 \pm 23}{10}$

$x = \dfrac{17 + 23}{10} = \dfrac{40}{10} = 4$ or $x = \dfrac{17 - 23}{10} = \dfrac{-6}{10} = -\dfrac{3}{5}$

13. Opening: down since $a < 0$

Vertex: $x = \dfrac{-b}{2a} = \dfrac{-6}{2(-1)} = \dfrac{-6}{-2} = 3$. Substituting this value for x, we obtain

$y = -3^2 + 6 \cdot 3 - 8 = -9 + 18 - 8 = 1$ as the y-coordinate of the vertex.

The vertex is therefore $(3, 1)$.

x-intercepts: Set $y = 0$. We get $0 = -x^2 + 6x - 8$ and by using the quadratic formula, we obtain

$x = \dfrac{-(6) \pm \sqrt{(6)^2 - 4(-1)(-8)}}{2(-1)} = \dfrac{-6 \pm \sqrt{36 - 32}}{-2} = \dfrac{-6 \pm \sqrt{4}}{-2} = \dfrac{-6 \pm 2}{-2}$

$x = \dfrac{-6 + 2}{-2} = \dfrac{-4}{-2} = 2$ or $x = \dfrac{-6 - 2}{-2} = \dfrac{-8}{-2} = 4$

The x-intercepts of the graph are therefore $(2, 0)$ and $(4, 0)$.

y-intercept: Set $x = 0$. Thus $y = -0^2 + 6 \cdot 0 - 8 = -8$. The y-intercept is therefore $(0, -8)$.

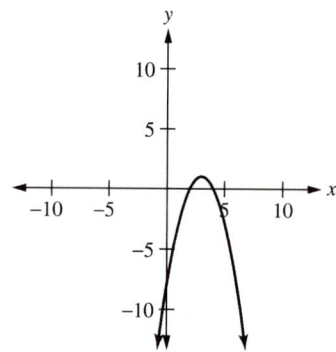

15. Opening: down since $a < 0$

   Vertex: $x = \dfrac{-b}{2a} = \dfrac{-8}{2(-4)} = \dfrac{-8}{-8} = 1$. Substituting this value for x, we obtain

   $y = -4 \cdot 1^2 + 8 \cdot 1 + 5 = -4 + 8 + 5 = 9$ as the y-coordinate of the vertex.

   The vertex is therefore $(1, 9)$.

   x-intercepts: Set $y = 0$. We get $0 = -4x^2 + 8x + 5$ and by using the quadratic formula, we obtain

   $$x = \dfrac{-(8) \pm \sqrt{(8)^2 - 4(-4)(5)}}{2(-4)} = \dfrac{-8 \pm \sqrt{64 + 80}}{-8} = \dfrac{-8 \pm \sqrt{144}}{-8} = \dfrac{-8 \pm 12}{-8}$$

   $$x = \dfrac{-8 + 12}{-8} = \dfrac{4}{-8} = -\dfrac{1}{2} \quad \text{or} \quad x = \dfrac{-8 - 12}{-8} = \dfrac{-20}{-8} = \dfrac{5}{2}$$

   The x-intercepts of the graph are therefore $\left(-\dfrac{1}{2}, 0\right)$ and $\left(\dfrac{5}{2}, 0\right)$.

   y-intercept: Set $x = 0$. Thus $y = -4 \cdot 0^2 + 8 \cdot 0 + 5 = 5$. The y-intercept is therefore $(0, 5)$.

   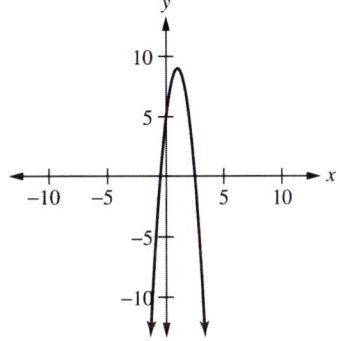

17. Opening: up since $a > 0$

   Vertex: $x = \dfrac{-b}{2a} = \dfrac{-(-4)}{2(4)} = \dfrac{4}{8} = \dfrac{1}{2}$. Substituting this value for x, we obtain

   $y = 4 \cdot \left(\dfrac{1}{2}\right)^2 - 4 \cdot \dfrac{1}{2} - 2 = 1 - 2 - 2 = -3$ as the y-coordinate of the vertex.

   The vertex is therefore $\left(\dfrac{1}{2}, -3\right)$.

   x-intercepts: Set $y = 0$. We get $0 = 4x^2 - 4x - 2$ and by using the quadratic formula, we obtain

   $$x = \dfrac{-(-4) \pm \sqrt{(-4)^2 - 4(4)(-2)}}{2(4)} = \dfrac{4 \pm \sqrt{16 + 32}}{8} = \dfrac{4 \pm \sqrt{48}}{8} = \dfrac{4 \pm 4\sqrt{3}}{8} = \dfrac{1 \pm \sqrt{3}}{2}$$

   $$x = \dfrac{1 + \sqrt{3}}{2} \approx 1.4 \quad \text{or} \quad x = \dfrac{1 - \sqrt{3}}{2} = \approx -0.4$$

   The x-intercepts of the graph are therefore $\left(\dfrac{1 + \sqrt{3}}{2}, 0\right)$ and $\left(\dfrac{1 - \sqrt{3}}{2}, 0\right)$.

   y-intercept: Set $x = 0$. Thus $y = 4 \cdot 0^2 - 4 \cdot 0 - 2 = -2$. The y-intercept is therefore $(0, -2)$.

17. (continued)

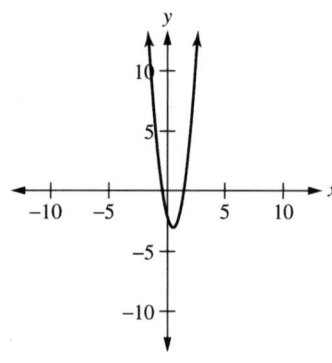

19. Opening: up since $a > 0$

    Vertex: $x = \dfrac{-b}{2a} = \dfrac{-(7)}{2(3)} = \dfrac{-7}{6} = -\dfrac{7}{6}$. Substituting this value for x, we obtain

    $$y = 3 \cdot \left(-\dfrac{7}{6}\right)^2 + 7 \cdot \left(-\dfrac{7}{6}\right) - 6 = \dfrac{49}{12} + \left(-\dfrac{49}{6}\right) - 6 = -\dfrac{121}{12}$$ as the y-coordinate of the vertex.

    The vertex is therefore $\left(-\dfrac{7}{6}, -\dfrac{121}{12}\right)$.

    x-intercepts: Set $y = 0$. We get $0 = 3x^2 + 7x - 6$ and by using the quadratic formula, we obtain

    $$x = \dfrac{-(7) \pm \sqrt{(7)^2 - 4(3)(-6)}}{2(3)} = \dfrac{-7 \pm \sqrt{49 + 72}}{6} = \dfrac{-7 \pm \sqrt{121}}{6} = \dfrac{-7 \pm 11}{6} =$$

    $$x = \dfrac{-7 + 11}{6} = \dfrac{4}{6} = \dfrac{2}{3} \text{ or } x = \dfrac{-7 - 11}{6} = -3$$

    The x-intercepts of the graph are therefore $\left(\dfrac{2}{3}, 0\right)$ and $(-3, 0)$.

    y-intercept: Set $x = 0$. Thus $y = 3 \cdot 0^2 + 7 \cdot 0 - 6 = -6$. The y-intercept is therefore $(0, -6)$.

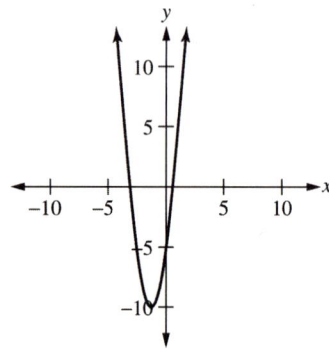

21. Opening: down since $a < 0$

Vertex: $x = \dfrac{-b}{2a} = \dfrac{-(7)}{2(-1)} = \dfrac{-7}{-2} = \dfrac{7}{2}$. Substituting this value for x, we obtain

$$y = -\left(\dfrac{7}{2}\right)^2 + 7 \cdot \left(\dfrac{7}{2}\right) - 12 = -\dfrac{49}{4} + \dfrac{49}{2} - 12 = \dfrac{1}{4}$$ as the y-coordinate of the vertex.

The vertex is therefore $\left(\dfrac{7}{2}, \dfrac{1}{4}\right)$.

x-intercepts: Set $y = 0$. We get $0 = -x^2 + 7x - 12$ and by using the quadratic formula, we obtain

$$x = \dfrac{-(7) \pm \sqrt{(7)^2 - 4(-1)(-12)}}{2(-1)} = \dfrac{-7 \pm \sqrt{49 - 48}}{-2} = \dfrac{-7 \pm \sqrt{1}}{-2} = \dfrac{-7 \pm 1}{-2}$$

$$x = \dfrac{-7+1}{-2} = \dfrac{-6}{-2} = 3 \text{ or } x = \dfrac{-7-1}{-2} = \dfrac{-8}{-2} = 4$$

The x-intercepts of the graph are therefore $(3, 0)$ and $(4, 0)$.

y-intercept: Set $x = 0$. Thus $y = -1 \cdot 0^2 + 7 \cdot 0 - 12 = -12$. The y-intercept is therefore $(0, -12)$.

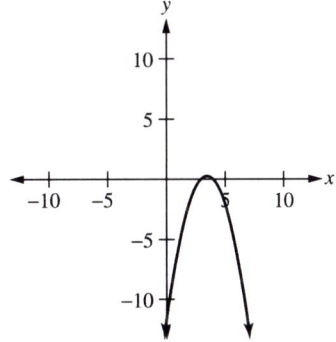

23. a) Week 2: $-0.125 \cdot 2^2 + 1.75 \cdot 2 + 2 = -0.5 + 3.5 + 2 = 5$;
    Week 4: $-0.125 \cdot 4^2 + 1.75 \cdot 4 + 2 = -2 + 7 + 2 = 7$
    Week 6: $-0.125 \cdot 6^2 + 1.75 \cdot 6 + 2 = -4.5 + 10.5 + 2 = 8$

    b) We need to find the vertex of $A = -0.125x^2 + 1.75x + 2$. $x = \dfrac{-b}{2a} = \dfrac{-(1.75)}{2(-0.125)} = \dfrac{-1.75}{-0.25} = 7$

    Substituting this value for x, we obtain
    $A = -0.125 \cdot (7)^2 + 1.75 \cdot (7) + 2 = -6.125 + 12.25 + 2 = 8.125$.

25. a) Month 1: $P = 30 \cdot 1^2 - 220 \cdot 1 - 10 = 30 - 220 - 10 = -200$
    Month 2: $P = 30 \cdot 2^2 - 220 \cdot 2 - 10 = 120 - 440 - 10 = -330$
    Month 3: $P = 30 \cdot 3^2 - 220 \cdot 3 - 10 = 270 - 660 - 10 = -400$

    b) We need to find the vertex of $P = 30x^2 - 220x - 10$. $x = \dfrac{-b}{2a} = \dfrac{-(-220)}{2(30)} = \dfrac{220}{60} = \dfrac{11}{3} \approx 3.7$

    implies at the end of the fourth month. Substituting this 4 for x, we obtain
    $P = 30 \cdot (4)^2 - 220 \cdot (4) - 10 = 480 - 880 - 10 = -410$, a cumulative loss of $410.

142 Chapter 6: Algebraic Models

25. (continued)

c) We need to determine when the cumulative profit is zero. This corresponds to solving the equation $0 = 30x^2 - 220x - 10$ and by using the quadratic formula, we obtain

$$x = \frac{-(-220) \pm \sqrt{(-220)^2 - 4(30)(-10)}}{2(30)} = \frac{220 \pm \sqrt{48400 + 1200}}{60} = \frac{220 \pm \sqrt{49600}}{60} = \frac{220 \pm 40\sqrt{31}}{60}$$

Since one of these values is negative, it can be disregarded. The value $x \approx 7.4$ implies that by the end of the eighth month, one can expect to show a cumulative profit.

d) Month 10: $P = 30 \cdot 10^2 - 220 \cdot 10 - 10 = 3{,}000 - 2{,}200 - 10 = \$790$

e) Month 9: $P = 30 \cdot 9^2 - 220 \cdot 9 - 10 = 2{,}430 - 1{,}980 - 10 = 440$
790−440=$350

27. We need to solve the equation $48 = 2.5x^2 - 9.5x + 48$ or $0 = 2.5x^2 - 9.5x$ by using the quadratic formula, we obtain

$$x = \frac{-(-9.5) \pm \sqrt{(-9.5)^2 - 4(2.5)(0)}}{2(2.5)} = \frac{9.5 \pm \sqrt{90.25 - 0}}{5} = \frac{9.5 \pm \sqrt{90.25}}{5} = \frac{9.5 \pm 9.5}{5}$$

$$x = \frac{9.5 + 9.5}{5} = \frac{19}{5} = 3.8 \text{ or } n = \frac{9.5 - 9.5}{5} = \frac{0}{5} = 0$$

We can expect this to occur by the fourth week.

29. a) Vertex: $t = \dfrac{-b}{2a} = \dfrac{-0}{2(-16)} = \dfrac{0}{-32} = 0$. Substituting this value for t, we obtain

$H = 160 - 16 \cdot 0^2 = 160 - 0 = 160$ as the second coordinate of the vertex, $(0, 160)$

t-intercepts: We need to solve $0 = 160 - 16t^2$ and by using the quadratic formula, we obtain

$$t = \frac{-(0) \pm \sqrt{(0)^2 - 4(-16)(160)}}{2(-16)} = \frac{0 \pm \sqrt{10240}}{-32} = \frac{\pm 32\sqrt{10}}{-32} = \pm\sqrt{10} \approx \pm 3.16$$

The t-intercepts of the graph are therefore $\left(-\sqrt{10}, 0\right)$ and $\left(\sqrt{10}, 0\right)$.

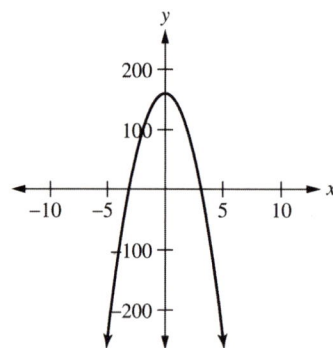

b) t (time) cannot be negative.

c) Approximately after 3.16 seconds

31. a) Vertex: $t = \dfrac{-b}{2a} = \dfrac{-100}{2(-16)} = \dfrac{-100}{-32} = \dfrac{25}{8} = 3.125$. Substituting this value for t, we obtain

$d = 100 \cdot \dfrac{25}{8} - 16 \cdot \left(\dfrac{25}{8}\right)^2 = \dfrac{625}{2} - \dfrac{625}{4} = \dfrac{625}{4} = 156.25$ as the second coordinate of the vertex, $(3.125, 156.25)$.

t-intercepts: We need to solve $0 = 100t - 16t^2$ and by using the quadratic formula, we obtain

$t = \dfrac{-(100) \pm \sqrt{(100)^2 - 4(-16)(0)}}{2(-16)} = \dfrac{-100 \pm \sqrt{10000}}{-32} = \dfrac{-100 \pm 100}{-32}$

$t = \dfrac{-100 - 100}{-32} = \dfrac{-200}{-32} = \dfrac{25}{4} = 6.25$ or $t = \dfrac{-100 + 100}{-32} = \dfrac{0}{-32} = 0$.

The t-intercepts of the graph are therefore $(6.25, 0)$ and $(0, 0)$.

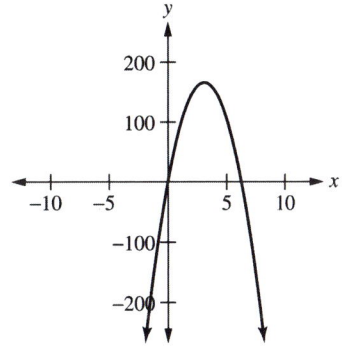

b) t (time) cannot be negative.

c) After 3.125 seconds

d) After 6.25 seconds

33. The crate begins falling a point above the ground and picks up speed as it falls. A linear equation would not be appropriate, because the rate of change of the distance of the crate above the ground is not constant.

35. a) Vertex: $t = \dfrac{-b}{2a} = \dfrac{-8}{2(0.15)} = \dfrac{-8}{0.3} = -\dfrac{80}{3} \approx -26.7$. Substituting this value for t, we obtain

$d = 0.15\left(-\dfrac{80}{3}\right)^2 + 8 \cdot \left(-\dfrac{80}{3}\right) = \dfrac{320}{3} + \dfrac{-640}{3} = -\dfrac{320}{3} \approx -106.7$ as the second coordinate of the vertex, $\left(-\dfrac{80}{3}, -\dfrac{320}{3}\right)$.

t-intercepts: We need to solve $0 = 0.15t^2 + 8t$ and by using the quadratic formula, we obtain

$t = \dfrac{-(8) \pm \sqrt{(8)^2 - 4(0.15)(0)}}{2(0.15)} = \dfrac{-8 \pm \sqrt{64 - 0}}{0.30} = \dfrac{-8 \pm \sqrt{64}}{0.30} = \dfrac{-8 \pm 8}{0.30}$

$t = \dfrac{-8 - 8}{0.30} \approx -53.3$ or $t = \dfrac{-8 + 8}{0.30} = 0$.

The t-intercepts of the graph are therefore approximately $(-53.3, 0)$ and $(0, 0)$

35. (continued)

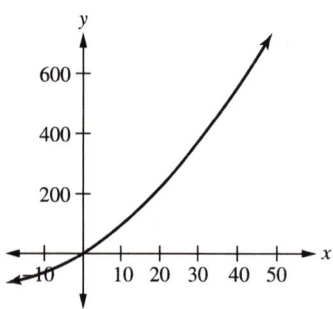

b) t (time) cannot be negative.

c) Considering only non-negative values, the runner is running more slowly when t is close to 0.

d) We need to solve $100 = 0.15t^2 + 8t$ or $0 = 0.15t^2 + 8t - 100$ and by using the quadratic formula, we obtain

$$t = \frac{-(8) \pm \sqrt{(8)^2 - 4(0.15)(-100)}}{2(0.15)} = \frac{-8 \pm \sqrt{64+60}}{0.30} = \frac{-8 \pm \sqrt{124}}{0.30}$$

$$t = \frac{-8 - \sqrt{124}}{0.30} \approx -63.79 \quad \text{or} \quad t = \frac{-8 + \sqrt{124}}{0.30} \approx 10.45$$

Since we disregard the negative solution, it would take approximately 10.45 seconds for the runner to finish the race.

37. No solution provided

Section 6.4   Exponential Equations And Growth

1. A is the amount after compounding, P is the principle, r is the interest rate, and n is the number of years the money is compounded.

3. The initial population and the growth rate.

5. It reverses the operation of raising 10 to a power.

7. When the rate of growth of a quantity is proportional to the amount present.

9. $A = 1,000(1+0.05)^1 = 1,000 \cdot 1.05 = \$1,050$

11. $A = 4,000(1+0.025)^1 = 4,000 \cdot 1.025 = \$4,100$

13. $A = 5,000(1+0.05)^5 = 5,000 \cdot 1.05^5 \approx \$6,381.40$

15. $A = 4,000(1+0.08)^2 = 4,000 \cdot 1.08^2 = \$4,665.60$

17. $A = 10,000(1+0.03)^8 = 10,000 \cdot 1.03^8 \approx \$12,667.70$

19. $A = 15,000(1+0.035)^{12} = 15,000 \cdot 1.035^{12} \approx \$22,666.02$

21. $P(1+r)^n = 1,030(1+0.016)^9 = 1,030 \cdot 1.016^9 \approx 1,188.2$ (million)

23. $P(1+r)^n = 12(1+0.023)^{20} = 12 \cdot 1.023^{20} \approx 18.9$ (million)

25. $P(1+r)^n = 1,273(1+0.0088)^9 = 1,273 \cdot 1.0088^9 \approx 1,377.4$ (million)

27. $5^x = 20$
    $\log 5^x = \log 20$
    $x \log 5 = \log 20$
    $x = \dfrac{\log 20}{\log 5} \approx \dfrac{1.30103}{0.69897} \approx 1.86$

29. $3^x = 10$
    $\log 3^x = \log 10$
    $x \log 3 = \log 10$
    $x = \dfrac{\log 10}{\log 3} \approx \dfrac{1}{0.47712} \approx 2.10$

31. $10^x = 3.2$
    $\log 10^x = \log 3.2$
    $x \log 10 = \log 3.2$
    $x = \dfrac{\log 3.2}{\log 10} \approx \dfrac{0.50515}{1} \approx 0.51$

33. $(3.4)^x = 6.85$
    $\log(3.4)^x = \log 6.85$
    $x \log 3.4 = \log 6.85$
    $x = \dfrac{\log 6.85}{\log 3.4} \approx \dfrac{0.83569}{0.53148} \approx 1.57$

35. $P_{4+1} = [1 + (0.03) \cdot (1 - 0.36)] \cdot 0.36$
    $P_5 = [1 + (0.03) \cdot 0.64] \cdot 0.36$
    $P_5 = [1 + 0.0192] \cdot 0.36$
    $P_5 = 1.0192 \cdot 0.36$
    $P_5 \approx 0.3669$

    At the end of the fifth year, the population is at approximately 36.69% of its maximum capacity.

37. $P_{8+1} = [1 + (0.045) \cdot (1 - 0.72)] \cdot 0.72$
    $P_9 = [1 + (0.045) \cdot 0.28] \cdot 0.72$
    $P_9 = [1 + 0.0126] \cdot 0.72$
    $P_9 = 1.0126 \cdot 0.72$
    $P_9 \approx 0.7291$

    At the end of the ninth year, the population is at approximately 72.91% of its maximum capacity.

39. $P_{0+1} = [1 + (0.08) \cdot (1 - 0.30)] \cdot 0.30$
    $P_1 = [1 + (0.08) \cdot 0.70] \cdot 0.30$
    $P_1 = [1 + 0.056] \cdot 0.30$
    $P_1 = 1.056 \cdot 0.30$
    $P_1 = 0.3168$

    $P_{1+1} = [1 + (0.08) \cdot (1 - 0.3168)] \cdot 0.3168$
    $P_2 = [1 + (0.08) \cdot 0.6832] \cdot 0.3168$
    $P_2 = [1 + 0.054656] \cdot 0.3168$
    $P_2 = 1.054656 \cdot 0.3168$
    $P_2 \approx 0.3341150208$

    $P_{2+1} \approx [1 + (0.08) \cdot (1 - 0.3341150208)] \cdot 0.3341150208$
    $P_3 \approx [1 + (0.08) \cdot 0.6658849792] \cdot 0.3341150208$
    $P_3 \approx [1 + 0.053270798336] \cdot 0.3341150208$
    $P_3 \approx 1.053270798336 \cdot 0.3341150208$
    $P_3 \approx 0.3519 \approx 35.2\%$

41. $P_{0+1} = [1+(0.10) \cdot (1-0.25)] \cdot 0.25$
$P_1 = [1+(0.10) \cdot 0.75] \cdot 0.25$
$P_1 = [1+0.075] \cdot 0.25$
$P_1 = 1.075 \cdot 0.25$
$P_1 = 0.26875$

$P_{1+1} = [1+(0.10) \cdot (1-0.26875)] \cdot 0.26875$
$P_2 = [1+(0.10) \cdot 0.73125] \cdot 0.26875$
$P_2 = [1+0.073125] \cdot 0.26875$
$P_2 = 1.073125 \cdot 0.26875$
$P_2 \approx 0.28840234375$

$P_{2+1} \approx [1+(0.10) \cdot (1-0.28840234375)] \cdot 0.28840234375$
$P_3 \approx [1+(0.10) \cdot 0.71159765625] \cdot 0.28840234375$
$P_3 \approx [1+0.071159765625] \cdot 0.28840234375$
$P_3 \approx 1.071159765625 \cdot 0.28840234375$
$P_3 \approx 0.3089 \approx 30.9\%$

43. $P_{0+1} = [1+(0.045) \cdot (1-0.60)] \cdot 0.60$
$P_1 = [1+(0.045) \cdot 0.40] \cdot 0.60$
$P_1 = [1+0.018] \cdot 0.60$
$P_1 = 1.018 \cdot 0.60$
$P_1 = 0.6108$

$P_{1+1} = [1+(0.045) \cdot (1-0.6108)] \cdot 0.6108$
$P_2 = [1+(0.045) \cdot 0.3892] \cdot 0.6108$
$P_2 = [1+0.017514] \cdot 0.6108$
$P_2 = 1.017514 \cdot 0.6108$
$P_2 = 0.6214975512$

$P_{2+1} \approx [1+(0.045) \cdot (1-0.6214975512)] \cdot 0.6214975512$
$P_3 \approx [1+(0.045) \cdot 0.3785024488] \cdot 0.6214975512$
$P_3 \approx [1+0.017032610196] \cdot 0.6214975512$
$P_3 \approx 1.017032610196 \cdot 0.6214975512$
$P_3 \approx 0.632 = 63.2\%$

45. $P_{0+1} = [1+(0.085) \cdot (1-0.50)] \cdot 0.50$
$P_1 = [1+(0.085) \cdot 0.50] \cdot 0.50$
$P_1 = [1+0.0425] \cdot 0.50$
$P_1 = 1.0425 \cdot 0.50$
$P_1 = 0.52125$

$P_{1+1} = [1+(0.085) \cdot (1-0.52125)] \cdot 0.52125$
$P_2 = [1+(0.085) \cdot 0.47875] \cdot 0.52125$
$P_2 = [1+0.04069375] \cdot 0.52125$
$P_2 = 1.04069375 \cdot 0.52125$
$P_2 \approx 0.542461617188$

$P_{2+1} \approx [1+(0.085) \cdot (1-0.542461617188)] \cdot 0.542461617188$
$P_3 \approx [1+(0.085) \cdot 0.457538382813] \cdot 0.542461617188$
$P_3 \approx [1+0.038890762539] \cdot 0.542461617188$
$P_3 \approx 1.038890762539 \cdot 0.542461617188$
$P_3 \approx 0.564 = 56.4\%$

47. $20,000 = 10,000(1+0.05)^n$

$2 = 1.05^n$

$\log 2 = \log 1.05^n$

$\log 2 = n \cdot \log 1.05$

$\dfrac{\log 2}{\log 1.05} = n$

$n \approx \dfrac{0.30103}{0.02119} \approx 14.21$ years

49. $562 = 281(1+0.012)^n$

$2 = 1.012^n$

$\log 2 = \log 1.012^n$

$\log 2 = n \cdot \log 1.012$

$\dfrac{\log 2}{\log 1.012} = n$

$n \approx \dfrac{0.30103}{0.00518} \approx 58.11$ years

By the year 2059, the population should double.

51. $1,320(1+(-0.005))^{15} = 1,320 \cdot 0.995^{15} \approx 1,224.4$ (million)

53. $50 = 100(1+(-0.0035))^n$

$\dfrac{1}{2} = 0.9965^n$

$\log \dfrac{1}{2} = \log 0.9965^n$

$\log \dfrac{1}{2} = n \cdot \log 0.9965$

$\dfrac{\log \dfrac{1}{2}}{\log 0.9965} = n$

$n \approx \dfrac{-0.30103}{-0.00152} \approx 197.70$

after approximately 198 years

55. $500(1+(-0.15))^3 = 500 \cdot 0.85^3 \approx 307$ mg

57. After 5 hours: $500(1+(-0.15))^5 \approx 222$

After 6 hours: $500(1+(-0.15))^6 \approx 189$

After 7 hours: $500(1+(-0.15))^7 \approx 160$

After 8 hours: $500(1+(-0.15))^8 \approx 136$

Therefore after approximately 8 hours the patient should stop feeling numbness.

59. $P_0 = \dfrac{200}{800} = 0.25$

$P_{0+1} = [1+(0.18)\cdot(1-0.25)]\cdot 0.25$

$P_1 = [1+(0.18)\cdot 0.75]\cdot 0.25$

$P_1 = [1+0.135]\cdot 0.25$

$P_1 = 1.135 \cdot 0.25$

$P_1 = 0.28375$

$\Rightarrow 0.28375 \cdot 800 \approx 227$ fish

$P_{1+1} = [1+(0.18)\cdot(1-0.28375)]\cdot 0.28375$

$P_2 = [1+(0.18)\cdot 0.71625]\cdot 0.28375$

$P_2 = [1+0.128925]\cdot 0.28375$

$P_2 = 1.128925 \cdot 0.28375$

$P_2 \approx 0.32033246875$

$\Rightarrow 0.32033246875 \cdot 800 \approx 256$ fish

$P_{2+1} \approx [1+(0.18)\cdot(1-0.32033246875)]\cdot 0.32033246875$

$P_3 \approx [1+(0.18)\cdot 0.67966753125]\cdot 0.32033246875$

$P_3 \approx [1+0.122340155625]\cdot 0.32033246875$

$P_3 \approx 1.122340155625 \cdot 0.32033246875$

$P_3 \approx 0.359521992829$

$\Rightarrow 0.359521992829 \cdot 800 \approx 288$ fish

59. (continued)

$$P_{3+1} \approx [1+(0.18)\cdot(1-0.359521992829)]\cdot 0.359521992829$$
$$P_4 \approx [1+(0.18)\cdot 0.640478007171]\cdot 0.359521992829$$
$$P_4 \approx [1+0.115286041291]\cdot 0.359521992829$$
$$P_4 \approx 1.11528604129\cdot 0.359521992829$$
$$P_4 \approx 0.400969860139$$
$$\Rightarrow 0.400969860139\cdot 800 \approx 321 \text{ fish}$$

The population will exceed 300 by the end of the fourth year. Fishing can begin toward the end of the fourth year.

61. By trial and error, let's first try 150.

$$P_0 = \frac{150}{1000} = 0.15$$
$$P_1 = [1+(0.25)\cdot(1-0.15)]\cdot 0.15 = 0.181875$$
$$0.181875\cdot 1000 \approx 182 \text{ fish}$$
$$P_2 = [1+(0.25)\cdot(1-0.181875)]\cdot 0.181875 \approx 0.219074121094$$
$$0.219074121094\cdot 1000 \approx 219 \text{ fish}$$
$$P_3 = [1+(0.25)\cdot(1-0.219074121094)]\cdot 0.219074121094 \approx 0.261844283734$$
$$0.261844283734\cdot 1000 \approx 262 \text{ fish}$$

150 fish is not enough to have 300 fish by the end of the third year (beginning of fourth).
Let's now try 200.

$$P_0 = \frac{200}{1000} = 0.2$$
$$P_1 = [1+(0.25)\cdot(1-0.2)]\cdot 0.2 = 0.24$$
$$0.24\cdot 1000 = 240 \text{ fish}$$
$$P_2 = [1+(0.25)\cdot(1-0.24)]\cdot 0.24 = 0.2856$$
$$0.2856\cdot 1000 \approx 286 \text{ fish}$$
$$P_3 = [1+(0.25)\cdot(1-0.2856)]\cdot 0.2856 = 0.33660816$$
$$0.33660816\cdot 1000 \approx 337 \text{ fish}$$

200 fish would be the smallest amount Carl could purchase.

Section 6.5 Proportions And Variation

1. The ratio of gas to the distance this week was compared with the ratio of gas to distance for an upcoming trip.

3. Step 1: Find the constant of variation, k. Step 2: Use k to rewrite the variation equation and then substitute values to find the desired quantity.

5. y decreases.

7. Rewrite $24:x=18:3$ as $\dfrac{24}{x}=\dfrac{18}{3}$.

$$\dfrac{24}{x}=\dfrac{18}{3}$$
$$24\cdot 3 = x\cdot 18$$
$$72 = 18x$$
$$4 = x$$
$$x = 4$$

9. $\dfrac{50}{4}=\dfrac{x}{5}$

$$50\cdot 5 = 4\cdot x$$
$$250 = 4x$$
$$\dfrac{250}{4} = x$$
$$x = 62.5$$

11. Rewrite $x:12=3:2$ as $\dfrac{x}{12}=\dfrac{3}{2}$.

$$\dfrac{x}{12}=\dfrac{3}{2}$$
$$x\cdot 2 = 12\cdot 3$$
$$2x = 36$$
$$x = 18$$

13. $\dfrac{30}{40}=\dfrac{27}{x}$

$$30\cdot x = 40\cdot 27$$
$$30x = 1{,}080$$
$$x = 36$$

15. $y = kx$; $y = 5x$
$37.5 = k\cdot 7.5$; $y = 5\cdot 13$
$\dfrac{37.5}{7.5} = k$; $y = 65$
$k = 5$

17. $r = \dfrac{k}{s}$; $r = \dfrac{8}{s}$
$12 = \dfrac{k}{2/3}$; $r = \dfrac{8}{8}$
$12\cdot \dfrac{2}{3} = k$; $r = 1$
$k = 8$

19. $a = kb^2$; $a = \dfrac{4}{9}b^2$
$16 = k\cdot 6^2$; $a = \dfrac{4}{9}\cdot 15^2$
$16 = k\cdot 36$; $a = \dfrac{4}{9}\cdot 225$
$\dfrac{16}{36} = k$; $a = 100$
$k = \dfrac{4}{9}$

21. $D = \dfrac{k}{C}$; $D = \dfrac{3/2}{C}$
$3\cdot 2 = 4k$; $D = \dfrac{3/2}{24}$
$6 = 4k$
$k = \dfrac{6}{4} = \dfrac{3}{2}$; $D = \dfrac{3}{2}\cdot\dfrac{1}{24} = \dfrac{1}{16}$

23. $r = kxy$; $r = 1.25xy$
$12.5 = k\cdot 2\cdot 5$; $r = 1.25\cdot 8\cdot 2.5$
$12.5 = k\cdot 10$; $r = 25$
$\dfrac{12.5}{10} = k$
$k = 1.25$

25. $y = kwx^2$; $y = \dfrac{7}{2}wx^2$
$504 = k\cdot 4\cdot 6^2$; $y = \dfrac{7}{2}\cdot 10\cdot 14^2$
$504 = k\cdot 4\cdot 36$; $y = \dfrac{7}{2}\cdot 10\cdot 196$
$504 = k\cdot 144$; $y = 6{,}860$
$\dfrac{504}{144} = k$
$k = \dfrac{7}{2}$

27. $y = k\dfrac{w}{x}$; $y = \dfrac{20}{3}\cdot\dfrac{w}{x}$
$4 = k\dfrac{6}{10}$; $y = \dfrac{20}{3}\cdot\dfrac{3}{15}$
$40 = 6k$
$\dfrac{40}{6} = k$; $y = \dfrac{4}{3}$
$k = \dfrac{20}{3}$

150   Chapter 6: Algebraic Models

29.
$$y = k\frac{x^2 w}{z}$$
$$15 = k\frac{2.5^2 \cdot 8}{20}$$
$$300 = k(6.25 \cdot 8)$$
$$300 = 50k$$
$$\frac{300}{50} = k$$
$$k = 6$$

$$y = 6 \cdot \frac{x^2 w}{z}$$
$$y = 6 \cdot \frac{8^2 \cdot 7}{14}$$
$$y = 6 \cdot \frac{64 \cdot 7}{14}$$
$$y = 6 \cdot \frac{64}{2}$$
$$y = 6 \cdot 32 = 192$$

In the solutions to Exercises 31 – 39, your proportion could be set up differently.

31.
$$\frac{150}{6} = \frac{65}{x}$$
$$150 \cdot x = 6 \cdot 65$$
$$150x = 390$$
$$x = \frac{390}{150}$$
$$x = 2.6 \text{ mg}$$

33.
$$\frac{30}{48} = \frac{x}{100}$$
$$30 \cdot 100 = 48 \cdot x$$
$$3{,}000 = 48x$$
$$\frac{3{,}000}{48} = x$$
$$x = 62.5 \text{ mph}$$

35. Garth's front lawn is $120 \cdot 40 = 4{,}800$ square feet.
$$\frac{1}{2{,}000} = \frac{x}{4{,}800}$$
$$1 \cdot 4{,}800 = 2{,}000 \cdot x$$
$$4{,}800 = 2{,}000x$$
$$\frac{4{,}800}{2{,}000} = x$$
$$x = 2.4$$

Garth must buy 3 bags.

Note: The 25 is not figured into the proportion to the left since we were looking for the number of bags as opposed to the number of pounds of fertilizer.

39. The conference room is $18 \cdot 27 = 486$ square feet. The room that needs to be carpeted is $24 \cdot 33 = 792$ square feet.
$$\frac{864}{486} = \frac{x}{792}$$
$$864 \cdot 792 = 486 \cdot x$$
$$684{,}288 = 486x$$
$$\frac{684{,}288}{486} = x$$
$$x = \$1{,}408$$

41. $\dfrac{\text{number of tagged bald eagles in population}}{\text{number of bald eagles in population}} = \dfrac{\text{number of tagged bald eagles in sample}}{\text{number of bald eagles in sample}}$

$$\dfrac{400}{n} = \dfrac{8}{240}$$
$$400 \cdot 240 = n \cdot 8$$
$$96,000 = 8n$$
$$\dfrac{96,000}{8} = n$$
$$n = 12,000$$

43. $\dfrac{\text{number of tagged grizzly bears in population}}{\text{number of grizzly bears in population}} = \dfrac{\text{number of tagged grizzly bears in sample}}{\text{number of grizzly bears in sample}}$

$$\dfrac{55}{1,000} = \dfrac{n}{95}$$
$$55 \cdot 95 = 1,000 \cdot n$$
$$5,225 = 1,000n$$
$$\dfrac{5,225}{1,000} = n$$
$$n = 5.225$$

5 or 6 tagged grizzly bears could be expected.

4 or 5 tagged largemouth bass could be expected.

45. $\dfrac{5,250}{1,470} = \dfrac{1,554}{x}$

$5,250 \cdot x = 1,470 \cdot 1,554$

$5,250x = 2,284,380$

$x = 435.12$

The total number of students is less than the number on the dean's list. He was not consistent in the way he set up the two ratios in the proportion.

47. Let t be the amount of tax in dollars.
Let c be the cost of the item in dollars.

$t = kc$

$2.34 = k \cdot 45$

$\dfrac{2.34}{45} = k$

$k = 0.052$

$t = 0.052c$

$t = 0.052 \cdot 65$

$t = \$3.38$

49. Let l be the length of the spring in inches.
Let f be the force in pounds.

$l = kf$

$6 = k \cdot 8$

$\dfrac{6}{8} = k$

$k = \tfrac{3}{4}$

$l = \tfrac{3}{4}f$

$10 = \tfrac{3}{4}f$

$10 \cdot \tfrac{4}{3} = f$

$f = 13\tfrac{1}{3}$ pounds

152   Chapter 6: Algebraic Models

51. Let i be the amount of illumination.
Let d be the distance in feet.

$$i = \frac{k}{d^2}$$

$i' = \frac{k}{4^2}$, where i' is the desired amount of illumination

$$i' = \frac{k}{16}$$

$$16i' = k$$

$$k = 16i'$$

$$i = \frac{16i'}{d^2}$$

$$\frac{i'}{2} = \frac{16i'}{d^2}$$

$$d^2 = 32$$

$$d = \sqrt{32} = 4\sqrt{2} \approx 5.66 \text{ feet}$$

53. Let a be the monthly attendance at the pool in people.
Let t be the temperature in degrees.
Let r be the number of rain days in a month.

$$a = k\frac{t}{r}$$

$$3,200 = k \cdot \frac{88}{8}$$

$$3,200 = 11k$$

$$\frac{3,200}{11} = k$$

$$k = \frac{3,200}{11}$$

$$a = \frac{3,200}{11} \cdot \frac{t}{r}$$

$$a = \frac{3,200}{11} \cdot \frac{92}{12}$$

$$a = \frac{3,200}{11} \cdot \frac{23}{3}$$

$$a = \frac{73,600}{33} \approx 2,230 \text{ people}$$

55. Let p be the pressure in pounds per square inch.
Let v be the volume in cubic inches.

$$p = \frac{k}{v}$$

$$4 = \frac{k}{120}$$

$$4 \cdot 120 = k$$

$$k = 480$$

$$p = \frac{480}{v}$$

$$p = \frac{480}{75}$$

$$p = 6.4 \text{ psi}$$

57. In $m:n$, we are comparing m objects with n objects. In $n:(m+n)$, we are comparing the n objects with the total of all the objects.

59. The strength doubles.

$$s = k\frac{(2w) \cdot d^2}{l} = 2\left(k\frac{w \cdot d^2}{l}\right)$$

61. The strength triples.

$$s = k\frac{w \cdot (3d)^2}{3l} = k\frac{9w \cdot d^2}{3l} = 3\left(k\frac{w \cdot d^2}{l}\right)$$

Section 6.6   Functions

1. f is the function that relates c, the cost of gasoline, to n, the number of miles driven.

3. $(3,4)$ and $(3,8)$.

5. Two different ordered pairs having the same first coordinate.

7. "y equals f of x."

9. "g of x equals 8."

11. $f(x) = 3x + 4$
$f(2) = 3(2) + 4$
$f(2) = 6 + 4$
$f(2) = 10$

13. $h(a) = 3a^2 - 2a + 5$
    $h(5) = 3(5)^2 - 2(5) + 5$
    $h(5) = 3(25) - 10 + 5$
    $h(5) = 75 - 10 + 5$
    $h(5) = 70$

15. $g(t) = \sqrt{t-4}$
    $g(40) = \sqrt{(40)-4}$
    $g(40) = \sqrt{36}$
    $g(40) = 6$

17. $f(x) = \sqrt{x-9}$
    $f(5) = \sqrt{(5)-9}$
    $f(5) = \sqrt{-4}$
    Does not exist.

19. $\{3,4,5,6,7,9\}$

21. 8

23. 9

25. Does not exist.

27. Does not exist.

29. –2

31. Not a function

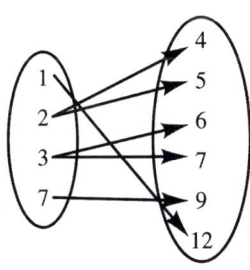

33. Function

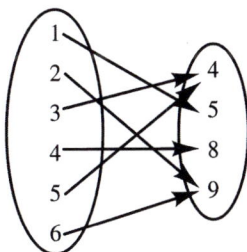

35. Function;

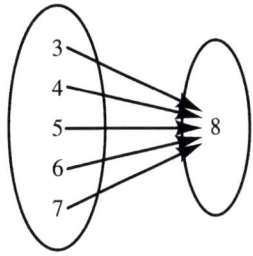

37. Yes; $f(x) = 5x + 3$
    $f(-4) = 5(-4) + 3 = -20 + 3 = -17$
    $f(0) = 5(0) + 3 = 0 + 3 = 3$
    $f(2) = 5(2) + 3 = 10 + 3 = 13$

39. No; $\{-2,-2\},\{-2,2\}$

41. Yes; $f(x) = \sqrt{x+3}$

    $f(-7) = \sqrt{(-7)+3} = \sqrt{-4}$; Does not exist.

    $f(0) = \sqrt{(0)+3} = \sqrt{3}$
    $f(6) = \sqrt{(6)+3} = \sqrt{9} = 3$

43. No; $\{0,-1\},\{0,1\}$

45. Function, passes vertical line test.

47. Not a function, fails vertical line test.

49. $y = 2x^2 - 5x + 3 \Rightarrow f(x) = 2x^2 - 5x + 3$

51. $y = 3x^2 - 5x \Rightarrow f(x) = 3x^2 - 5x$

53. Yes

55. No, a movie may have many stars.

57. No, a person may have many people in their immediate family.

59. No, some TV programs may appear several nights per week.

61. $d(t) = 104t + 64$
    $d(14) = 104(14) + 64$
    $d(14) = 1456 + 64$
    $d(14) = 1520$

63. $f(t) = 1.25t + 131.95$
    $f(3) = 1.25(3) + 131.95$
    $f(3) = 3.75 + 131.95$
    $f(3) = 135.7$

67. $A(n) = 97n + 975$
    $A(5) = 97(5) + 975$
    $A(5) = 485 + 975$
    $A(5) = 1460$

65. $S(n) = -13n^2 + 169n$
    $S(6.5) = -13(6.5)^2 + 169(6.5)$
    $S(6.5) = -13(42.25) + 1098.5$
    $S(6.5) = -549.25 + 1098.5$
    $S(6.5) = 549.25$

69. $(f+g)(x) = f(x) + g(x) = (2x+1) + (x^2+3) = x^2 + 2x + 4$

    $(f+g)(0) = (0)^2 + 2(0) + 4 = 0 + 0 + 4 = 4$
    $(f+g)(3) = (3)^2 + 2(3) + 4 = 9 + 6 + 4 = 19$
    $(f+g)(5) = (5)^2 + 2(5) + 4 = 25 + 10 + 4 = 39$

71. $(f \cdot g)(x) = f(x) \cdot g(x) = (2x+1) \cdot (x^2+3) = 2x^3 + x^2 + 6x + 3$

    $(f \cdot g)(0) = 2(0)^3 + (0)^2 + 6(0) + 3 = 0 + 0 + 0 + 3 = 3$
    $(f \cdot g)(2) = 2(2)^3 + (2)^2 + 6(2) + 3 = 16 + 4 + 12 + 3 = 35$
    $(f \cdot g)(4) = 2(4)^3 + (4)^2 + 6(4) + 3 = 128 + 16 + 24 + 3 = 171$

73. $(h \cdot (f+g))(x) = h(x) \cdot (f+g)(x)$

    $(h \cdot (f+g))(x) = h(x) \cdot (f+g)(x) = (x-5) \cdot ((2x+1) + (x^2+3)) = (x-5)(2x+1) + (x-5) \cdot (x^2+3)$
    $(h \cdot f + h \cdot g)(x) = h(x) \cdot f(x) + h(x) \cdot g(x) = (x-5) \cdot (2x+1) + (x-5) \cdot (x^2+3)$
    Thus, $(h \cdot (f+g))(x) = (h \cdot f + h \cdot g)(x)$

Chapter 6 Review Exercises

1. a) $\dfrac{2}{3}x + 2 = \dfrac{1}{6}x + 4$

    $6 \cdot \left[\dfrac{2}{3}x + 2\right] = \left[\dfrac{1}{6}x + 4\right] \cdot 6$

    $4x + 12 = x + 24$
    $3x + 12 = 24$
    $3x = 12$
    $x = 4$

   b) $0.3x - 2 = 3.5x - 0.4$
    $10 \cdot [0.3x - 2] = [3.5x - 0.4] \cdot 10$
    $3x - 20 = 35x - 4$
    $-32x - 20 = -4$
    $-32x = 16$
    $x = \dfrac{16}{-32} = -\dfrac{1}{2}$

2. $$A = P(1+rt)$$
$$A = P + \text{Pr}t$$
$$A - P = \text{Pr}t$$
$$\frac{A-P}{Pt} = \frac{\text{Pr}t}{Pt} = r$$
$$r = \frac{A-P}{Pt}$$

3. a) Let h be the number of hours Minxia works.
Let s be Minxia's salary for the week.
$$s = 40 \cdot 5 + (h - 40)(5)(2)$$
$$s = 200 + 10(h - 40)$$
b) $s = 200 + 10(46 - 40) = 200 + 10 \cdot 6 = 200 + 60 = \$260$

4. Let c be the number of hours Anton works in the camera shop.
Let d be the number of hours he works as a data entry technician.
$$5.80c + 11.35d = 160.40$$

5. x-intercept (when $y = 0$)
$$3x + 5 \cdot 0 = 20$$
$$3x + 0 = 20$$
$$3x = 20$$
$$x = \frac{20}{3} \Rightarrow \left(6\frac{2}{3}, 0\right)$$
y-intercept (when $x=0$)
$$3 \cdot 0 + 5y = 20$$
$$0 + 5y = 20$$
$$5y = 20$$
$$y = 4 \Rightarrow (0, 4)$$

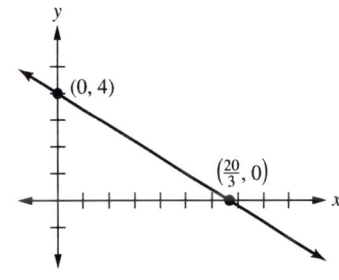

6. $m = \dfrac{y_2 - y_1}{x_2 - x_1} = \dfrac{8-5}{6-2} = \dfrac{3}{4}$

7. Line 1 is a; Line 2 is d; Line 3 is b; Line 4 is c.

8. Let m be the number of months.
$$29 + 11m = 39 + 8.5m$$
$$29 + 2.5m = 39$$
$$2.5m = 10$$
$$m = 4$$
after the fourth month or in the fifth month

9. when there is a constant rate of change between two variables

10. $m = \dfrac{y_2 - y_1}{x_2 - x_1} = \dfrac{9-4}{6-3} = \dfrac{5}{3}$
$$y = mx + b$$
$$y = \frac{5}{3}x + b$$
$$4 = \frac{5}{3} \cdot 3 + b$$
$$4 = 5 + b$$
$$-1 = b$$
$$y = \frac{5}{3}x - 1$$

11. Let x be the number of years after 1980.
    Let y be the percentage (in decimal form) of the U.S. population that were foreign born.
    The given information corresponds to the two ordered pairs $(0, 0.062)$ and $(20, 0.104)$.

    $$m = \frac{y_2 - y_1}{x_2 - x_1} = \frac{0.104 - 0.062}{20 - 0} = \frac{0.042}{20} = 0.0021$$

    $y = mx + b$
    $y = 0.0021x + b$
    $0.062 = 0.0021 \cdot 0 + b$
    $0.062 = 0 + b$
    $0.062 = b$
    $y = 0.0021x + 0.062$
    $y = 0.0021 \cdot 40 + 0.062 = 0.084 + 0.062 = 0.146$

    14.6% estimated foreign born in 2020

12. a technique by which we find a linear equation that bests models a set of data

13. a) Opening: down since $a < 0$

    b) Vertex: $x = \frac{-b}{2a} = \frac{-(-10)}{2(-2)} = \frac{10}{-4} = -\frac{5}{2}$. Substituting this value for x, we obtain

    $$y = -2 \cdot \left(-\frac{5}{2}\right)^2 - 10 \cdot \left(-\frac{5}{2}\right) - 8 = -\frac{25}{2} + 25 - 8 = \frac{9}{2}$$ as the y-coordinate of the vertex.

    The vertex is therefore $\left(-\frac{5}{2}, \frac{9}{2}\right)$.

    c) x-intercepts: Set $y = 0$. We get $0 = -2x^2 - 10x - 8$ and by using the quadratic formula, we obtain

    $$x = \frac{-(-10) \pm \sqrt{(-10)^2 - 4(-2)(-8)}}{2(-2)} = \frac{10 \pm \sqrt{100 - 64}}{-4} = \frac{10 \pm \sqrt{36}}{-4} = \frac{10 \pm 6}{-4}$$

    $$x = \frac{10 + 6}{-4} = \frac{16}{-4} = -4 \text{ or } x = \frac{10 - 6}{-4} = \frac{4}{-4} = -1$$

    The x-intercepts of the graph are therefore $(-4, 0)$ and $(-1, 0)$.

    d) y-intercept: Set $x = 0$. Thus $y = -2 \cdot 0^2 - 10 \cdot 0 - 8 = -8$. The y-intercept is therefore $(0, -8)$.

    e)

    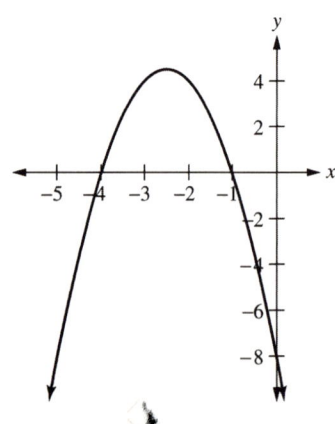

14. a technique by which we find a quadratic equation that bests models a set of data

15. We need to find the first component of the vertex of $A = -0.125x^2 + 2x + 1.125$.
$$x = \frac{-b}{2a} = \frac{-2}{2(-0.125)} = \frac{-2}{-0.25} = 8 \text{ implies in week 8.}$$

16. $A = 10,000(1+0.048)^5 = 10,000 \cdot 1.048^5 \approx \$12,641.72$

17. $20,000 = 10,000(1+0.048)^n$
$2 = 1.048^n$
$\log 2 = \log 1.048^n$
$\log 2 = n \cdot \log 1.048$
$\dfrac{\log 2}{\log 1.048} = n$
$n \approx \dfrac{0.30103}{0.02036} \approx 14.79$

It will double by the 15th year.

18. $P_{3+1} = [1+(0.03) \cdot (1-0.50)] \cdot 0.50$
$P_4 = [1+(0.03) \cdot 0.50] \cdot 0.50$
$P_4 = [1+0.015] \cdot 0.50$
$P_4 = 1.015 \cdot 0.50$
$P_4 = 0.5075$

19. In a logistic model we adjust the rate of growth as the population grows. The larger the population, the smaller the rate of growth.

20. a) Rewrite $25:8 = x:2$ as $\dfrac{25}{8} = \dfrac{x}{2}$.

$\dfrac{25}{8} = \dfrac{x}{2}$
$25 \cdot 2 = 8 \cdot x$
$50 = 8x$
$\dfrac{50}{8} = x$
$x = \dfrac{25}{4} = 6.25$

b) $\dfrac{30}{4} = \dfrac{x}{5}$
$30 \cdot 5 = 4 \cdot x$
$150 = 4x$
$\dfrac{150}{4} = x$
$x = \dfrac{75}{2} = 37.5$

21. Your proportion could be set up differently.
$\dfrac{3.5}{840} = \dfrac{x}{1,500}$
$3.5 \cdot 1,500 = 840 \cdot x$
$5,250 = 840x$
$\dfrac{5,250}{840} = x$
$x = 6.25$ gal

22. $$\frac{\text{number of tagged penguins in population}}{\text{number of penguins in population}} = \frac{\text{number of tagged penguins in sample}}{\text{number of penguins in sample}}$$

$$\frac{180}{n} = \frac{12}{55}$$
$$180 \cdot 55 = n \cdot 12$$
$$9,900 = 12n$$
$$\frac{9,900}{12} = n$$
$$n = 825$$

23. $y = \dfrac{k}{x}$ $\quad y = \dfrac{42}{x}$

$14 = \dfrac{k}{3}$ $\quad y = \dfrac{42}{18}$

$14 \cdot 3 = k$ $\quad y = \dfrac{7}{3}$

$k = 42$

24. $d = k\dfrac{a^2 b}{c}$ $\quad d = 32 \cdot \dfrac{a^2 b}{c}$

$144 = k\dfrac{3^2 \cdot 7}{14}$ $\quad d = 32 \cdot \dfrac{8^2 \cdot 11}{16}$

$2,016 = k(9 \cdot 7)$ $\quad d = 32 \cdot \dfrac{64 \cdot 11}{16}$

$2,016 = 63k$ $\quad d = 32 \cdot (4 \cdot 11)$

$\dfrac{2,016}{63} = k$ $\quad d = 32 \cdot 44 = 1,408$

$k = 32$

25. The strength of the beam stays the same.

$$s = k\frac{(2w) \cdot d^2}{2l} = k\frac{w \cdot d^2}{l}$$

26. a) $f(x) = 2x - 3$
$f(5) = 2(5) - 3$
$f(5) = 10 - 3$
$f(5) = 7$

b) $g(t) = 2t^2 + 4$
$g(6) = 2(6)^2 + 4$
$g(6) = 2(36) + 4$
$g(6) = 72 + 4$
$g(6) = 76$

27. a) $\{1, 2, 3, 4, 5, 7\}$
b) $\{-2, -1, 4, 6, 8\}$
c) -1
d) 3 and 7

28. a) Function
b) Not a function; $\{0,0\}, \{0,1\}$
c) Not a function; $\{-1,-1\}, \{-1,1\}$
d) Function

29. a) Not a function; fails vertical line test.
b) Function; passes vertical line test.

30. a) Function
b) Function

Chapter 6 Test

1. a) $\dfrac{3}{4}x + 5 = \dfrac{2}{3}x + 4$

$12 \cdot \left[\dfrac{3}{4}x + 5\right] = \left[\dfrac{2}{3}x + 4\right] \cdot 12$

$9x + 60 = 8x + 48$

$x + 60 = 48$

$x = -12$

b) $0.25x + 4 = 1.5x - 0.2$

$100 \cdot [0.25x + 4] = [1.5x - 0.2] \cdot 100$

$25x + 400 = 150x - 20$

$400 = 125x - 20$

$420 = 125x$

$x = \dfrac{420}{125} = 3.36$

2. $X = a(1+b)$ $X = a(1+b)$
   $\dfrac{X}{a} = \dfrac{a(1+b)}{a}$ $X = a + ab$
   $\dfrac{X}{a} = 1+b$ or $X - a = ab$
   $b = \dfrac{X}{a} - 1$ $ab = X - a$
   $b = \dfrac{X-a}{a}$

3. $m = \dfrac{y_2 - y_1}{x_2 - x_1} = \dfrac{12-5}{8-3} = \dfrac{7}{5}$

4. x-intercept (when $y=0$)
   $5x - 4 \cdot 0 = 10$
   $5x - 0 = 10$
   $5x = 10$
   $x = \dfrac{10}{5} \Rightarrow (2, 0)$

   y-intercept (when $x=0$)
   $5 \cdot 0 - 4y = 10$
   $0 - 4y = 10$
   $-4y = 10$
   $y = \dfrac{10}{-4} \Rightarrow \left(-\dfrac{5}{2}, 0\right)$

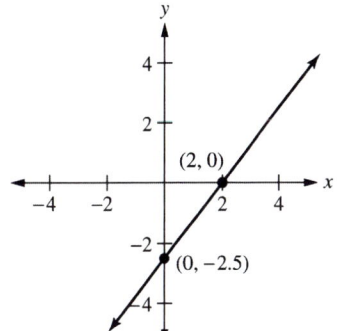

5. a) Let m be the number of minutes Brandon uses each month.
      Let c be Brandon's cost for one month.
      $c = 30 + 0.045(m - 1200)$

   b) $c = 30 + 0.045\big((1520) - 1200\big) = 30 + 0.045(320) = 30 + 14.4 = \$44.40$

6. 1 and b, 2 and d, 3 and a, 4 and c

7. Let r be the number of rentals.
   $40 + 1.5r = 22 + 2r$
   $18 = 0.5r$
   $r = \dfrac{18}{0.5}$
   $r = 36$
   after 36 rentals.

8. When the rate of change between the two variables is constant.

160   Chapter 6: Algebraic Models

9. $m = \dfrac{y_2 - y_1}{x_2 - x_1} = \dfrac{8-5}{6-2} = \dfrac{3}{4}$

$y = mx + b$

$y = \dfrac{3}{4}x + b$

$5 = \dfrac{3}{4} \cdot 2 + b$

$5 = \dfrac{3}{2} + b$

$\dfrac{7}{2} = b$

$y = \dfrac{3}{4}x + \dfrac{7}{2}$

10. Let p be the number of hours Shanaya works in the psychology lab.
    Let d be the number of hours she works in the dining hall.
    $8.25p + 6.30d = 137.70$

11. A technique to find a linear equation that bests models a set of data.

12. Let x be the number of years after 2002.
    Let y be the number (in millions) of the U.S. population that live in poverty.
    The given information corresponds to the two ordered pairs (0,34.6) and (1,35.9).
       (Note: The number living in poverty was up 1.3 million from 2002, so there are 35.9 − 1.3 = 34.6 million in 2002)

$m = \dfrac{y_2 - y_1}{x_2 - x_1} = \dfrac{35.9 - 34.6}{1 - 0} = \dfrac{1.3}{1} = 1.3$

$y = mx + b$

$y = 1.3x + b$

$34.6 = 1.3 \cdot 0 + b$

$34.6 = 0 + b$

$34.6 = b$

$y = 1.3x + 34.6$

$y = 1.3 \cdot 8 + 34.6 = 10.4 + 34.6 = 45$

The estimated number of people living in poverty in 2010 is 45 million.

13. a) opening down since $a < 0$

b) Vertex: $x = \dfrac{-b}{2a} = \dfrac{-(11)}{2(-1)} = \dfrac{-11}{-2} = \dfrac{11}{2}$. Substituting this value for x, we obtain

$y = -\left(\dfrac{11}{2}\right)^2 + 11 \cdot \left(\dfrac{11}{2}\right) - 24 = -\dfrac{121}{4} + \dfrac{121}{2} - 8 = -\dfrac{121}{4} + \dfrac{242}{4} - \dfrac{96}{4} = \dfrac{25}{4}$ as the y-coordinate of the vertex.

The vertex is therefore $\left(\dfrac{11}{2}, \dfrac{25}{4}\right)$.

13. (continued)

c) *x*-intercepts: Set $y = 0$. We get $0 = -x^2 + 11x - 24$ and by using the quadratic formula, we obtain

$$x = \frac{-(11) \pm \sqrt{(11)^2 - 4(-1)(-24)}}{2(-1)} = \frac{-11 \pm \sqrt{121 - 96}}{-2} = \frac{-11 \pm \sqrt{25}}{-2} = \frac{-11 \pm 5}{-2}$$

$$x = \frac{-11 + 5}{-2} = \frac{-6}{-2} = 3 \quad \text{or} \quad x = \frac{-11 - 5}{-2} = \frac{-16}{-2} = 8$$

The *x*-intercepts of the graph are therefore $(3, 0)$ and $(8, 0)$.

d) *y*-intercept: Set $x = 0$. Thus $y = -(0)^2 + 11 \cdot 0 - 24 = -24$. The *y*-intercept is therefore $(0, -24)$.

e)

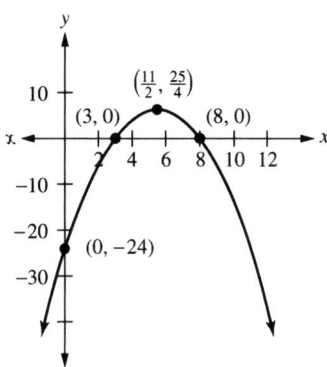

14. $A = 5{,}000(1 + 0.024)^8 = 5{,}000 \cdot 1.024^8 \approx \$6{,}044.63$

15. $10{,}000 = 5{,}000(1 + 0.024)^n$

    $2 = 1.024^n$

    $\log 2 = \log 1.024^n$

    $\log 2 = n \cdot \log 1.024$

    $\dfrac{\log 2}{\log 1.024} = n$

    $n \approx \dfrac{0.30103}{0.103} \approx 29.2$

    It will double just after 29 years.

16. quadratic regression

17. We need to find the first component of the vertex of $S = -1.2x^2 + 8x$.

    $x = \dfrac{-b}{2a} = \dfrac{-(8)}{2(-1.2)} = \dfrac{-8}{-2.4} \approx 3.33$, which implies during week 4.

18. $P_{8+1} = [1 + (0.02) \cdot (1 - 0.40)] \cdot 0.40$

    $P_9 = [1 + (0.02) \cdot 0.60] \cdot 0.40$

    $P_9 = [1 + 0.012] \cdot 0.40$

    $P_9 = 1.012 \cdot 0.40$

    $P_9 = 0.4048$

19. As the population increases, the environment cannot sustain the same growth rate, so the growth rate must be reduced.

20. $$\frac{\text{number of homeless in first sample}}{\text{number of homeless in population}} = \frac{\text{number of homeless from first sample in second sample}}{\text{total number of homeless in second sample}}$$

$$\frac{75}{x} = \frac{5}{120}$$
$$75 \cdot 120 = 5 \cdot x$$
$$9{,}000 = 5x$$
$$\frac{9{,}000}{5} = x$$
$$x = \frac{9{,}000}{5} = 1{,}800$$

21. a) Rewrite $40:6 = x:9$ as $\frac{40}{6} = \frac{x}{9}$.

$$\frac{40}{6} = \frac{x}{9}$$
$$40 \cdot 9 = 6 \cdot x$$
$$360 = 6x$$
$$\frac{360}{6} = x$$
$$x = 60$$

b) $$\frac{8}{3} = \frac{x}{10}$$
$$10 \cdot 8 = 3 \cdot x$$
$$80 = 3x$$
$$\frac{80}{3} = x$$
$$x = \frac{80}{3}$$

22. Your proportion could be set up differently.

$$\frac{x}{90} = \frac{4}{35}$$
$$35 \cdot x = 4 \cdot 90$$
$$35x = 360$$
$$x = \frac{360}{35} \approx 10.3$$

Curt Shilling could get 10 or 11 hits.

23. The strength of the beam would double.

$$s = k\frac{d^2}{l}$$
$$s_{new} = k\frac{(2d)^2}{(2l)} = k\frac{4d^2}{2l} = k\frac{2d^2}{l} = 2\left(k\frac{d^2}{l}\right)$$

24. $$y = \frac{k}{x}$$
$$8 = \frac{k}{2/3}$$
$$8 \cdot \left(\frac{2}{3}\right) = k$$
$$k = \frac{16}{3}$$

$$y = \frac{16/3}{x} = \frac{16}{3x}$$
$$y = \frac{16}{3(4)}$$
$$y = \frac{16}{12} = \frac{4}{3}$$

25. $$s = \frac{kxy}{t}$$
$$16 = \frac{k(2)(4)}{(6)}$$
$$16 = \frac{8 \cdot k}{6}$$
$$k = \frac{16 \cdot 6}{8} = 12$$

$$s = \frac{12xy}{t}$$
$$s = \frac{12(3)(5)}{(9)} = 20$$

26. a) $\{1, 2, 4, 5, 7, 8\}$
    b) $\{-3, -2, 4, 6, 8\}$
    c) -3
    d) 2 and 7

27. a) $f(x) = 3x - 8$
$f(5) = 3(5) - 8$
$f(5) = 15 - 8$
$f(5) = 7$

b) $g(t) = 5t^2 - 4$
$g(6) = 5(6)^2 - 4$
$g(6) = 5(36) - 4$
$g(6) = 180 - 4$
$g(6) = 176$

28. a) Not a function; $\{0, 5\}, \{0, 6\}$.
b) Function.
c) Not a function; $\{-1, -1\}, \{-1, 1\}$
d) Function.

29. a) Not a function; fails vertical line test.
b) Function; passes vertical line test.

30. a only

Of Further Interest: Dynamical Systems

1. All models in both examples are dynamical systems. That is, sequences of numbers of the form $A_0, A_1, A_2, \ldots$, where the value of A_{n+1} depends on the value of A_n.

2. A number a such that if $A_n = a$, then $A_{n+1} = a$. $\dfrac{5{,}000}{6} = 833.\overline{3}$

3. In Example 2, the equilibrium value $\dfrac{5{,}000}{6}$ was stable. In Example 3, the equilibrium value $\dfrac{5}{3}$ was unstable.

4. There are equations relating A_1 to A_0, A_2 to A_1, A_3 to A_2, and so on.

5. Although we began with values very close to $\dfrac{5}{3}$, as we computed values for the dynamical system, we got valued very far away from $\dfrac{5}{3}$.

6. If $-1 < m < 1$, then the equilibrium value will be stable.

7. $A_0 = 3$
$A_1 = 2 \cdot 3 - 1 = 6 - 1 = 5$
$A_2 = 2 \cdot 5 - 1 = 10 - 1 = 9$

8. $A_0 = 1$

$A_1 = 3 \cdot 1 + 2 = 3 + 2 = 5$
$A_2 = 3 \cdot 5 + 2 = 15 + 2 = 17$

9. $A_0 = -2$
$A_1 = -3 \cdot (-2) + 4 = 6 + 4 = 10$
$A_2 = -3 \cdot 10 + 4 = -30 + 4 = -26$
$A_3 = -3 \cdot (-26) + 4 = 78 + 4 = 82$

10. $A_0 = 2$
$A_1 = 2.5 \cdot 2 - 3 = 5 - 3 = 2$
$A_2 = 2.5 \cdot 2 - 3 = 5 - 3 = 2$
$A_3 = 2.5 \cdot 2 - 3 = 5 - 3 = 2$

11. $A_0 = 4$
$A_1 = 1.8 \cdot 4 - 2 = 7.2 - 2 = 5.2$
$A_2 = 1.8 \cdot 5.2 - 2 = 9.36 - 2 = 7.36$
$A_3 = 1.8 \cdot 7.36 - 2 = 13.248 - 2 = 11.248$
$A_4 = 1.8 \cdot 11.248 - 2 = 20.2464 - 2 = 18.2464$

12. $A_0 = 1.5$
    $A_1 = -0.8 \cdot 1.5 - 2 = -1.2 - 2 = -3.2$
    $A_2 = -0.8 \cdot (-3.2) - 2 = 2.56 - 2 = 0.56$
    $A_3 = -0.8 \cdot 0.56 - 2 = -0.448 - 2 = -2.448$
    $A_4 = -0.8 \cdot (-2.448) - 2 = 1.9584 - 2 = -0.0416$

13. In $A_{n+1} = 2A_n + 3$, substitute a for both A_{n+1} and A_n.
    $a = 2a + 3$
    $-a = 3$
    $a = -3$
    Since $2 > 1$, the equilibrium value is unstable.

14. In $A_{n+1} = 4A_n - 5$, substitute a for both A_{n+1} and A_n.
    $a = 4a - 5$
    $-3a = -5$
    $a = \dfrac{5}{3}$
    Since $4 > 1$, the equilibrium value is unstable.

15. In $B_{n+1} = 0.25B_n + 4$, substitute a for both B_{n+1} and B_n.
    $a = 0.25a + 4$
    $0.75a = 4$
    $a = \dfrac{4}{0.75} = \dfrac{400}{75} = \dfrac{16}{3}$
    Since $-1 < 0.25 < 1$, the equilibrium value is stable.

16. In $B_{n+1} = 0.10B_n - 2$, substitute a for both B_{n+1} and B_n.
    $a = 0.10a - 2$
    $0.90a = -2$
    $a = \dfrac{-2}{0.9} = -\dfrac{20}{9}$
    Since $-1 < 0.10 < 1$, the equilibrium value is stable.

17. $A_{n+1} = 1.05A_n$ where $n = 0, 1, 2, \ldots$ and $A_0 = 1,000$
    $A_1 = 1.05 \cdot 1,000 = 1,050$
    $A_2 = 1.05 \cdot 1,050 = 1,102.5$

    You would have \$1,102.50 at the end of 2 years.

18. $A_{n+1} = 1.06A_n$ where $n = 0, 1, 2, \ldots$ and $A_0 = 1,500$
    $A_1 = 1.06 \cdot 1,500 = 1,590$
    $A_2 = 1.06 \cdot 1,590 = 1,685.40$
    $A_3 = 1.06 \cdot 1,685.40 = 1,786.524$

    You would have \$1,786.52 at the end of 3 years.

19. $P_{n+1} = [1+(0.08)\cdot(1-P_n)]\cdot P_n$ where $n = 0,1,2,...$ and $P_0 = 0.30$

$P_1 = [1+(0.08)\cdot(1-0.30)]\cdot 0.30$
$P_1 = [1+(0.08)\cdot 0.70]\cdot 0.30$
$P_1 = [1+0.056]\cdot 0.30$
$P_1 = 1.056\cdot 0.30$
$P_1 = 0.3168$

$P_2 = [1+(0.08)\cdot(1-0.3168)]\cdot 0.3168$
$P_2 = [1+(0.08)\cdot 0.6832]\cdot 0.3168$
$P_2 = [1+0.054656]\cdot 0.3168$
$P_2 = 1.054656\cdot 0.3168$
$P_2 \approx 0.3341150208$

Approximately 33.4% of the lemur population's maximum capacity will be attained at the end of two years.

20. $P_{n+1} = [1+(0.12)\cdot(1-P_n)]\cdot P_n$ where $n = 0,1,2,...$ and $P_0 = 0.20$

$P_1 = [1+(0.12)\cdot(1-0.20)]\cdot 0.20$
$P_1 = [1+(0.12)\cdot 0.80]\cdot 0.20$
$P_1 = [1+0.096]\cdot 0.20$
$P_1 = 1.096\cdot 0.20$
$P_1 = 0.2192$

$P_2 = [1+(0.12)\cdot(1-0.2192)]\cdot 0.2192$
$P_2 = [1+(0.12)\cdot 0.7808]\cdot 0.2192$
$P_2 = [1+0.093696]\cdot 0.2192$
$P_2 = 1.093696\cdot 0.2192$
$P_2 \approx 0.2397381632$

Approximately 24.0% of the lemur population's maximum capacity will be attained at the end of two years.

21. $D_{n+1} = 0.60\cdot D_n + 250$ where $n = 0,1,2,...$ and $D_0 = 0$

$D_1 = 0.60\cdot 0 + 250 = 0 + 250 = 250$
$D_2 = 0.60\cdot 250 + 250 = 150 + 250 = 400$
$D_3 = 0.60\cdot 400 + 250 = 240 + 250 = 490$

490 milligrams of antibiotic will be in your bloodstream after three doses.

22. $D_{n+1} = 0.25\cdot D_n + 1{,}000$ where $n = 0,1,2,...$ and $D_0 = 0$

$D_1 = 0.25\cdot 0 + 1{,}000 = 0 + 1{,}000 = 1{,}000$
$D_2 = 0.25\cdot 1{,}000 + 1{,}000 = 250 + 1{,}000$
$\quad = 1{,}250$
$D_3 = 0.25\cdot 1{,}250 + 1{,}000 = 312.5 + 1{,}000$
$\quad = 1{,}312.5$

1,312.5 milligrams of antibiotic will be in your bloodstream after three doses.

23. $C_k = (0.99988)^k \cdot C_0 = 0.99988^k$ since $C_0 = 1$ where $k = 0, 1, 2, ...$

$0.90 = 0.99988^k$
$\log 0.90 = \log 0.99988^k$
$\log 0.90 = k\cdot \log 0.99988$
$\dfrac{\log 0.90}{\log 0.99988} = k$

$k \approx \dfrac{-0.04575749}{-0.00005212} \approx 877.93$

The fossilized bone is approximately 900 years old.

24. $C_k = (0.99988)^k \cdot C_0 = 0.99988^k$ since $C_0 = 1$ where $k = 0, 1, 2, ...$

$$0.60 = 0.99988^k$$
$$\log 0.60 = \log 0.99988^k$$
$$\log 0.60 = k \cdot \log 0.99988 \quad \Rightarrow \quad k \approx \frac{-0.22184875}{-0.00005212} \approx 4,256.50$$
$$\frac{\log 0.60}{\log 0.99988} = k$$

The fossilized plant is approximately 4,300 years old.

25. $P_0 = \dfrac{500}{1,000} = 0.5$ and 80 turkeys represents $\dfrac{80}{1,000} = 0.08$ or 8% of the maximum population.

$P_1 = [1 + (0.10) \cdot (1 - 0.5)] \cdot 0.5$
$P_1 = [1 + (0.10) \cdot 0.5] \cdot 0.5$
$P_1 = [1 + 0.05] \cdot 0.5$
$P_1 = 1.05 \cdot 0.5$
$P_1 = 0.525$
$\Rightarrow 0.525 - 0.08 = 0.445$

$P_2 = [1 + (0.10) \cdot (1 - 0.445)] \cdot 0.445$
$P_2 = [1 + (0.10) \cdot 0.555] \cdot 0.445$
$P_2 = [1 + 0.0555] \cdot 0.445$
$P_2 = 1.0555 \cdot 0.445$
$P_2 = 0.4696975$
$\Rightarrow 0.4696975 - 0.08 = 0.3896975$

$P_3 = [1 + (0.10) \cdot (1 - 0.3896975)] \cdot 0.3896975$
$P_3 = [1 + (0.10) \cdot 0.6103025] \cdot 0.3896975$
$P_3 = [1 + 0.06103025] \cdot 0.3896975$
$P_3 = 1.06103025 \cdot 0.3896975$
$P_3 \approx 0.413480835849$
$\Rightarrow 0.413480835849 - 0.08 = 0.333480835849$

There would be approximately 333 turkeys at the end of the three years.

26. $P_0 = \dfrac{750}{1000} = 0.75$ and 100 turkeys represents $\dfrac{100}{1000} = 0.10$ or 10% of the maximum population.

$P_1 = [1 + (0.10) \cdot (1 - 0.75)] \cdot 0.75$
$P_1 = [1 + (0.10) \cdot 0.25] \cdot 0.75$
$P_1 = [1 + 0.025] \cdot 0.75$
$P_1 = 1.025 \cdot 0.75$
$P_1 = 0.76875$
$\Rightarrow 0.76875 - 0.10 = 0.66875$

$P_2 = [1 + (0.10) \cdot (1 - 0.66875)] \cdot 0.66875$
$P_2 = [1 + (0.10) \cdot 0.33125] \cdot 0.66875$
$P_2 = [1 + 0.033125] \cdot 0.66875$
$P_2 = 1.033125 \cdot 0.66875$
$P_2 \approx 0.69090234375$
$\Rightarrow 0.69090234375 - 0.10 = 0.59090234375$

$P_3 \approx [1 + (0.10) \cdot (1 - 0.59090234375)] \cdot 0.59090234375$
$P_3 \approx [1 + (0.10) \cdot 0.40909765625] \cdot 0.59090234375$
$P_3 \approx [1 + 0.040909765625] \cdot 0.59090234375$
$P_3 \approx 1.040909765625 \cdot 0.59090234375$
$P_3 \approx 0.61507602014$
$\Rightarrow 0.61507602014 - 0.10 = 0.51507602014$

There would be approximately 515 turkeys at the end of the three years.

Chapter 7
MODELING WITH SYSTEMS OF LINEAR EQUATIONS AND INEQUALITIES: What's The Best Way To Do It?

Section 7.1 Systems Of Linear Equations

1. The lines can intersect in a single point as in Example 1; they can be parallel, with no points in common, as in Example 2; they can be the same line, as in Example 3.

3. The numbers are easier to work with.

5. When solving, there is a unique value for x or y.

7. When solving, there is a true statement, such as $0 + 0 = 0$.

9. The solution is $(4, 2)$.

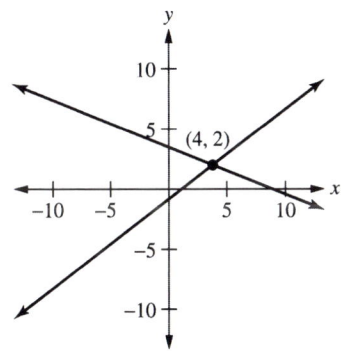

$$2 \cdot 4 + 5 \cdot 2 = 8 + 10 = 18$$
$$-3 \cdot 4 + 4 \cdot 2 = -12 + 8 = -4$$

11. The solution is $(3, 0)$.

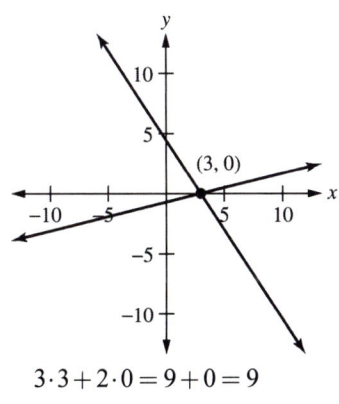

$$3 \cdot 3 + 2 \cdot 0 = 9 + 0 = 9$$
$$-1 \cdot 3 + 2 \cdot 0 = -3 + 0 = -3$$

13. The solution is $(2, 5)$.

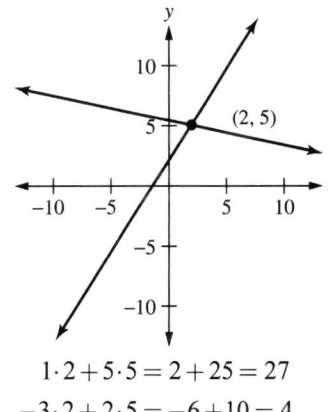

$$1 \cdot 2 + 5 \cdot 5 = 2 + 25 = 27$$
$$-3 \cdot 2 + 2 \cdot 5 = -6 + 10 = 4$$

15. The solution is $(1, 5)$.

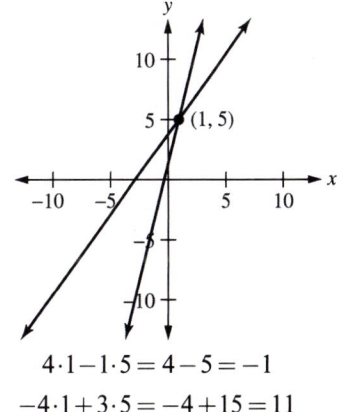

$$4 \cdot 1 - 1 \cdot 5 = 4 - 5 = -1$$
$$-4 \cdot 1 + 3 \cdot 5 = -4 + 15 = 11$$

17. The solution is $(4,-2)$.

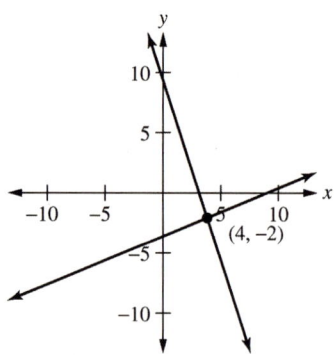

$2 \cdot 4 - 5 \cdot (-2) = 8 - (-10) = 18$
$3 \cdot 4 + 1 \cdot (-2) = 12 + (-2) = 10$

19. The solution is $(1,-1)$.

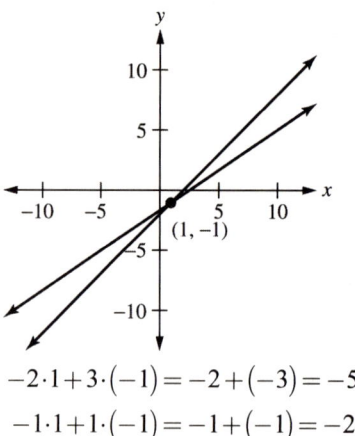

$-2 \cdot 1 + 3 \cdot (-1) = -2 + (-3) = -5$
$-1 \cdot 1 + 1 \cdot (-1) = -1 + (-1) = -2$

21. $(1,4)$

$2 \cdot [-3x + 2y] = [5] \cdot 2$
$6x + 4y = 22$

$-6x + 4y = 10$
$\underline{6x + 4y = 22}$

$0 + 8y = 32$
$y = 4$

$-3x + 2 \cdot 4 = 5$
$-3x + 8 = 5$
$-3x = -3$
$x = 1$

23. $(4,5)$

$-4x + 2y = -6$
$2 \cdot [2x + y] = [13] \cdot 2$

$-4x + 2y = -6$
$\underline{4x + 2y = 26}$

$0 + 4y = 20$
$y = 5$

$2x + 5 = 13$
$2x = 8$
$x = 4$

25. $(-6,-2)$

$3x - 2y = -14$
$2 \cdot [-4x + y] = [22] \cdot 2$

$3x - 2y = -14$
$\underline{-8x + 2y = 44}$

$-5x + 0 = 30$
$x = -6$

$-6 \cdot 3 - 2y = -14$
$-18 - 2y = -14$
$-2y = 4$
$y = -2$

27. no solution

$8x - 2y = -2$
$2 \cdot [-4x + y] = [3] \cdot 2$

$8x - 2y = -2$
$\underline{-8x + 2y = 6}$

$0 + 0 = 4$

29. infinite number of solutions

$4 \cdot [x - y] = [-2] \cdot 4$
$-4x + 4y = 8$

$4x - 4y = -8$
$\underline{-4x + 4y = 8}$

$0 + 0 = 0$

31. no solution

$2 \cdot [6x - 9y] = [8] \cdot 2$
$3 \cdot [-4x + 6y] = [10] \cdot 3$

$12x - 18y = 16$
$\underline{-12x + 18y = 30}$

$0 + 0 = 46$

33. $\left(\frac{1}{4}, \frac{1}{2}\right)$

$$12x - 8y = -1$$
$$6 \cdot [-2x + 5y] = [2] \cdot 6$$

$$12x - 8y = -1$$
$$\underline{-12x + 30y = 12}$$

$$0 + 22y = 11$$
$$y = \tfrac{11}{22} = \tfrac{1}{2}$$
$$12x - 8 \cdot \tfrac{1}{2} = -1$$
$$12x - 4 = -1$$
$$12x = 3$$
$$x = \tfrac{3}{12} = \tfrac{1}{4}$$

35. $\left(\frac{1}{5}, \frac{2}{5}\right)$

$$4 \cdot [3x - 9y] = [-3] \cdot 4$$
$$3 \cdot [-4x + 2y] = [0] \cdot 3$$

$$12x - 36y = -12$$
$$\underline{-12x + 6y = 0}$$

$$0 - 30y = -12$$
$$y = \tfrac{-12}{-30} = \tfrac{2}{5}$$
$$-4x + 2 \cdot \tfrac{2}{5} = 0$$
$$5 \cdot \left[-4x + \tfrac{4}{5}\right] = [0] \cdot 5$$
$$-20x + 4 = 0$$
$$-20x = -4$$
$$x = \tfrac{-4}{-20} = \tfrac{1}{5}$$

37. Let w be the number of games won.
    Let l be the number of games lost.
$$w + l = 162$$
$$w = l + 48$$

$$\begin{array}{ll} w + l = 162 & 105 + l = 162 \\ \underline{w - l = 48} & l = 57 \\ 2w + 0 = 210 & \\ w = 105 & \end{array}$$

105 wins and 57 losses

39. 4 bagels and 5 ounces of cream cheese

Nutrition	Amount per Bagel (b)	Amount per Ounce of Cream Cheese (c)	Desired
Calcium	30 mg	25 mg	245 mg
Iron	2 mg	0.4 mg	10 mg

$$\begin{array}{l} 30b + 25c = 245 \\ 10 \cdot [2b + 0.4c] = [10] \cdot 10 \end{array} \Rightarrow \begin{array}{l} 2 \cdot [30b + 25c] = [245] \cdot 2 \\ (-3) \cdot [20b + 4c] = [100] \cdot (-3) \end{array} \Rightarrow$$

$$\begin{array}{l} 60b + 50c = 490 \\ \underline{-60b - 12c = -300} \end{array} \Rightarrow 38c = 190 \Rightarrow c = 5$$

$$2b + 0.4 \cdot 5 = 10 \Rightarrow 2b + 2 = 10 \Rightarrow 2b = 8 \Rightarrow b = 4$$

170   Chapter 7: Modeling with Systems of Equations and Inequalities

41. Let c be the number of computers sold.
Let p be the pay.
$p = 225 + 45c$ Best Deal Electronics
$p = 400 + 20c$ Circuit Town

$$\begin{array}{c} (-1)\cdot[-45c+p]=[225]\cdot(-1) \\ -20c+p=400 \end{array} \Rightarrow \begin{array}{c} 45c-p=-225 \\ -20c+p=400 \end{array} \Rightarrow 25c=175 \Rightarrow c=7$$

At 7 systems, he makes $p = 400 + 20 \cdot 7 = 400 + 140 = \$540$ at either store. After that, he earns more at Best Deal because he earns more commission per computer.

43. Let w be the number of games won.
Let l be the number of games lost.

$$\begin{array}{c} w+l=34 \\ w=l+2 \end{array} \Rightarrow \begin{array}{c} w+l=34 \\ w-l=2 \end{array} \Rightarrow 2w=36 \Rightarrow w=18 \text{ thus } 18+l=34 \Rightarrow l=16$$

18 wins and 16 losses

45. Let A be the number of passengers at Atlanta's Hartsfield International (in millions).
Let C be the number of passengers at Chicago's O'Hare International (in millions).

$$\begin{array}{c} A+C=148 \\ A=C+10 \end{array} \Rightarrow \begin{array}{c} A+C=148 \\ A-C=10 \end{array} \Rightarrow 2A=158 \Rightarrow A=79 \text{ thus } 79+C=148 \Rightarrow A=69$$

Atlanta's Hartsfield International has 79 million and Chicago's O'Hare International has 69 million.

47. $11 per hour

Supply		Demand	
Price	Tutors Supplied	Price	Tutors Demanded
8	9	8	30
15	37	15	9

Supply:

$\text{slope} = \dfrac{37-9}{15-8} = \dfrac{28}{7} = 4$

$y = mx + b$
$y = 4x + b$
$9 = 4 \cdot 8 + b$
$9 = 32 + b \Rightarrow b = -23$
$y = 4x - 23 \Rightarrow -4x + y = -23$

Demand:

$\text{slope} = \dfrac{9-30}{15-8} = \dfrac{-21}{7} = -3$

$y = mx + b$
$y = -3x + b$
$30 = -3 \cdot 8 + b$
$30 = -24 + b \Rightarrow b = 54$
$y = -3x + 54 \Rightarrow 3x + y = 54$

We must now solve the following system.

$$-4x + y = -23 \quad \text{(supply)}$$
$$3x + y = 54 \quad \text{(demand)}$$

$$\begin{array}{c} -4x+y=-23 \\ (-1)\cdot[3x+y]=[54]\cdot(-1) \end{array} \Rightarrow \begin{array}{c} -4x+y=-23 \\ -3x-y=-54 \end{array} \Rightarrow -7x = -77 \Rightarrow x = 11$$

49. $34 per book

Supply		Demand	
Price	Books Supplied	Price	Books Demanded
30	60	30	95
42	120	42	50

Supply:

$$\text{slope} = \frac{120-60}{42-30} = \frac{60}{12} = 5$$

$y = mx + b$
$y = 5x + b$
$60 = 5 \cdot 30 + b$
$60 = 150 + b \Rightarrow b = -90$
$y = 5x - 90 \Rightarrow -5x + y = -90$

Demand:

$$\text{slope} = \frac{50-95}{42-30} = \frac{-45}{12} = -\frac{15}{4}$$

$y = mx + b$
$y = -\frac{15}{4}x + b$
$95 = -\frac{15}{4} \cdot 30 + b$
$95 = -\frac{225}{2} + b \Rightarrow b = \frac{415}{2}$
$y = -\frac{15}{4}x + \frac{415}{2} \Rightarrow 15x + 4y = 830$

We must now solve the following system.

$$-5x + y = -90 \quad \text{(supply)}$$
$$15x + 4y = 830 \quad \text{(demand)}$$

$$\begin{array}{l}(-4) \cdot [-5x+y] = [-90] \cdot (-4) \\ 15x + 4y = 830\end{array} \Rightarrow \begin{array}{l} 20x - 4y = 360 \\ 15x + 4y = 830 \end{array} \Rightarrow 35x = 1190 \Rightarrow x = 34$$

51. $(-1, 3)$

$x - 2y = -7 \Rightarrow x = 2y - 7$
$3x + 5y = 12$
$3(2y - 7) + 5y = 12$
$6y - 21 + 5y = 12$
$11y - 21 = 12$
$11y = 33$
$y = 3$
$x = 2 \cdot 3 - 7 = 6 - 7 = -1$

53. $(-1, 4)$

$-2x + y = 6 \Rightarrow y = 2x + 6$
$-2x + 3y = 14$
$-2x + 3(2x + 6) = 14$
$-2x + 6x + 18 = 14$
$4x + 18 = 14$
$4x = -4$
$x = -1$
$y = 2 \cdot (-1) + 6 = -2 + 6 = 4$

55. no solution

$x + y = 6$
$2x = 8 - 2y \Rightarrow x = 4 - y$
$(4 - y) + y = 6$
$4 - y + y = 6$
$4 = 6$ This is a false statement.

57. infinite number of solutions

$x - y = 4$
$3y = 3x - 12 \Rightarrow y = x - 4$
$x - (x - 4) = 4$
$x - x + 4 = 4$
$4 = 4$ This is a true statement.

172   Chapter 7: Modeling with Systems of Equations and Inequalities

59. a) $\begin{array}{l}(-2)\cdot[2x+3y+z]=[10]\cdot(-2) \\ x-y+2z=7\end{array} \Rightarrow \begin{array}{l}-4x-6y-2z=-20 \\ x-y+2z=7\end{array} \Rightarrow -3x-7y-13$

b) $\begin{array}{l}2x+3y+z=10 \\ 3x+y-z=4\end{array} \Rightarrow 5x+4y=14$

61. The first and the second equation:

$\begin{array}{l}x-3y+2z=6 \\ x+y-2z=2\end{array} \Rightarrow \frac{1}{2}\cdot[2x-2y]=[8]\cdot\frac{1}{2} \Rightarrow x-y=4$

The first and the third equation:

$\begin{array}{l}x-3y+2z=6 \\ (-2)[3x+2y+z]=[13]\cdot(-2)\end{array} \Rightarrow \begin{array}{l}x-3y+2z=6 \\ -6x-4y-2z=-26\end{array} \Rightarrow -5x-7y=-20$

Solving for x and y:

$\begin{array}{l}5\cdot[x-y]=[4]\cdot 5 \\ -5x-7y=-20\end{array} \Rightarrow \begin{array}{l}5x-5y=20 \\ -5x-7y=-20\end{array} \Rightarrow -12y=0 \Rightarrow y=0$

$x-0=4 \Rightarrow x=4$

Solving for z:

$4-3\cdot 0+2z=6 \Rightarrow 4-0+2z=6 \Rightarrow 4+2z=6 \Rightarrow 2z=2 \Rightarrow z=1$

The solution to the system is therefore $(4,0,1)$.

63. The first and the second equation:

$\begin{array}{l}x-y-z=2 \\ 4x+y+2z=24\end{array} \Rightarrow 5x+z=26$

The first and the third equation:

Since we eliminated y above, we should do the same in the next grouping.

$\begin{array}{l}x-y-z=2 \\ x+y-z=6\end{array} \Rightarrow \frac{1}{2}\cdot[2x-2z]=[8]\cdot\frac{1}{2} \Rightarrow x-z=4$

Solving for x and z:

$\begin{array}{l}5x+z=26 \\ x-z=4\end{array} \Rightarrow 6x=30 \Rightarrow x=5$

$5-z=4 \Rightarrow -z=-1 \Rightarrow z=1$

Solving for y:

$5-y-1=2 \Rightarrow 4-y=2 \Rightarrow -y=-2 \Rightarrow y=2;$ The solution to the system is therefore $(5,2,1)$.

65. and 67. No solutions provided

Section 7.2 Systems Of Linear Inequalities

1. It is a half-plane.

3. There were an infinite number of solutions. There are an infinite number of servings of shrimp and broccoli that satisfy the nutritional requirements.

5. (0, 0)

Section 7.2: Systems of Linear Inequalities   173

7. a and d

   a) $3\cdot 3+4\cdot 5\overset{?}{\geq} 2 \Rightarrow 9+20\overset{?}{\geq} 2 \Rightarrow 29\geq 2$; True

   b) $3\cdot 1+4\cdot(-2)\overset{?}{\geq} 2 \Rightarrow 3-8\overset{?}{\geq} 2 \Rightarrow -5\geq 2$; False

   c) $3\cdot 0+4\cdot 0\overset{?}{\geq} 2 \Rightarrow 0+0\overset{?}{\geq} 2 \Rightarrow 0\geq 2$; False

   d) $3\cdot(-4)+4\cdot 6\overset{?}{\geq} 2 \Rightarrow -12+24\overset{?}{\geq} 2 \Rightarrow 12\geq 2$; True

9. c and d

   a) $5\cdot 3-3\cdot 2\overset{?}{<} 2 \Rightarrow 15-6\overset{?}{<} 2 \Rightarrow 9<2$; False

   b) $5\cdot 1-3\cdot 1\overset{?}{<} 2 \Rightarrow 5-3\overset{?}{<} 2 \Rightarrow 2<2$; False

   c) $5\cdot 0-3\cdot 0\overset{?}{<} 2 \Rightarrow 0-0\overset{?}{<} 2 \Rightarrow 0<2$; True

   d) $5\cdot(-2)-3\cdot(-3)\overset{?}{<} 2 \Rightarrow -10+9\overset{?}{<} 2 \Rightarrow -1<2$; True

11. $3\cdot 0+4\cdot 0\overset{?}{\geq} 12 \Rightarrow 0+0\overset{?}{\geq} 12 \Rightarrow 0\geq 12$; False

    Shade the side that does not contain $(0,0)$.

    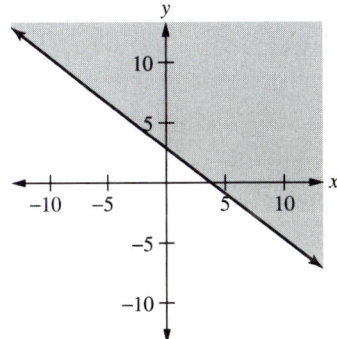

13. $2\cdot 0-4\cdot 0\overset{?}{<} 12 \Rightarrow 0-0\overset{?}{<} 12 \Rightarrow 0<12$; True

    Shade the side that contains $(0,0)$.

    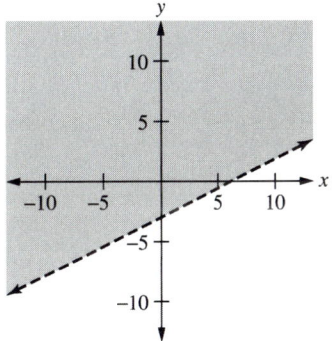

15. $0\overset{?}{\geq} 3\cdot 0-9 \Rightarrow 0\overset{?}{\geq} 0-9 \Rightarrow 0\geq -9$; True

    Shade the side that contains $(0,0)$.

    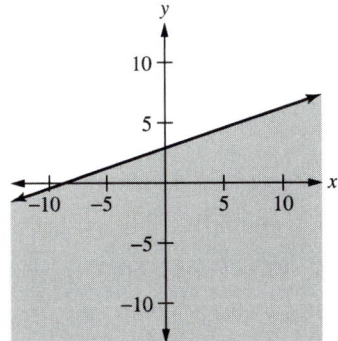

17. $4\cdot 0-8\overset{?}{<} 2\cdot 0 \Rightarrow 0-8\overset{?}{<} 0 \Rightarrow -8<0$; True

    Shade the side that contains $(0,0)$.

    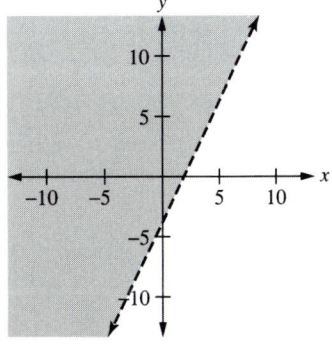

19. Testing $(0,0)$ in the first inequality, we get $2\cdot 0 - 3\cdot 0 = 0$ which is greater than -5, so $(0,0)$ is not a solution. Testing $(0,0)$ in the second inequality, we get $0 - 2\cdot 0 = 0$ which is greater than -8, so $(0,0)$ is not a solution. To solve for the corner point, we solve the system.

$$\begin{matrix} 2x-3y=-5 \\ x-2y=-8 \end{matrix} \Rightarrow \begin{matrix} 2x-3y=-5 \\ (-2)\cdot[x-2y]=[-8]\cdot(-2) \end{matrix} \Rightarrow \begin{matrix} 2x-3y=-5 \\ -2x+4y=16 \end{matrix} \Rightarrow y=11$$

$$2x - 3\cdot 11 = -5 \Rightarrow 2x - 33 = -5 \Rightarrow 2x = 28 \Rightarrow x = 14$$

Thus the corner point is $(14, 11)$.

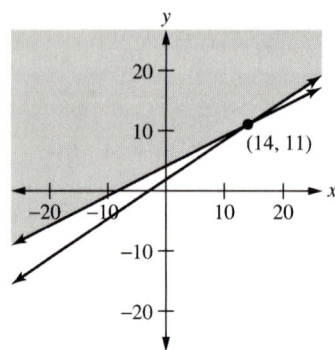

21. Testing $(0,0)$ in the first inequality, we get $2\cdot 0 - 0 = 0$ which is less than 3, so $(0,0)$ is not a solution. Testing $(0,0)$ in the second inequality, we get $0 - 0 = 0$ which is greater than -1, so $(0,0)$ is not a solution. To solve for the corner point, we solve the system.

$$\begin{matrix} 2x-y=3 \\ x-y=-1 \end{matrix} \Rightarrow \begin{matrix} 2x-y=3 \\ (-1)\cdot[x-y]=[-1]\cdot(-1) \end{matrix} \Rightarrow \begin{matrix} 2x-y=3 \\ -x+y=1 \end{matrix} \Rightarrow x=4$$

$$4 - y = -1 \Rightarrow -y = -5 \Rightarrow y = 5$$

Thus the corner point is $(4, 5)$.

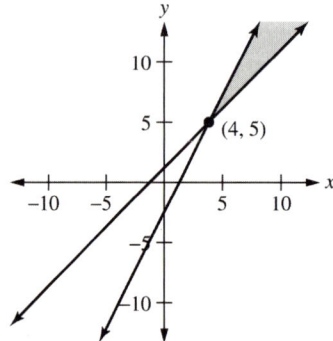

23. Testing $(0,0)$ in the first inequality, we get $-4 \cdot 0 + 3 \cdot 0 = 0$ which is less than 23, so $(0,0)$ is a solution. Testing $(0,0)$ in the second inequality, we get $3 \cdot 0 \stackrel{?}{>} 19 - 5 \cdot 0 \Rightarrow 0 \stackrel{?}{>} 19 - 0 \Rightarrow 0 > 19$ which is false, so $(0,0)$ is not a solution. To solve for the corner point, we solve the system.

$$\begin{matrix} -4x+3y=23 \\ 3x=19-5y \end{matrix} \Rightarrow \begin{matrix} 3 \cdot [-4x+3y] = [23] \cdot 3 \\ 4 \cdot [3x+5y] = [19] \cdot 4 \end{matrix} \Rightarrow \begin{matrix} -12x+9y=69 \\ 12x+20y=76 \end{matrix} \Rightarrow 29y=145 \Rightarrow y=5$$

$$3x = 19 - 5 \cdot 5 \Rightarrow 3x = 19 - 25 \Rightarrow 3x = -6 \Rightarrow x = -2$$

Thus the corner point is $(-2, 5)$.

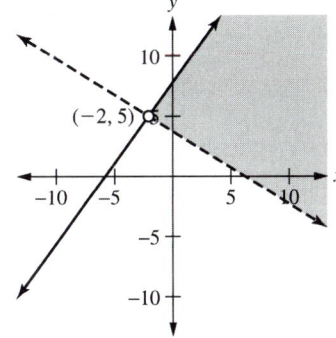

25. Testing $(0,0)$ in the first inequality, we get $3 \cdot 0 + 5 \cdot 0 = 0$ which is less than 32, so $(0,0)$ is a solution. Testing $(0,0)$ in the second inequality, we get $0 \geq 4$ which is false, so $(0,0)$ is not a solution. To solve for the corner point, we solve the system.

$$3x+5y=32$$
$$y=4$$

$$3x + 5 \cdot 4 = 32 \Rightarrow 3x + 20 = 32 \Rightarrow 3x = 12 \Rightarrow x = 4$$

Thus the corner point is $(4, 4)$.

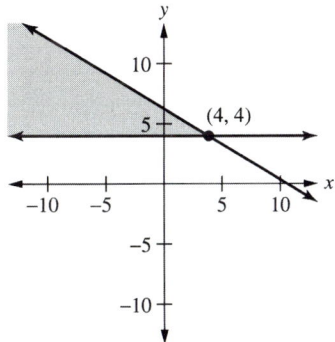

27. Testing $(0,0)$ in the first inequality, we get $2\cdot 0+5\cdot 0=0$ which is less than 26, so $(0,0)$ is not a solution. Testing $(0,0)$ in the second inequality, we get $0<8$ which is true, so $(0,0)$ is a solution. To solve for the corner point, we solve the system.

$$2x+5y=26$$
$$x=8$$
$$2\cdot 8+5y=26 \Rightarrow 16+5y=26 \Rightarrow 5y=10 \Rightarrow y=2$$

Thus the corner point is $(8,2)$.

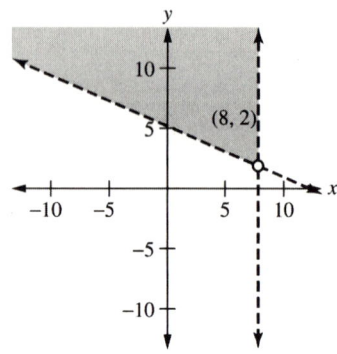

29. There are no points in which the regions given by the two inequalities intersect.

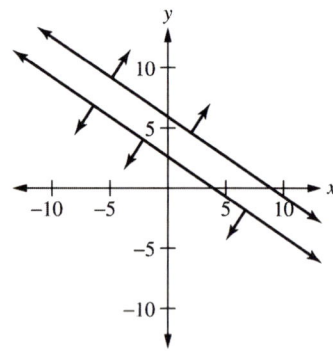

31. There are no points in which the regions given by the two inequalities intersect.

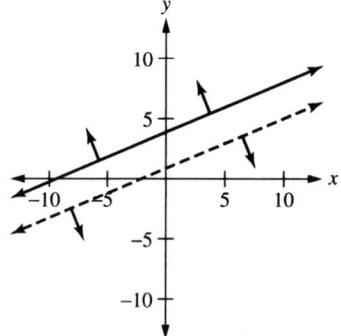

33. Since the number of entertainment centers cannot be negative, we will consider restrictions in the first quadrant given by the following table.

	Athens (a)	Barcelona (b)	Available	Inequalities
fancy molding in feet	4	15	360	$4a+15b \leq 360$
time in hours	4	3	120	$4a+3b \leq 120$

The origin, $(0,0)$, an a-intercept, $(30,0)$, and a b-intercept, $(0,24)$, are all corner points. To find the fourth corner point we solve the following system.

$$4a+15b=360$$
$$4a+3b=120$$

$$\begin{array}{c} 4a+15b=360 \\ (-1)\cdot[4a+3b]=[120]\cdot(-1) \end{array} \Rightarrow \begin{array}{c} 4a+15b=360 \\ -4a-3b=-120 \end{array} \Rightarrow 12b=240 \Rightarrow b=20$$

$4a+3\cdot 20=120 \Rightarrow 4a+60=120 \Rightarrow 4a=60 \Rightarrow a=15$ The fourth corner point is $(15,20)$.

33. (continued)

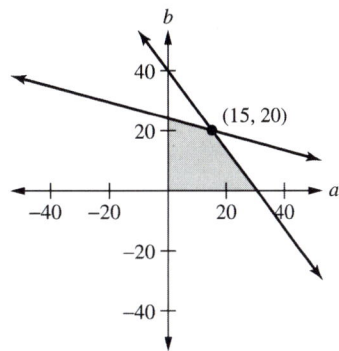

35. Let h be the number of hats.
    Let t be the number of T-shirts.

    Since the number of items cannot be negative, we will consider restrictions in the first quadrant given by the following.

    $$30h + 20t \leq 600$$

    The origin, $(0,0)$, the h-intercept, $(20,0)$, and the t-intercept, $(0,30)$, are all corner points.

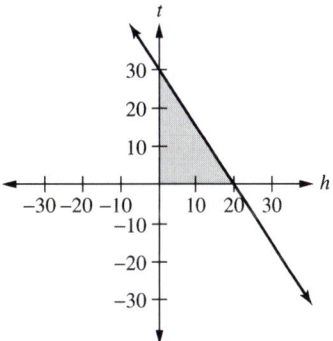

37. Let b be the amount invested in bonds in dollars.
    Let m be the amount invested in mutual funds in dollars.

    Since the amount invested cannot be negative, we will consider restrictions in the first quadrant given by the following.

    $$b + m \leq 18,000$$
    $$b \geq 2m$$

    The origin, $(0,0)$, and a b-intercept, $(18000,0)$, are corner points. To find the third corner point we solve the following system.

    $$b + m = 18,000$$
    $$b = 2m$$

    $2m + m = 18,000 \Rightarrow 3m = 18,000 \Rightarrow m = 6,000$

    $b = 2 \cdot 6,000 = 12,000$ The third corner point is $(12000, 6000)$.

37. (continued)

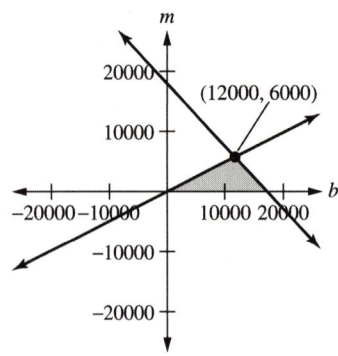

39. Since the amount of nutritional supplements cannot be negative, we will consider restrictions in the first quadrant given by the following table.

	PowerUp (p)	StressTabs (s)	Required	Inequalities
Niacin in mg	30	40	180	$30p + 40s \geq 180$
Vitamin C in mg	200	400	1,600	$200p + 400s \geq 1,600$

A p-intercept, $(8,0)$, and an s-intercept, $(0, 4\frac{1}{2})$, are both corner points. To find the third corner point we solve the following system.

$$30p + 40s = 180$$
$$200p + 400s = 1,600$$

$$\begin{array}{c}(-10)\cdot[30p+40s]=[180]\cdot(-10) \\ 200p+400s=1,600\end{array} \Rightarrow \begin{array}{c}-300p-400s=-1,800 \\ 200p+400s=1,600\end{array} \Rightarrow -100p=-200 \Rightarrow p=2$$

$30 \cdot 2 + 40s = 180 \Rightarrow 60 + 40s = 180 \Rightarrow 40s = 120 \Rightarrow s = 3$ The third corner point is $(2,3)$.

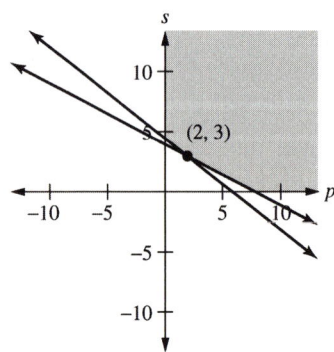

41. Let b be the number of bagels.
    Let c be the number of ounces of cream cheese.

    Since the number of bagels or amount of cream cheese cannot be negative, we will consider restrictions in the first quadrant given by the following.
    $$200b + 100c \leq 700$$
    $$b \geq 2$$

    Two b-intercepts, $(2,0)$ and $\left(3\tfrac{1}{2},0\right)$, are all corner points. To find the third corner point we solve the following system.
    $$200b + 100c = 700$$
    $$b = 2$$

    $200 \cdot 2 + 100c = 700 \Rightarrow 400 + 100c = 700 \Rightarrow 100c = 300 \Rightarrow c = 3$ The third corner point is $(2,3)$.

    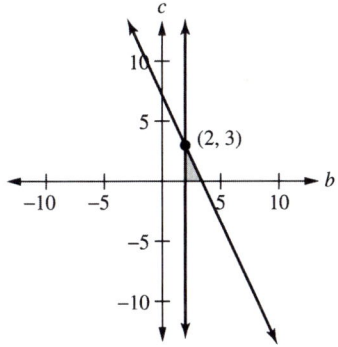

43. Let r be the amount spent on radio advertising in dollars.
    Let n be the amount spent on newspaper advertising in dollars.

    Since the amount spent on advertising cannot be negative, we will consider restrictions in the first quadrant given by the following.
    $$100r + 300n \leq 2,400$$
    $$r \geq 3n$$

    The origin, $(0,0)$, and an r-intercept, $(24,0)$, are corner points. To find the third corner point we solve the following system.
    $$100r + 300n = 2,400$$
    $$r = 3n$$

    $100 \cdot 3n + 300n = 2,400 \Rightarrow 300n + 300n = 2,400 \Rightarrow 600n = 2,400 \Rightarrow n = 4$

    $r = 3 \cdot 4 = 12$ The third corner point is $(12,4)$.

180 Chapter 7: Modeling with Systems of Equations and Inequalities

43. (continued)

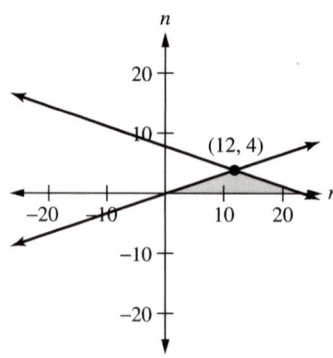

45. The constraints $x \geq 0$ and $y \geq 0$ imply that we are limited to quadrant I.
Let the equation that corresponds to the inequality $2x+3y \leq 25$ be ℓ_1.
Let the equation that corresponds to the inequality $5y \geq 20+x$ be ℓ_2.
Let the equation that corresponds to the inequality $y \leq x+5$ be ℓ_3.

Two of the corner points are y-intercepts, namely $(0,4)$ and $(0,5)$. To find the other two corner points we need to find the point of intersection for ℓ_1 and ℓ_2 as well as for ℓ_1 and ℓ_3.

ℓ_1 and ℓ_2:

$\begin{array}{l} 2x+3y=25 \\ 5y=20+x \end{array} \Rightarrow \begin{array}{l} 2x+3y=25 \\ 2\cdot[-x+5y]=[20]\cdot 2 \end{array} \Rightarrow \begin{array}{l} 2x+3y=25 \\ -2x+10y=40 \end{array} \Rightarrow 13y=65 \Rightarrow y=5$

$5\cdot 5 = 20+x \Rightarrow 25 = 20+x \Rightarrow 5 = x$ A corner point is $(5,5)$.

ℓ_1 and ℓ_3:

$\begin{array}{l} 2x+3y=25 \\ y=x+5 \end{array} \Rightarrow 2x+3(x+5)=25 \Rightarrow 2x+3x+15=25 \Rightarrow 5x+15=25 \Rightarrow 5x=10 \Rightarrow x=2$

$y = 2+5 = 7$ A corner point is $(2,7)$.

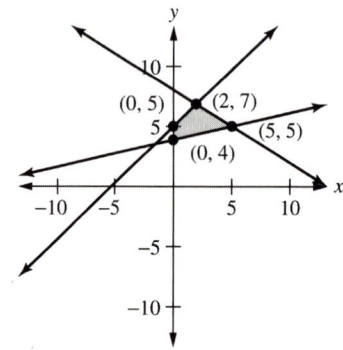

47. The constraints $x \geq 0$ and $y \geq 0$ imply that we are limited to quadrant I.

Let the equation that corresponds to the inequality $x - 2y \leq 8$ be ℓ_1.
Let the equation that corresponds to the inequality $2x + y \leq 19$ be ℓ_2.
Let the equation that corresponds to the inequality $x - y \geq 5$ be ℓ_3.
One corner point is the x-intercept of ℓ_3, $(5,0)$. Another is the x-intercept of ℓ_1, $(8,0)$. To find the other two corner points we need to find the point of intersection for ℓ_1 and ℓ_2 as well as for ℓ_2 and ℓ_3.

ℓ_1 and ℓ_2:
$$\begin{array}{l} x - 2y = 8 \\ 2x + y = 19 \end{array} \Rightarrow \begin{array}{l} x - 2y = 8 \\ 2\cdot[2x+y] = [19]\cdot 2 \end{array} \Rightarrow \begin{array}{l} x - 2y = 8 \\ 4x + 2y = 38 \end{array} \Rightarrow 5x = 46 \Rightarrow x = \frac{46}{5}$$

$\frac{46}{5} - 2y = 8 \Rightarrow -2y = 8 - \frac{46}{5} = \frac{40}{5} - \frac{46}{5} = -\frac{6}{5} \Rightarrow y = \frac{3}{5}$ A corner point is $\left(\frac{46}{5}, \frac{3}{5}\right)$.

ℓ_2 and ℓ_3:
$$\begin{array}{l} 2x + y = 19 \\ x - y = 5 \end{array} \Rightarrow 3x = 24 \Rightarrow x = 8$$

$8 - y = 5 \Rightarrow -y = -3 \Rightarrow y = 3$ A corner point is $(8, 3)$.

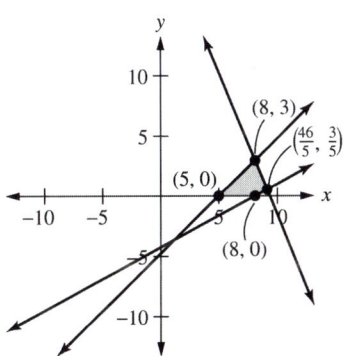

For Exercises 49 and 51, we consider how the origin $(0,0)$ relates in each of the equations. We will replace the blank with either $\leq$ or $\geq$, depending on whether we want the statement to be true or false.

Case A: For $3y = 43 - 5x$ we have $3\cdot 0 \overset{?}{\_} 43 - 5\cdot 0$ or $0 \overset{?}{\_} 43 - 0$ or $0 \overset{?}{\_} 43$.

Case B: For $11y = 27 - 2x$ we have $11\cdot 0 \overset{?}{\_} 27 - 2\cdot 0$ or $0 \overset{?}{\_} 27 - 0$ or $0 \overset{?}{\_} 27$.

Case C: For $3y = 1 + 2x$ we have $3\cdot 0 \overset{?}{\_} 1 + 2\cdot 0$ or $0 \overset{?}{\_} 1 + 0$ or $0 \overset{?}{\_} 1$.

49. In Region 1, when we look at the location of the origin, $(0,0)$, Case A should be true, Case B should be false and Case C should be false. This yields the following set of inequalities.

$$3y \leq 43 - 5x$$
$$11y \geq 27 - 2x$$
$$3y \geq 1 + 2x$$

51. In Region 3, when we look at the location of the origin, $(0,0)$, Case A should be true, Case B should be false and Case C should be true. This yields the following set of inequalities.

$$3y \leq 43 - 5x$$
$$11y \geq 27 - 2x$$
$$3y \leq 1 + 2x$$

Chapter Review Exercises

1. $(3,-5)$

$$4x-3y=27$$
$$(-4)\cdot[x+2y]=[-7]\cdot(-4)$$

$$\begin{aligned} 4x-3y &= 27 \\ -4x-8y &= 28 \\ \hline 0-11y &= 55 \\ y &= -5 \end{aligned}$$

$$\begin{aligned} x+2\cdot(-5) &= -7 \\ x-10 &= -7 \\ x &= 3 \end{aligned}$$

2. a) Case 2; no solutions
   b) Case 3; infinite number of solutions
   c) Case 1; one, unique solution

3. Let u be the number of points scored by the U.S. team.
   Let p be the number of points by the opponent.

$$u+p=123$$
$$u=p+17$$

$$\begin{aligned} u+p &= 123 \\ u-p &= 17 \\ \hline 2u+0 &= 140 \\ u &= 70 \end{aligned}$$

$$\begin{aligned} 70+p &= 123 \\ p &= 53 \end{aligned}$$

U.S.: 70; Opponent: 53

4. Let f be the number of hours worked at the fast food restaurant.
   Let p be the number of hours worked as a personal trainer.

$$f = p+7$$
$$5.85f + 15p = 145.20$$

$$\begin{array}{c} f = p+7 \\ 100\cdot[5.85f+15p] = [145.20]\cdot 100 \end{array} \Rightarrow \begin{array}{c} f = p+7 \\ 585f + 1,500p = 14,520 \end{array}$$

Using substitution, $585(p+7)+1,500p=14,520 \Rightarrow 585p+4,095+1,500p=14,520 \Rightarrow$
$2,085p+4,095=14,520 \Rightarrow 2,085p=10,425 \Rightarrow p=5$

Aaron worked 5 hours as a personal trainer and $f = 5+7 = 12$ hours at the fast food restaurant.

5. $6\cdot 0 + 5\cdot 0 \overset{?}{>} 20 \Rightarrow 0+0 \overset{?}{>} 20 \Rightarrow 0 > 20$; False. Shade the side that does not contain $(0,0)$.

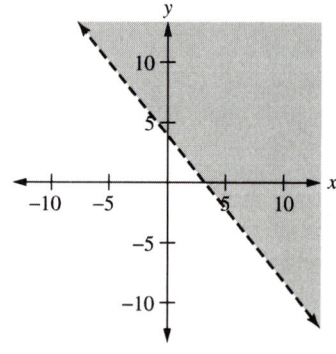

6. Testing $(0,0)$ in the first inequality, we get $2\cdot 0+5\cdot 0\overset{?}{\geq} 24 \Rightarrow 0+0\overset{?}{\geq} 24 \Rightarrow 0\geq 24$ which is false, so $(0,0)$ is not a solution. Testing $(0,0)$ in the second inequality, we get $2\cdot 0\overset{?}{\leq} 2\cdot 0+18 \Rightarrow 0\overset{?}{\leq} 0+18 \Rightarrow 0\leq 18$ which is true, so $(0,0)$ is a solution. To solve for the corner point, we solve the system.

$$2x+5y=24$$
$$2y=2x+18$$

$$2x+5y=24$$
$$-2x+2y=18$$
$$\overline{}$$
$$0+7y=42$$
$$y=6$$

$$2x+5\cdot 6=24 \Rightarrow 2x+30=24 \Rightarrow$$
$$2x=-6 \Rightarrow x=-3$$

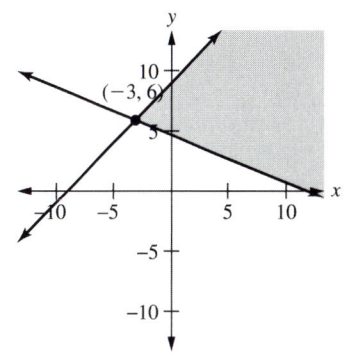

Thus the corner point is $(-3,6)$.

7. Let t be the number of T-shirts.
Let h be the number of hats.

Since the number of hats and T-shirts cannot be negative, we will consider restrictions in the first quadrant given by the following.

$$4t+2h\leq 180$$
$$h\geq t+15$$

The h-intercepts, $(0,90)$ and $(0,15)$, are corner points. To find the third corner point we solve the following system.

$$4t+2h=180$$
$$h=t+15$$

$$4t+2(t+15)=180$$
$$4t+2t+30=180$$
$$6t+30=180$$
$$6t=150$$
$$t=25$$

$$h=25+15=40$$

The third corner point is $(25,40)$.

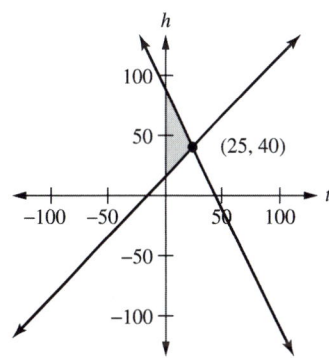

Chapter Test

1. $(2,-3)$

   $(3)\cdot[3x+4y]=[-6]\cdot(3)$
   $(4)\cdot[2x-3y]=[13]\cdot(4)$

   $9x+12y=-18$
   $8x-12y=52$
   ___
   $17x+0=34$
   $x=2$

   $3(2)+4y=-6$
   $6+4y=-6$
   $4y=-12$
   $y=-3$

2. a) Case 3; infinite number of solutions
   b) Case 1; one unique solution
   c) Case 2; no solution

3. Let c be the number of children.
   Let p be the number of adults.

   $c+p=76$
   $9c+18p=927$

   $(-9)\cdot[c+p]=76\cdot(-9)$
   $9c+18p=927$

   $-9c-9p=-684$
   $9c+18p=927$
   ___
   $9p=243$
   $p=27$

   $c+27=76$
   $c=49$

   49 children; 27 adults

4. Let f be the number of hours Shandra worked at the fast food restaurant.
   Let s be the number of hours she worked giving swimming lessons.

   $\begin{aligned}s&=f+5\\5.65f+8.50s&=212.30\end{aligned} \Rightarrow \begin{aligned}s&=f+5\\100\cdot[5.65f+8.50s]&=[212.30]\cdot100\end{aligned} \Rightarrow \begin{aligned}s&=f+5\\565f+850s&=21,230\end{aligned}$

   Using substitution, $565f+850(f+5)=21,230 \Rightarrow 565f+850f+4,250=21,230 \Rightarrow$

   $1,415f+4,250=21,230 \Rightarrow 1,415f=16,980 \Rightarrow f=12$

   Shandra worked 12 hours at the restaurant and $s=12+5=17$ hours giving swimming lessons.

5. $4\cdot0-3\cdot0\overset{?}{\geq}8 \Rightarrow 0-0\overset{?}{\geq}8 \Rightarrow 0\geq8$; False. Shade the side that does not contain $(0,0)$.

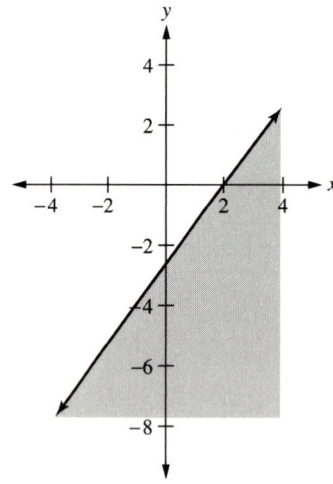

6. Testing $(0,0)$ in the first inequality yields $0 \stackrel{?}{\leq} -(0) + 14 \Rightarrow 0 \stackrel{?}{\leq} 14 \Rightarrow 0 \leq 14$ which is true, so $(0,0)$ is a solution. Testing $(0,0)$ in the second inequality yields $2 \cdot 0 - 5 \cdot 0 \stackrel{?}{\geq} -14 \Rightarrow 0 \stackrel{?}{\geq} -14 \Rightarrow 0 \geq 14$ which is true, so $(0,0)$ is a solution. Testing $(0,0)$ in the third inequality yields $5 \cdot 0 - 2 \cdot 0 \stackrel{?}{\geq} 7 \Rightarrow 0 \stackrel{?}{\geq} 7 \Rightarrow 0 \geq 7$ which is false, so $(0,0)$ is not a solution. To solve for the corner point, we solve the following systems.

First and second

$-2 \cdot [x+y] = 14 \cdot -2$
$2x - 5y = -14$

$-2x - 2y = -28$
$2x - 5y = -14$
───────────
$-7y = -42$
$y = 6$

$x + 6 = 14$
$x = 8$

Corner point is (6,8)

First and third

$2 \cdot [x+y] = 14 \cdot 2$
$5x - 2y = 7$

$2x + 2y = 28$
$5x - 2y = 7$
───────────
$7x = 35$
$x = 5$

$5 + y = 14$
$y = 9$

Corner point is (5,9)

Second and third

$-2 \cdot [2x - 5y] = -14 \cdot -2$
$5 \cdot [5x - 2y] = 7 \cdot 5$

$-4x + 10y = 28$
$25x - 10y = 35$
───────────
$21x + 0 = 63$
$x = 3$

$5 \cdot 3 - 2y = 7$
$15 - 2y = 7$
$-2y = -8$
$y = 4$

Corner point is (3,4)

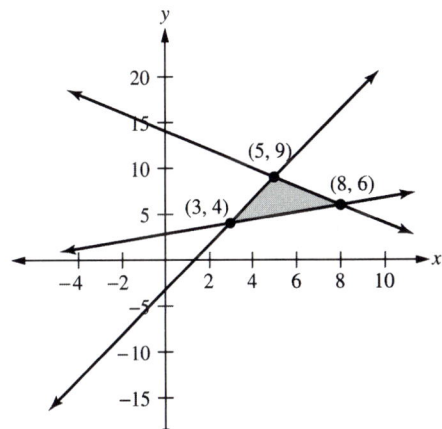

7. Let s be the number of small boxes.
   Let l be the number of large boxes.

   Since the number of boxes cannot be negative, we will consider restrictions in the first quadrant given by the following.

   $2s + 6l \leq 90$ (Amount of wood)
   $s + 2l \leq 40$ (Amount of time)

   The points, $(40, 0)$ and $(0, 15)$, are corner points. To find the third corner point we solve the following system.

   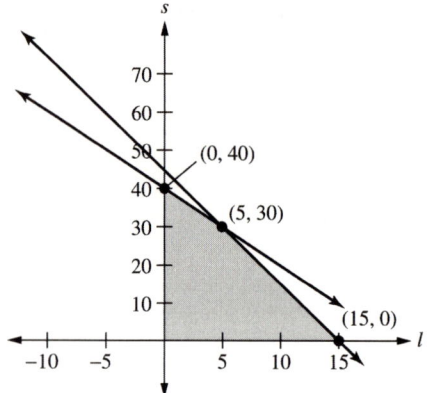

   $2s + 6l = 90$
   $-2 \cdot [s + 2l] = 40 \cdot -2$

   $\begin{aligned} 2s + 6l &= 90 \\ -2s - 4l &= -80 \\ \hline 2l &= 10 \\ l &= 5 \end{aligned}$ $\begin{aligned} s + 2 \cdot 5 &= 40 \\ s + 10 &= 40 \\ s &= 30 \end{aligned}$

   The third corner point is $(30, 5)$.

Of Further Interest: Linear Programming

1. In Example 1, the maximum value of the quantity $2x - 5y$ within the solution set for the system of inequalities is being found.

2. The maximum and minimum values of the objective function are located at the corner points.

3. The constraints are on copper, $2s + 16c \leq 160$, and on labor, $1s + 3c \leq 40$.

4. The variables, the constraints, and the objective function.

5. By solving the system of equations whose intersection creates each corner.

6. First, graph the system of constraints, second, find the corner points; third, evaluate the objective function at the corner points.

7. The origin, $(0,0)$, is a corner point. Another is the y-intercept of $x+2y=16$, namely $(0,8)$. Another is the x-intercept of $4x+y=24$, namely $(6,0)$.

A fourth corner point can be found by finding the point of intersection between $x+2y=16$ and $2x+y=14$.

$$\begin{array}{c} x+2y=16 \\ (-2)\cdot[2x+y]=[14]\cdot(-2) \end{array} \Rightarrow \begin{array}{c} x+2y=16 \\ -4x-2y=-28 \end{array} \Rightarrow -3x=-12 \Rightarrow x=4$$

$$2\cdot 4+y=14 \Rightarrow 8+y=14 \Rightarrow y=6$$

This corner point is $(4,6)$.

A fifth corner point can be found by finding the point of intersection between $4x+y=24$ and $2x+y=14$.

$$\begin{array}{c} 4x+y=24 \\ (-1)\cdot[2x+y]=[14]\cdot(-1) \end{array} \Rightarrow \begin{array}{c} 4x+y=24 \\ -2x-y=-14 \end{array} \Rightarrow 2x=10 \Rightarrow x=5$$

$$2\cdot 5+y=14 \Rightarrow 10+y=14 \Rightarrow y=4$$

This corner point is $(5,4)$.

8. The origin, $(0,0)$, is a corner point. Another is the y-intercept of $5y=20+x$, namely $(0,4)$. Another is the x-intercept of $y=x-5$, namely $(5,0)$.

A fourth corner point can be found by finding the point of intersection between $2x+3y=25$ and $5y=20+x$.

$$\begin{array}{c} 2x+3y=25 \\ -x+5y=20 \end{array} \Rightarrow \begin{array}{c} x+2y=16 \\ (-2)\cdot[2x+y]=[14]\cdot(-2) \end{array} \Rightarrow \begin{array}{c} 2x+3y=25 \\ 2\cdot[-x+5y]=[20]\cdot 2 \end{array} \Rightarrow 13y=65 \Rightarrow y=5$$

$$-x+5\cdot 5=20 \Rightarrow -x+25=20 \Rightarrow -x=-5 \Rightarrow x=5$$

This corner point is $(5,5)$.

A fifth corner point can be found by finding the point of intersection between $2x+3y=25$ and $y=x-5$.

$$2x+3(x-5)=25 \Rightarrow 2x+3x-15=25 \Rightarrow 5x-15=25 \Rightarrow 5x=40 \Rightarrow x=8$$

$$y=8-5=3$$

This corner point is $(8,3)$.

188   Chapter 7: Modeling with Systems of Equations and Inequalities

9. The origin, $(0,0)$, is a corner point. Another is the y-intercept of $x+4y=24$, namely $(0,6)$. Another is the x-intercept of $3x+2y=22$, namely $\left(\frac{22}{3},0\right)$.

A fourth corner point can be found by finding the point of intersection between $x+y=8$ and $x+4y=24$.

$$\begin{matrix} x+y=8 \\ (-1)[x+4y]=[24]\cdot(-1) \end{matrix} \Rightarrow \begin{matrix} x+y=8 \\ -x-4y=-24 \end{matrix} \Rightarrow -3y=-16 \Rightarrow y=\frac{16}{3}$$

$$x+\tfrac{16}{3}=8 \Rightarrow x=8-\tfrac{16}{3}=\tfrac{24}{3}-\tfrac{16}{3}=\tfrac{8}{3}$$

This corner point is $\left(\frac{8}{3},\frac{16}{3}\right)$.

A fifth corner point can be found by finding the point of intersection between $x+y=8$ and $3x+2y=22$.

$$\begin{matrix} (-2)\cdot[x+y]=[8]\cdot(-2) \\ 3x+2y=22 \end{matrix} \Rightarrow \begin{matrix} -2x-2y=-16 \\ 3x+2y=22 \end{matrix} \Rightarrow x=6$$

$$6+y=8 \Rightarrow y=2$$

This corner point is $(6,2)$.

10. A corner point can be found by finding the point of intersection between $6y=14+5x$ and $2y=23-2x$.

$$\begin{matrix} -5x+6y=14 \\ 2x+2y=23 \end{matrix} \Rightarrow \begin{matrix} -5x+6y=14 \\ (-3)\cdot[2x+2y]=[23]\cdot(-3) \end{matrix} \Rightarrow \begin{matrix} -5x+6y=14 \\ -6x-6y=-69 \end{matrix} \Rightarrow -11x=-55 \Rightarrow x=5$$

$$2y=23-2\cdot5 \Rightarrow 2y=23-10 \Rightarrow 2y=13 \Rightarrow y=\tfrac{13}{2}$$

This corner point is $\left(5,\frac{13}{2}\right)$.

A second corner point can be found by finding the point of intersection between $6y=14+5x$ and $13y=56-2x$.

$$\begin{matrix} -5x+6y=14 \\ 2x+13y=56 \end{matrix} \Rightarrow \begin{matrix} 2\cdot[-5x+6y]=[14]\cdot2 \\ 5\cdot[2x+13y]=[56]\cdot5 \end{matrix} \Rightarrow \begin{matrix} -10x+12y=28 \\ 10x+65y=280 \end{matrix} \Rightarrow 77y=308 \Rightarrow y=4$$

$$2x+13\cdot4=56 \Rightarrow 2x+52=56 \Rightarrow 2x=4 \Rightarrow x=2$$

This corner point is $(2,4)$.

A third corner point can be found by finding the point of intersection between $13y=56-2x$ and $2y=23-2x$.

$$\begin{matrix} 2x+13y=56 \\ 2x+2y=23 \end{matrix} \Rightarrow \begin{matrix} (-1)\cdot[2x+13y]=[56]\cdot(-1) \\ 2x+2y=23 \end{matrix} \Rightarrow \begin{matrix} -2x-13y=-56 \\ 2x+2y=23 \end{matrix} \Rightarrow -11y=-33 \Rightarrow y=3$$

$$2x+2\cdot3=23 \Rightarrow 2x+6=23 \Rightarrow 2x=17 \Rightarrow x=\tfrac{17}{2}$$

This corner point is $\left(\frac{17}{2},3\right)$.

11. The origin, $(0,0)$, is a corner point. Another is the y-intercept of $x+2y=16$, namely $(0,8)$. Another is the x-intercept of $2x+y=16$, namely $(8,0)$.

A fourth corner point can be found by finding the point of intersection between $x+y=10$ and $x+2y=16$.

$$\begin{array}{c} x+y=10 \\ (-1)[x+2y]=[16]\cdot(-1) \end{array} \Rightarrow \begin{array}{c} x+y=10 \\ -x-2y=-16 \end{array} \Rightarrow -y=-6 \Rightarrow y=6$$

$$x+6=10 \Rightarrow x=4$$

This corner point is $(4,6)$.

A fifth corner point can be found by finding the point of intersection between $x+y=10$ and $2x+y=16$.

$$\begin{array}{c} x+y=10 \\ (-1)[2x+y]=[16]\cdot(-1) \end{array} \Rightarrow \begin{array}{c} x+y=10 \\ -2x-y=-16 \end{array} \Rightarrow -x=-6 \Rightarrow x=6$$

$$6+y=10 \Rightarrow y=4$$

This corner point is $(6,4)$.

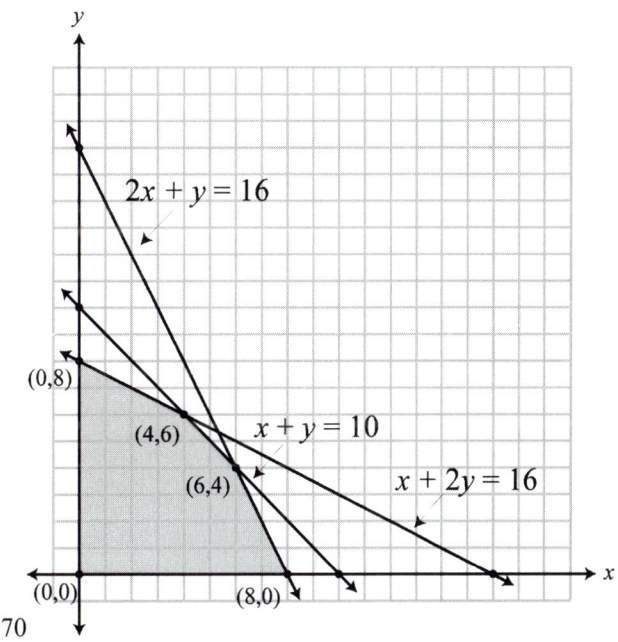

Corner point	Value of $P=3x+4y$	
$(0,0)$	$3\cdot 0+4\cdot 0=0+0=0$	
$(0,8)$	$3\cdot 0+4\cdot 8=0+32=32$	
$(8,0)$	$3\cdot 8+4\cdot 0=24+0=24$	
$(4,6)$	$3\cdot 4+4\cdot 6=12+24=36$	⇐ Maximum
$(6,4)$	$3\cdot 6+4\cdot 4=18+16=34$	

Maximum is 36 at $(4,6)$.

12. The origin, $(0,0)$, is a corner point. Another is the y-intercept of $x-3y=-9$, namely $(0,3)$. Another is the x-intercept of $3x+y=18$, namely $(6,0)$.

A fourth corner point can be found by finding the point of intersection between $x+2y=11$ and $x-3y=-9$.

$$\begin{array}{c} x+2y=11 \\ (-1)[x-3y]=[-9]\cdot(-1) \end{array} \Rightarrow \begin{array}{c} x+2y=11 \\ -x+3y=9 \end{array} \Rightarrow 5y=20 \Rightarrow y=4$$

$$x-3\cdot 4=-9 \Rightarrow x-12=-9 \Rightarrow x=3$$

This corner point is $(3,4)$.

A fifth corner point can be found by finding the point of intersection between $x+2y=11$ and $3x+y=18$.

$$\begin{array}{c} x+2y=11 \\ (-2)[3x+y]=[18]\cdot(-2) \end{array} \Rightarrow \begin{array}{c} x+2y=11 \\ -6x-2y=-36 \end{array} \Rightarrow -5x=-25 \Rightarrow x=5$$

$$3\cdot 5+y=18 \Rightarrow 15+y=18 \Rightarrow y=3$$

This corner point is $(5,3)$.

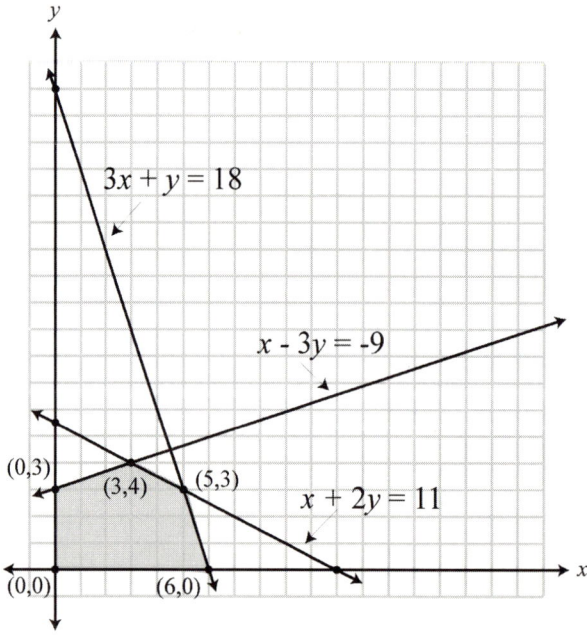

Corner point	Value of $P=x+y$	
$(0,0)$	$0+0=0$	
$(0,3)$	$0+3=3$	
$(6,0)$	$6+0=6$	
$(3,4)$	$3+4=7$	
$(5,3)$	$5+3=8$	⇐ Maximum

Maximum is 8 at $(5,3)$.

13. The origin, $(0,0)$, is a corner point. Another is the y-intercept of $x-6y=-18$, namely $(0,3)$. Another is the x-intercept of $x-3y=5$, namely $(5,0)$.

A fourth corner point can be found by finding the point of intersection between $3x+2y=26$ and $x-3y=5$.

$$\begin{array}{c} 3x+2y=26 \\ (-3)[x-3y]=[5]\cdot(-3) \end{array} \Rightarrow \begin{array}{c} 3x+2y=26 \\ -3x+9y=-15 \end{array} \Rightarrow 11y=11 \Rightarrow y=1$$

$$x-3\cdot 1=5 \Rightarrow x-3=5 \Rightarrow x=8$$

This corner point is $(8,1)$.

A fifth corner point can be found by finding the point of intersection between $3x+2y=26$ and $x-6y=-18$.

$$\begin{array}{c} 3x+2y=26 \\ (-3)\cdot[x-6y]=[-18]\cdot(-3) \end{array} \Rightarrow \begin{array}{c} 3x+2y=26 \\ -3x+18y=54 \end{array} \Rightarrow 20y=80 \Rightarrow y=4$$

$$x-6\cdot 4=-18 \Rightarrow x-24=-18 \Rightarrow x=6$$

This corner point is $(6,4)$.

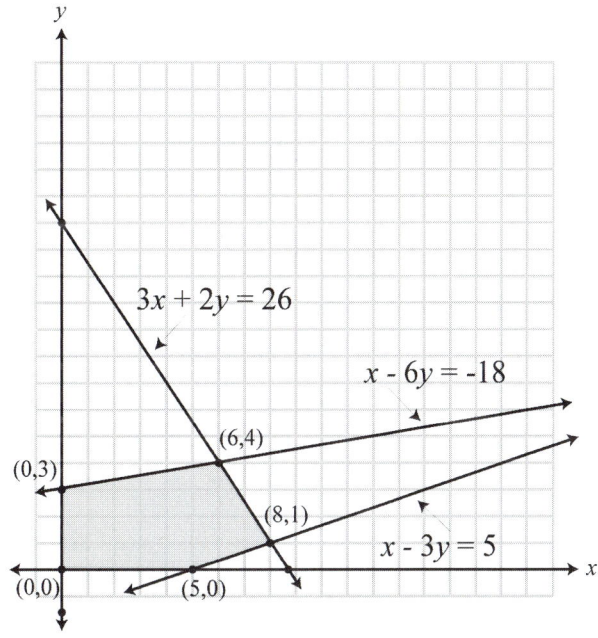

Corner point	Value of $P=5x-y$	
$(0,0)$	$5\cdot 0-0=0-0=0$	
$(0,3)$	$5\cdot 0-3=0-3=-3$	$\Leftarrow$ Minimum
$(5,0)$	$5\cdot 5-0=25-0=25$	
$(8,1)$	$5\cdot 8-1=40-1=39$	
$(6,4)$	$5\cdot 6-4=30-4=26$	

Minimum is -3 at $(0,3)$.

192   Chapter 7: Modeling with Systems of Equations and Inequalities

14. The origin, $(0,0)$, is a corner point. Another is the y-intercept of $x+5y=40$, namely $(0,8)$. Another is the x-intercept of $4x+y=32$, namely $(8,0)$.

A fourth corner point can be found by finding the point of intersection between $3x+2y=29$ and $4x+y=32$.

$$\begin{array}{c} 3x+2y=29 \\ (-2)[4x+y]=[32]\cdot(-2) \end{array} \Rightarrow \begin{array}{c} 3x+2y=29 \\ -8x-2y=-64 \end{array} \Rightarrow -5x=-35 \Rightarrow x=7$$

$$4\cdot 7+y=32 \Rightarrow 28+y=32 \Rightarrow y=4$$

This corner point is $(7,4)$.

A fifth corner point can be found by finding the point of intersection between $3x+2y=29$ and $x+5y=40$.

$$\begin{array}{c} 3x+2y=29 \\ (-3)\cdot[x+5y]=[40]\cdot(-3) \end{array} \Rightarrow \begin{array}{c} 3x+2y=29 \\ -3x-15y=-120 \end{array} \Rightarrow -13y=-91 \Rightarrow y=7$$

$$x+5\cdot 7=40 \Rightarrow x+35=40 \Rightarrow x=5$$

This corner point is $(5,7)$.

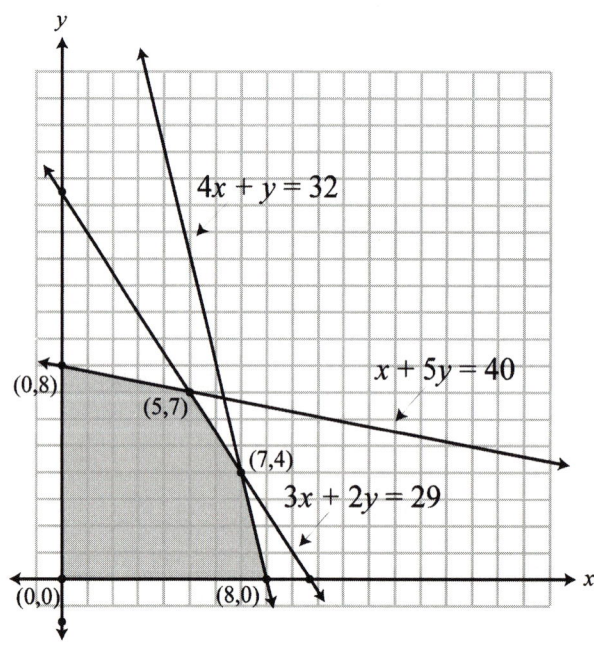

Corner point	Value of $P=x+3y$	
$(0,0)$	$0+3\cdot 0=0+0=0$	$\Leftarrow$ Minimum
$(0,8)$	$0+3\cdot 8=0+24=24$	
$(8,0)$	$8+3\cdot 0=8+0=8$	
$(7,4)$	$7+3\cdot 4=7+12=19$	
$(5,7)$	$5+3\cdot 7=5+21=26$	

Minimum is 0 at $(0,0)$.

15. A corner point can be found by finding the point of intersection between $y=3$ and $x-y=-1$.

$$x-3=-1 \Rightarrow x=2$$

This corner point is $(2,3)$.

A second corner point can be found by finding the point of intersection between $y=3$ and $2x+y=21$.

$$2x+3=21 \Rightarrow 2x=18 \Rightarrow x=9$$

This corner point is $(9,3)$.

A third corner point can be found by finding the point of intersection between $y=5$ and $x-y=-1$.

$$x-5=-1 \Rightarrow x=4$$

This corner point is $(4,5)$.

A fourth corner point can be found by finding the point of intersection between $y=5$ and $2x+y=21$.

$$2x+5=21 \Rightarrow 2x=16 \Rightarrow x=8$$

This corner point is $(8,5)$.

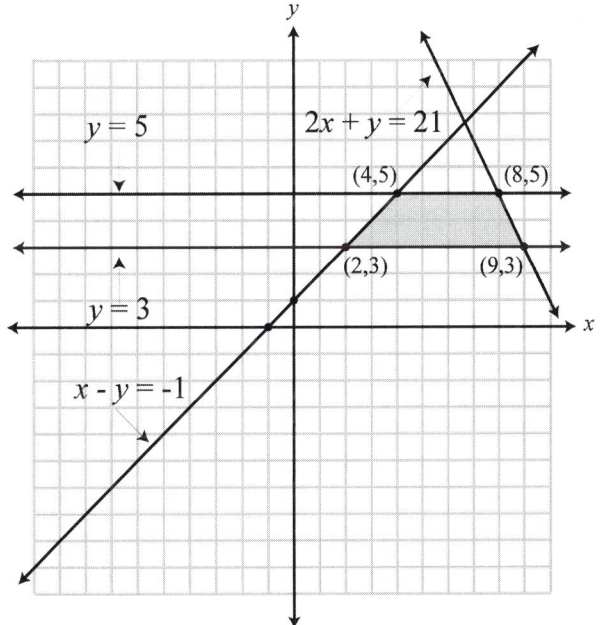

Corner point	Value of $P=20x-12y$	
$(2,3)$	$20 \cdot 2 - 12 \cdot 3 = 40 - 36 = 4$	$\Leftarrow$ Minimum
$(9,3)$	$20 \cdot 9 - 12 \cdot 3 = 180 - 36 = 144$	
$(4,5)$	$20 \cdot 4 - 12 \cdot 5 = 80 - 60 = 20$	
$(8,5)$	$20 \cdot 8 - 12 \cdot 5 = 160 - 60 = 100$	

Minimum is 4 at $(2,3)$.

16. A corner point can be found by finding the point of intersection between $x=6$ and $x-y=-2$.

$$6-y=-2 \Rightarrow -y=-8 \Rightarrow y=8$$

This corner point is $(6,8)$.

A second corner point can be found by finding the point of intersection between $x=6$ and $x+4y=14$.

$$6+4y=14 \Rightarrow 4y=8 \Rightarrow y=2$$

This corner point is $(6,2)$.

A third point can be found by finding the point of intersection between $x=2$ and $x-y=-2$.

$$2-y=-2 \Rightarrow -y=-4 \Rightarrow y=4$$

This corner point is $(2,4)$.

A fourth point can be found by finding the point of intersection between $x=2$ and $x+4y=14$.

$$2+4y=14 \Rightarrow 4y=12 \Rightarrow y=3$$

This corner point is $(2,3)$.

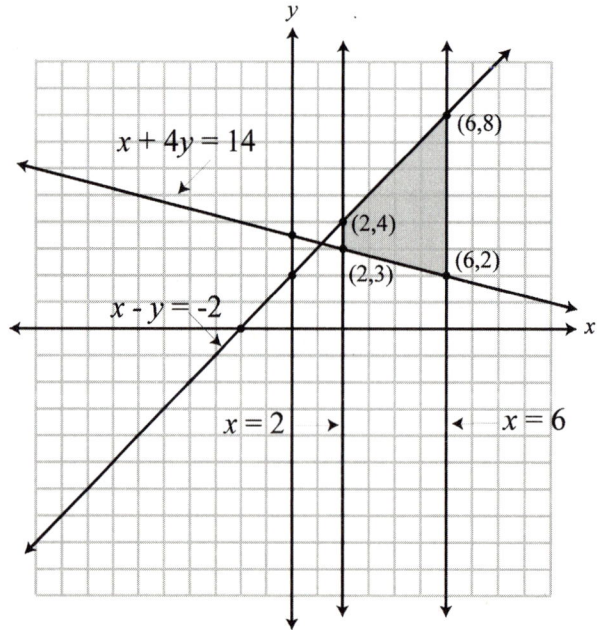

Corner point	Value of $P=10x-6y$	
$(6,8)$	$10\cdot 6 - 6\cdot 8 = 60 - 48 = 12$	
$(6,2)$	$10\cdot 6 - 6\cdot 2 = 60 - 12 = 48$	⇐ Maximum
$(2,4)$	$10\cdot 2 - 6\cdot 4 = 20 - 24 = -4$	
$(2,3)$	$10\cdot 2 - 6\cdot 3 = 20 - 18 = 2$	

Maximum is 48 at $(6,2)$.

17. The corner point can be found by finding the point of intersection between $x - 2y = -2$ and $x + y = 4$.

$$\begin{array}{c} x - 2y = -2 \\ 2 \cdot [x + y] = [4] \cdot 2 \end{array} \Rightarrow \begin{array}{c} x - 2y = -2 \\ 2x + 2y = 8 \end{array} \Rightarrow 3x = 6 \Rightarrow x = 2$$

$$2 + y = 4 \Rightarrow y = 2$$

The corner point is $(2, 2)$.

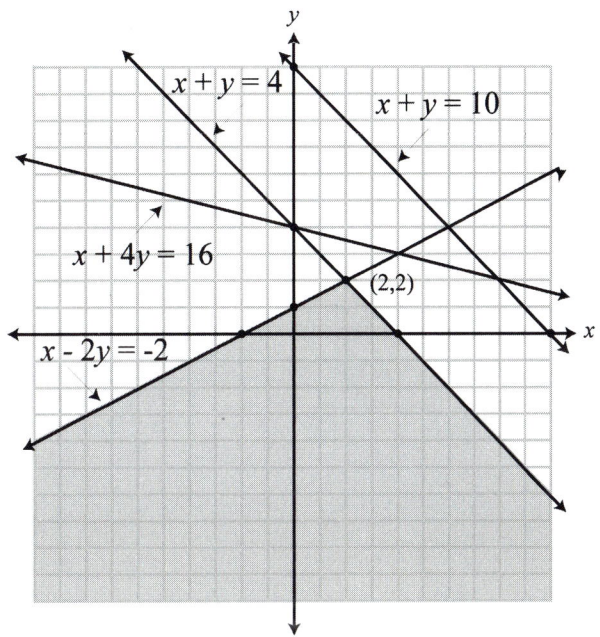

Corner point	Value of $P = 5x + 12y$	
$(2, 2)$	$5 \cdot 2 + 12 \cdot 2 = 10 + 24 = 34$	$\Leftarrow$ Maximum

Maximum is 34 at $(2, 2)$.

18. A corner point is the x-intercept of $3x - y = 12$, namely $(4, 0)$. Another is the x-intercept of $3x + y = 30$, namely $(10, 0)$.

A third corner point can be found by finding the point of intersection between $x - y = 2$ and $3x - y = 12$.

$$\begin{array}{c} (-1)[x - y] = [2] \cdot (-1) \\ 3x - y = 12 \end{array} \Rightarrow \begin{array}{c} -x + y = -2 \\ 3x - y = 12 \end{array} \Rightarrow 2x = 10 \Rightarrow x = 5$$

$$5 - y = 2 \Rightarrow -y = -3 \Rightarrow y = 3$$

This corner point is $(5, 3)$.

A fourth corner point can be found by finding the point of intersection between $x - y = 2$ and $3x + y = 30$.

18. (continued)

$$x - y = 2$$
$$3x + y = 30$$
$$\Rightarrow 4x = 32 \Rightarrow x = 8$$

$$8 - y = 2 \Rightarrow -y = -6 \Rightarrow y = 6$$

This corner point is $(8, 6)$.

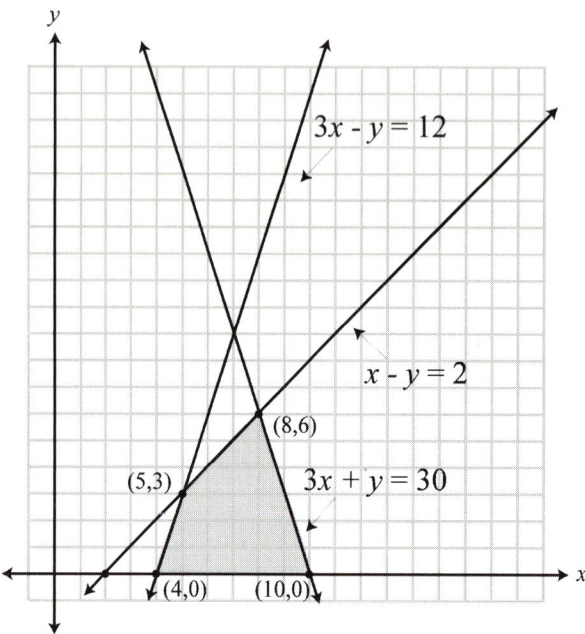

Corner point	Value of $P = x + y$	
$(4, 0)$	$4 + 0 = 4$	
$(10, 0)$	$10 + 0 = 10$	
$(5, 3)$	$5 + 3 = 8$	
$(8, 6)$	$8 + 6 = 14$	⇐ Maximum

Maximum is 14 at $(8, 6)$.

19. A corner point can be found by finding the point of intersection between $6x - 7y = -6$ and $3x + 4y = 42$.

$$\begin{array}{c} 6x - 7y = -6 \\ (-2) \cdot [3x + 4y] = [42] \cdot (-2) \end{array} \Rightarrow \begin{array}{c} 6x - 7y = -6 \\ -6x - 8y = -84 \end{array} \Rightarrow -15y = -90 \Rightarrow y = 6$$

$$3x + 4 \cdot 6 = 42 \Rightarrow 3x + 24 = 42 \Rightarrow 3x = 18 \Rightarrow x = 6$$

This corner point is $(6, 6)$.

19. (continued)

A second corner point can be found by finding the point of intersection between $6x-7y=-6$ and $6x-22y=-51$.

$$\begin{array}{c} 6x-7y=-6 \\ (-1)\cdot[6x-22y]=[-51]\cdot(-1) \end{array} \Rightarrow \begin{array}{c} 6x-7y=-6 \\ -6x+22y=51 \end{array} \Rightarrow 15y=45 \Rightarrow y=3$$

$$6x-7\cdot 3=-6 \Rightarrow 6x-21=-6 \Rightarrow 6x=15 \Rightarrow x=\tfrac{15}{6}=\tfrac{5}{2}$$

This corner point is $\left(\tfrac{5}{2},3\right)$.

A third corner point can be found by finding the point of intersection between $6x-22y=-51$ and $3x+4y=42$.

$$\begin{array}{c} 6x-22y=-51 \\ (-2)\cdot[3x+4y]=[42]\cdot(-2) \end{array} \Rightarrow \begin{array}{c} 6x-22y=-51 \\ -6x-8y=-84 \end{array} \Rightarrow -30y=-135 \Rightarrow y=\tfrac{135}{30}=\tfrac{9}{2}$$

$$3x+4\cdot\tfrac{9}{2}=42 \Rightarrow 3x+18=42 \Rightarrow 3x=24 \Rightarrow x=8$$

This corner point is $\left(8,\tfrac{9}{2}\right)$.

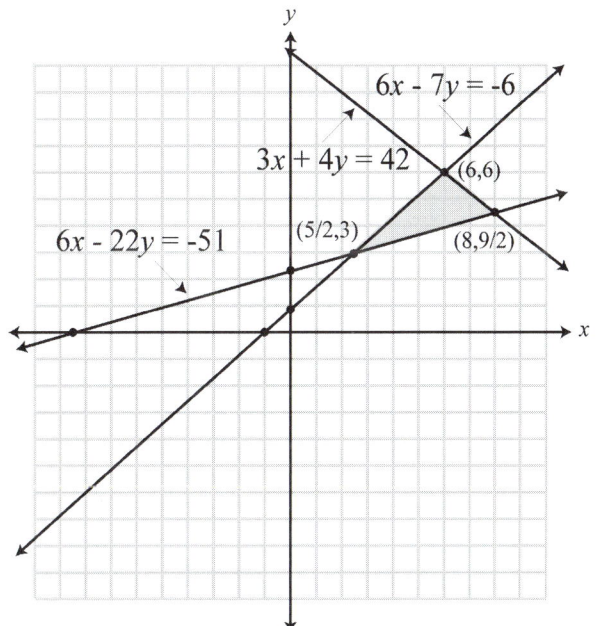

Corner point	Value of $P=4x+2y$	
$(6,6)$	$4\cdot 6+2\cdot 6=24+12=36$	
$\left(\tfrac{5}{2},3\right)$	$4\cdot\tfrac{5}{2}+2\cdot 3=10+6=16$	⇐ Minimum
$\left(8,\tfrac{9}{2}\right)$	$4\cdot 8+2\cdot\tfrac{9}{2}=32+9=41$	

Minimum is 16 at $\left(\tfrac{5}{2},3\right)$.

198   Chapter 7: Modeling with Systems of Equations and Inequalities

20. A corner point can be found by finding the point of intersection between $3x+5y=61$ and $x+y=13$.

$$\begin{array}{c} 3x+5y=61 \\ (-3)\cdot[x+y]=[13]\cdot(-3) \end{array} \Rightarrow \begin{array}{c} 3x+5y=61 \\ -3x-3y=-39 \end{array} \Rightarrow 2y=22 \Rightarrow y=11$$

$$x+11=13 \Rightarrow x=2$$

This corner point is $(2,11)$.

A second corner point can be found by finding the point of intersection between $2x+y=22$ and $x+y=13$.

$$\begin{array}{c} 2x+y=22 \\ (-1)\cdot[x+y]=[13]\cdot(-1) \end{array} \Rightarrow \begin{array}{c} 2x+y=22 \\ -x-y=-13 \end{array} \Rightarrow x=9$$

$$9+y=13 \Rightarrow y=4$$

This corner point is $(9,4)$.

A third corner point can be found by finding the point of intersection between $3x+5y=61$ and $2x+y=22$.

$$\begin{array}{c} 3x+5y=61 \\ (-5)\cdot[2x+y]=[22]\cdot(-5) \end{array} \Rightarrow \begin{array}{c} 3x+5y=61 \\ -10x-5y=-110 \end{array} \Rightarrow -7x=-49 \Rightarrow x=7$$

$$2\cdot 7+y=22 \Rightarrow 14+y=22 \Rightarrow y=8$$

This corner point is $(7,8)$.

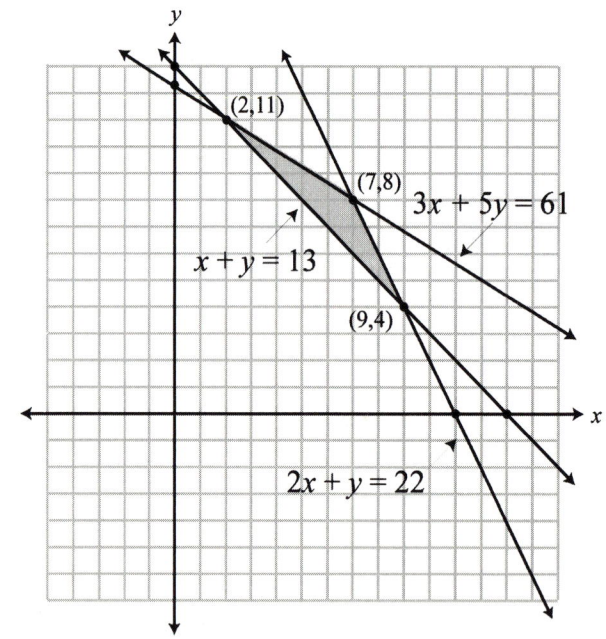

Corner point	Value of $P=3x+6y$	
$(2,11)$	$3\cdot 2+6\cdot 11=6+66=72$	
$(9,4)$	$3\cdot 9+6\cdot 4=27+24=51$	⇐ Minimum
$(7,8)$	$3\cdot 7+6\cdot 8=21+48=69$	

Minimum is 51 at $(9,4)$.

21. Let x be the number of Athens entertainment centers.
    Let y be the number of Barcelona entertainment centers.

    Since the number of entertainment centers cannot be negative, we will consider restrictions in the first quadrant given by the following table.

	Athens (x)	Barcelona (y)	Available	Inequalities
Fancy molding in feet	4	15	360	$4x+15y \leq 360$
Time in hours	4	3	120	$4x+3y \leq 120$

    The constraints-objective are therefore as follows.

    Maximize $P = 9x + 12y$ subject to the constraints

    $$4x + 15y \leq 360$$
    $$4x + 3y \leq 120$$
    $$x \geq 0, y \geq 0$$

    The origin, $(0,0)$, an x-intercept, $(30,0)$, and a y-intercept, $(0,24)$, are all corner points. To find the fourth corner point we solve the following system.

    $$4x + 15y = 360$$
    $$4x + 3y = 120$$

    $$\begin{array}{c} 4x+15y=360 \\ (-1)\cdot[4x+3y]=[120]\cdot(-1) \end{array} \Rightarrow \begin{array}{c} 4x+15y=360 \\ -4x-3y=-120 \end{array} \Rightarrow 12y = 240 \Rightarrow y = 20$$

    $4x + 3 \cdot 20 = 120 \Rightarrow 4x + 60 = 120 \Rightarrow 4x = 60 \Rightarrow x = 15$ The fourth corner point is $(15, 20)$

    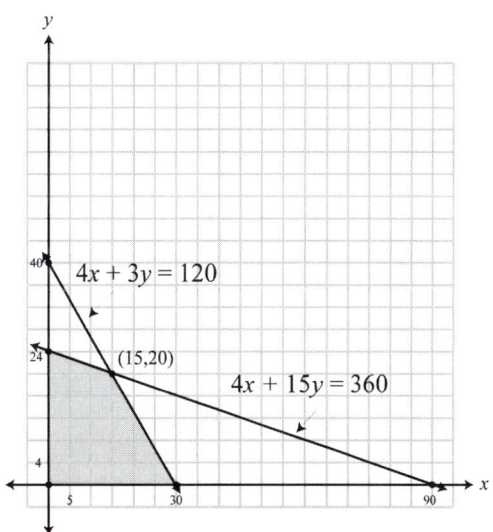

Corner point	Value of $P = 9x + 12y$	
$(0,0)$	$9 \cdot 0 + 12 \cdot 0 = 0 + 0 = 0$	
$(30,0)$	$9 \cdot 30 + 12 \cdot 0 = 270 + 0 = 270$	
$(0,24)$	$9 \cdot 0 + 12 \cdot 24 = 0 + 288 = 288$	
$(15,20)$	$9 \cdot 15 + 12 \cdot 20 = 135 + 240 = 375$	⇐ Maximum

    Scott should manufacture 15 Athens and 20 Barcelona entertainment centers to produce a maximum profit of $375.

22. Let x be the amount invested in pharmaceuticals in dollars.
    Let y be the amount invested in communications in dollars.

    The constraints-objective are therefore as follows.

    $$\text{Maximize } P = 0.075x + 0.125y \text{ subject to the constraints}$$

    $$x + y \leq 3,000$$
    $$x \geq 3y$$
    $$x \geq 0, y \geq 0$$

    The origin, $(0,0)$, and an x-intercept, $(3000,0)$, are corner points. To find the third corner point we solve the following system.

    $$x + y = 3,000$$
    $$x = 3y$$

    $$3y + y = 3,000 \Rightarrow 4y = 3,000 \Rightarrow y = 750$$

    $x = 3 \cdot 750 = 2,250$ The third corner point is $(2250, 750)$.

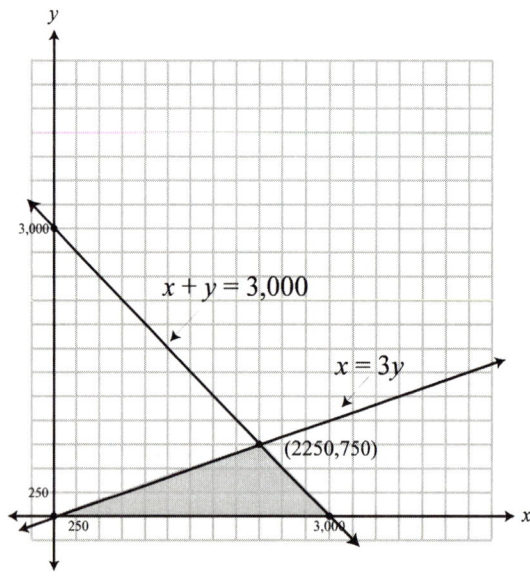

Corner point	Value of $P = 0.075x + 0.125y$	
$(0,0)$	$0.075 \cdot 0 + 0.125 \cdot 0 = 0 + 0 = 0$	
$(3000, 0)$	$0.075 \cdot 3000 + 0.125 \cdot 0 = 225 + 0 = 225$	
$(2250, 750)$	$0.075 \cdot 2250 + 0.125 \cdot 750 = 168.75 + 93.75 = 262.50$	⇐ Maximum

Gina should invest $2,250 in pharmaceuticals and $750 in communications to realize a maximum return on the investment of $262.50.

23. Let x be the number of hats.
    Let y be the number of T-shirts.

    The constraints-objective are therefore as follows.

    $$\text{Maximize } P = 4x + 3y \text{ subject to the constraints}$$

    $$10x + 9y \leq 140$$
    $$y \geq 2x$$
    $$x \geq 0, y \geq 0$$

The origin, $(0,0)$, and a y-intercept, $\left(0, \frac{140}{9}\right)$, are corner points. To find the third corner point we solve the following system.

$$10x + 9y = 140$$
$$y = 2x$$

$$10x + 9(2x) = 140 \Rightarrow 10x + 18x = 140 \Rightarrow 28x = 140 \Rightarrow x = 5$$

$y = 2 \cdot 5 = 10$ The third corner point is $(5, 10)$.

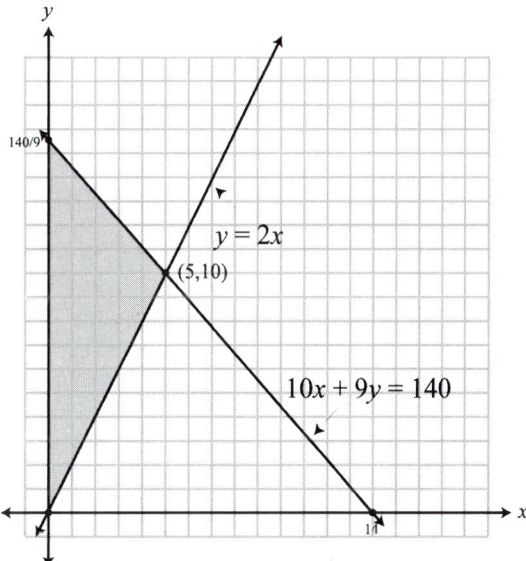

Corner point	Value of $P = 4x + 3y$	
$(0,0)$	$4 \cdot 0 + 3 \cdot 0 = 0 + 0 = 0$	
$\left(0, \frac{140}{9}\right)$	$4 \cdot 0 + 3 \cdot \frac{140}{9} = 0 + \frac{140}{3} = \frac{140}{3} \approx 46.67$	
$(5, 10)$	$4 \cdot 5 + 3 \cdot 10 = 20 + 30 = 50$	$\Leftarrow$ Maximum

Nicole should make 5 hats and 10 T-shirts for a maximum profit of $50.

Note: Although in this solution we considered a corner point of $\left(0, \frac{140}{9}\right)$, in a real-life situation we would only consider whole numbers.

24. Let x be the amount of time spent on plain text slide in minutes.
Let y be the amount of time spent on graphics slides in minutes.

The constraints-objective are therefore as follows.

$$\text{Maximize } P = x + 2.30y \text{ subject to the constraints}$$

$$10x + 15y \leq 350$$
$$x \geq y + 5$$
$$x \geq 0, y \geq 0$$

x-intercepts, $(5,0)$ and $(35,0)$, are corner points. To find the third corner point we solve the following system.

$$10x + 15y = 350$$
$$x = y + 5$$

$10(y+5) + 15y = 350 \Rightarrow 10y + 50 + 15y = 350 \Rightarrow 25y + 50 = 350 \Rightarrow 25y = 300 \Rightarrow y = 12$

$x = 12 + 5 = 17$ The third corner point is $(17, 12)$.

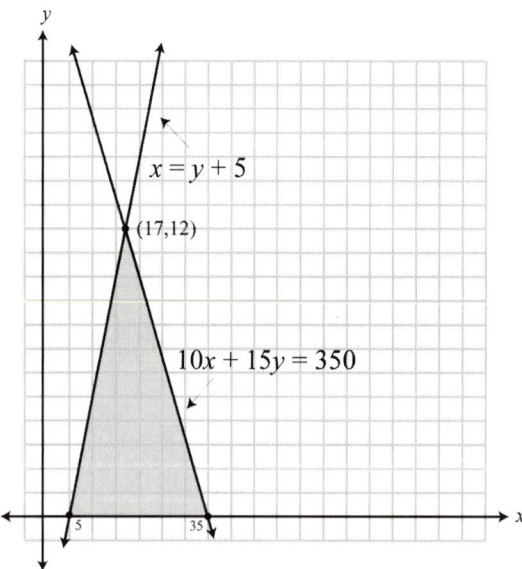

Corner point	Value of $P = x + 2.30y$	
$(5,0)$	$5 + 2.30 \cdot 0 = 5 + 0 = 5$	
$(35,0)$	$35 + 2.30 \cdot 0 = 35 + 0 = 35$	
$(17,12)$	$17 + 2.30 \cdot 12 = 17 + 27.60 = 44.60$	⇐ Maximum

Jaleel should make 17 plain slides and 12 graphics slides for a maximum profit of $44.60.

25. Let x be the number of PowerUp tablets.
Let y be the number of StressTabs tablets.

Since the amount of nutritional supplements cannot be negative, we will consider restrictions in the first quadrant given by the following table.

	PowerUp (x)	StressTabs (y)	Required	Inequalities
Niacin in mg	30	40	180	$30x + 40y \geq 180$
Vitamin C in mg	200	400	1,600	$200x + 400y \geq 1,600$

The constraints-objective are therefore as follows.

Minimize $P = 0.09x + 0.13y$ subject to the constraints

$$30x + 40y \geq 180$$
$$200x + 400y \geq 1,600$$
$$x + y \leq 8$$
$$x \geq 0, y \geq 0$$

An x-intercept, $(8,0)$, and two y-intercept, $\left(0, 4\frac{1}{2}\right)$ and $(0,8)$, are all corner points. To find the fourth corner point we solve the following system.

$$30x + 40y = 180$$
$$200x + 400y = 1,600$$

$\begin{aligned}(-10) \cdot [30x + 40y] &= [180] \cdot (-10) \\ 200x + 400y &= 1,600\end{aligned} \Rightarrow \begin{aligned}-300x - 400y &= -1,800 \\ 200x + 400y &= 1,600\end{aligned} \Rightarrow -100x = -200 \Rightarrow x = 2$

$30 \cdot 2 + 40y = 180 \Rightarrow 60 + 40y = 180 \Rightarrow 40y = 120 \Rightarrow y = 3$ The fourth corner point is $(2,3)$.

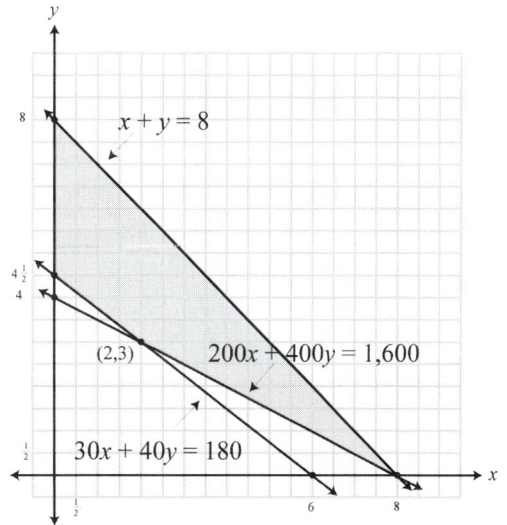

Corner point	Value of $P = 0.09x + 0.13y$	
$(8,0)$	$0.09 \cdot 8 + 0.13 \cdot 0 = 0.72 + 0 = 0.72$	
$\left(0, 4\frac{1}{2}\right)$	$0.09 \cdot 0 + 0.13 \cdot \frac{9}{2} = 0 + 0.585 \approx 0.59$	
$(0,8)$	$0.09 \cdot 0 + 0.13 \cdot 8 = 0 + 1.04 = 1.04$	
$(2,3)$	$0.09 \cdot 2 + 0.13 \cdot 3 = 0.18 + 0.39 = 0.57$	⇐ Minimum

Raphael should take 2 PowerUp tablets and 3 StressTabs for a minimum cost of $0.57.

26. Let x be the number of plates.
Let y be the number of cups.

Since the number of plates or cups cannot be negative, we will consider restrictions in the first quadrant, as well as the number of the two kinds of pieces being at most 28, and those given by the following table.

	Plates (x)	Cups (y)	Available	Inequalities
Time in minutes	10	20	$8 \cdot 60 = 480$	$10x + 20y \leq 480$
Clay in pounds	1	0.5	20	$x + 0.5y \leq 20$ or $10x + 5y \leq 200$

The constraints-objective are therefore as follows.

$$\text{Maximize } P = 12x + 8y \text{ subject to the constraints}$$

$$10x + 20y \leq 480$$
$$10x + 5y \leq 200$$
$$x + y \leq 28$$
$$x \geq 0, y \geq 0$$

The origin, $(0,0)$, an x-intercept, $(20,0)$, and a y-intercept, $(0,24)$, are all corner points.

To find the fourth corner point we solve the following system.

$$10x + 20y = 480$$
$$x + y = 28$$

$$\begin{array}{c} 10x + 20y = 480 \\ (-10) \cdot [x+y] = [28] \cdot (-10) \end{array} \Rightarrow \begin{array}{c} 10x + 20y = 480 \\ -10x - 10y = -280 \end{array} \Rightarrow 10y = 200 \Rightarrow y = 20$$

$x + 20 = 28 \Rightarrow x = 8$ The fourth corner point is $(8, 20)$.

To find the fifth corner point we solve the following system.

$$10x + 5y = 200$$
$$x + y = 28$$

$$\begin{array}{c} 10x + 5y = 200 \\ (-10) \cdot [x+y] = [28] \cdot (-10) \end{array} \Rightarrow \begin{array}{c} 10x + 5y = 200 \\ -10x - 10y = -280 \end{array} \Rightarrow -5y = -80 \Rightarrow y = 16$$

$x + 16 = 28 \Rightarrow x = 12$ The fifth corner point is $(12, 16)$.

26. (continued)

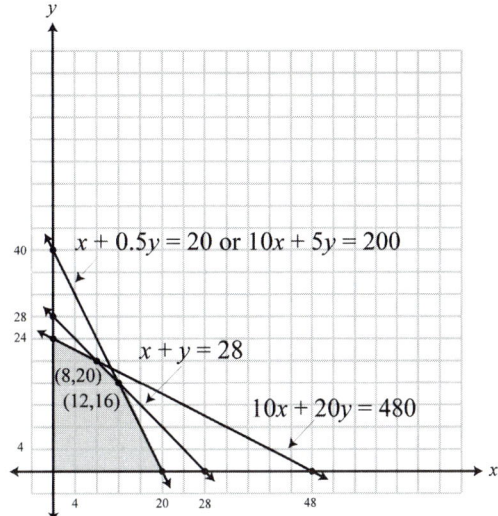

Corner point	Value of $P = 12x + 8y$	
$(0,0)$	$12 \cdot 0 + 8 \cdot 0 = 0 + 0 = 0$	
$(20,0)$	$12 \cdot 20 + 8 \cdot 0 = 240 + 0 = 240$	
$(0,24)$	$12 \cdot 0 + 8 \cdot 24 = 0 + 192 = 192$	
$(8,20)$	$12 \cdot 8 + 8 \cdot 20 = 96 + 160 = 256$	
$(12,16)$	$12 \cdot 12 + 8 \cdot 16 = 144 + 128 = 272$	⇐ Maximum

Joleen should make 12 plates and 16 cups for a maximum profit of $272.

27. Let x be the number of lizards.
Let y be the number of frogs.

Since the number of animals cannot be negative, we will consider restrictions in the first quadrant given by the following table.

	Lizards (x)	Frogs (y)	Available	Inequalities
Living space in square feet	2	2	400	$2x + 2y \leq 400$
Cost to feed in dollars	6	1	600	$6x + y \leq 600$

The constraints-objective are therefore as follows.

Maximize $P = 17x + 6y$ subject to the constraints

$$2x + 2y \leq 400$$
$$6x + y \leq 600$$
$$x \geq 0, y \geq 0$$

27. (continued)

The origin, $(0,0)$, an x-intercept, $(100,0)$, and a y-intercept, $(0,200)$, are all corner points. To find the fourth corner point we solve the following system.

$$2x + 2y = 400$$
$$6x + y = 600$$

$$\begin{array}{c} 2x + 2y = 400 \\ (-2) \cdot [6x + y] = [600] \cdot (-2) \end{array} \Rightarrow \begin{array}{c} 2x + 2y = 400 \\ -12x - 2y = -1,200 \end{array} \Rightarrow -10x = -800 \Rightarrow x = 80$$

$6 \cdot 80 + y = 600 \Rightarrow 480 + y = 600 \Rightarrow y = 120$ The fourth corner point is $(80, 120)$.

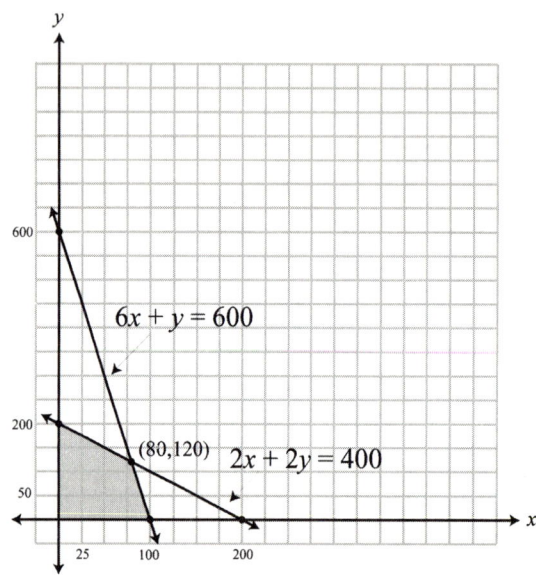

Corner point	Value of $P = 17x + 6y$	
$(0,0)$	$17 \cdot 0 + 6 \cdot 0 = 0 + 0 = 0$	
$(100,0)$	$17 \cdot 100 + 6 \cdot 0 = 1,700 + 0 = 1,700$	
$(0,200)$	$17 \cdot 0 + 6 \cdot 200 = 0 + 1,200 = 1,200$	
$(80,120)$	$17 \cdot 80 + 6 \cdot 120 = 1,360 + 720 = 2,080$	$\Leftarrow$ Maximum

The Reptile Farm should host 80 lizards and 120 frogs for a maximum profit of \$2,080.

28. Let x be the number of two-bedroom apartments.
    Let y be the number of three-bedroom apartments.

    The constraints-objective are as follows.

    Maximize $P = x + 2y$ subject to the constraints

    $$600x + 700y \geq 26,000$$
    $$600x + 700y \leq 30,000$$
    $$y \geq x$$
    $$x \geq 0, y \geq 0$$

28. (continued)

y-intercepts, $\left(0,37\frac{1}{7}\right)$ and $\left(0,42\frac{6}{7}\right)$, are corner points. To find the third corner point we solve the following system.

$$600x + 700y = 26,000$$
$$y = x$$

$$600y + 700y = 26,000 \Rightarrow 1,300y = 26,000 \Rightarrow y = 20$$

$20 = x \Rightarrow x = 20$ The third corner point is $(20, 20)$.

To find the fourth corner point we solve the following system.

$$600x + 700y = 30,000$$
$$y = x$$

$$600y + 700y = 30,000 \Rightarrow 1,300y = 26,000 \Rightarrow y = 20$$

$23\frac{1}{13} = x \Rightarrow x = 23\frac{1}{13}$ The fourth corner point is $\left(23\frac{1}{13}, 23\frac{1}{13}\right)$.

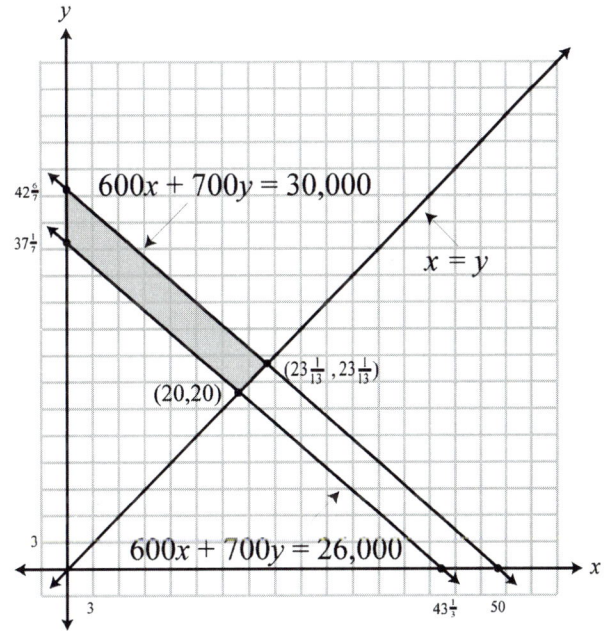

Corner point	Value of $P = x + 2y$	
$\left(0, 37\frac{1}{7}\right)$	$0 + 2 \cdot \frac{260}{7} = 0 + \frac{520}{7} = \frac{520}{7} \approx 74.3$	
$\left(0, 42\frac{6}{7}\right)$	$0 + 2 \cdot \frac{300}{7} = 0 + \frac{600}{7} = \frac{600}{7} \approx 85.7$	
$(20, 20)$	$20 + 2 \cdot 20 = 20 + 40 = 60$	$\Leftarrow$ Minimum
$\left(23\frac{1}{13}, 23\frac{1}{13}\right)$	$\frac{300}{13} + 2 \cdot \frac{300}{13} = \frac{300}{13} + \frac{600}{13} = \frac{900}{13} \approx 69.2$	

They should develop 20 apartments of each type for a minimum increase of 60 children.

Note: Although in this solution we considered corner points of $\left(0, 37\frac{1}{7}\right), \left(0, 42\frac{6}{7}\right)$, and $\left(23\frac{1}{13}, 23\frac{1}{13}\right)$, in a real-life situation we would only consider whole numbers. Also, the evaluated objective function would be rounded down to represent whole children.

29. – 32. No solutions provided

Chapter 8
GEOMETRY: Ancient And Modern Mathematics Embrace

Section 8.1 Lines, Angles, And Circles

1. The postulates are specific assumptions about geometric objects. The axioms are more general assumptions that can apply to objects beyond geometry.

3. a) 2 and 6; 3 and 7   b) there are no others

5. The sum of supplementary angles is 180°; the sum of complementary angles is 90°.

7. "Alternate" means that the angles are on opposite sides of the transversal. "Exterior" means that the angles are outside the parallel lines.

9. 7 and 9; 8 and 10; 2 and 5; 1 and 6; 3 and 4   13. 10; 8

11. 3 and 7

15. 9 and 10; 7 and 10; 7 and 8; 8 and 9; 3 and 10; 4 and 10; 4 and 8; 3 and 8

17. True

19. False; Complementary angles are two angles whose sum of the measures is 90 degrees. The two angles do not each have to measure 45 degrees.

21. True

23. False; For example, a 30° angle is the complement of a 60° angle and the supplement of a 150° angle.

25. False; For example, the supplement of a 30° angle is a 150° angle.

27. e and d 29. b and f; c and g

31. complement: $90° - 30° = 60°$; supplement: $180° - 30° = 150°$

33. complement: none; supplement: $180° - 120° = 60°$

35. complement: $90° - 51.2° = 38.8°$; supplement: $180° - 51.2° = 128.8°$

37. $m\angle a = 180° - 36° = 144°$; $m\angle b = 36°$; $m\angle c = 144°$

39. $m\angle a = 45°$; $m\angle b = 180° - 45° = 135°$; $m\angle c = 45°$

41. $m\angle a = 90° - 38° = 52°$; $m\angle b = 90°$; $m\angle c = 180° - 52° = 128°$

43. $\dfrac{\text{measure of central angle}}{360°} = \dfrac{\text{arc length}}{\text{circumference}}$

$\dfrac{90°}{360°} = \dfrac{a}{24 \text{ feet}}$

$90 \cdot 24 = 360a$

$2{,}160 = 360a$

$a = \dfrac{2{,}160}{360} = 6$

Arc AB has length 6 feet.

45. $\dfrac{\text{measure of central angle}}{360°} = \dfrac{\text{arc length}}{\text{circumference}}$

$\dfrac{m}{360°} = \dfrac{4 \text{ m}}{12 \text{ m}}$

$12m = 4 \cdot 360$

$12m = 1{,}440$

$m = 120$

The measure of the central angle, $\angle ACB$, is $120°$.

47. $\dfrac{\text{measure of central angle}}{360°} = \dfrac{\text{arc length}}{\text{circumference}}$

$\dfrac{30°}{360°} = \dfrac{100 \text{ mm}}{c}$

$30c = 360 \cdot 100$

$30c = 36{,}000$

$c = 1{,}200$

The circumference of the circle is 1,200 millimeters.

49. This is possible. If the two lines are perpendicular, then all four angles are right angles. Each angle measures $90°$.

51. No more than two could be obtuse. Because vertical angles are equal, if there were a third obtuse angle then the fourth would also have to be obtuse, giving a sum of more than $360°$.

53. $90 - x$

55. $3(90 - x)$

57. The angle is $30°$.

$180 - x = 2(90 - x) + 30$

$180 - x = 180 - 2x + 30$

$180 - x = 210 - 2x$

$180 + x = 210$

$x = 30$

59. The angle is $75°$.

$(180 - x) + (90 - x) = 120$

$180 - x + 90 - x = 120$

$270 - 2x = 120$

$-2x = -150$

$x = 75$

61. 12

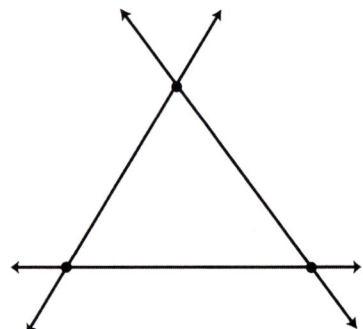

63. $\dfrac{\text{measure of central angle}}{360°} = \dfrac{\text{arc length}}{\text{circumference}}$

$\dfrac{m}{360°} = \dfrac{1{,}000 \text{ miles}}{25{,}000 \text{ miles}}$

$25{,}000m = 360 \cdot 1{,}000$

$25{,}000m = 360{,}000$

$m = 14.4$

The measure of the angle should have been $14.4°$.

65. No solution provided

67. Yes, if the angles both measure 45°.

69.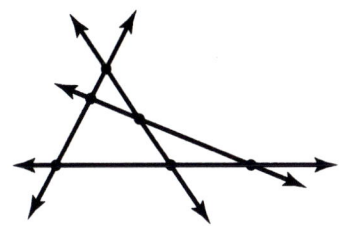

71.

Number of lines	Most number of angles formed
2	$4 = 4 \cdot 1$
3	$12 = 4 \cdot 3 = 4 \cdot (1+2)$
4	$24 = 4 \cdot 6 = 4 \cdot (1+2+3)$
5	$40 = 4 \cdot 10 = 4 \cdot (1+2+3+4)$
⋮	⋮
10	$4 \cdot (1+2+3+4+5+6+7+8+9) = 4 \cdot 45 = 180$

Section 8.2 Polygons

1. Alternate interior angles are equal; $m\angle 1 = m\angle 4$, $m\angle 3 = m\angle 5$.

3. If a convex polygon has n sides, the sum of its interior angles is $(n-2) \times 180°$. $\angle BZY$ is the supplement of $\angle XZV$, so its measure is $180° - 108° = 72°$.

5. Isosceles triangles have two equal sides; equilateral triangles have 3 equal sides.

7. A convex polygon with n sides has an interior angle sum of $(n-2) \times 180°$.

9. True

11. False; A trapezoid has at least one pair of parallel sides where a parallelogram must have two pairs of parallel sides.

13. True;
$$\frac{(10-2) \cdot 180°}{10} = \frac{8 \cdot 180°}{10} = \frac{1,440°}{10} = 144°$$

15. False; Consider a rectangle and a square.

17. False; The number of sides must be a natural number.
$$(n-2) \cdot 180° = 400°$$
$$n - 2 = \frac{400°}{180°} = \frac{20}{9}$$
$$n = 4\frac{2}{9}$$

19. polygon

21. not a polygon; not made of line segments

23. No. A scalene triangle has no sides equal.

25. Both have four equal sides. A rhombus does not have to have right interior angles.

27. $m\angle A + m\angle B + m\angle C = 180°$
$m\angle A + 2m\angle A + 3m\angle A = 180°$
$6m\angle A = 180°$
$m\angle A = 30°$

$m\angle B = 2m\angle A = 2 \cdot 30° = 60°$
$m\angle C = 3m\angle A = 3 \cdot 30° = 90°$

29. $m\angle A + m\angle B + m\angle C = 180°$
$m\angle A + (m\angle A + 10) + (2m\angle A - 10) = 180°$
$4m\angle A = 180°$
$m\angle A = 45°$

$m\angle B = m\angle A + 10 = 45 + 10 = 55°$
$m\angle C = 2m\angle A - 10 = 2 \cdot 45 - 10 = 90 - 10 = 80°$

31. Divide the hexagon into four triangles. The angle sum is $4 \cdot 180° = 720°$.

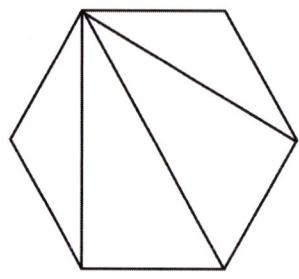

33. $\dfrac{(20-2) \cdot 180°}{20} = \dfrac{18 \cdot 180°}{20} = \dfrac{3,240°}{20} = 162°$

35. $160° = \dfrac{(n-2) \cdot 180°}{n}$
$160n = (n-2) \cdot 180$
$160n = 180n - 360$
$-20n = -360$
$n = \dfrac{-360}{-20} = 18$

37. $m\angle E = 40°$
$\dfrac{EH}{8} = \dfrac{21}{24}$
$24(EH) = 21 \cdot 8$
$24(EH) = 168$
$EH = \dfrac{168}{24} = 7$

39. $m\angle I = 35°$
$\dfrac{HI}{120} = \dfrac{105}{150}$
$150(HI) = 105 \cdot 120$
$150(HI) = 12,600$
$HI = \dfrac{12,600}{150} = 84$

41. Let x be the length of the pond.
$\dfrac{x}{270+60} = \dfrac{35}{60}$
$\dfrac{x}{330} = \dfrac{35}{60}$
$60x = 330 \cdot 35$
$60x = 11,550$
$x = \dfrac{11,550}{60} = 192.5$ feet

43.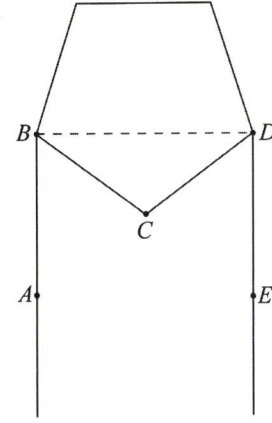

Label the points as shown. Draw a line connecting B to D. Angles BDE and ABD will be right angles. Find the measure of angle BCD.

$$m\angle BCD = \frac{(5-2)\cdot 180°}{5} = \frac{3\cdot 180°}{5} = \frac{540°}{5} = 108°$$

Since triangle BCD is isosceles, $m\angle BDC = \frac{180° - 108°}{2} = \frac{72°}{2} = 36°$. Also since $\angle BDC$ and $\angle CDE$ are complementary, the desired angles each measure $90° - 36° = 54°$.

45. The triangles formed by the sides and supports of the smaller and larger gazebos are similar, so the corresponding sides are proportional. Let x be the length of the long support

$$\frac{x}{12} = \frac{17.4}{10}$$
$$10x = 12 \cdot 17.4$$
$$10x = 208.8$$
$$x = \frac{208.8}{10} = 20.88 \text{ ft}$$

47. Let x be the length from the base of the hedge to the edge of the garden.

$$\frac{x}{8} = \frac{10}{4}$$
$$4x = 8 \cdot 10$$
$$4x = 80$$
$$x = \frac{80}{4} = 20 \text{ ft}$$

49. It is possible to construct. Additional answers may vary.

51. It is not possible to construct. The angle sum would be greater than $180°$.

53. No solution provided

55. The measure of the interior angles gets larger. For very large n, the interior angles have measures close to 180°. For example, evaluate the expression $\frac{(n-2)\cdot 180°}{n}$ with n=10, 100, 1000, 10000, etc.

n	$\frac{[(n-2)\cdot 180°]}{n}$	simplified
10	$\frac{(10-2)\cdot 180°}{10}$	144°
100	$\frac{(100-2)\cdot 180°}{100}$	176.4°
1,000	$\frac{(1,000-2)\cdot 180°}{1,000}$	179.64°
10,000	$\frac{(10,000-2)\cdot 180°}{10,000}$	179.964°
⋮	⋮	⋮

57. Construct the level (see figure) so that the lengths AB and BC are equal and so that $m\angle EDB$ is equal to $m\angle BED$. Mark a vertical line at M, the midpoint of DE. The level is resting on level ground when the string lines up with the mark made at point M.

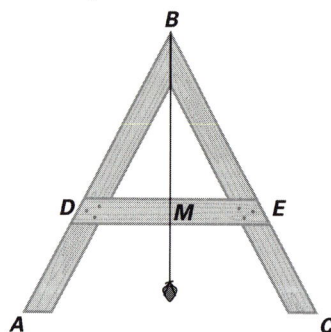

59. a) Triangles are rigid. The scaffolding containing triangles will not shift.
    b) The rectangles in this scaffolding can deform into nonrectangular parallelograms, causing the scaffolding to collapse.

Section 8.3 Perimeter And Area

1. In both cases, the area is the height times the base.

3. The area of the trapezoid is the sum of the areas of two triangles.

5. rectangle, parallelogram, triangle, trapezoid

7. We had the length of the hypotenuse of $\triangle TMC$ and needed the length of another leg to find the height h.

9. $A = l \cdot w = 16 \cdot 10 = 160 \text{ ft}^2$ 11. $A = h \cdot b = 7 \cdot 20 = 140 \text{ in}^2$

13. $A = \frac{1}{2} \cdot (b_1 + b_2) \cdot h = \frac{1}{2} \cdot (22 + 14) \cdot 6 = \frac{1}{2} \cdot (36) \cdot 6 = 18 \cdot 6 = 108 \text{ cm}^2$

15. $A = \frac{1}{2} \cdot h \cdot b = \frac{1}{2} \cdot 6 \cdot 24 = 3 \cdot 24 = 72 \text{ yd}^2$

19. $r = \frac{d}{2} = \frac{8}{2} = 4 \text{ m}$

$A = \pi r^2 \approx 3.14 \cdot 4^2 = 3.14 \cdot 16 = 50.24 \text{ m}^2$

17. $A = \pi r^2 \approx 3.14 \cdot 5^2 = 3.14 \cdot 25 = 78.5 \text{ cm}^2$

21. Entire area: $A = l \cdot w = 20 \cdot 10 = 200 \text{ cm}^2$

Unshaded area: $A = \frac{1}{2} \cdot h \cdot b = \frac{1}{2} \cdot 10 \cdot 20 = 5 \cdot 20 = 100 \text{ cm}^2$

Shaded area: $200 - 100 = 100 \text{ cm}^2$

23. Entire area: $A = \frac{1}{2} \cdot (b_1 + b_2) \cdot h = \frac{1}{2} \cdot (20 + 12) \cdot 7 = \frac{1}{2} \cdot (32) \cdot 7 = 16 \cdot 7 = 112 \text{ m}^2$

Unshaded area: $A = \frac{1}{2} \cdot h \cdot b = \frac{1}{2} \cdot 7 \cdot 20 = \frac{7}{2} \cdot 20 = 70 \text{ m}^2$

Shaded area: $112 - 70 = 42 \text{ m}^2$

25. Entire area: $A = l \cdot w = 8 \cdot (2 + 2) = 8 \cdot 4 = 32 \text{ m}^2$

Unshaded area: $A = \pi r^2 \approx 3.14 \cdot 2^2 = 3.14 \cdot 4 = 12.56 \text{ m}^2$

$A = \pi r^2 \approx 3.14 \cdot 2^2 = 3.14 \cdot 4 = 12.56 \text{ m}^2$

Shaded area: $32 - 12.56 - 12.56 = 6.88 \text{ m}^2$

27. Entire area: $A = l \cdot w = 2 \cdot 2 = 4 \text{ m}^2$

Unshaded area: $A = \pi r^2 \approx 3.14 \cdot \left(\frac{2}{2}\right)^2 = 3.14 \cdot 1 = 3.14 \text{ m}^2$

Shaded area: $4 - 3.14 = 0.86 \text{ m}^2$

For Exercises 29 and 31, we need to find the height of the parallelogram. Since the area of a parallelogram is height times base, the height is therefore 6 inches.

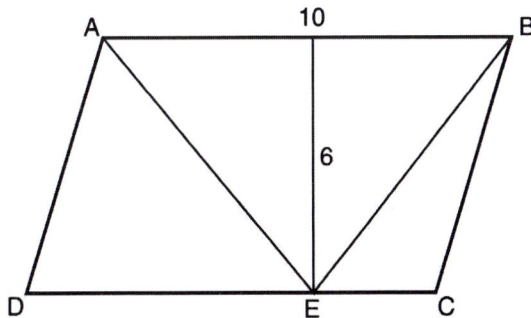

29. The area of $\triangle ABE$ is $A = \frac{1}{2} \cdot h \cdot b = \frac{1}{2} \cdot 6 \cdot 10 = 30$ square inches.

31. It is given that the area of $\triangle BEC$ is 6 square inches. The area of $\triangle ABE$ is $A = \frac{1}{2} \cdot h \cdot b = \frac{1}{2} \cdot 6 \cdot 10 = 30$ square inches. So, the area of $\triangle ADE$ is $60 - 6 - 30 = 24$ square inches. The base of $\triangle ADE$, DE, must satisfy $24 = \frac{1}{2} \cdot 6 \cdot b \Rightarrow 24 = 3b \Rightarrow b = 8$ inches. The area of trapezoid $ABED$ is $A = \frac{1}{2} \cdot (b_1 + b_2) \cdot h = \frac{1}{2} \cdot (8 + 10) \cdot 6 = \frac{1}{2} \cdot (18) \cdot 6 = 9 \cdot 6 = 54$ square inches.

or

To find the area of trapezoid $ABED$, we add the areas of $\triangle ADE$ and $\triangle ABE$. Thus, the area of trapezoid $ABED$ is $24 + 30 = 54$ square inches.

For Exercises 33 and 35, we need to find the areas $\triangle EDB$ and $\triangle AEF$. Since the area of $\triangle BAE$ is 30 square yards, $\triangle EDB$ also has an area of 30 square yards. This is a result of the right angles that yields these two triangles congruent. Since the area of trapezoid $ABDF$ is 66 square yards, the area of $\triangle AEF$ is $66 - 30 - 30 = 6$ square yards. Also, lengths AB and DE are 10 yards each because they must satisfy the relation $30 = \frac{1}{2} \cdot 6 \cdot b \Rightarrow 30 = 3b \Rightarrow b = 10$.

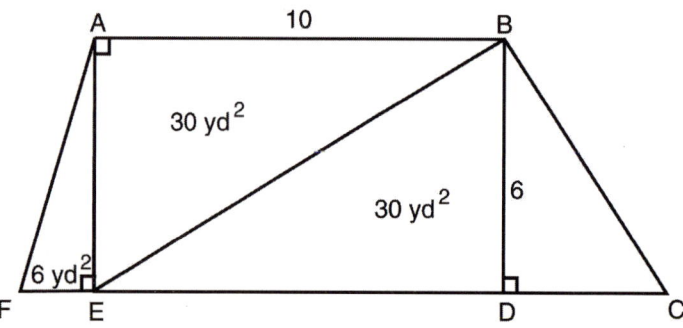

33. Since the area of $\triangle BDC$ is twice $\triangle AEF$, the area of $\triangle BDC$ is $2 \cdot 6 = 12$ square yards.

35. From Exercise 33, we have the area of $\triangle BDC$ as 12 square yards. The base of $\triangle BDC$, CD, must satisfy $12 = \frac{1}{2} \cdot 6 \cdot b \Rightarrow 12 = 3b \Rightarrow b = 4$ yards. The area of trapezoid $ABCE$ is $A = \frac{1}{2} \cdot (b_1 + b_2) \cdot h = \frac{1}{2} \cdot [(10 + 4) + 10] \cdot 6 = \frac{1}{2} \cdot (24) \cdot 6 = 12 \cdot 6 = 72$ square yards.

or

To find the area of trapezoid $ABCE$, we add the areas of $\triangle BAE$, $\triangle BDE$, and $\triangle BDC$. Thus the area of trapezoid $ABCE$ is $30 + 30 + 12 = 72$ square yards.

37. 7 in²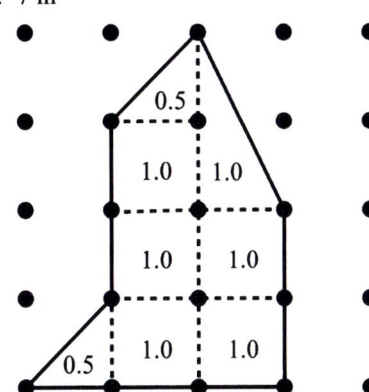

39. 3.5 in²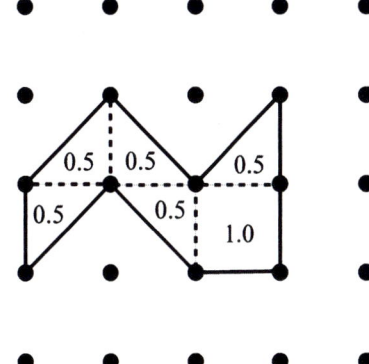

41. $s = \frac{1}{2}(a+b+c) = \frac{1}{2}(15+6+18) = \frac{39}{2}$

$A = \sqrt{s(s-a)(s-b)(s-c)} = \sqrt{\frac{39}{2}\left(\frac{39}{2}-15\right)\left(\frac{39}{2}-6\right)\left(\frac{39}{2}-18\right)} = \sqrt{\frac{39}{2} \cdot \frac{9}{2} \cdot \frac{27}{2} \cdot \frac{3}{2}} = \sqrt{\frac{28,431}{16}}$

$A \approx 42.15 \text{ cm}^2$

43. $s = \frac{1}{2}(a+b+c) = \frac{1}{2}(10+15+12) = \frac{37}{2}$

$A = \sqrt{s(s-a)(s-b)(s-c)} = \sqrt{\frac{37}{2}\left(\frac{37}{2}-10\right)\left(\frac{37}{2}-15\right)\left(\frac{37}{2}-12\right)} = \sqrt{\frac{37}{2} \cdot \frac{17}{2} \cdot \frac{7}{2} \cdot \frac{13}{2}} = \sqrt{\frac{57,239}{16}}$

$A \approx 59.81 \text{ m}^2$

45. $s = \frac{1}{2}(a+b+c) = \frac{1}{2}(19+7+18) = \frac{44}{2} = 22$

$A = \sqrt{s(s-a)(s-b)(s-c)} = \sqrt{22(22-19)(22-7)(22-18)} = \sqrt{22 \cdot 3 \cdot 15 \cdot 4} = \sqrt{3,960} \approx 62.93 \text{ cm}^2$

$A = \frac{1}{2} \cdot h \cdot b$

$\sqrt{3,960} = \frac{1}{2} \cdot h \cdot 19$

$h = \frac{2\sqrt{3,960}}{19} \approx 6.62 \text{ cm}$

47. $a^2 + b^2 = c^2$

$5^2 + 12^2 = x^2$

$25 + 144 = x^2$

$169 = x^2$

$x = \sqrt{169} = 13 \text{ m}$

49. $a^2 + b^2 = c^2$

$x^2 + 11^2 = 13^2$

$x^2 + 121 = 169$

$x^2 = 48$

$x = \sqrt{48} \approx 6.93 \text{ yd}$

218 Chapter 8: Geometry

51. We need to first find the height, h, of the triangle using the Pythagorean theorem.

$12^2 + h^2 = 13^2$
$144 + h^2 = 169$
$h^2 = 25$
$h = \sqrt{25} = 5$ in

$A = \dfrac{1}{2} \cdot h \cdot b$

$A = \dfrac{1}{2} \cdot 5 \cdot (11+12)$

$A = \dfrac{1}{2} \cdot 5 \cdot 23 = \dfrac{5}{2} \cdot 23 = \dfrac{115}{2} = 57.5$ in^2

53. We need to first find the height, h, of the triangle using the Pythagorean theorem.

$10^2 + h^2 = 12^2$
$100 + h^2 = 144$
$h^2 = 44$
$h = \sqrt{44} = 2\sqrt{11} \approx 6.63$ ft

$A = \dfrac{1}{2} \cdot h \cdot b$

$A = \dfrac{1}{2} \cdot 2\sqrt{11} \cdot (10+10)$

$A = \sqrt{11} \cdot 20 = 20\sqrt{11} \approx 66.33$ ft^2

55. The distance from home plate to second base represents the hypotenuse of an isosceles right triangle. Let d represent this distance.

$$90^2 + 90^2 = d^2$$
$$8{,}100 + 8{,}100 = d^2$$
$$16{,}200 = d^2$$
$$d = \sqrt{16{,}200} \approx 127.28 \text{ ft}$$

57. AB is 2 m.

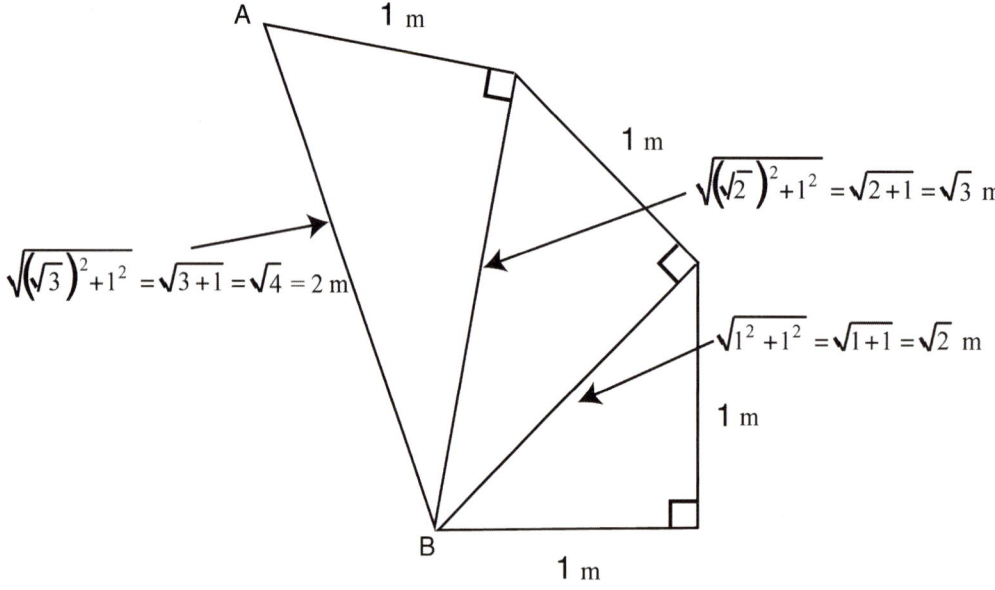

59. area

61. perimeter

63. We need to sum the areas of a rectangle and a semi-circle (half of a circle).

    Area of rectangle: $A = h \cdot b = 12 \cdot 6 = 72$ ft^2

    Area of semi-circle: $A = \frac{1}{2}\pi\ r^2 \approx \frac{1}{2} \cdot 3.14 \cdot \left(\frac{6}{2}\right)^2 = \frac{1}{2} \cdot 3.14 \cdot 3^2 = \frac{1}{2} \cdot 3.14 \cdot 9 = 14.13$ ft^2

    Total area: 72 + 14.13 = 86.13 ft^2

65. $A = \pi\ r^2$

    $50 \approx 3.14 \cdot r^2$

    $\dfrac{50}{3.14} \approx r^2$

    $r \approx \sqrt{\dfrac{50}{3.14}} \approx 4.00$

    The radius should be approximately 4 ft.

67. False; it quadruples

    Area of original circle: $A_0 = \pi\ r^2$

    Area of circle with double radius: $A_d = \pi\ (2r)^2 = \pi \cdot 4 \cdot r^2 = 4 \cdot \pi\ r^2 = 4A_0$

69. There are two rectangular pieces. Each of these has an area of $4 \cdot 100 = 400$ m^2.
    The two ends combined would form a washer-like figure in which we need to find the area of the shaded region.

    Outer Area: $A_0 = \pi\ r^2$

    $A_0 \approx 3.14 \cdot \left(\dfrac{20}{2}\right)^2$

    $A_0 \approx 3.14 \cdot 10^2$

    $A_0 \approx 3.14 \cdot 100$

    $A_0 \approx 314$ m^2

    $A_I \approx 3.14 \left(\dfrac{20 - 4 - 4}{2}\right)^2$

    $A_I \approx 3.14 \cdot 6^2$

    $A_I \approx 3.14 \cdot 36$

    $A_I \approx 113.04$ m^2

    Inner Area: $A_I = \pi\ r^2$
    Surface area of track: 400 + 400 + 314 − 113.04 = 1,000.96 m^2

71. a) The circumference of a circle with an 18 foot radius is $C = 2 \cdot \pi\ r \approx 2 \cdot 3.14 \cdot 18 = 113.04$ ft. The amount of fencing we need would be $\dfrac{72}{360} = \dfrac{1}{5}$ of this or approximately 22.61 ft.

    b) The area of a circle with an 18-foot radius is $A = \pi\ r^2 \approx 3.14 \cdot 18^2 = 3.14 \cdot 324 = 1,017.36$ ft^2. One fifth of this is about 203.47 ft^2

220 Chapter 8: Geometry

73. The area of quadrilateral (parallelogram) WXYZ is one-half of the area of rectangle ABCD. Explanations may vary.

Let b be the base of rectangle ABCD and h the height. Although it is more common to use length and width, we will use base and height in order to mark the triangles clearly. The area of rectangle ABCD will be $A = h \cdot b$. The sides of the four congruent triangles will be as marked.

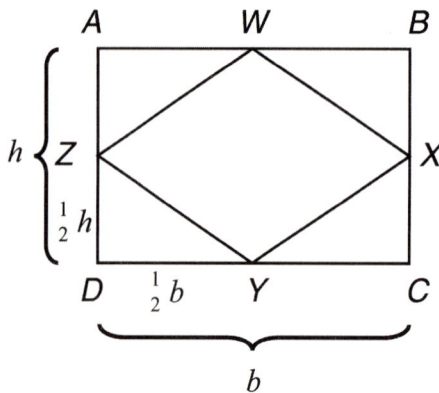

The area of each of the four congruent triangles will be $\dfrac{1}{2} \cdot \left(\dfrac{1}{2}h\right) \cdot \left(\dfrac{1}{2}b\right) = \dfrac{1}{8}hb$. The area of the four congruent triangles together will be $4 \cdot \left(\dfrac{1}{8}hb\right) = \dfrac{1}{2}hb$. Thus the area of quadrilateral (parallelogram) WXYZ will be $h \cdot b - \dfrac{1}{2}hb = \dfrac{1}{2}hb$ or one-half the area of rectangle ABCD.

75. The figure that will have the maximum area will be a square with dimensions 50 feet by 50 feet.

77. We need to first find the approximate radius of the open area.

$A = \pi r^2$

$11{,}304 \approx 3.14 \cdot r^2$

$\dfrac{11{,}304}{3.14} \approx r^2$

$r \approx \sqrt{\dfrac{11{,}304}{3.14}} = 60 \text{ ft}$

We are next concerned with a washer-like figure in which we need to find the area of the shaded region.

Outer Area: $A_o = \pi r^2$

$A_o \approx 3.14 \cdot (180 + 60)^2$

$A_o \approx 3.14 \cdot 240^2$

$A_o \approx 3.14 \cdot 57{,}600$

$A_o \approx 180{,}864 \text{ ft}^2$

Inner Area: $11{,}304 \text{ ft}^2$

Outer Area – Inner Area = $180{,}864 - 11{,}304 = 169{,}560 \text{ ft}^2$

Each of the lots will be one-eighth of this area or about $21{,}195 \text{ ft}^2$

79. $A = (2r)^2 - \pi r^2 = 4r^2 - \pi r^2 = (4 - \pi)r^2$

Section 8.4 Volume And Surface Area

1. Both are the area of the base times the height.

3. The radius of the can that minimized its surface area is between 0.5 and 0.6 of a foot.

5. Its volume is less that the volume of a cylinder with the radius r and height 2r, so it is some fractional part of $2\pi r^3$, namely $\frac{4}{3}\pi r^3$.

7. a) Surface area: $S = 2lw + 2lh + 2wh = 2 \cdot 6 \cdot 4 + 2 \cdot 6 \cdot 5 + 2 \cdot 4 \cdot 5 = 48 + 60 + 40 = 148$ cm^2
   b) Volume: $V = lwh = 6 \cdot 4 \cdot 5 = 120$ cm^3

9. a) Surface area: $S = \pi r\sqrt{r^2 + h^2} \approx 3.14 \cdot 3 \cdot \sqrt{3^2 + 8^2} = 9.42\sqrt{9 + 64} = 9.42\sqrt{73} \approx 80.48$ in^2
   b) Volume: $V = \frac{1}{3}\pi r^2 h \approx \frac{1}{3} \cdot 3.14 \cdot 3^2 \cdot 8 = \frac{3.14 \cdot 9 \cdot 8}{3} = \frac{226.08}{3} = 75.36$ in^3

11. a) Surface area: $S = 2\pi rh + 2\pi r^2 \approx 2 \cdot 3.14 \cdot 5 \cdot 8 + 2 \cdot 3.14 \cdot 5^2 = 251.2 + 157 = 408.2$ ft^2
    b) Volume: $V = \pi r^2 h \approx 3.14 \cdot 5^2 \cdot 8 = 3.14 \cdot 25 \cdot 8 = 628$ ft^3

13. a) Surface area: $S = 4\pi r^2 \approx 4 \cdot 3.14 \cdot 20^2 = 12.56 \cdot 400 = 5{,}024$ cm^2
    b) Volume: $V = \frac{4}{3}\pi r^3 \approx \frac{4}{3} \cdot 3.14 \cdot 20^3 = \frac{4 \cdot 3.14 \cdot 8{,}000}{3} = \frac{100{,}480}{3} \approx 33{,}493.33$ cm^3

15. Volume = area of the base times height = $25 \cdot 3 = 75$ ft^3

17. Area of base = $A = \frac{1}{2} \cdot (b_1 + b_2) \cdot h = \frac{1}{2} \cdot (11 + 7) \cdot 4 = \frac{1}{2} \cdot (18) \cdot 4 = 9 \cdot 4 = 36$ in^2
    Volume = area of the base times height = $36 \cdot 5 = 180$ in^3

19. $s = \frac{1}{2}(a + b + c) = \frac{1}{2}(5 + 5 + 4) = \frac{14}{2} = 7$
    Area of base applying Heron's formula is
    $A = \sqrt{s(s-a)(s-b)(s-c)} = \sqrt{7(7-5)(7-5)(7-4)} = \sqrt{7 \cdot 2 \cdot 2 \cdot 3} = \sqrt{84} \approx 9.17$ m^2
    Volume = area of the base times height = $\sqrt{84} \cdot 2 \approx 18.33$ m^3

21. Area of base = $\frac{45}{360}\pi r^2 \approx \frac{1}{8} \cdot 3.14 \cdot 8^2 = \frac{3.14 \cdot 64}{8} = \frac{200.96}{8} = 25.12$ in^2
    Volume = area of the base times height $\approx 25.12 \cdot 3 = 75.36$ in^3

222   Chapter 8: Geometry

23. We need to first determine the volume of the punch bowl and the ladle.

   Punch bowl volume: $V = \frac{1}{2} \cdot \left(\frac{4}{3}\pi r^3\right) \approx \frac{4}{6} \cdot 3.14 \cdot 9^3 = \frac{4 \cdot 3.14 \cdot 729}{6} = \frac{9,156.24}{6} = 1,526.04$ in^3

   Ladle volume: $V = \frac{1}{2} \cdot \left(\frac{4}{3}\pi r^3\right) \approx \frac{4}{6} \cdot 3.14 \cdot 2^3 = \frac{4 \cdot 3.14 \cdot 8}{6} = \frac{100.48}{6} \approx 16.75$ in^3

   We then divide the volume of the punch bowl by the volume of the ladle. $\frac{1,526.04}{16.75} \approx 91.11$

   There are approximately 91 ladles of punch.

25. $12^3 = 1,728$

27. We need to first determine the volume of the wheelbarrow.

   Area of base = $A = \frac{1}{2} \cdot (b_1 + b_2) \cdot h = \frac{1}{2} \cdot (2+3) \cdot 1 = \frac{1}{2} \cdot (5) \cdot 1 = \frac{5}{2}$ ft^2

   Volume = area of the base times height = $\frac{5}{2} \cdot \frac{5}{2} = \frac{25}{4} = 6.25$ ft^3

   The volume of the wheelbarrow is then multiplied by 16. $\frac{25}{4} \cdot 16 = 100$ ft^3

   She needs 100 ft^3 of stone and this is approximately $\frac{100}{27} \approx 3.7$ yd^3 of stone.

29. Volume of the rectangular cake: $V = lwh = 14 \cdot 9 \cdot 3 = 378$ in^3
    Volume of the round cake: $V = \pi r^2 h \approx 3.14 \cdot 5^2 \cdot 4 = 3.14 \cdot 25 \cdot 4 = 314$ in^3
    The rectangular cake has more volume.

31. a) $V = \pi r^2 h \approx 3.14 \cdot 10^2 \cdot 7 = 3.14 \cdot 100 \cdot 7 = 2,198$ in^3
    b) $V = \pi r^2 h \approx 3.14 \cdot 12^2 \cdot 5 = 3.14 \cdot 144 \cdot 5 = 2,260.8$ in^3

33. a) $V = \frac{1}{3}\pi r^2 h \approx \frac{1}{3} \cdot 3.14 \cdot 6^2 \cdot 5 = \frac{3.14 \cdot 36 \cdot 5}{3} = \frac{565.2}{3} = 188.4$ in^3
    b) $V = \frac{1}{3}\pi r^2 h \approx \frac{1}{3} \cdot 3.14 \cdot 8^2 \cdot 3 = \frac{3.14 \cdot 64 \cdot 3}{3} = \frac{602.88}{3} = 200.96$ in^3

35. We get a larger increase by increasing the radius because in the formula for the volume of a cylinder, we square the radius. This is not the case in general, however.

37. Let d be the diameter of the can. The height of the can is $3d$; the circumference of the can is $d \approx 3.14d$, which is larger.

39. Because the volume of the tin can is to be 2 cubic feet, we set $\pi r^2 h = 2$. Dividing both sides of this equation by πr^2, we get $h = \dfrac{2}{\pi r^2}$. Thus, the height depends on the choice of the radius. Recall the formula for finding the surface area of a tin can is $S = 2\pi rh + 2\pi r^2$. We can substitute $\dfrac{2}{\pi r^2}$ for h in this equation to get $S = 2\pi r \dfrac{2}{\pi r^2} + 2\pi r^2 = \dfrac{4\pi r}{\pi r^2} + 2\pi r^2 = \dfrac{4}{r} + 2\pi r^2$. Try various values for r.

Radius: r	Surface Area: $S = \dfrac{4}{r} + 2\pi r^2$	
0.1	40.0628	
0.2	20.2512	decreasing
0.3	13.8985	decreasing
0.4	11.0048	decreasing
0.5	9.5700	decreasing
0.6	8.9275	decreasing
0.7	8.7915	decreasing
0.8	9.0192	increasing

The table indicates that we should start somewhere between 0.7 and 0.8. In actuality, the value should be between 0.6 and 0.7, so it would be best to start with a larger interval such as 0.6 to 0.8.

The exact value is $r = \dfrac{1}{\sqrt[3]{\pi}} \approx 0.6828$ ft.

41. We must first find the volume of the concrete floor in cubic feet. To find the volume, we need to find the base area; that is, the area of the regular hexagon. The regular hexagon can be divided up into 6 equilateral triangles. To find the area of each of these triangles, we can either apply Heron's formula or use the Pythagorean theorem to find the height and then apply the area formula for a triangle. In this solution, we will use Heron's formula.

$$s = \dfrac{1}{2}(a+b+c) = \dfrac{1}{2}(8+8+8) = \dfrac{24}{2} = 12$$

Area of base applying Heron's formula is

$$A = \sqrt{s(s-a)(s-b)(s-c)} = \sqrt{12(12-8)(12-8)(12-8)} = \sqrt{12 \cdot 4 \cdot 4 \cdot 4} = \sqrt{768} = 16\sqrt{3} \approx 27.71 \text{ ft}^2$$

Since there are 6 of these triangles, we have the area of the hexagon to be $16\sqrt{3} \cdot 6 = 96\sqrt{3} \approx 166.28$ ft^2.

Since the height of 6 inches is $\dfrac{1}{2}$ foot, we have

Volume = area of the base times height = $96\sqrt{3} \cdot \dfrac{1}{2} = 48\sqrt{3} \approx 83.14$ ft^3. He must purchase 84 cubic feet of concrete.

224   Chapter 8: Geometry

43. The volume of the earth is approximately:

$$V = \frac{4}{3}\pi r^3 \approx \frac{4}{3} \cdot 3.14 \cdot \left(\frac{7{,}920}{2}\right)^3 = \frac{4}{3} \cdot 3.14 \cdot 3{,}960^3 = \frac{4}{3} \cdot 3.14 \cdot 62{,}099{,}136{,}000 = 259{,}988{,}382{,}720 \text{ mi}^3.$$

The volume of the moon is therefore approximately $\dfrac{259{,}988{,}382{,}720}{49} \approx 5{,}305{,}885{,}361.63$.

$$V = \frac{4}{3}\pi r^3$$
$$5{,}305{,}885{,}361.63 \approx \frac{4}{3} \cdot 3.14 \cdot r^3$$
$$1{,}267{,}329{,}306.12 \approx r^3$$
$$r \approx \sqrt[3]{1{,}267{,}329{,}306.12}$$
$$r \approx 1{,}082.17 \text{ mi}$$

The diameter is therefore approximately $2 \cdot 1{,}082.17 \approx 2{,}164$ miles.

45. The volume is four times as large because in the formula of the volume of a cone, the radius is squared.

$$V = \frac{1}{3}\pi r^2 h = \frac{1}{3}\pi \cdot (2r)^2 \cdot h = \frac{1}{3}\pi \cdot 4 \cdot r^2 h = 4 \cdot \left(\frac{1}{3}\pi r^2 h\right)$$

47. The smaller cubes have more surface area. For example, compare the surface area of 1 3-inch cube with the surface of 27 1-inch cubes.

49. The volume of the $4 \times 4 \times \frac{1}{4}$ hamburger is $4 \cdot 4 \cdot \frac{1}{4} = 4 \text{ in}^3$. The volume of the "tri-burger" will also be 4 in³. This volume can be found from base area times height, where the height is ¼ in and the base is an equilateral triangle. Since the height is ¼ inch, the base area is 16 in². Since each side of an equilateral triangle has the same measure, let this measure be a. The perimeter of this triangle will be $3a$.

Applying Heron's formula we have $s = \dfrac{1}{2}(3a) = \dfrac{3a}{2}$ and

$$16 = \sqrt{\frac{3a}{2}\left(\frac{3a}{2} - a\right)\left(\frac{3a}{2} - a\right)\left(\frac{3a}{2} - a\right)}$$
$$16 = \sqrt{\frac{3a}{2} \cdot \frac{a}{2} \cdot \frac{a}{2} \cdot \frac{a}{2}}$$
$$16 = \sqrt{\frac{3a^4}{16}}$$
$$256 = \frac{3a^4}{16}$$
$$4{,}096 = 3a^4$$
$$\frac{4{,}096}{3} = a^4$$
$$a = \left(\frac{4{,}096}{3}\right)^{1/4} \approx 6.08 \text{ in.}$$

51. In this exercise, we are cutting off a cone that has a radius of $\frac{r}{2}$ and a height of $\frac{h}{2}$. The original cone has a surface area of $S = \pi r\sqrt{r^2+h^2}$. The surface area that we are cutting off is

$$S = \pi \left(\frac{r}{2}\right)\sqrt{\left(\frac{r}{2}\right)^2 + \left(\frac{h}{2}\right)^2} = \pi \frac{r}{2}\sqrt{\frac{r^2}{4} + \frac{h^2}{4}} = \pi \frac{r}{2}\sqrt{\frac{r^2+h^2}{4}} = \frac{1}{4}\pi r\sqrt{r^2+h^2}.$$

The remaining surface area is therefore $\pi r\sqrt{r^2+h^2} - \frac{1}{4}\pi r\sqrt{r^2+h^2} = \frac{3}{4}\pi r\sqrt{r^2+h^2}$.

53. No solution provided

Section 8.5 The Metric System And Dimensional Analysis

1. Moving three columns to the right in the table gave us $10^3 = 1,000$ as many objects.

3. $\dfrac{1.6 \text{ kilometers}}{1 \text{ mile}}$

5. 10

7. More decimeters because decimeters are smaller than hectometers.

9. h; $\dfrac{1}{4}$ pound $\times \dfrac{454 \text{ grams}}{1 \text{ pound}} = 113.5$ grams $= 1,135$ dg

11. e; $15 \text{ feet} = 15 \text{ feet} \times \dfrac{1 \text{ yard}}{3 \text{ feet}} = 5 \text{ yards} \times \dfrac{1 \text{ meter}}{1.0936 \text{ yards}} \approx 4.572 \text{ meters} = 4,572 \text{ mm}$ or

$15 \text{ feet} = 15 \text{ feet} \times \dfrac{12 \text{ inches}}{1 \text{ foot}} = 180 \text{ inches} \times \dfrac{1 \text{ meter}}{39.37 \text{ inches}} \approx 4.572 \text{ meters} = 4,572 \text{ mm}$

13. g; $6 \text{ inches} \times \dfrac{1 \text{ meter}}{39.37 \text{ inches}} \approx 0.1524$ meters $= 15.24$ cm

15. a; $8.5 \text{ inches} \times \dfrac{1 \text{ meter}}{39.37 \text{ inches}} \approx 0.2159$ meters $= 0.02159$ dam

17. d; $6 \text{ ounces} = 6 \text{ ounces} \times \dfrac{1 \text{ quart}}{32 \text{ ounces}} = 0.1875 \text{ quarts} \times \dfrac{1 \text{ liter}}{1.0567 \text{ quarts}} \approx 0.177$ liters $= 1.77$ dl

19. 24,000 dl

21. 34.5 hm

23. 8,500 dg

25. 1,800 g

27. 4.5 m

29. 350 dl

31. Because dekagrams are much larger than milligrams, we would need fewer dekagrams. Therefore, we would move the decimal point to the left.

33. The prefix "centi" reminds us of 100 and "deci" reminds us of ten. Therefore, "centi" means hundredths and "deci" means tenths.

The explanations to Exercises 35 – 41 may vary.

35. b; A pencil is about seven inches long. Since 39.37 inches is about one meter or 100 centimeters, one inch is about $\frac{100}{39.37} \approx 2.54$ cm. So, seven inches is approximately 17.78 cm.

37. c; 6 kg = 6,000g $\times \frac{1 \text{ pound}}{454 \text{ grams}} \approx 13.2$ pounds.

39. c; A bottle of wine might be a little less than a quart (or a liter).

41. b; The dog might be about two feet tall or about $\frac{2}{3}$ of a meter.

43. 1,460 grams $\times \frac{1 \text{ pound}}{454 \text{ grams}} \approx 3.22$ pounds

45. 27 gallons $\times \frac{4 \text{ quarts}}{1 \text{ gallon}} = 108$ quarts $\times \frac{1 \text{ liter}}{1.0567 \text{ quarts}} \approx 102.20$ liters

47. 4 yards $\times \frac{1 \text{ meter}}{1.0936 \text{ yards}} \approx 3.6576$ meters = 365.76 centimeters

49. 10,000 milliliters = 10 liters $\times \frac{1.0567 \text{ quarts}}{1 \text{ liter}} \approx 10.57$ quarts

51. 514 decimeters = 51.4 meters $\times \frac{1.0936 \text{ yards}}{1 \text{ meter}} \approx 56.21$ yards

53. 47 pounds $\times \frac{454 \text{ grams}}{1 \text{ pound}} = 21,338$ grams = 21.338 kilograms ≈ 21.34 kilograms

55. 507,820 milligrams = 507.82 grams $\times \frac{1 \text{ pound}}{454 \text{ grams}} \approx 1.12$ pounds

57. 480 dekagrams = 4,800 grams $\times \frac{1 \text{ pound}}{454 \text{ grams}} \approx 10.57$ pounds

59. 3,500 cm = 35 meters $\times \frac{1.0936 \text{ yards}}{1 \text{ meter}} \approx 38.28$ yards

61. 45,000 kg = 45,000,000 grams $\times \frac{1 \text{ pound}}{454 \text{ grams}} \approx 99,118.94$ pounds $\times \frac{1 \text{ ton}}{2,000 \text{ pounds}} \approx 49.56$ tons

63. $0.24 \text{ gallons} \times \dfrac{4 \text{ quarts}}{1 \text{ gallon}} = 0.96 \text{ quarts} \times \dfrac{1 \text{ liter}}{1.0567 \text{ quarts}} \approx 0.91 \text{ liters}$

65. $400 \text{ yards} \times \dfrac{1 \text{ meter}}{1.0936 \text{ yards}} \approx 365.8 \text{ meters} = 36.58 \text{ dekameters}$

67. Don't give him an inch.

69. It is first down and ten yards to go.

71. a) $V = lwh = 8 \cdot 5 \cdot 4 = 160$ cubic meters

    b) $160 \text{ cubic meters} \times \dfrac{1,000 \text{ liters}}{1 \text{ cubic meter}} = 160,000 \text{ liters}$

    c) $160,000 \text{ liters} \times \dfrac{1 \text{ kilogram}}{1 \text{ liter}} = 160,000 \text{ kilograms}$

73. $V = lwh = 40 \cdot 20 \cdot 6 = 4,800 \text{ cubic feet} \times \dfrac{1 \text{ cubic yard}}{3 \times 3 \times 3 \text{ cubic feet}} \approx 177.7778 \text{ cubic yards}$

    We now convert to cubic meters.

    $177.7778 \text{ cubic yards} \times \dfrac{1 \text{ cubic meter}}{1.0936 \times 1.0936 \times 1.0936 \text{ cubic yards}} \approx 135.9258 \text{ cubic meters}$

    We now convert to kiloliters.

    $135.9258 \text{ cubic meters} \times \dfrac{1 \text{ kiloliter}}{1 \text{ cubic meter}} \approx 135.93 \text{ kiloliters}$

75. $\dfrac{\$2.75}{1 \text{ kilogram}} \times \dfrac{1 \text{ kilogram}}{1,000 \text{ grams}} = \dfrac{\$2.75}{1,000 \text{ grams}} \times \dfrac{454 \text{ grams}}{1 \text{ pound}} = \dfrac{\$1,248.5}{1,000 \text{ pounds}} \approx \$1.25 \text{ per pound}$

77. Since $P = 2l + 2w$, we have $2 \cdot 62 + 2 \cdot 35 = 124 + 70 = 194$ feet of fencing. We need to convert this to meters. $194 \text{ feet} \times \dfrac{1 \text{ yard}}{3 \text{ feet}} \approx 64.6667 \text{ yards} \times \dfrac{1 \text{ meter}}{1.0936 \text{ yards}} \approx 59.13 \text{ meters}$. She will need to buy 60 meters.

79. $\dfrac{30 \text{ miles}}{1 \text{ gallon}} \times \dfrac{1 \text{ gallon}}{4 \text{ quarts}} \times \dfrac{1.0567 \text{ quarts}}{1 \text{ liter}} \times \dfrac{1.609 \text{ km}}{1 \text{ mile}} \approx \dfrac{51.0069 \text{ km}}{4 \text{ liters}} \approx 12.75 \text{ kilometers per liter}$

81. $65°$ C

    $149 = \dfrac{9}{5}C + 32$

    $117 = \dfrac{9}{5}C$

    $585 = 9C$

    $65 = C$

83. $140°$ F

    $F = \dfrac{9}{5} \cdot 60 + 32$

    $F = 9 \cdot 12 + 32$

    $F = 108 + 32$

    $F = 140$

228   Chapter 8: Geometry

85. 68° F

$F = \dfrac{9}{5} \cdot 20 + 32$

$F = 9 \cdot 4 + 32$

$F = 36 + 32$

$F = 68$

87. 45° C

$113 = \dfrac{9}{5} C + 32$

$81 = \dfrac{9}{5} C$

$405 = 9C$

$45 = C$

89. One hectare $= 100 \text{ m} \cdot 100 \text{ m} = 10{,}000$ square meters

91. The land that Thiep purchased is $0.75 \text{ km} \cdot 1.2 \text{ km} = 0.9 \text{ km}^2$. It was shown in Exercise 90 that one square kilometer is equal to 100 hectares. So, Thiep purchased 90 hectares.

93. – 99. No solutions provided

Section 8.6 Geometric Symmetry And Tessellations

1. There are more rigid motions that can be applied to the star so that its beginning and ending position are the same.

3. The measure of an interior angle for the polygon is greater than 120°, so if we have three or more polygons sharing a vertex, the angle sum at the vertex will exceed 360°.

5. 360°; the measure of an interior angle of a regular hexagon is 120°, which divides 360° evenly, so there can be three hexagons at a vertex. The measure of an interior angle of a regular pentagon is 108°, which does not divide 360° evenly, so the tessellation is impossible.

7. and 9.

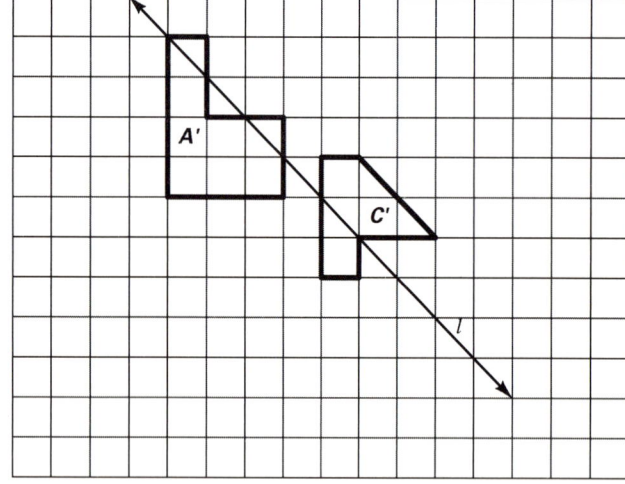

11.

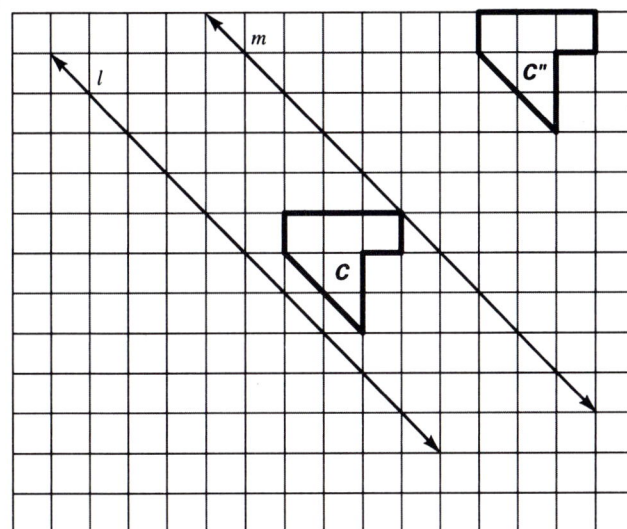

13.

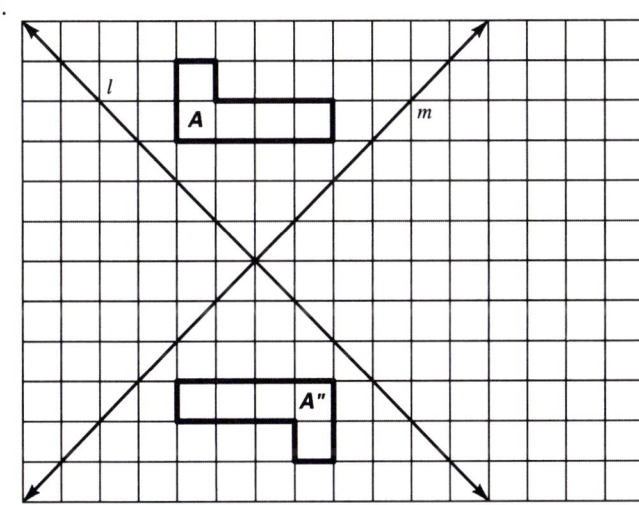

15. a) yes b) The effect is the same as if we performed a translation.

17.

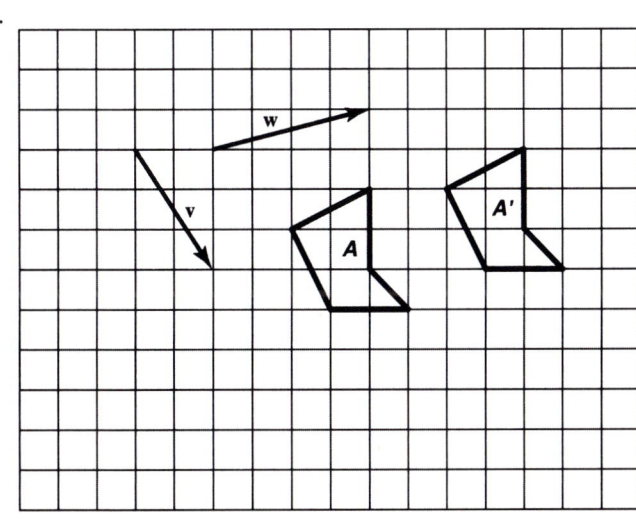

19.

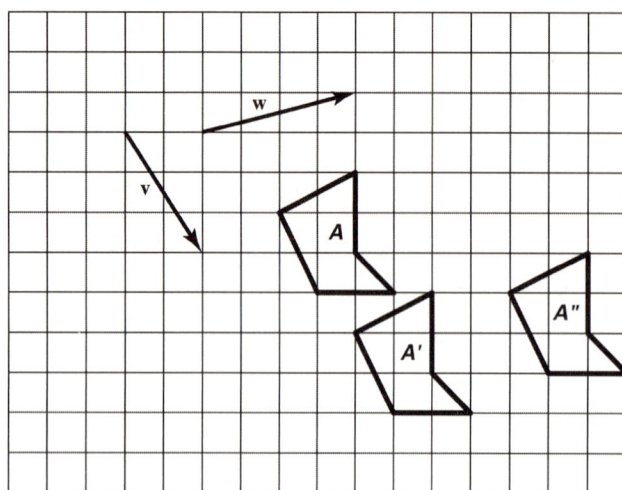

21.

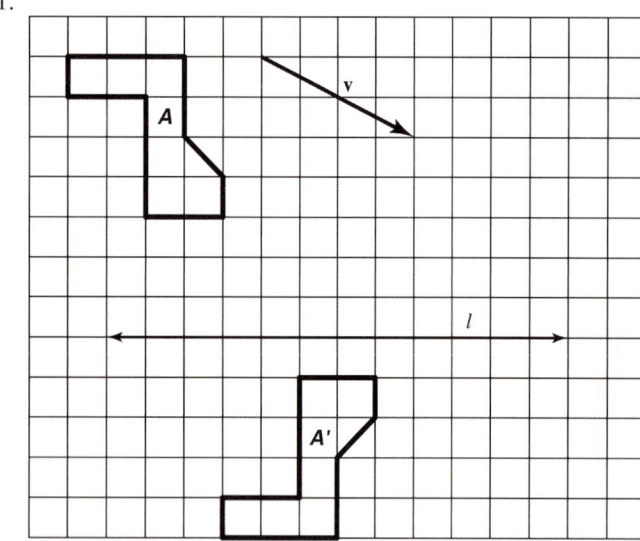

23.

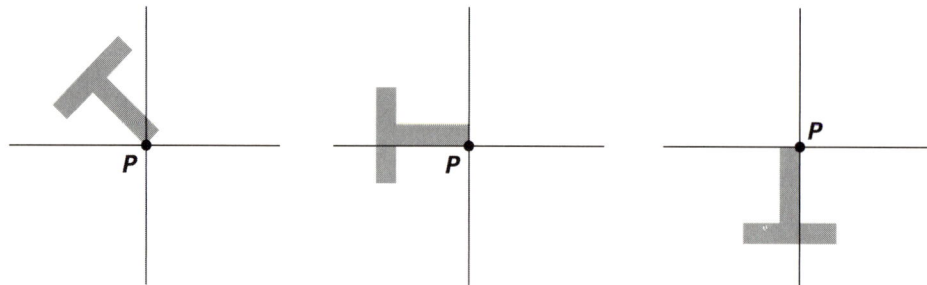

25.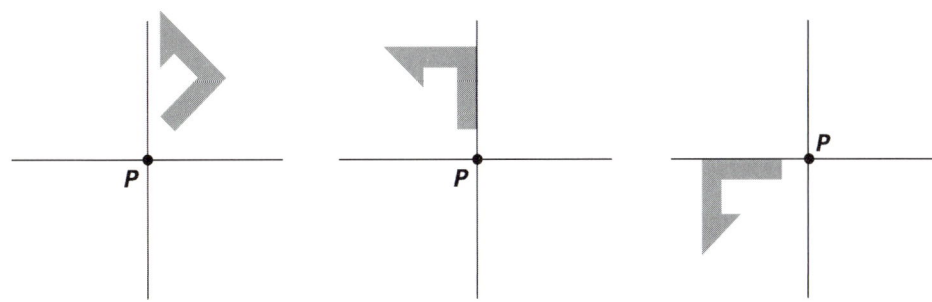

27. $(12-2) \cdot 180 = 10 \cdot 180 = 1,800°$

$\dfrac{1,800}{12} = 150°$

29. Using the interior angles of a regular pentagon, we cannot obtain an angle sum of $360°$ around a point.

31.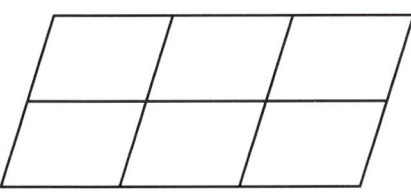

33. (c), (d)

35. (a), (e)

37. Reflectional symmetries: about a vertical line, about a horizontal line, and two diagonal lines

Rotational symmetries: $90°$, $180°$, and $270°$

39. Reflectional symmetries: about a vertical line

Rotational symmetries: none

41. Reflectional symmetries: about a vertical line
Rotational symmetries: none

43. Each interior angle of an equilateral triangle is $\dfrac{(3-2) \cdot 180°}{3} = \dfrac{180°}{3} = 60°$.

Each interior angle of a regular hexagon is $\dfrac{(6-2) \cdot 180°}{6} = \dfrac{4 \cdot 180°}{6} = \dfrac{720°}{6} = 120°$.

At the points where the vertices of the equilateral triangles and the regular hexagons meet, the angle sum is $120° + 60° + 120° + 60° = 360°$.

45. Each interior angle of an equilateral triangle is $\dfrac{(3-2) \cdot 180°}{3} = \dfrac{180°}{3} = 60°$.

Each interior angle of a square is $\dfrac{(4-2) \cdot 180°}{4} = \dfrac{360°}{4} = 90°$.

Each interior angle of a regular hexagon is $\dfrac{(6-2) \cdot 180°}{6} = \dfrac{4 \cdot 180°}{6} = \dfrac{720°}{6} = 120°$.

At the points where the vertices of the equilateral triangle, squares, and the regular hexagon meet, the angle sum is $60° + 90° + 120° + 90° = 360°$.

232   Chapter 8: Geometry

47. Because the sum of the interior angles of a convex quadrilateral is 360°, you can always arrange four copies of the quadrilateral around a point to make an angle sum of 360°.

49. In the figure, we constructed the perpendicular bisectors of segments AA' and CC'. The point where these perpendicular bisectors meet is the center of rotation.

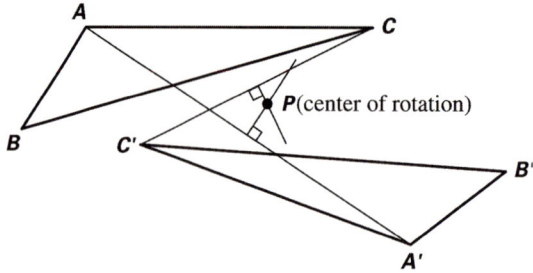

Chapter Review Exercises

1. a) b, g
    b) d, e

2. $m\angle a = 42°$; $m\angle b = 180° - 42° = 138°$;
    $m\angle c = 180° - 90° - 42° = 48°$

3. $\dfrac{(18-2) \cdot 180°}{18} = \dfrac{16 \cdot 180°}{18} = \dfrac{2{,}880°}{18} = 160°$

4. $m\angle E = 45°$

    $\dfrac{EH}{8} = \dfrac{10}{12}$
    $12(EH) = 10 \cdot 8$
    $12(EH) = 80$
    $EH = \dfrac{80}{12} = \dfrac{20}{3}$

    $\dfrac{FG}{8} = \dfrac{6}{12}$
    $12(FG) = 6 \cdot 8$
    $12(FG) = 48$
    $FG = \dfrac{48}{12} = 4$

5. a) $A = \dfrac{1}{2} \cdot (b_1 + b_2) \cdot h = \dfrac{1}{2} \cdot (20 + 12) \cdot 5 = \dfrac{1}{2} \cdot (32) \cdot 5 = 16 \cdot 5 = 80 \text{ cm}^2$

    b) $A = \dfrac{1}{2} \cdot h \cdot b = \dfrac{1}{2} \cdot 4 \cdot 10 = 2 \cdot 10 = 20 \text{ in}^2$

6. a) Entire area: $A = \dfrac{1}{2} \cdot (b_1 + b_2) \cdot h = \dfrac{1}{2} \cdot (10 + 6) \cdot 4 = \dfrac{1}{2} \cdot (16) \cdot 4 = 8 \cdot 4 = 32 \text{ m}^2$

    Unshaded area: $A = \dfrac{1}{2} \cdot h \cdot b = \dfrac{1}{2} \cdot 4 \cdot 10 = 2 \cdot 10 = 20 \text{ m}^2$

    Shaded area: $32 - 20 = 12 \text{ m}^2$

    b) Semi-circle's area: $A = \dfrac{1}{2} \pi \, r^2 \approx \dfrac{1}{2} \cdot 3.14 \cdot \left(\dfrac{8}{2}\right)^2 = \dfrac{1}{2} \cdot 3.14 \cdot 4^2 = \dfrac{1}{2} \cdot 3.14 \cdot 16 = 25.12 \text{ ft}^2$

    Triangle's area: $A = \dfrac{1}{2} \cdot h \cdot b = \dfrac{1}{2} \cdot 4 \cdot 8 = 2 \cdot 8 = 16 \text{ ft}^2$

    Shaded area: $25.12 + 16 = 41.12 \text{ ft}^2$

7. a) $s = \frac{1}{2}(a+b+c) = \frac{1}{2}(10+8+4) = \frac{22}{2} = 11$

$A = \sqrt{s(s-a)(s-b)(s-c)} = \sqrt{11(11-10)(11-8)(11-4)} = \sqrt{11 \cdot 1 \cdot 3 \cdot 7} = \sqrt{231} \approx 15.2 \text{ cm}^2$

b) $\quad A = \frac{1}{2} \cdot h \cdot b$

$\sqrt{231} = \frac{1}{2} \cdot h \cdot 10$

$\sqrt{231} = 5h$

$h = \frac{\sqrt{231}}{5} \approx 3.04 \text{ cm}$

8. There are two rectangular pieces. Each of these has an area of $5 \cdot 100 = 500 \text{ m}^2$.
    The two ends combined would form a washer-like figure in which we need to find the area of the shaded region.

    Outer Area: $\quad A_o = \pi r^2$

    $A_o \approx 3.14 \cdot \left(\frac{20}{2}\right)^2$

    $A_o \approx 3.14 \cdot 10^2$

    $A_o \approx 3.14 \cdot 100$

    $A_o \approx 314 \text{ m}^2$

    Inner Area: $\quad A_I = \pi r^2$

    $A_I \approx 3.14 \left(\frac{20-5-5}{2}\right)^2$

    $A_I \approx 3.14 \cdot 5^2$

    $A_I \approx 3.14 \cdot 25$

    $A_I \approx 78.5 \text{ m}^2$

    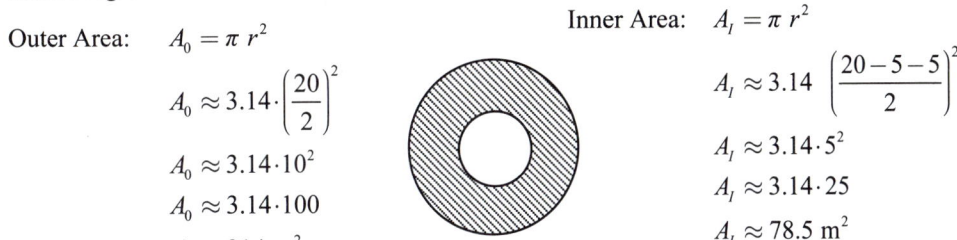

    Surface area of track: $500 + 500 + 314 - 78.5 = 1{,}235.5 \text{ m}^2$

9. a) Area of base $= \frac{60}{360} \pi r^2 \approx \frac{1}{6} \cdot 3.14 \cdot 6^2 = \frac{3.14 \cdot 36}{6} = \frac{113.04}{6} = 18.84 \text{ in}^2$

    Volume = area of the base times height $\approx 18.84 \cdot 2 = 37.68 \text{ in}^3$

    b) Area of base $= A = \frac{1}{2} \cdot (b_1 + b_2) \cdot h = \frac{1}{2} \cdot (12+10) \cdot 5 = \frac{1}{2} \cdot (22) \cdot 5 = 11 \cdot 5 = 55 \text{ cm}^2$

    Volume = area of the base times height $= 55 \cdot 6 = 330 \text{ cm}^3$

10. We need to first determine the volume of the punch bowl and the cylindrical glass.

    Punch bowl volume: $V = \frac{1}{2} \cdot \left(\frac{4}{3}\pi r^3\right) \approx \frac{4}{6} \cdot 3.14 \cdot 9^3 = \frac{4 \cdot 3.14 \cdot 729}{6} = \frac{9{,}156.24}{6} = 1{,}526.04 \text{ in}^3$

    Glass volume: $V = \pi r^2 h \approx 3.14 \cdot \left(\frac{3}{2}\right)^2 \cdot 3 = 3.14 \cdot \frac{9}{4} \cdot 3 = 21.195 \text{ in}^3$

    We then divide the volume of the punch bowl by the volume of the glass. $\frac{1{,}526.04}{21.195} = 72$

    There are approximately 72 glasses of punch.

11. The volume is four times as large, because in the formula of the volume of a cone, the radius is squared.

    $V = \frac{1}{3}\pi r^2 h = \frac{1}{3} \cdot \pi \cdot (2r)^2 \cdot h = \frac{1}{3}\pi \cdot 4 \cdot r^2 h = 4 \cdot \left(\frac{1}{3}\pi r^2 h\right)$

12. a) 3.5 meters b) 43,150 centigrams c) 38,600 deciliters

234   Chapter 8: Geometry

13. 514 decimeters = 51.4 meters $\times \dfrac{1.0936 \text{ yards}}{1 \text{ meter}} \approx 56.21$ yards

14. 2.1 kiloliters = 2,100 liters $\times \dfrac{1.0567 \text{ quarts}}{1 \text{ liter}} = 2,219.07$ quarts

15.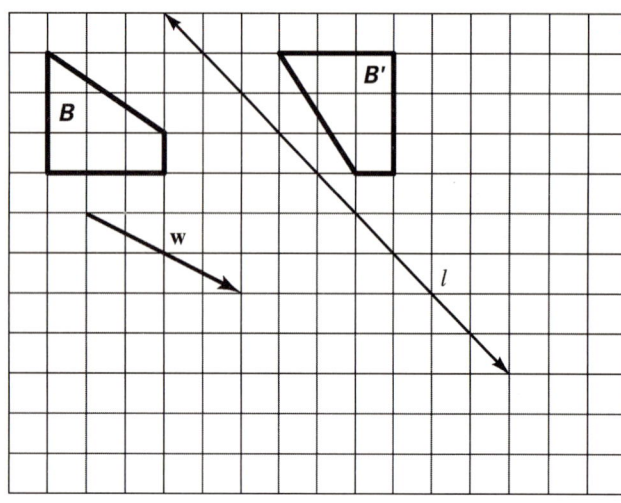

16. a) c, e b) d

17. Reflectional symmetries: about a vertical line and about a horizontal line
    Rotational symmetries: 180°

18.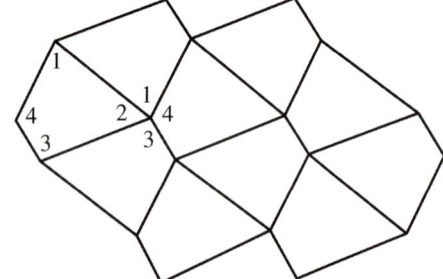

Chapter Test

1. a) vertical angles
   b) corresponding angles
   c) alternate exterior angles
   d) alternate interior angles

2. $12 \cdot \left(\dfrac{(12-2) \cdot 180°}{12} \right) = \cancel{12} \cdot \left(\dfrac{(12-2) \cdot 180°}{\cancel{12}} \right) = (10) \cdot 180° = 1800°$

3. $m\angle c = 40°$; $m\angle b = 180° - 40° = 140°$;
   $m\angle a = 180° - \left(180° - 40° - 90°\right) = 180° - 50° = 130°$

4. a) $A = \frac{1}{2} \cdot (b_1 + b_2) \cdot h = \frac{1}{2} \cdot (15+12) \cdot 2 = \frac{1}{2} \cdot (27) \cdot 2 = \frac{1}{2} \cdot 2 \cdot (27) = 27 \text{ in}^2$

   b) $A = \frac{1}{2} \cdot h \cdot b = \frac{1}{2} \cdot 4 \cdot 16 = 2 \cdot 16 = 32 \text{ cm}^2$

5. a) Entire area: $A = \frac{1}{2} \cdot (b_1 + b_2) \cdot h = \frac{1}{2} \cdot (20+13) \cdot 4 = \frac{1}{2} \cdot (33) \cdot 4 = \frac{1}{2} \cdot 4 \cdot (33) = 2 \cdot 33 = 66 \text{ m}^2$

   Unshaded area: $A = \frac{1}{2} \cdot h \cdot b = \frac{1}{2} \cdot 4 \cdot 20 = 2 \cdot 20 = 40 \text{ m}^2$

   Shaded area: $66 - 40 = 26 \text{ m}^2$

   b) Semi-circle's area: $A = \frac{1}{2} \pi r^2 \approx \frac{1}{2} \cdot 3.14 \cdot 3^2 = \frac{1}{2} \cdot 3.14 \cdot 9 = 14.13 \text{ ft}^2$

   Triangle's area: $A = \frac{1}{2} \cdot h \cdot b = \frac{1}{2} \cdot 3 \cdot 6 = \frac{1}{2} \cdot 18 = 9 \text{ ft}^2$

   Shaded area: $14.13 - 9 = 5.13 \text{ ft}^2$

6. $m\angle F = 35°$

   $\frac{FG}{12} = \frac{4}{6}$ $\qquad$ $\frac{GH}{10} = \frac{4}{6}$

   $6(FG) = 4 \cdot 12$ $\qquad$ $6(GH) = 10 \cdot 4$

   $6(FG) = 48$ $\qquad$ $6(GH) = 40$

   $FG = \frac{48}{6} = 8$ $\qquad$ $GH = \frac{40}{6} = \frac{20}{3}$

7. a) The pool consists of a rectangle with area $(16 - 4 - 4) \cdot 20 = 160 \text{ ft}^2$. The two ends combine to form a circle with area $\pi(4)^2 = \pi(16) = 16\pi \text{ ft}^2$. So, the surface area of the pool is : 160 ft² + 16 ft² = or approximately 210.24 ft².

   b) Volume = area of the base times height = $210.24 \cdot 3 = 630.72 \text{ ft}^2$.

   c) There are two rectangular pieces. Each of these has an area of $4 \cdot 20 = 80 \text{ ft}^2$.
   The two ends combined would form a washer-like figure in which we need to find the area of the shaded region.

   Outer Area: $\quad A_O = \pi \ r^2$ $\qquad\qquad$ Inner Area: $\quad A_I = \pi \ r^2$

   $A_O \approx 3.14 \cdot \left(\frac{16}{2}\right)^2$ $\qquad\qquad$ $A_I \approx 3.14 \ \left(\frac{16-4-4}{2}\right)^2$

   $A_O \approx 3.14 \cdot 8^2$ $\qquad\qquad$ $A_I \approx 3.14 \cdot 4^2$

   $A_O \approx 3.14 \cdot 64$ $\qquad\qquad$ $A_I \approx 3.14 \cdot 16$

   $A_O \approx 200.96 \text{ ft}^2$ $\qquad\qquad$ $A_I \approx 50.24 \text{ ft}^2$

   Surface area of track: $80 + 80 + 200.96 - 50.24 = 310.72 \text{ ft}^2$

8. a) $s = \frac{1}{2}(a+b+c) = \frac{1}{2}(6+9+12) = \frac{27}{2} = 13.5$

$A = \sqrt{s(s-a)(s-b)(s-c)} = \sqrt{13.5(13.5-6)(13.5-9)(13.5-12)}$
$= \sqrt{(13.5) \cdot (7.5) \cdot (4.5) \cdot (1.5)} = \sqrt{683.4375} \approx 26.1 \text{ ft}^2$

b) $A = \frac{1}{2} \cdot h \cdot b$

$\sqrt{683.4375} = \frac{1}{2} \cdot h \cdot 12$

$\sqrt{683.4375} = 6h$

$h = \frac{\sqrt{683.4375}}{6} \approx 4.4 \text{ ft}$

9. We need to first determine the volume of the spherical tank.

Spherical tank volume: $V = \left(\frac{4}{3}\pi r^3\right) \approx \frac{4}{3} \cdot 3.14 \cdot 15^3 = \frac{4 \cdot 3.14 \cdot 3375}{3} = \frac{42,390}{3} = 14,130 \text{ ft}^3$

Let h be the height of the cylindrical tank.

$V = \pi r^2 h \Rightarrow 14,130 \approx 3.14 \cdot (15)^2 h \Rightarrow$

$h = 3.14 \cdot 225h \Rightarrow 706.5h = 14,130 \Rightarrow$

$h = \frac{14,130}{706.5} = 20 \text{ ft}$

The cylindrical tank should be approximately 20 ft high..

10. a) Area of base: $s = \frac{1}{2}(a+b+c) = \frac{1}{2}(6+6+8) = \frac{20}{2} = 10$

$A = \sqrt{s(s-a)(s-b)(s-c)} = \sqrt{10(10-6)(10-6)(10-8)} =$
$\sqrt{10 \cdot 4 \cdot 4 \cdot 2} = \sqrt{320} \approx 17.9 \text{ cm}^2$

Volume = area of the base times height $\approx 17.9 \cdot 4 = 71.6 \text{ cm}^3$

b) Area of base = $A = \frac{1}{2} \cdot (b_1 + b_2) \cdot h = \frac{1}{2} \cdot (10+6) \cdot 2 = \frac{1}{2} \cdot (16) \cdot 2 = 8 \cdot 2 = 16 \text{ in}^2$

Volume = area of the base times height = $16 \cdot 3 = 48 \text{ in}^3$

11. a) 24 meters b) 3,460 milligrams c) 2,140 centiliters

12. The volume is four times as large, because in the formula for the surface area of a sphere, $A = 4\pi r^2$, the radius is squared.

If the radius is doubled: $A = 4\pi r^2 = 4\pi(2r)^2 = 4\pi(4r^2) = 4\pi(4r^2) = 4(4\pi r^2)$

13. 18 yards $\times \frac{1 \text{ meter}}{1.0936 \text{ yards}} \approx 16.459$ meters = 164.59 decimeters

14. 2,614.35 quarts $\times \frac{1 \text{ liter}}{1.0567 \text{ quarts}} \approx 2,474$ liters = 2.474 kiloliters

15. a) d, e, and f b) b and c

16.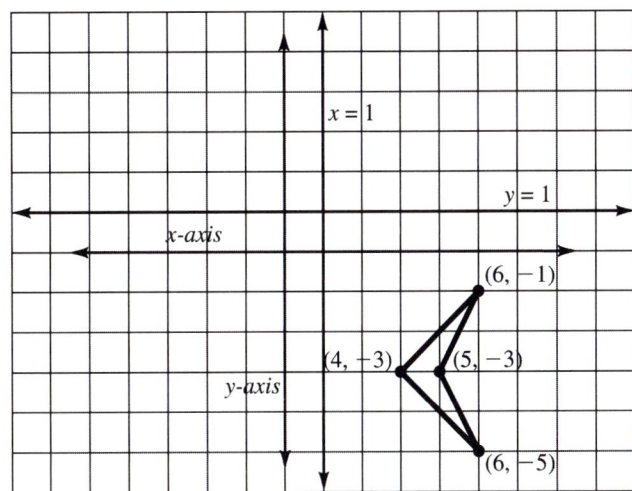

17. Reflectional symmetries about lines through *AE, BF, CG, DH*; rotational symmetries through $0°, 90°, 180°,$ and $270°$.

18.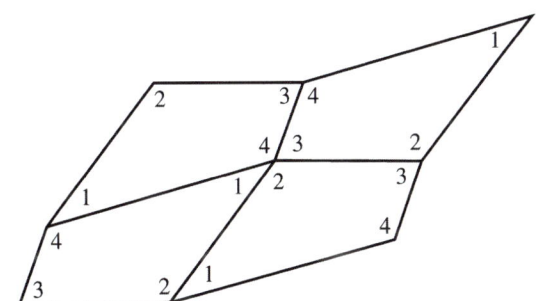

Of Further Interest: Fractals

1. As the image of the cloud is magnified, the same type of pattern appears.

2. Small portions of these objects repeat the same pattern as the original objects.

3. $3^D = 4$; the log function

4. The length of the curve at step $n + 1$ is 4/3 the length at step n, so as n increases, the length of the curve will continuously increase and exceed any fixed number.

5. The area of the gasket at step $n + 1$ is 3/4 the area at step n, so as n increases, the area of the gasket will continually decrease and be less than any fixed positive number.

6. The fractal dimension of a figure provides an idea of how much space the figure fills.

7. $4^5 = 1,024$

8. $\left(\dfrac{4}{3}\right)^5 = \dfrac{1,024}{243} \approx 4.21$

9. $\left(\dfrac{4}{3}\right)^{10} = \dfrac{1,048,576}{59,049} \approx 17.76$

10. False

11. a)

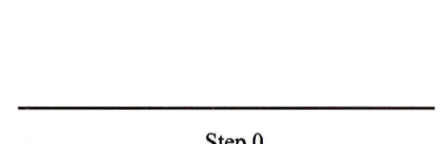
Step 0

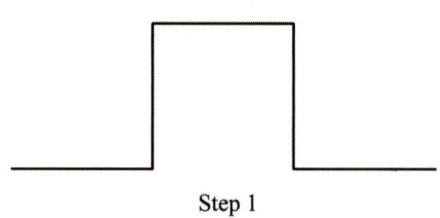
Step 1

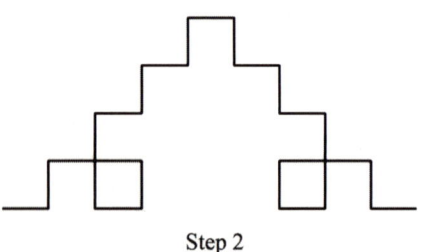
Step 2

b) Length in step 5: $\left(\dfrac{5}{3}\right)^5 = \dfrac{3{,}125}{243} \approx 12.86$

12. a)

Step 0

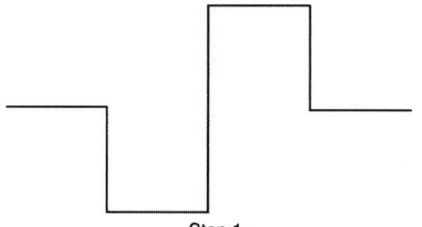
Step 1

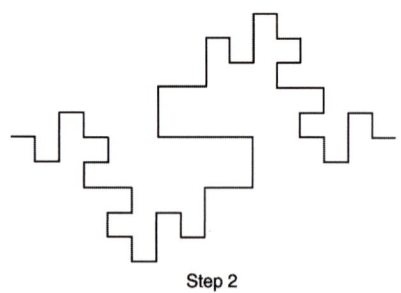
Step 2

b) Length in step 5: $\left(\dfrac{8}{4}\right)^5 = 2^5 = 32$

13. $4^D = 6$
$\log 4^D = \log 6$
$D \log 4 = \log 6$
$D = \dfrac{\log 6}{\log 4} \approx 1.29$

14. $4^D = 12$
$\log 4^D = \log 12$
$D \log 4 = \log 12$
$D = \dfrac{\log 12}{\log 4} \approx 1.79$

15. $3^D = 5$
$\log 3^D = \log 5$
$D \log 3 = \log 5$
$D = \dfrac{\log 5}{\log 3} \approx 1.46$

16. $4^D = 8$
$\log 4^D = \log 8$
$D \log 4 = \log 8$
$D = \dfrac{\log 8}{\log 4} = 1.5$

17.

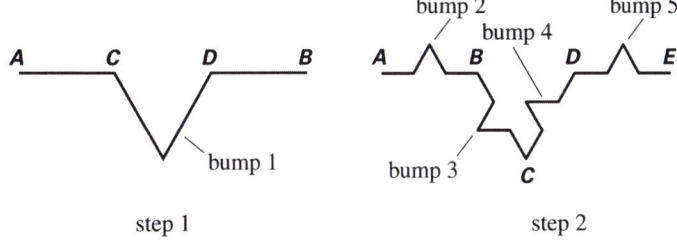

18. The dimension is the same as the Koch Curve, 1.26.

19.

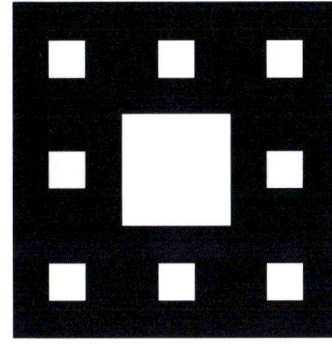

step 2

20.

step 2

21. 5^n

22. 7^n

23. At step 10: $\left(\dfrac{3}{4}\right)^{10} = \dfrac{59,049}{1,048,576} \approx 0.06$

　　At step n: $\left(\dfrac{3}{4}\right)^n$

24. At step 10: $\left(\dfrac{8}{9}\right)^{10} = \dfrac{1,073,741,824}{3,486,784,401} \approx 0.31$

　　At step n: $\left(\dfrac{8}{9}\right)^n$

25. – 26. No solutions provided

Chapter 9
LEGISLATIVE APPORTIONMENT: How Do We Measure Fairness?

Section 9.1 Understanding Apportionment

1. After the integer parts of the number of representatives each state deserves have been allocated, additional representatives are signed to the states with the largest fractional parts.

3. A reallocation of the extra tenths will not be done at a later time.

5. When measuring the fairness of an apportionment, it is important to take into account the size of the states involved.

7. a) 24 is the integer part; 0.098 is the fractional part b) 5 is the integer part; 0.02 is the fractional part

9.

State	Number of customer %	Step 1: Determine exact number of seats deserved	Step 2: Assign integer parts	Examine fractional parts	Step 3: Assign 1 additional seat
California	56,000 (37.6%)	0.376 × 11 = 4.136	4	0.136	4
Arizona	52,000 (34.9%)	0.349 × 11 = 3.839	3	0.839	4
Nevada	41,000 (27.5%)	0.275 × 11 = 3.025	3	0.025	3
Total	149,000 (100%)		11	10		11

11.

Guild	Number of members %	Step 1: Determine exact number of booths	Step 2: Assign integer parts	Examine fractional parts	Step 3: Assign 2 additional booths
Painters	87 (46.8%)	0.468 × 31 = 14.508	14	0.508	14
Sculptures	46 (24.7%)	0.247 × 31 = 7.657	7	0.657	8
Weavers	53 (28.5%)	0.285 × 31 = 8.835	8	0.835	9
Total	186 (100%)		31	29		31

13.

State	Number of residents %	Step 1: Determine exact number of Representative	Step 2: Assign integer parts	Examine fractional parts	Step 3: Assign 1 additional Representative
Alabama	4,041 (37.3%)	0.373 × 19 = 7.087	7	0.087	7
Mississippi	2,573 (23.7%)	0.237 × 19 = 4.503	4	0.503	5
Louisiana	4,220 (39.0%)	0.390 × 19 = 7.410	7	0.410	7
Total	10,834 (100%)		19	18		19

15. a) 353.7
    b) 353.785

17. 12.6, 34.2, 25.4, 11.7, 16.1

19. 34.79, 11.20, 23.90, 12.03, 2.99, 15.09

Original Number	Number truncated to hundredths	Discarded portion
34.789	34.78	0.009
11.2	11.20	0.000
23.897	23.89	0.007
12.034	12.03	0.004
2.987	2.98	0.007
15.093	15.09	0.003
100.000	99.97	

21. $\text{average constituency} = \dfrac{\text{population}}{\text{number of representatives}} = \dfrac{177{,}408}{3} = 59{,}136$

23. A is more poorly represented.

$\text{average constituency of A} = \dfrac{\text{population}}{\text{number of representatives}} = \dfrac{27{,}600}{16} = 1{,}725$

$\text{average constituency of B} = \dfrac{\text{population}}{\text{number of representatives}} = \dfrac{23{,}100}{14} = 1{,}650$

The absolute unfairness is $1{,}725 - 1{,}650 = 75$

25. $\text{average constituency of Naxxon} = \dfrac{\text{stockholders}}{\text{number of members}} = \dfrac{4{,}700}{5} = 940$

$\text{average constituency of Aroco} = \dfrac{\text{stockholders}}{\text{number of members}} = \dfrac{3{,}700}{4} = 925$

The absolute unfairness is $940 - 925 = 15$.

27. $\text{average constituency of Colorado} = \dfrac{\text{population}}{\text{number of representatives}} = \dfrac{2{,}200{,}000}{5} = 440{,}000$

$\text{average constituency of Delaware} = \dfrac{\text{population}}{\text{number of representatives}} = \dfrac{600{,}000}{1} = 600{,}000$

The absolute unfairness is $600{,}000 - 440{,}000 = 160{,}000$

$\text{relative unfairness} = \dfrac{\text{absolute unfairness}}{\text{smaller average consituency}} = \dfrac{160{,}000}{440{,}000} \approx 0.364$

29. a)

Division	Number of employees %	Step 1: Determine exact number of council members	Step 2: Assign integer parts	Examine fractional parts	Step 3: Assign 2 additional members
(V)ideo	140 (54.9%)	$0.549 \times 12 = 6.588$	6	0.588	7
(C)omputers	85 (33.3%)	$0.333 \times 12 = 3.996$	3	0.996	4
(B)usiness Products	30 (11.8%)	$0.118 \times 12 = 1.416$	1	0.416	1
Total	255 (100%)	12	10		12

b)

Division	Number of employees %	Step 1: Determine exact number of council members	Step 2: Assign integer parts	Examine fractional parts	Step 3: Assign 1 additional member
(V)ideo	140 (54.9%)	$0.549 \times 13 = 7.137$	7	0.137	7
(C)omputers	85 (33.3%)	$0.333 \times 13 = 4.329$	4	0.329	4
(B)usiness Products	30 (11.8%)	$0.118 \times 13 = 1.534$	1	0.534	2
Total	255 (100%)	13	12		13

Division	Number of employees %	Step 1: Determine exact number of council members	Step 2: Assign integer parts	Examine fractional parts	Step 3: Assign 2 additional members
(V)ideo	140 (54.9%)	$0.549 \times 14 = 7.686$	7	0.686	8
(C)omputers	85 (33.3%)	$0.333 \times 14 = 4.662$	4	0.662	5
(B)usiness Products	30 (11.8%)	$0.118 \times 14 = 1.652$	1	0.652	1
Total	255 (100%)	14	12		14

c) Business Products loses a seat when the council size is increased from 13 to 14.

31. 12.5, 31.1, 36.0, 15.7, 4.7

Original Number	Number truncated to tenths	Discarded portion
12.45	12.4	0.05
31.11	31.1	0.01
36.02	36.0	0.02
15.73	15.7	0.03
4.69	4.6	0.09
100.00	99.8	

33. 101.453, 96.001, 33.980, 67.133

Original Number	Number truncated to thousandths	Discarded portion
101.45328	101.453	0.00028
96.00059	96.000	0.00059
33.98023	33.980	0.00023
67.1329	67.132	0.0009
298.56700	298.565	

244 Chapter 9: Legislative Apportionment

35. a) average constituency of Naxxon = $\dfrac{\text{stockholders}}{\text{number of members}} = \dfrac{4,700}{6} \approx 783$

   average constituency of Aroco = $\dfrac{\text{stockholders}}{\text{number of members}} = \dfrac{3,700}{4} = 925$

   The absolute unfairness is 925 − 783 = 142.

   relative unfairness = $\dfrac{\text{absolute unfairness}}{\text{smaller average consituency}} = \dfrac{142}{783} \approx 0.181$

   b) average constituency of Naxxon = $\dfrac{\text{stockholders}}{\text{number of members}} = \dfrac{4,700}{5} = 940$

   average constituency of Aroco = $\dfrac{\text{stockholders}}{\text{number of members}} = \dfrac{3,700}{5} = 740$

   The absolute unfairness is 940 − 740 = 200.

   relative unfairness = $\dfrac{\text{absolute unfairness}}{\text{smaller average consituency}} = \dfrac{200}{740} \approx 0.270$

   c) Giving the extra representative to Naxxon will result in a smaller unfairness.

Section 9.2   The Huntington–Hill Apportionment Principle

1. A smaller relative unfairness occurs if the next representative is given to B.

3. It was assigned to the company with the highest Huntington–Hill number.

5. It minimizes relative unfairness.

7. a) average constituency of Musicians' Guild = $\dfrac{\text{members}}{\text{number of representatives}} = \dfrac{908}{5} = 181.6$

   average constituency of Artists' Alliance = $\dfrac{\text{members}}{\text{number of representatives}} = \dfrac{633}{3} = 211$

   relative unfairness = $\dfrac{\text{larger average consituency} - \text{smaller average consituency}}{\text{smaller average consituency}}$

   $= \dfrac{211 - 181.6}{181.6} \approx 0.162$

   Note: If you round the average constituency of the Musicians' Guild to the nearest whole person, it would be approximately 182. This implies the relative unfairness would be approximately 0.159.

7. (continued)

   b) $\text{average constituency of Musicians' Guild} = \dfrac{\text{members}}{\text{number of representatives}} = \dfrac{908}{4} = 227$

   $\text{average constituency of Artists' Alliance} = \dfrac{\text{members}}{\text{number of representatives}} = \dfrac{633}{4} \approx 158$

   $\text{relative unfairness} = \dfrac{\text{larger average consituency} - \text{smaller average consituency}}{\text{smaller average consituency}}$

   $= \dfrac{227 - 158}{158} \approx 0.437$

   Note: If you don't round the average constituency of the Artists' Alliance to the nearest whole person, it would be 158.25. The relative unfairness would be approximately 0.434.

   c) Giving the additional representative to Musicians' Guild will result in a smaller unfairness.

9. a) $\text{average number of passengers on red line} = \dfrac{\text{passengers}}{\text{number of cars}} = \dfrac{405}{10} = 40.5$

   $\text{average number of passengers on blue line} = \dfrac{\text{passengers}}{\text{number of cars}} = \dfrac{287}{7} = 41$

   $\text{relative unfairness} = \dfrac{\text{larger average} - \text{smaller average}}{\text{smaller average}} = \dfrac{41 - 40.5}{40.5} \approx 0.012$

   b) $\text{average number of passengers on red line} = \dfrac{\text{passengers}}{\text{number of cars}} = \dfrac{405}{9} = 45$

   $\text{average number of passengers on blue line} = \dfrac{\text{passengers}}{\text{number of cars}} = \dfrac{287}{8} = 35.875$

   $\text{relative unfairness} = \dfrac{\text{larger average} - \text{smaller average}}{\text{smaller average}} = \dfrac{45 - 35.875}{35.875} \approx 0.254$

   c) Giving the additional car to red line will result in a smaller unfairness.

11. If Alabama receives the additional representative

    $\text{average constituency of Alabama} = \dfrac{\text{population}}{\text{number of representatives}} = \dfrac{4{,}221{,}826}{8} \approx 527{,}728$

    $\text{average constituency of New York} = \dfrac{\text{population}}{\text{number of representatives}} = \dfrac{17{,}990{,}778}{31} \approx 580{,}348$

    $\text{relative unfairness} = \dfrac{\text{larger average consituency} - \text{smaller average consituency}}{\text{smaller average consituency}}$

    $= \dfrac{580{,}348 - 527{,}728}{527{,}728} \approx 0.100$

246 Chapter 9: Legislative Apportionment

11. (continued)

    If New York receives the additional representative

    $$\text{average constituency of Alabama} = \frac{\text{population}}{\text{number of representatives}} = \frac{4{,}221{,}826}{7} = 603{,}118$$

    $$\text{average constituency of New York} = \frac{\text{population}}{\text{number of representatives}} = \frac{17{,}990{,}778}{32} \approx 562{,}212$$

    $$\text{relative unfairness} = \frac{\text{larger average consituency} - \text{smaller average consituency}}{\text{smaller average consituency}}$$

    $$= \frac{603{,}118 - 562{,}212}{562{,}212} \approx 0.073$$

    Giving the additional representative to New York will result in a smaller unfairness.

13. If Colorado receives the additional representative

    $$\text{average constituency of Colorado} = \frac{\text{population}}{\text{number of representatives}} = \frac{3{,}294{,}473}{7} = 470{,}639$$

    $$\text{average constituency of Delaware} = \frac{\text{population}}{\text{number of representatives}} = \frac{666{,}168}{1} = 666{,}168$$

    $$\text{relative unfairness} = \frac{\text{larger average consituency} - \text{smaller average consituency}}{\text{smaller average consituency}}$$

    $$= \frac{666{,}168 - 470{,}639}{470{,}639} \approx 0.415$$

    If Delaware receives the additional representative

    $$\text{average constituency of Colorado} = \frac{\text{population}}{\text{number of representatives}} = \frac{3{,}294{,}473}{6} \approx 549{,}079$$

    $$\text{average constituency of Delaware} = \frac{\text{population}}{\text{number of representatives}} = \frac{666{,}168}{2} = 333{,}084$$

    $$\text{relative unfairness} = \frac{\text{larger average consituency} - \text{smaller average consituency}}{\text{smaller average consituency}}$$

    $$= \frac{549{,}079 - 333{,}084}{333{,}084} \approx 0.648$$

    Giving the additional representative to Colorado will result in a smaller unfairness.

15. Huntington-Hill number $= \dfrac{(\text{population } X)^2}{x(x+1)} = \dfrac{908^2}{4 \cdot 5} = 41{,}223.2$

17. Huntington-Hill number $= \dfrac{(\text{population } X)^2}{x(x+1)} = \dfrac{287^2}{7 \cdot 8} \approx 1{,}470.9$

19. Huntington-Hill number $= \dfrac{(\text{population } X)^2}{x(x+1)} = \dfrac{4.2^2}{7 \cdot 8} \approx 0.315$

Section 9.2: The Huntington–Hill Apportionment Principle

21. Huntington-Hill number $= \dfrac{(\text{population } X)^2}{x(x+1)} = \dfrac{3.3^2}{6 \cdot 7} \approx 0.259$

23. Huntington-Hill number for Indiana $= \dfrac{(\text{population } X)^2}{x(x+1)} = \dfrac{5.5^2}{10 \cdot 11} = 0.275$

    Huntington-Hill number for Illinois $= \dfrac{(\text{population } X)^2}{x(x+1)} = \dfrac{11.4^2}{20 \cdot 21} \approx 0.309$

    Since $0.275 < 0.309$, Illinois should receive the additional representative.

25. Huntington-Hill number for Alaska $= \dfrac{(\text{population } X)^2}{x(x+1)} = \dfrac{0.6^2}{1 \cdot 2} = 0.180$

    Huntington-Hill number for New Hampshire $= \dfrac{(\text{population } X)^2}{x(x+1)} = \dfrac{1.1^2}{2 \cdot 3} \approx 0.202$

    Huntington-Hill number for Wyoming $= \dfrac{(\text{population } X)^2}{x(x+1)} = \dfrac{0.5^2}{1 \cdot 2} = 0.125$

    Since $0.125 < 0.180 < 0.202$, New Hampshire should receive the additional representative.

27. Huntington-Hill number for Musicians' Guild $= \dfrac{(\text{population } X)^2}{x(x+1)} = \dfrac{908^2}{4 \cdot 5} \approx 41{,}223$

    Huntington-Hill number for Artists' Alliance $= \dfrac{(\text{population } X)^2}{x(x+1)} = \dfrac{633^2}{3 \cdot 4} \approx 33{,}391$

    Huntington-Hill number for Actors' Coalition $= \dfrac{(\text{population } X)^2}{x(x+1)} = \dfrac{420^2}{2 \cdot 3} = 29{,}400$

    Since $29400 < 33391 < 41223$, the Musicians' Guild should receive the additional representative.

29. Huntington-Hill number for red line $= \dfrac{(\text{population } X)^2}{x(x+1)} = \dfrac{405^2}{9 \cdot 10} = 1{,}822.5$

    Huntington-Hill number for blue line $= \dfrac{(\text{population } X)^2}{x(x+1)} = \dfrac{287^2}{7 \cdot 8} = 1{,}470.875$

    Huntington-Hill number for yellow line $= \dfrac{(\text{population } X)^2}{x(x+1)} = \dfrac{156^2}{3 \cdot 4} = 2{,}028$

    Since $1{,}470.875 < 1{,}822.5 < 2{,}028$ the yellow line should receive the additional car.

248   Chapter 9: Legislative Apportionment

31. Huntington-Hill number for Musicians' Guild $= \dfrac{(\text{population } X)^2}{x(x+1)} = \dfrac{908^2}{5 \cdot 6} \approx 27{,}482$

   Huntington-Hill number for Artists' Alliance $= \dfrac{(\text{population } X)^2}{x(x+1)} = \dfrac{633^2}{3 \cdot 4} \approx 33{,}391$

   Huntington-Hill number for Actors' Coalition $= \dfrac{(\text{population } X)^2}{x(x+1)} = \dfrac{420^2}{2 \cdot 3} = 29{,}400$

   Since $27{,}482 < 29{,}400 < 33{,}391$ the Artists' Alliance should receive the second additional representative.

33. According to the assignment of the 11 representatives in Exercise 27, the Musicians' Guild has 5; the Artists' Alliance has 4; and the Actors Coalition has 2. This is consistent with the Hamilton apportionment method

Groups	Number of members %	Step 1: Determine exact number of representatives	Step 2: Assign integer parts	Examine fractional parts	Step 3: Assign additional representative
Musicians' Guild	908 (46.30%)	0.4630 × 11 = 5.0930	5	0.0930	5
Artists' Alliance	633 (32.28%)	0.3228 × 11 = 3.5508	3	0.5508	4
Actors' Coalition	420 (21.42%)	0.2142 × 11 = 2.3562	2	0.3562	2
Total	1961 (100%)	11	10		11

35. The apportionment principle allows us to assign representatives one at a time, at each stage, giving the next representative to the state that most deserves it. We make the assignments so as to minimize the relative unfairness of the apportionment.

37. This would be similar to the situation that Michigan has 15 representatives and Wisconsin has 9 and we want to assign one more representative.

   Huntington-Hill number for Michigan $= \dfrac{(\text{population } X)^2}{x(x+1)} = \dfrac{9.3^2}{15 \cdot 16} \approx 0.360$

   Huntington-Hill number for Wisconsin $= \dfrac{(\text{population } X)^2}{x(x+1)} = \dfrac{X^2}{9 \cdot 10} = \dfrac{X^2}{90}$

   $0.360 < \dfrac{X^2}{90} \Rightarrow 32.4 < X^2 \Rightarrow 5.6921 < X$ Therefore, Wisconsin's population needs to be approximately 5.7 million in order to take a seat away from Michigan.

39. No solution provided

Section 9.3   Applications Of The Apportionment Principle

1. Naxxon had the largest Huntington–Hill number.

3. Calculate a table of Huntington–Hill numbers.

5. We use the entries in row 1. Since the carpenters have the largest number, they get the fourth seat.

7. We look at the entries that read 104.2, 162.0, and 160.2. Since the plumbers have the largest number, they get the fifth seat.

9. We use the entries in row 1. Since the Carney building has the largest number, they get the fourth seat.

11. We look at the entries that read 264.5, 181.5, and 308.2. Since the Carney building has the largest number, they get the sixth seat.

13. Theater, Music, and Dance each get one representative in no particular order. The remaining seven assignments in order would be MTMTMTD.

15. a)

Boroughs	Number of residents %	Step 1: Determine exact number of representatives	Step 2: Assign integer parts	Examine fractional parts	Step 3: Assign 2 additional representatives
Alsace	23,000 (32.4%)	$0.324 \times 10 = 3.24$	3	0.24	3
Bradford	34,000 (47.9%)	$0.479 \times 10 = 4.79$	4	0.79	5
Cambria	14,000 (19.7%)	$0.197 \times 10 = 1.97$	1	0.97	2
Total	71,000 (100%)	10	8		10

b) Make the following table.

Current Representation	(A)lsace	(B)radford	(C)ambria
1	$\dfrac{23^2}{1 \cdot 2} = 264.5$	$\dfrac{34^2}{1 \cdot 2} = 578$	$\dfrac{14^2}{1 \cdot 2} = 98$
2	$\dfrac{23^2}{2 \cdot 3} \approx 88.2$	$\dfrac{34^2}{2 \cdot 3} \approx 192.7$	$\dfrac{14^2}{2 \cdot 3} \approx 32.7$
3	$\dfrac{23^2}{3 \cdot 4} \approx 44.1$	$\dfrac{34^2}{3 \cdot 4} \approx 96.3$	$\dfrac{14^2}{3 \cdot 4} \approx 16.3$
4	$\dfrac{23^2}{4 \cdot 5} \approx 26.5$	$\dfrac{34^2}{4 \cdot 5} = 57.8$	$\dfrac{14^2}{4 \cdot 5} = 9.8$

Alsace, Bradford, and Cambria each get one representative in no particular order. The remaining seven assignments in order would be BABCBAB. So Alsace should have 3 representatives, Bradford should have 5, and Cambria should have 2, according to the Huntington–Hill apportionment principle.

17. Make the following table.

Current Representation	Utah	Idaho	Oregon
1	$\dfrac{1.7^2}{1\cdot 2}=1.445$	$\dfrac{1.0^2}{1\cdot 2}=0.500$	$\dfrac{2.8^2}{1\cdot 2}=3.920$
2	$\dfrac{1.7^2}{2\cdot 3}\approx 0.482$	$\dfrac{1.0^2}{2\cdot 3}\approx 0.167$	$\dfrac{2.8^2}{2\cdot 3}\approx 1.307$
3	$\dfrac{1.7^2}{3\cdot 4}\approx 0.241$	$\dfrac{1.0^2}{3\cdot 4}\approx 0.083$	$\dfrac{2.8^2}{3\cdot 4}\approx 0.653$
4	$\dfrac{1.7^2}{4\cdot 5}=0.145$	$\dfrac{1.0^2}{4\cdot 5}=0.050$	$\dfrac{2.8^2}{4\cdot 5}=0.392$

Utah, Idaho, and Oregon each get one representative in no particular order. If we assume "UT" represents Utah, "ID" represents Idaho, and "OR" represents Oregon then the remaining seven assignments in order would be OR, UT, OR, OR, ID, UT, OR. So Utah would have 3 representatives, Idaho would have 2, and Oregon would have 5 according to the Huntington–Hill method.

19. Make the following table.

Current Representation	Arkansas	Kansas	Nebraska
1	$\dfrac{2.4^2}{1\cdot 2}=2.880$	$\dfrac{2.5^2}{1\cdot 2}=3.125$	$\dfrac{1.6^2}{1\cdot 2}=1.280$
2	$\dfrac{2.4^2}{2\cdot 3}=0.960$	$\dfrac{2.5^2}{2\cdot 3}\approx 1.042$	$\dfrac{1.6^2}{2\cdot 3}\approx 0.427$
3	$\dfrac{2.4^2}{3\cdot 4}=0.480$	$\dfrac{2.5^2}{3\cdot 4}\approx 0.521$	$\dfrac{1.6^2}{3\cdot 4}\approx 0.213$
4	$\dfrac{2.4^2}{4\cdot 5}=0.288$	$\dfrac{2.5^2}{4\cdot 5}\approx 0.313$	$\dfrac{1.6^2}{4\cdot 5}=0.128$
5	$\dfrac{2.4^2}{5\cdot 6}=0.192$	$\dfrac{2.5^2}{5\cdot 6}\approx 0.208$	$\dfrac{1.6^2}{5\cdot 6}\approx 0.085$

Arkansas, Kansas, and Nebraska each get one representative in no particular order. If we assume "AR" represents Arkansas, "KS" represents Kansas, and "NE" represents Nebraska then the remaining eight assignments in order would be KS, AR, NE, KS, AR, KS, AR, NE. So Arkansas would have 4 representatives, Kansas would have 4, and Nebraska would have 3 according to the Huntington–Hill method.

21. BB; We need to find one more entry for row 6 for Business. That entry is $\dfrac{40^2}{6\cdot 7}\approx 38.1$. By comparing 45.0, 33.3, and 38.1 we can see that the twelfth representative should be assigned to H. Humanities would have 4 representatives, Science would have 2, and Business would have 5.

23. Make the following table.

Current Representation	Region 1	Region 2	Region 3
1	$\dfrac{123^2}{1\cdot 2} \approx 7565$	$\dfrac{44^2}{1\cdot 2} = 968$	$\dfrac{79^2}{1\cdot 2} \approx 3121$
2	$\dfrac{123^2}{2\cdot 3} \approx 2522$	$\dfrac{44^2}{2\cdot 3} \approx 323$	$\dfrac{79^2}{2\cdot 3} \approx 1040$
3	$\dfrac{123^2}{3\cdot 4} \approx 1261$	$\dfrac{44^2}{3\cdot 4} \approx 161$	$\dfrac{79^2}{3\cdot 4} \approx 520$
4	$\dfrac{123^2}{4\cdot 5} \approx 756$	$\dfrac{44^2}{4\cdot 5} \approx 97$	$\dfrac{79^2}{4\cdot 5} \approx 312$

Region 1, region 2, and region 3 each get one representative in no particular order. If we assume "A" represents region 1, "B" represents region 2, and "C" represents region 3, then the remaining four assignments in order would be ACAA. So region 1 would have 4 representatives, region 2 would have 1, and region 3 would have 2, according to the Huntington–Hill method.

25. Make the following table.

Current Representation	High-impact	Low-impact	Jazzercise	Step
1	$\dfrac{58^2}{1\cdot 2} = 1682$	$\dfrac{32^2}{1\cdot 2} = 512$	$\dfrac{11^2}{1\cdot 2} \approx 61$	$\dfrac{39^2}{1\cdot 2} = 761$
2	$\dfrac{58^2}{2\cdot 3} \approx 561$	$\dfrac{32^2}{2\cdot 3} \approx 171$	$\dfrac{11^2}{2\cdot 3} \approx 20$	$\dfrac{39^2}{2\cdot 3} \approx 254$

High-Impact aerobics, Low-Impact aerobics, Jazzercise, and Step Exercise each get one class in no particular order. The remaining 2 classes would be High-Impact aerobics and Step Exercise, according to the Huntington–Hill method.

27. – 31. No solutions provided

Section 9.4 Other Paradoxes And Apportionment Methods

1. The standard divisor is the total population divided by the number of representatives; the standard quota is the state's population divided by the standard divisor.

3. by trial and error

5. Jefferson rounds the modified quotas down; Adams rounds them up; Webster rounds them in the usual way.

7. a) Standard Divisor $= \dfrac{\text{Total Population}}{\text{Number of Representatives Allocated}} = \dfrac{100{,}000}{8} = 12{,}500$

   b) Standard Quota $= \dfrac{\text{State's Population}}{\text{Standard Divisor}} = \dfrac{27{,}000}{12{,}500} = 2.16$

   c) 2.16 rounded down is 2.

9. a) Standard Divisor = $\dfrac{\text{Total Population}}{\text{Number of Representatives Allocated}} = \dfrac{220{,}000}{11} = 20{,}000$

   b) Standard Quota = $\dfrac{\text{State's Population}}{\text{Standard Divisor}} = \dfrac{92{,}000}{20{,}000} = 4.6$

   c) 4.6 rounded down is 4.

11. a) Standard Divisor = $\dfrac{\text{Total Population}}{\text{Number of Representatives Allocated}} = \dfrac{120{,}000}{15} = 8{,}000$

   b) Standard Quota = $\dfrac{\text{State's Population}}{\text{Standard Divisor}} = \dfrac{36{,}800}{8{,}000} = 4.6$

   c) 4.6 rounded up is 5.

13. a) Standard Divisor = $\dfrac{\text{Total Population}}{\text{Number of Representatives Allocated}} = \dfrac{134{,}400}{28} = 4{,}800$

   b) Standard Quota = $\dfrac{\text{State's Population}}{\text{Standard Divisor}} = \dfrac{56{,}160}{4{,}800} = 11.7$

   c) 11.7 rounded up is 12.

15. a) Standard Divisor = $\dfrac{\text{Total Population}}{\text{Number of Representatives Allocated}} = \dfrac{504{,}000}{18} = 28{,}000$

   b) Standard Quota = $\dfrac{\text{State's Population}}{\text{Standard Divisor}} = \dfrac{74{,}200}{28{,}000} = 2.65$

   c) 2.65 rounded is 3.

17. a) Standard Divisor = $\dfrac{\text{Total Population}}{\text{Number of Representatives Allocated}} = \dfrac{102{,}000}{12} = 8{,}500$

   b) Standard Quota = $\dfrac{\text{State's Population}}{\text{Standard Divisor}} = \dfrac{19{,}975}{8{,}500} = 2.35$

   c) 2.35 rounded is 2.

19. Because the Jefferson method rounds the modified quotas down, we often need larger modified quotas and hence smaller modified divisors.

21. She needs to try a larger modified divisor.

23. The total population considered is $56{,}000 + 52{,}000 + 41{,}000 = 149{,}000$.

   Standard Divisor = $\dfrac{149{,}000}{11} \approx 13{,}545.45$; California's Standard Quota = $\dfrac{56{,}000}{13{,}545.45} \approx 4.134$;

   Nevada's Standard Quota = $\dfrac{41{,}000}{13{,}545.45} \approx 3.027$; Arizona's Standard Quota = $\dfrac{52{,}000}{13{,}545.45} \approx 3.839$

25. With the standard divisor of 13,545.45, a modified divisor is 14,000.

	California	Nevada	Arizona
Number of Customers	56,000	41,000	52,000
Standard Quota	4.134	3.027	3.839
Modified Quota	$\dfrac{56,000}{14,000} = 4$	$\dfrac{41,000}{14,000} \approx 2.93$	$\dfrac{52,000}{14,000} \approx 3.71$
Round Modified Quota Up	4	3	4
			Total = 11

27. The total number of workers considered is $213 + 273 + 178 = 664$.

    Standard Divisor $= \dfrac{664}{20} = 33.2$; Performers' Standard Quota $= \dfrac{213}{33.2} \approx 6.416$;

    Food Workers' Standard Quota $= \dfrac{273}{33.2} \approx 8.223$;

    Maintenance Workers' Standard Quota $= \dfrac{178}{33.2} \approx 5.361$

29. With the standard divisor of 33.2, a modified divisor is 35.

	Performers	Food Workers	Maintenance
Number of Workers	213	273	178
Standard Quota	6.416	8.223	5.361
Modified Quota	$\dfrac{213}{35} \approx 6.09$	$\dfrac{273}{35} = 7.8$	$\dfrac{178}{35} \approx 5.09$
Round Modified Quota Up	7	8	6
			Total = 21

This total was too high, so try a modified divisor of 36.

	Performers	Food Workers	Maintenance
Number of Workers	213	273	178
Modified Quota	$\dfrac{213}{36} \approx 5.92$	$\dfrac{273}{36} \approx 7.58$	$\dfrac{178}{36} \approx 4.94$
Round Modified Quota Up	6	8	5
			Total = 19

This total was too low, so try a modified divisor of 35.5.

	Performers	Food Workers	Maintenance
Number of Workers	213	273	178
Modified Quota	$\dfrac{213}{35.5} = 6$	$\dfrac{273}{35.5} \approx 7.69$	$\dfrac{178}{35.5} \approx 5.01$
Round Modified Quota Up	6	8	6
			Total = 20

254   Chapter 9: Legislative Apportionment

31. The total number of members considered is $25 + 18 + 29 + 31 = 103$.

    Standard Divisor $= \dfrac{103}{20} = 5.15$; Electricians Union's Standard Quota $= \dfrac{25}{5.15} \approx 4.854$;

    Plumber Union's Standard Quota $= \dfrac{18}{5.15} \approx 3.495$; Painters Union's Standard Quota $= \dfrac{29}{5.15} \approx 5.631$;

    Carpenters Union's Standard Quota $= \dfrac{31}{5.15} \approx 6.019$

33. With the standard divisor of 5.15, a modified divisor is 5.5.

	Electricians	Plumbers	Painters	Carpenters
Number of Members	25	18	29	31
Standard Quota	4.85	3.50	5.63	6.02
Modified Quota	$\dfrac{25}{5.5} \approx 4.55$	$\dfrac{18}{5.5} \approx 3.27$	$\dfrac{29}{5.5} \approx 5.27$	$\dfrac{31}{5.5} \approx 5.64$
Round Modified Quota Up	5	4	6	6
				Total = 21

    This total was too high, so try a modified divisor of 6.

	Electricians	Plumbers	Painters	Carpenters
Number of Members	25	18	29	31
Modified Quota	$\dfrac{25}{6} \approx 4.17$	$\dfrac{18}{6} = 3$	$\dfrac{29}{6} \approx 4.83$	$\dfrac{31}{6} \approx 5.17$
Round Modified Quota Up	5	3	5	6
				Total = 19

    This total was too low, so try a modified divisor of 5.8.

	Electricians	Plumbers	Painters	Carpenters
Number of Members	25	18	29	31
Modified Quota	$\dfrac{25}{5.8} \approx 4.31$	$\dfrac{18}{5.8} \approx 3.10$	$\dfrac{29}{5.8} = 5$	$\dfrac{31}{5.8} \approx 5.34$
Round Modified Quota Up	5	4	5	6
				Total = 20

35. The total number of graduate students considered is $30 + 20 + 17 + 14 = 81$.

    Standard Divisor $= \dfrac{81}{19} \approx 4.263$; Fiction Majors' Standard Quota $= \dfrac{30}{4.263} \approx 7.037$;

    Poetry Majors' Standard Quota $= \dfrac{20}{4.263} \approx 4.692$;

    Technical Writing Majors' Standard Quota $= \dfrac{17}{4.263} \approx 3.988$;

    Media Writers Majors' Standard Quota $= \dfrac{14}{4.26} \approx 3.284$

37. With the standard divisor of 4.263 a modified divisor is 4.3.

	Fiction	Poetry	Technical	Media
Number of Students	30	20	17	14
Standard Quota	7.037	4.692	3.988	3.284
Modified Quota	$\frac{30}{4.3} \approx 6.98$	$\frac{20}{4.3} \approx 4.65$	$\frac{17}{4.3} \approx 3.95$	$\frac{14}{4.3} \approx 3.26$
Round Modified Quota Up	7	5	4	4
				Total = 20

This total was too high, so try a modified divisor of 5.

	Fiction	Poetry	Technical	Media
Number of Students	30	20	17	14
Modified Quota	$\frac{30}{5} = 6$	$\frac{20}{5} = 4$	$\frac{17}{5} = 3.4$	$\frac{14}{5} = 2.8$
Round Modified Quota Up	6	4	4	3
				Total = 17

This total was too low, so try a modified divisor of 4.8.

	Fiction	Poetry	Technical	Media
Number of Students	30	20	17	14
Modified Quota	$\frac{30}{4.8} = 6.25$	$\frac{20}{4.8} \approx 4.17$	$\frac{17}{4.8} \approx 3.54$	$\frac{14}{4.8} \approx 2.92$
Round Modified Quota Up	7	5	4	3
				Total = 19

39. The total number of incidents considered is $107 + 65 + 43 + 59 + 27 = 301$.

   Standard Divisor $= \frac{301}{13} \approx 23.154$; Region 1's Standard Quota $= \frac{107}{23.154} \approx 4.621$;

   Region 2's Standard Quota $= \frac{65}{23.154} \approx 2.807$; Region 3's Standard Quota $= \frac{43}{23.154} \approx 1.857$;

   Region 4's majors' Standard Quota $= \frac{59}{23.154} \approx 2.548$; Region 5's Standard Quota $= \frac{27}{23.154} \approx 1.166$

41. With the standard divisor of 23.154, a modified divisor is 25.

	Region 1	Region 2	Region 3	Region 4	Region 5
Number of Incidents	107	65	43	59	27
Standard Quota	4.621	2.807	1.857	2.548	1.166
Modified Quota	$\frac{107}{25}=4.28$	$\frac{65}{25}=2.6$	$\frac{43}{25}=1.72$	$\frac{59}{25}=2.36$	$\frac{27}{25}=1.08$
Round Modified Quota Up	5	3	2	3	2
					Total = 15

This total was too high, so try a modified divisor is 30.

	Region 1	Region 2	Region 3	Region 4	Region 5
Number of Incidents	107	65	43	59	27
Modified Quota	$\frac{107}{30}\approx 3.57$	$\frac{65}{30}\approx 2.17$	$\frac{43}{30}\approx 1.43$	$\frac{59}{30}\approx 1.97$	$\frac{27}{30}=0.9$
Round Modified Quota Up	4	3	2	2	1
					Total = 12

This total was too low, so try a modified divisor is 27.

	Region 1	Region 2	Region 3	Region 4	Region 5
Number of Incidents	107	65	43	59	27
Modified Quota	$\frac{107}{27}\approx 3.96$	$\frac{65}{27}\approx 2.41$	$\frac{43}{27}\approx 1.59$	$\frac{59}{27}\approx 2.19$	$\frac{27}{27}=1$
Round Modified Quota Up	4	3	2	3	1
					Total = 13

43. The total number of students considered is $56+29+11+4=100$.

Standard Divisor $=\frac{100}{6}\approx 16.667$; High-Impact's Standard Quota $=\frac{56}{16.667}\approx 3.36$;

Low-Impact's Standard Quota $=\frac{29}{16.667}\approx 1.74$;

Jazzercise's Standard Quota $=\frac{11}{16.667}\approx 0.66$;

Step Exercise's Standard Quota $=\frac{4}{16.667}\approx 0.24$

45. With the standard divisor of 16.67 a modified divisor of 27.5.

	High Impact	Low Impact	Jazzercise	Step Exercise
Number of Students	56	29	11	4
Standard Quota	3.36	1.74	0.66	0.24
Modified Quota	$\frac{56}{27.5} \approx 2.04$	$\frac{29}{27.5} \approx 1.05$	$\frac{11}{27.5} = 0.4$	$\frac{4}{27.5} \approx 0.15$
Round Modified Quota Up	3	2	1	1
				Total = 7

This total was too high, so try a modified divisor of 29.

	High Impact	Low Impact	Jazzercise	Step Exercise
Number of Students	56	29	11	4
Modified Quota	$\frac{56}{29} \approx 1.93$	$\frac{29}{29} = 1$	$\frac{11}{29} \approx 0.38$	$\frac{4}{29} \approx 0.14$
Round Modified Quota Up	2	1	1	1
				Total = 5

This total was too low, so try a modified divisor of 28

	High Impact	Low Impact	Jazzercise	Step Exercise
Number of Students	56	29	11	4
Modified Quota	$\frac{56}{28} = 2$	$\frac{29}{28} \approx 1.04$	$\frac{11}{28} \approx 0.39$	$\frac{4}{28} \approx 0.14$
Round Modified Quota Up	2	2	1	1
				Total = 6

47. We must first find how the ten representatives were originally apportioned according to the Hamilton method. The standard divisor $= \dfrac{10{,}000}{10} = 1{,}000$.

State	Number of residents	Standard Quota	Integer parts	Fractional Parts	Assign 2 additional representatives
A	570	$\dfrac{570}{1{,}000} = 0.570$	0	0.570	1
B	2,557	$\dfrac{2{,}557}{1{,}000} = 2.557$	2	0.557	2
C	6,873	$\dfrac{6{,}873}{1{,}000} = 6.873$	6	0.873	7
Total	10,000		8		10

Determine the apportionment ten years later. The standard divisor $= \dfrac{10{,}080}{10} = 1{,}008$.

State	Number of residents	Standard Quota	Integer parts	Fractional Parts	Assign 2 additional representatives
A	590	$\dfrac{590}{1{,}008} \approx 0.585$	0	0.585	0
B	2,617	$\dfrac{2{,}617}{1{,}008} \approx 2.596$	2	0.596	3
C	6,873	$\dfrac{6{,}873}{1{,}008} \approx 6.818$	6	0.818	7
Total	10,080		8		10

Rate of increase in A's population $= \dfrac{20}{570} \approx 3.5\%$;

Rate of increase in B's population $= \dfrac{60}{2{,}557} \approx 2.3\%$

Yes, because A's population has increased faster than B's, but A loses a representative to B in the reapportionment.

49. We must first find how the 150 security personnel were originally apportioned according to the Hamilton method. The standard divisor $= \dfrac{150{,}000}{150} = 1{,}000$.

Airport	Number of passengers	Standard Quota	Integer parts	Fractional Parts	Assign 2 security personnel
Allenport	120,920	$\dfrac{120{,}920}{1{,}000} = 120.920$	120	0.920	121
Bakerstown	5,550	$\dfrac{5{,}550}{1{,}000} = 5.550$	5	0.550	6
Columbia City	23,530	$\dfrac{23{,}530}{1{,}000} = 23.530$	23	0.530	23
Total	150,000		148		150

Determine the apportionment one year later. The standard divisor $= \dfrac{151{,}000}{150} \approx 1006.67$.

Airport	Number of passengers	Standard Quota	Integer parts	Fractional Parts	Assign 2 security personnel
Allenport	121,420	$\dfrac{121{,}420}{1{,}006.67} \approx 120.615$	120	0.615	121
Bakerstown	5,650	$\dfrac{5{,}650}{1{,}006.67} \approx 5.613$	5	0.613	5
Columbia City	23,930	$\dfrac{23{,}930}{1{,}006.67} \approx 23.771$	23	0.771	24
Total	151,000		148		150

Rate of increase at Columbia City $= \dfrac{23{,}930 - 23{,}530}{23{,}530} = \dfrac{400}{23{,}530} \approx 1.7\%$

Rate of increase at Bakerstown $= \dfrac{5{,}650 - 5{,}550}{5{,}550} = \dfrac{100}{5{,}550} \approx 1.8\%$

Yes, because the number of passengers per week at Bakerstown has increased faster than at Columbia City, but Bakerstown loses a security guard to Columbia City.

51. Consider the apportionment among only states A and B. The standard divisor $= \dfrac{9{,}770}{100} = 97.7$.

State	Number of residents	Standard Quota	Integer parts	Fractional Parts	Assign 1 additional computer labs
A	2,000	$\dfrac{2{,}000}{97.7} \approx 20.471$	20	0.471	20
B	7,770	$\dfrac{7{,}770}{97.7} \approx 79.529$	79	0.529	**80**
Total	9,770		99		100

Now consider approximately the number of representatives C would get with the standard divisor of 97.7. C would get $\dfrac{600}{97.7} \approx 6.141$ representatives.

Determine the apportionment with 6 additional committee members.

The standard divisor $= \dfrac{10{,}370}{106} \approx 97.83$.

State	Number of residents	Standard Quota	Integer parts	Fractional Parts	Assign 1 additional computer labs
A	2,000	$\dfrac{2{,}000}{97.83} \approx 20.444$	20	0.444	21
B	7,770	$\dfrac{7{,}770}{97.83} \approx 79.423$	79	0.423	**79**
C	600	$\dfrac{600}{97.83} \approx 6.133$	6	0.133	6
Total	10,370		105		106

Yes, when C's representatives are added to the commission, B loses a representative to A.

53. Consider the apportionment among only states A, B, and C. The standard divisor $= \dfrac{16,200}{56} \approx 289.29$.

State	Number of residents	Standard Quota	Integer parts	Fractional Parts	Assign 1 additional computer labs
A	8,700	$\dfrac{8,700}{289.29} \approx 30.074$	30	0.074	30
B	4,300	$\dfrac{4,300}{289.29} \approx 14.864$	14	0.864	15
C	3,200	$\dfrac{3,200}{289.29} \approx 11.062$	11	0.062	11
Total	16,200		55		56

Now consider approximately the number of representatives D would get with the standard divisor of 289.29. D would get $\dfrac{6,500}{289.29} \approx 22.469$ representatives.

Determine the apportionment with 22 additional committee members.

The standard divisor $= \dfrac{22,700}{78} \approx 291.03$.

State	Number of residents	Standard Quota	Integer parts	Fractional Parts	Assign 3 additional computer labs
A	8,700	$\dfrac{8,700}{291.03} \approx 29.894$	29	0.894	30
B	4,300	$\dfrac{4,300}{291.03} \approx 14.775$	14	0.775	15
C	3,200	$\dfrac{3,200}{291.03} \approx 10.995$	10	0.995	11
D	6,500	$\dfrac{6,500}{291.03} \approx 22.334$	22	0.334	22
Total	22,700		75		78

No, when D's representatives are added to the commission, there is no change in A's, B's, or C's apportionment.

55. Total Population $= 700 + 1,500 + 820 + 4,530 + 2,200 + 550 = 10,300$

Standard Divisor $= \dfrac{\text{Total Population}}{\text{Number of Representatives Allocated}} = \dfrac{10,300}{200} = 51.5$

State	Number of residents	Standard Quota	Lower Quota	Upper Quota
A	700	$\frac{700}{51.5} \approx 13.592$	13	14
B	1,500	$\frac{1,500}{51.5} \approx 29.126$	29	30
C	820	$\frac{820}{51.5} \approx 15.922$	15	16
D	4,530	$\frac{4,530}{51.5} \approx 87.961$	87	**88**
E	2,200	$\frac{2,200}{51.5} \approx 42.718$	42	43
F	550	$\frac{550}{51.5} \approx 10.680$	10	11

55. (continued)

Jefferson apportionment method: With the standard divisor of 51.5 a modified divisor is 50.5.

State	Population	Modified Quota	Round Modified Quota Down
A	700	$\frac{700}{50.5} \approx 13.861$	13
B	1,500	$\frac{1,500}{50.5} \approx 29.703$	29
C	820	$\frac{820}{50.5} \approx 16.238$	16
D	4,530	$\frac{4,530}{50.5} \approx 89.703$	**89**
E	2,200	$\frac{2,200}{50.5} \approx 43.564$	43
F	550	$\frac{550}{50.5} \approx 10.891$	10
Total	10,300		200

Adams apportionment method: With the standard divisor of 51.5, a modified divisor is 52.1.

State	Population	Modified Quota	Round Modified Quota Up
A	700	$\frac{700}{52.1} \approx 13.436$	14
B	1,500	$\frac{1,500}{52.1} \approx 28.791$	29
C	820	$\frac{820}{52.1} \approx 15.739$	16
D	4,530	$\frac{4,530}{52.1} \approx 86.948$	87
E	2,200	$\frac{2,200}{52.1} \approx 42.226$	43
F	550	$\frac{550}{52.1} \approx 10.557$	11
Total	10,300		200

Using the Jefferson method, D would receive 89 representatives, which is greater than its upper quota.

57. Total Population $= 1{,}400 + 2{,}000 + 1{,}500 + 9{,}000 + 4{,}300 + 500 = 18{,}700$

$$\text{Standard Divisor} = \frac{\text{Total Population}}{\text{Number of Representatives Allocated}} = \frac{18{,}700}{500} = 37.4$$

State	Number of residents	Standard Quota	Lower Quota	Upper Quota
A	1,400	$\frac{1{,}400}{37.4} \approx 37.433$	37	38
B	2,000	$\frac{2{,}000}{37.4} \approx 53.476$	53	54
C	1,500	$\frac{1{,}500}{37.4} \approx 40.107$	40	41
D	9,000	$\frac{9{,}000}{37.4} \approx 240.642$	**240**	**241**
E	4,300	$\frac{4{,}300}{37.4} \approx 114.973$	114	115
F	500	$\frac{500}{37.4} \approx 13.369$	13	14

Jefferson apportionment method: With the standard divisor of 37.4 a modified divisor is 37.1.

State	Population	Modified Quota	Round Modified Quota Down
A	1,400	$\frac{1{,}400}{37.1} \approx 37.736$	37
B	2,000	$\frac{2{,}000}{37.1} \approx 53.908$	53
C	1,500	$\frac{1{,}500}{37.1} \approx 40.431$	40
D	9,000	$\frac{9{,}000}{37.1} \approx 242.588$	**242**
E	4,300	$\frac{4{,}300}{37.1} \approx 115.903$	115
F	500	$\frac{500}{37.1} \approx 13.477$	13
Total	18,700		500

Adams apportionment method: With the standard divisor of 37.4 a modified divisor is 37.7.

State	Population	Modified Quota	Round Modified Quota Up
A	1,400	$\frac{1{,}400}{37.7} \approx 37.135$	38
B	2,000	$\frac{2{,}000}{37.7} \approx 53.050$	54
C	1,500	$\frac{1{,}500}{37.7} \approx 39.788$	40
D	9,000	$\frac{9{,}000}{37.7} \approx 238.727$	**239**
E	4,300	$\frac{4{,}300}{37.7} \approx 114.058$	115
F	500	$\frac{500}{37.7} \approx 13.263$	14
Total	18,700		500

Using the Jefferson method, D would receive 242 representatives, which is greater than its upper quota. Using the Adams method, D would receive 239 representatives, which is less than its lower quota.

59. smallest 926; largest 940

61. smallest 1,058; largest 1,066

63. No solution provided

Chapter Review Exercises

1. When adding a member to the legislature and without changing populations, a state loses a representative.

2.

Division	Number of members %	Step 1: Determine exact number of seats	Step 2: Assign integer parts	Step 3: Examine fractional parts	Step 3: Assign additional seats
Revolutionary War	560 (28.00%)	$0.2800 \times 11 = 3.0800$	3	0.0800	3
Civil War	524 (26.20%)	$0.2620 \times 11 = 2.8820$	2	0.8820	3
World War I	431 (21.55%)	$0.2155 \times 11 = 2.3705$	2	0.3705	2
World War II	485 (24.25%)	$0.2425 \times 11 = 2.6675$	2	0.6675	3
Total	2,000 (100%)	11	9		11

3. No solution provided

4. a) B is more poorly represented.

    $$\text{average constituency of A} = \frac{\text{population}}{\text{number of representatives}} = \frac{935}{5} = 187 \text{ (thousand)}$$

    $$\text{average constituency of B} = \frac{\text{population}}{\text{number of representatives}} = \frac{2{,}343}{11} = 213 \text{ (thousand)}$$

    The absolute unfairness is $213 - 187 = 26$ (thousand).

    b) $\text{relative unfairness} = \frac{\text{absolute unfairness}}{\text{smaller average consituency}} = \frac{26}{187} \approx 0.139$

5. 11.1, 27.3, 36.9, 3.0, 21.7

Original Number	Number truncated to tenths	Discarded portion
11.045	11.0	0.045
27.333	27.3	0.033
36.901	36.9	0.001
3.006	3.0	0.006
21.715	21.7	0.015
100.000	99.9	

6. a) Huntington-Hill number for South Carolina $= \frac{(\text{population } X)^2}{x(x+1)} = \frac{3.5^2}{6 \cdot 7} \approx 0.292$

    Huntington-Hill number for Maryland $= \frac{(\text{population } X)^2}{x(x+1)} = \frac{4.1^2}{8 \cdot 9} \approx 0.233$

    b) Since $0.233 < 0.292$, South Carolina should receive the additional representative.

7. Once a representative is assigned, it is not reassigned at a later date.

8. Red, blue, yellow, and green routes each get one bus in no particular order. The remaining nine buses in order would be given to GRBGRGBRY. So the red route should have 4 buses, the blue route should have 3, the yellow route should have 2, and the green route should have 4 according to the Huntington–Hill apportionment principle.

9. Make the following table.

Current Representation	Tae Bo	Karate	Weight	Tae-Chi
1	$\dfrac{66^2}{1\cdot 2} = 2178$	$\dfrac{39^2}{1\cdot 2} \approx 761$	$\dfrac{18^2}{1\cdot 2} \approx 162$	$\dfrac{23^2}{1\cdot 2} = 265$
2	$\dfrac{66^2}{2\cdot 3} = 726$	$\dfrac{39^2}{2\cdot 3} \approx 254$	$\dfrac{18^2}{2\cdot 3} = 54$	$\dfrac{23^2}{2\cdot 3} \approx 88$
3	$\dfrac{66^2}{3\cdot 4} = 363$	$\dfrac{39^2}{3\cdot 4} \approx 127$	$\dfrac{18^2}{3\cdot 4} = 27$	$\dfrac{23^2}{3\cdot 4} \approx 44$
4	$\dfrac{66^2}{4\cdot 5} \approx 218$	$\dfrac{39^2}{4\cdot 5} \approx 76$	$\dfrac{18^2}{4\cdot 5} \approx 16$	$\dfrac{23^2}{4\cdot 5} \approx 26$

   Tae-Bo, karate, weight training, and Tae-Chi each get one class in no particular order. The remaining 4 classes in order would be Tae-Bo, karate, Tae-Bo, and Tae-Bo, according to the Huntington–Hill method. So Tae-Bo would have 4 classes, karate would have 2, weight training would have 1, and Tae-Chi would have 1.

10. Total Responses $= 8 + 64 + 11 + 31 = 114$

    Standard Divisor $= \dfrac{\text{Total Responses}}{\text{Number of Classes}} = \dfrac{114}{8} = 14.25$

    Jefferson apportionment method: With the standard divisor of 14.25, a modified divisor is 11.

Class	Responses	Modified Quota	Round Modified Quota Down
High-Impact aerobics	8	$\dfrac{8}{11} \approx 0.727$	0
Low-Impact aerobics	64	$\dfrac{64}{11} \approx 5.818$	5
Jazzercise	11	$\dfrac{11}{11} = 1$	1
Step Exercise	31	$\dfrac{31}{11} \approx 2.818$	2
Total	114		8

11. Total Responses $= 8 + 64 + 11 + 31 = 114$

    Standard Divisor $= \dfrac{\text{Total Responses}}{\text{Number of Classes}} = \dfrac{114}{8} = 14.25$

    Adams apportionment method: With the standard divisor of 14.25, a modified divisor is 16.

11. (continued)

Class	Responses	Modified Quota	Round Modified Quota Up
High-Impact aerobics	8	$\frac{8}{16} = 0.5$	1
Low-Impact aerobics	64	$\frac{64}{16} = 4$	4
Jazzercise	11	$\frac{11}{16} \approx 0.688$	1
Step Exercise	31	$\frac{31}{16} \approx 1.938$	2
Total	114		8

12. Total Responses $= 8 + 64 + 11 + 31 = 114$

    Standard Divisor $= \dfrac{\text{Total Responses}}{\text{Number of Classes}} = \dfrac{114}{8} = 14.25$

    Webster apportionment method: With the standard divisor of 14.25, a modified divisor is 15.

Class	Responses	Modified Quota	Round Modified Quota
High-Impact aerobics	8	$\frac{8}{15} \approx 0.533$	1
Low-Impact aerobics	64	$\frac{64}{15} \approx 4.267$	4
Jazzercise	11	$\frac{11}{15} \approx 0.733$	1
Step Exercise	31	$\frac{31}{15} \approx 2.067$	2
Total	114		8

13. We must first find how the 150 security personnel were originally apportioned according to the Hamilton method. The standard divisor $= \dfrac{180,600}{150} = 1,204$.

Airport	Number of passengers	Standard Quota	Integer parts	Fractional Parts	Assign 2 security personnel
A	80,500	$\frac{80,500}{1,204} \approx 66.860$	66	0.860	67
B	6,800	$\frac{6,800}{1,204} \approx 5.648$	5	0.648	6
C	93,300	$\frac{93,300}{1,204} \approx 77.492$	77	0.492	77
Total	180,600		148		150

13. (continued)

Determine the apportionment one year later. The standard divisor $= \dfrac{196,410}{150} \approx 1,309.4$.

Airport	Number of passengers	Standard Quota	Integer parts	Fractional Parts	Assign 2 security personnel
A	87,700	$\dfrac{87,700}{1,309.4} \approx 66.977$	66	0.977	67
B	7,410	$\dfrac{7410}{1,309.4} \approx 5.659$	5	0.659	6
C	101,300	$\dfrac{101,300}{1,309.4} \approx 77.364$	77	0.364	77
Total	196,410		148		150

No, the apportionment stayed the same.

14. The new states paradox occurs when a new state is added, and its share of seats is added to the legislature causing a change in the allocation of seats previously given to another state.

15. Jefferson's, Adams', and Webster's

Chapter Test

1. When adding a member to the legislature and without changing populations, a state loses a representative.

2. a) D is more poorly represented.

    average constituency of C $= \dfrac{\text{population}}{\text{number of representatives}} = \dfrac{1,640}{8} = 205$ (thousand)

    average constituency of D $= \dfrac{\text{population}}{\text{number of representatives}} = \dfrac{1,863}{9} = 207$ (thousand)

    The absolute unfairness is $207 - 205 = 2$ (thousand).

   b) relative unfairness $= \dfrac{\text{absolute unfairness}}{\text{smaller average consituency}} = \dfrac{2}{205} \approx 0.0098$

3.

Division	Number of members %	Step 1: Determine exact number of seats	Step 2: Assign integer parts	Examine fractional parts	Step 3: Assign additional seats
Art	47 (23.86%)	$0.2386 \times 8 = 1.9088$	1	0.9088	2
Music	111 (56.34%)	$0.5634 \times 8 = 4.5072$	4	0.5072	4
Theater	39 (19.80%)	$0.1980 \times 8 = 1.5840$	1	0.5840	2
Total	197 (100%)	8	6		8

4. If the number of representatives increases, representatives assigned earlier are reassigned.

268 Chapter 9: Legislative Apportionment

5. a) Huntington – Hill number for Kentucky $= \dfrac{(\text{population } X)^2}{x(x+1)} = \dfrac{4.041^2}{6 \cdot 7} \approx 0.389$

  Huntington – Hill number for Arizona $= \dfrac{(\text{population } X)^2}{x(x+1)} = \dfrac{5.13^2}{8 \cdot 9} \approx 0.366$

  b) Since $0.366 < 0.389$, Arizona should receive the additional representative.

6. 13.05, 24.89, 47.20, 4.02, 10.84

Original Number	Number truncated to hundredths	Discarded portion
13.0472	13.04	0.0072
24.895	24.89	0.005
47.2045	47.20	0.0045
4.0165	4.01	0.0065
10.8368	10.83	0.0068
100.000	99.97	

7. Once a representative is assigned, it is not reassigned at a later date.

8. Mountain Village, City Sidewalks, Jungle Village, and Great Frontier routes each get one bus in no particular order. The remaining ten buses in order would be given to MCJMGCJMGC. So the Mountain Village route should have 4 buses, the City Sidewalks route should have 4, the Jungle Village route should have 3, and the Great Frontier route should have 3 according to the Huntington–Hill apportionment principle.

9. Make the following table.

Current Representation	Mathematics	Reading	Study Skills
1	$\dfrac{16^2}{1 \cdot 2} = 128$	$\dfrac{9^2}{1 \cdot 2} \approx 41$	$\dfrac{6^2}{1 \cdot 2} \approx 18$
2	$\dfrac{16^2}{2 \cdot 3} \approx 43$	$\dfrac{9^2}{2 \cdot 3} \approx 14$	$\dfrac{6^2}{2 \cdot 3} = 6$
3	$\dfrac{16^2}{3 \cdot 4} \approx 21$	$\dfrac{9^2}{3 \cdot 4} \approx 7$	$\dfrac{6^2}{3 \cdot 4} = 3$
4	$\dfrac{16^2}{4 \cdot 5} \approx 13$	$\dfrac{9^2}{4 \cdot 5} \approx 4$	$\dfrac{6^2}{4 \cdot 5} \approx 2$

Mathematics, Reading, and Study Skills each get one tutor in no particular order. The remaining 8 tutors in order would be, Mathematics, Mathematics, Reading, Mathematics, Study Skills, Reading, Mathematics, and Reading, according to the Huntington–Hill method. So Mathematics would have 5 tutors, Reading would have 4, and Study Skills would have 2.

10. Consider the apportionment among only airports A, B, and C.

$$\text{The standard divisor} = \frac{180,600}{150} = 1204.$$

Airport	Number of passengers	Standard Quota	Integer parts	Fractional Parts	Assign 1 additional security officer
A	80,500	$\frac{80,500}{1204} \approx 66.86$	66	0.86	**67**
B	6,800	$\frac{6,800}{1204} \approx 5.65$	5	0.65	**6**
C	93,300	$\frac{93,300}{1204} \approx 77.49$	77	0.49	77
Total	180,600		148		150

Now consider approximately the number of representatives D would get with the standard divisor of 1204. D would get $\frac{41,400}{1204} \approx 34.39$ representatives.

Determine the apportionment with 34 additional committee members.

$$\text{The standard divisor} = \frac{222,000}{184} \approx 1206.5.$$

Airport	Number of passengers	Standard Quota	Integer parts	Fractional Parts	Assign 1 additional security officer
A	80,500	$\frac{80,500}{1206.5} \approx 66.72$	66	0.72	**67**
B	6,800	$\frac{6,800}{1206.5} \approx 5.64$	5	0.64	**6**
C	93,300	$\frac{93,300}{1206.5} \approx 77.33$	77	0.33	77
D	41,400	$\frac{41,400}{1206.5} \approx 34.31$	34	0.31	34
Total	222,000		182		184

No, because when D is added, no airport loses any security personnel.

11. The population paradox occurs when state A grows faster that state B but, with no change in the number of representatives, A loses a representative to B.

12. Total Responses $= 47 + 32 + 41 + 21 = 141$

$$\text{Standard Divisor} = \frac{\text{Total Responses}}{\text{Number of Classes}} = \frac{141}{9} = 15.67$$

Webster apportionment method: With the standard divisor of 15.67, a modified divisor is 15.

270   Chapter 9: Legislative Apportionment

12. (continued)

Class	Responses	Modified Quota	Round Modified Quota
massage	47	$\frac{47}{15} \approx 3.133$	3
aromatherapy	32	$\frac{32}{15} \approx 2.133$	2
yoga	41	$\frac{41}{15} \approx 2.733$	3
meditation	21	$\frac{21}{15} = 1.4$	1
Total	141		9

13. Total Responses $= 47 + 32 + 41 + 21 = 141$

    Standard Divisor $= \dfrac{\text{Total Responses}}{\text{Number of Classes}} = \dfrac{141}{9} = 15.67$

    Jefferson apportionment method: With the standard divisor of 15.67, a modified divisor is 12.

Class	Responses	Modified Quota	Round Modified Quota Down
massage	47	$\frac{47}{12} = 3.917$	3
aromatherapy	32	$\frac{32}{12} \approx 2.667$	2
yoga	41	$\frac{41}{12} = 3.417$	3
meditation	21	$\frac{21}{12} = 1.75$	1
Total	141		9

14. Total Responses $= 47 + 32 + 41 + 21 = 141$

    Standard Divisor $= \dfrac{\text{Total Responses}}{\text{Number of Classes}} = \dfrac{141}{9} = 15.67$

    Adams apportionment method: With the standard divisor of 15.67, a modified divisor is 20.8.

Class	Responses	Modified Quota	Round Modified Quota Up
massage	47	$\frac{47}{20.8} \approx 2.260$	3
aromatherapy	32	$\frac{32}{20.8} \approx 1.538$	2
yoga	41	$\frac{41}{20.8} \approx 1.971$	2
meditation	21	$\frac{21}{20.8} \approx 1.010$	2
Total	141		9

15. only Hamilton's

Of Further Interest: Fair Division

1. Her estimate of the value of the painting was divided by 3, which was the number of heirs.

2. She had the highest bids.

3. The amount of the fair shares for Tom and Karen were subtracted from the amount that Bob and Ann paid the estate. This surplus was divided evenly among the heirs.

4. Each person has a different estimate of the value of the estate and does not know the others' estimates.

5. Their bids are too low.

6. Those who think that items in the estate are worth more receive the items and then must contribute cash to the estate that is then (partially) distributed to those who made low estimates of the items in the estate.

7. a) discrete
   b) continuous
   c) discrete

8. a) discrete
   b) discrete
   c) continuous

9.

	Greg ($\frac{1}{2}$)	Dharma ($\frac{1}{2}$)
Bid on car	$110,000	$85,000
Fair share of estate	$110,000/2 = $55,000	$85,000/2 = $42,500
Item obtained with the highest bid	Car	

10.

	Karl ($\frac{1}{2}$)	Fredrich ($\frac{1}{2}$)
Big on book	$27,000	$35,000
Fair share of estate	$27,000/2 = $13,500	$35,000/2 = $17,500
Item obtained with the highest bid		Book

11.

	Alex (40%)	Marianne (60%)
Bid on copyright	$20,000	$28,000
Fair share of copyright	$20,000 \times 0.40 = $8,000	$28,000 \times 0.60 = $16,800
Item obtained with the highest bid		Copyright

12.

	Matt (35%)	Tony (65%)
Bid on pizza stand	$120,000	$90,000
Fair share of pizza stand	$120,000 \times 0.35 = $42,000	$90,000 \times 0.65 = $58,500
Item obtained with the highest bid	Pizza stand	

13.

	Ed ($\frac{1}{3}$)	Al ($\frac{1}{3}$)	Jerry ($\frac{1}{3}$)
Bid on painting	$13,000	$8,000	$9,000
Bid on statue	$14,000	$13,000	$15,000
Total value	$27,000	$21,000	$24,000
Fair share of estate	$\frac{1}{3} \cdot 27000 = \$9,000$	$\frac{1}{3} \cdot 21000 = \$7,000$	$\frac{1}{3} \cdot 24000 = \$8,000$

14.

	Franz ($\frac{1}{4}$)	Ida ($\frac{1}{4}$)	Bill ($\frac{1}{4}$)	Monica ($\frac{1}{4}$)
Bid on computer stock	$75,000	$80,000	$70,000	$90,000
Bid on oil stock	$35,000	$40,000	$45,000	$40,000
Bid on pharmaceutical stock	$40,000	$30,000	$25,000	$35,000
Total value	$150,000	$150,000	$140,000	$165,000
Fair share of portfolio	$\frac{1}{4} \cdot 150000 = \$37,500$	$\frac{1}{4} \cdot 150000 = \$37,500$	$\frac{1}{4} \cdot 140000 = \$35,000$	$\frac{1}{4} \cdot 165000 = \$41,250$

15.

	Greg ($\frac{1}{2}$)	Dharma ($\frac{1}{2}$)
Pays to estate (+) or receives from estate (−)	$55,000 (+)	$42,500 (−)
Division of estate balance ($12,500)	$6,250 (−)	$6,250 (−)
Summary of cash	Pays $48,750	Receives $48,750

16.

	Karl ($\frac{1}{2}$)	Fredrich ($\frac{1}{2}$)
Pays to estate (+) or receives from estate (−)	$13,500 (−)	$17,500 (+)
Division of estate balance ($4,000)	$2,000 (−)	$2,000 (−)
Summary of cash	Receives $15,500	Pays $15,500

17.

	Alex (40%)	Marianne (60%)
Pays to pool (+) or receives from pool (−)	$8,000 (−)	$11,200 (+)
Division of pool balance ($3,200)	$1,280 (−)	$1,920 (−)
Summary of cash	Receives $9,280	Pays $9,280

18.

	Matt (35%)	Tony (65%)
Pays to pool (+) or receives from pool (−)	$78,000 (+)	$58,500 (−)
Division of pool balance ($19,500)	$19500 \times 0.35 = \$6,825$ (−)	$19,500 \times 0.65 = \$12,675$ (−)
Summary of cash	Pays $71,175	Receives $71,175

19.

	Ed $\left(\frac{1}{3}\right)$	Al $\left(\frac{1}{3}\right)$	Jerry $\left(\frac{1}{3}\right)$
Items obtained with highest bid	Painting		Statue
Pays to estate (+) or receives from estate (−)	13,000 − 9,000 $4,000.00 (+)$	$7,000.00 (−)$	15,000 − 8,000 $7,000.00 (+)$
Division of estate balance ($4,000)	$1,333.33 (−)$	$1,333.33 (−)$	$1,333.33 (−)$
Summary of cash	Pays $2,666.67	Receives $8,333.33	Pays $5,666.67

20.

	Franz $\left(\frac{1}{4}\right)$	Ida $\left(\frac{1}{4}\right)$	Bill $\left(\frac{1}{4}\right)$	Monica $\left(\frac{1}{4}\right)$
Stock obtained with highest bid	Pharmaceutical Stock		Oil Stock	Computer Stock
Pays to pool (+) or receives from pool (−)	$2,500 (+)$	$37,500 (−)$	$10,000 (+)$	$48,750 (+)$
Division of estate balance ($23,750)	$5,937.50 (−)$	$5,937.50 (−)$	$5,937.50 (−)$	$5,937.50 (−)$
Summary of cash	Receives $3,437.50	Receives $43,437.50	Pays $4,062.50	Pays $42,812.50

21.

	Betty $\left(\frac{1}{2}\right)$	Dennis $\left(\frac{1}{2}\right)$
Big on ring	$16,000	$18,000
Bid on desk	$ 4,500	$ 5,000
Bid on books	$ 4,000	$ 3,000
Total value	$24,500	$26,000
Fair share of estate	$12,250	$13,000

	Betty $\left(\frac{1}{2}\right)$	Dennis $\left(\frac{1}{2}\right)$
Item with highest bid	Books	Ring and Desk
Pays to pool (+) or receives from pool (−)	$8,250 (−)$	$10,000 (+)$
Division of estate balance ($1,750)	$ 875 (−)$	$ 875 (−)$
Summary of Cash	Receives $9,125	Pays $9,125

Betty receives the books, Dennis receives the ring and desk; Dennis pays Betty $9,125.

22.

	Matt $\left(\frac{1}{3}\right)$	Dani $\left(\frac{1}{3}\right)$	Christian $\left(\frac{1}{3}\right)$
Big on A	$170,000	$180,000	$160,000
Bid on B	$145,000	$150,000	$155,000
Bid on C	$200,000	$190,000	$210,000
Bid on D	$ 70,000	$ 50,000	$ 45,000
Total value	$585,000	$570,000	$570,000
Fair share of restaurants	$195,000	$190,000	$190,000

22. (continued)

	Matt $\left(\frac{1}{3}\right)$	Dani $\left(\frac{1}{3}\right)$	Christian $\left(\frac{1}{3}\right)$
Restaurant with highest bid	D	A	B and C
Pays to pool (+) or receives from pool (−)	$125,000 (−)	$10,000 (−)	$175,000 (+)
Division of restaurants' balance ($40,000)	$13,333.33 (−)	$13,333.33 (−)	$13,333.33 (−)
Summary of cash	Receives $138,333.33	Receives $23,333.33	Pays $161,666.67

Matt receives D, Dani A, Christian B and C. Matt receives $138.333.33, Dani receives $23,333.33, and Christian pays $161,666.67.

23. – 30. No solutions provided

Chapter 10
VOTING: Using Mathematics To Make Choices

Section 10.1 Voting Methods

1. Her eleven second-place votes gave her more points than Carim's one extra first-place vote.

3. Each received ½ point because the number that preferred T over B was the same as the number that preferred B over T.

5. when one candidate receives a majority of first-place votes

7. a) No. The total number of votes is 2,156 + 1,462 + 986 + 428 = 5,032; $2156 \div 5032 \approx 0.428$, which is less than 50%
   b) Edelson

9. Athletic facilities with 33 votes

11. Since Student Union Building had the least votes, it is removed. The new table would be

Preference	15	30	18	17	10	2
1st	A	D	A	C	C	A
2nd	C	C	D	D	A	C
3rd	D	A	C	A	D	D

    By combining identical columns, we have

Preference	17	30	18	17	10
1st	A	D	A	C	C
2nd	C	C	D	D	A
3rd	D	A	C	A	D

    Since Campus Security (C) has the fewest first-place votes, it is then eliminated

Preference	17	30	18	17	10
1st	A	D	A	D	A
2nd	D	A	D	A	D

    By combining identical columns, we have

Preference	45	47
1st	A	D
2nd	D	A

    Since Dining Facilities (D) now has the most first-place votes, it wins the election.

13. Comedy with 23 votes

15. Since Greek Tragedy had the least votes, it is removed. The new table would be

Preference	10	15	13	12	5	7
1st	C	D	C	M	M	C
2nd	M	M	D	D	C	M
3rd	D	C	M	C	D	D

By combining identical columns, we have

Preference	17	15	13	12	5
1st	C	D	C	M	M
2nd	M	M	D	D	C
3rd	D	C	M	C	D

Since Drama (D) has the fewest first-place votes, it is then eliminated.

Preference	17	15	13	12	5
1st	C	M	C	M	M
2nd	M	C	M	C	C

By combining identical columns, we have

Preference	30	32
1st	C	M
2nd	M	C

Since Mystery (M) now has the most first-place votes, it wins the election.

17. Technology with 15 votes

19. Since Families (F) had the least votes, it is removed. The new table would be

Preference	15	7	13	5	2
1st	T	E	G	P	E
2nd	P	P	E	G	G
3rd	E	G	P	E	P
4th	G	T	T	T	T

Since Poverty in Third World Countries (P) has the fewest first-place votes, it is then eliminated.

Preference	15	7	13	5	2
1st	T	E	G	G	E
2nd	E	G	E	E	G
3rd	G	T	T	T	T

By combining identical columns we have

Preference	15	9	18
1st	T	E	G
2nd	E	G	E
3rd	G	T	T

Since Environment (E) has the fewest first-place votes, it is then eliminated.

Preference	15	9	18
1st	T	G	G
2nd	G	T	T

By combining identical columns, we have

Preference	15	27
1st	T	G
2nd	G	T

Since Gender Roles (G) now has the most first-place votes, it wins the election.

21. Londram with 17 votes

23. Since E-Mall (E) had the least votes, it is removed. The new table would be

Preference	15	11	9	10	2
1st	L	F	S	T	L
2nd	F	S	L	F	S
3rd	S	L	F	L	F
4th	T	T	T	S	T

Since Securenet (S) has the fewest first-place votes, it is then eliminated.

Preference	15	11	9	10	2
1st	L	F	L	T	L
2nd	F	L	F	F	F
3rd	T	T	T	L	T

By combining identical columns, we have

Preference	26	11	10
1st	L	F	T
2nd	F	L	F
3rd	T	T	L

Since Techenium (T) has the fewest first-place votes, it is then eliminated.

Preference	26	11	10
1st	L	F	F
2nd	F	L	L

By combining identical columns, we have

Preference	26	21
1st	L	F
2nd	F	L

Since Londram (L) now has the most first-place votes, it wins the election.

25. The Borda count is being used. Three points for 1st, two points for 2nd, and one point for 3rd.

Player, Team	1st	2nd	3rd	Total	
Matt Leinart, USC	267	**211**	102	1,325	$(1325 - 267 \times 3 - 102 \times 1) \div 2 = 211$
Adrian Peterson, Oklahoma	**154**	180	175	997	$(997 - 180 \times 2 - 175 \times 1) \div 3 = 154$
Jason White, Oklahoma	171	149	**146**	957	$(957 - 171 \times 3 - 149 \times 2) \div 1 = 146$
Alex Smith, Utah	98	112	117	635	$98 \times 3 + 112 \times 2 + 117 \times 1 = 635$
Reggie Bush, USC	118	80	**83**	597	$(597 - 118 \times 3 - 80 \times 2) \div 1 = 83$

27. Answers may vary.
    The first one eliminated should be in last place. The second one eliminated should be in the second-to-last place. This process continues until the winner is determined.

29. 1st: Athletic Facilities, 2nd: Dining Facilities, 3rd: Campus Security, and 4th: Student Union Building

31. 1st: Dining Facilities, 2nd: Athletic Facilities, 3rd: Campus Security, and 4th: Student Union Building

33. 1st: Technology, 2nd: Gender Roles, 3rd: Environment, 4th: Poverty in Third World Countries, and 5th: Families

35. 1st: Gender Roles, 2nd: Technology, 3rd: Environment, 4th: Poverty in Third World Countries, and 5th: Families

278   Chapter 10: Voting

37. Answers may vary.
There are only two candidates to compare and assuming the two don't have the same number of votes, one will have more than the other and this will represent the majority.

39. Answers may vary.
The candidate that was eliminated was the one with the fewest (fewer) votes. Thus, the candidate that wins was the candidate with the majority of the votes.

41. The possible points awarded are 4, 3, 2, or 1. $(4+3+2+1) \cdot 20 = 10 \cdot 20 = 200$

43. If the candidate is in first place in all selections, then $5 \cdot 22 = 110$ is the Borda count.

45. Assume the four candidates are A, B, C, and D. The comparisons you need to make are A to B, A to C, A to D, B to C, B to D, and C to D. Since 1 point is awarded to each of the winners of these comparisons, there would be 6 points total.

47. Assume the five candidates are A, B, C, D and E. The comparisons you need to make are A to B, A to C, A to D, A to E, B to C, B to D, B to E, C to D, C to E, and D to E. Since 1 point is awarded to each of the winners of these comparisons and any individual candidate appears in a comparison only 4 times, then the maximum that could be earned is 4 points.

49. – 53. No solution provided

Section 10.2 Defects In Voting Methods

1. Example 1; the Borda method violates the majority criterion.

3. the plurality and pairwise comparison methods

5. pairwise comparison; independence-of-irrelevant-alternatives

7. Option B with a total of 149 points

Preference	1st place votes × 4 (points)	2nd place votes × 3 (points)	3rd place votes × 2 (points)	4th place votes × 1 (points)	Total Points
A	25×4	0×3	15×2	9×1	139
B	15×4	21×3	13×2	0×1	149
C	9×4	15×3	12×2	13×1	118
D	0×4	13×3	9×2	27×1	84

9. No solution provided

11.

Preference		1st place votes × 4 (points)	2nd place votes × 3 (points)	3rd place votes × 2 (points)	4th place votes × 1 (points)	Total Points
Lower the age to 18	(A)	3×4	18×3	10×2	14×1	100
Lower the age to 19	(B)	10×4	14×3	21×2	0×1	124
Lower the age to 20	(C)	22×4	10×3	0×2	13×1	131
Keep the age at 21	(D)	10×4	3×3	14×2	18×1	95

Option C is the winner by the Borda count. The Condorcet's criterion is not satisfied in this example because option B can beat all of the other options in a head-to-head competition, but C is the winner by the Borda count. 24 out of 45 prefer B to A, 23 out of 45 prefer B to C, and 32 out of 45 prefer B to D.

13.

Preference		1st place votes × 4 (points)	2nd place votes × 3 (points)	3rd place votes × 2 (points)	4th place votes × 1 (points)	Total Points
Atlanta	(A)	2×4	16×3	29×2	0×1	114
Boston	(B)	18×4	19×3	2×2	8×1	141
Chicago	(C)	23×4	2×3	8×2	14×1	128
Denver	(D)	4×4	10×3	8×2	25×1	87

Option B is the winner by the Borda count. The independence-of-irrelevant-alternatives criterion is not satisfied in this example because if option A were taken out, then the preference table would be

Preference	15	4	8	10	8	2
1st	C	D	C	B	B	C
2nd	B	B	D	D	C	B
3rd	D	C	B	C	D	D

According to the Borda count, option C would be declared the winner.

Preference		1st place votes × 3 (points)	2nd place votes × 2 (points)	3rd place votes × 1 (points)	Total Points
Chicago	(C)	25×3	8×2	14×1	105
Boston	(B)	18×3	21×2	8×1	104
Denver	(D)	4×3	18×2	25×1	73

15. Since option B (reduce expenditures on art and music programs) had the least votes, it is removed.

The new table would be

Preference	9	12	4	5	4	1
1st	C	D	C	A	A	C
2nd	A	A	D	D	C	A
3rd	D	C	A	C	D	D

By combining identical columns, we have

Preference	10	12	4	5	4
1st	C	D	C	A	A
2nd	A	A	D	D	C
3rd	D	C	A	C	D

Since option A (reduce sports programs) has the fewest first-place votes, it is then eliminated.

Preference	10	12	4	5	4
1st	C	D	C	D	C
2nd	D	C	D	C	D

By combining identical columns, we have

Preference	18	17
1st	C	D
2nd	D	C

Since option C (increase class size) now has the most first-place votes, it wins the election. The Condorcet's criterion is not satisfied in this example because option A (reduce sport programs) can beat all of the other options in a head-to-head competition, but C (increase class size) is the winner by the plurality-with-elimination method. 30 out of 35 prefer A to B, 21 out of 35 prefer A to C, and 19 out of 35 prefer A to D.

280   Chapter 10: Voting

17. We must compare a) A with B, b) A with C, and c) B with C.

   a) In comparing A with B, we ignore C. Thus, 13 prefer A over B, and 10+5=15 prefer B over A. Thus, we award 1 point to option B.
   d) In comparing A with C, we ignore B. Thus, 13 prefer A over C, and 10+5=15 prefer C over A. Thus, we award 1 point to option C.
   e) In comparing B with C, we ignore A. Thus, 13+10=23 prefer B over C, and 5 prefer C over B. Thus, we award 1 point to option B.

   Since B has 2 points and C has 1, option B is declared the winner.

   This example doesn't violate the majority criterion because the criterion states that <u>if</u> a majority of the voters rank a candidate as their first choice, <u>then</u> that candidate should win the election. The hypothesis of this conditional criterion was not satisfied. The majority (over half) did not vote for any one candidate.

19. Answers may vary.

Preference	5	4
1st	B	A
2nd	A	C
3rd	C	B

Preference	1st place votes × 3 (points)	2nd place votes × 2 (points)	3rd place votes × 1 (points)	Total Points
A	4×3	5×2	0×1	22
B	5×3	0×2	4×1	19
C	0×3	4×2	5×1	13

   Option A is the winner by the Borda count. Condorcet's criterion is not satisfied in this example because option B can beat all of the other options in a head-to-head competition, but A is the winner by Borda count. 5 out of 9 prefer B to A and 5 out of 9 prefer B to C.

21. and 23. No solutions provided

25. Answers may vary.

Preference	20	30	8	8	12
1st	C	D	A	A	B
2nd	A	A	D	B	C
3rd	B	B	B	C	A
4th	D	C	C	D	D

Since option B had the least votes, it is removed. The new table would be

Preference	20	30	8	8	12
1st	C	D	A	A	C
2nd	A	A	D	C	A
3rd	D	C	C	D	D

By combining identical columns, we have

Preference	32	30	8	8
1st	C	D	A	A
2nd	A	A	D	C
3rd	D	C	C	D

25. (continued)

Since option A had the least votes, it is removed. The new table would be

Preference	32	30	8	8
1st	C	D	D	C
2nd	D	C	C	D

By combining identical columns, we have

Preference	40	38
1st	C	D
2nd	D	C

Since option C now has the most first-place votes, it wins the election. The Condorcet's criterion is not satisfied in this example because option A can beat all of the other options in a head-to-head competition, but option C is the winner by the plurality-with-elimination method. 66 out of 78 prefer A to B, 46 out of 78 prefer A to C, and 48 out of 78 prefer A to D

27. Answers may vary.
Since in the plurality method, you are looking for the option that has the most first-place votes, if an option has the majority of the votes (over 50%) then this will constitute the most first-place votes.

29. No solution provided

31. Answers may vary.
If the majority (over 50%) of the voters rank an option first, then that option will be chosen over any other choice in a head-to-head competition and thus will win using the pairwise comparison method.

33. – 37. No solutions provided

39. Answers may vary.
Yes.

Preference	5	4
1st	A	B
2nd	B	C
3rd	C	A

Preference	1st place votes × 3 (points)	2nd place votes × 2 (points)	3rd place votes × 1 (points)	Total Points
A	5×3	0×2	4×1	19
B	4×3	5×2	0×1	22
C	0×3	4×2	5×1	13

Option B is the winner by the Borda count even though option A has the majority (approximately 56% of the vote).

41. No solution provided

Section 10.3 Weighted Voting Systems

1. 51 was the quota; 26, 26, 12, 12, 12, and 12 were the weights of the six voters.

3. They were critical members in all 11 winning coalitions.

5. Without that voter's support, a resolution cannot pass.

282 Chapter 10: Voting

7. a) The quota is 5.
   b) There are 5 voters. Each voter has 1 vote.
   c) There are no dictators.
   d) The sum of the votes is the same as the quota. Each voter has veto power.

9. a) The quota is 11.
   b) There are 4 voters. Voter A has 10 votes, voter B has 3, voter C has 4, and voter D has 5.
   c) There are no dictators.
   d) No voter has veto power.

11. a) The quota is 15.
    b) There are 5 voters. Voter A has 1 vote, voter B has 2, two voters (C and D) have 3, and voter E has 4. Notice that the sum of the votes is less than the quota. No resolutions can be passed.
    c) There are no dictators.
    d) No voter has veto power, because no resolutions can be passed.

13. a) The quota is 12.
    b) There are 4 voters. Voter A has 1 vote, voter B has 3, voter C has 5, and voter D has 7.
    c) There are no dictators.
    d) Two voters have veto power. Voters C and D each have veto power.

15. a) The quota is 25.
    b) There are 5 voters. Two voters (A and B) have 4 votes, voter C has 6, voter D has 7, and voter E has 9.
    c) There are no dictators.
    d) Three voters have veto power. Voters C, D, and E each have veto power.

17. a) The quota is 51.
    b) There are 5 voters. Three voters (A, B, and C) have 20 votes and two (D and E) have 10.
    c) There are no dictators.
    d) No voter has veto power.

19.

Coalition	Weight	
{A}	1	
{B}	3	
{C}	5	
{D}	7	
{A,B}	4	
{A,C}	6	
{A,D}	8	
{B,C}	8	
{B,D}	10	
{C,D}	12	Winning
{A,B,C}	9	
{A,B,D}	11	
{A,C,D}	13	Winning
{B,C,D}	15	Winning
{A,B,C,D}	16	Winning

21.

Coalition	Weight	
{A}	4	
{B}	4	
{C}	6	
{D}	7	
{E}	9	
{A,B}	8	
{A,C}	10	
{A,D}	11	
{A,E}	13	
{B,C}	10	
{B,D}	11	
{B,E}	13	
{C,D}	13	
{C,E}	15	
{D,E}	16	
{A,B,C}	14	
{A,B,D}	15	
{A,B,E}	17	
{A,C,D}	17	
{A,C,E}	19	
{A,D,E}	20	
{B,C,D}	17	
{B,C,E}	19	
{B,D,E}	20	
{C,D,E}	22	
{A,B,C,D}	21	
{A,B,C,E}	23	
{A,B,D,E}	24	
{A,C,D,E}	26	Winning
{B,C,D,E}	26	Winning
{A,B,C,D,E}	30	Winning

23.

Coalition	Weight		Critical Voters
{C,D}	12	Winning	C,D
{A,C,D}	13	Winning	C,D
{B,C,D}	15	Winning	C,D
{A,B,C,D}	16	Winning	C,D

25.

Coalition	Weight		Critical Voters
{A,C,D,E}	26	Winning	A,C,D,E
{B,C,D,E}	26	Winning	B,C,D,E
{A,B,C,D,E}	30	Winning	C,D,E

27. A simple majority is 6, since there are 11 total voters.

Coalition	Weight	
{P}	5	
{T}	4	
{S}	2	
{P,T}	9	Winning
{P,S}	7	Winning
{T,S}	6	Winning
{P,S,T}	11	Winning

29.

Coalition	Weight	
{A}	3	
{F}	4	
{T}	3	
{N}	2	
{A,F}	7	
{A,T}	6	
{A,N}	5	
{F,T}	7	
{F,N}	6	
{T,N}	5	
{A,F,T}	10	Winning
{A,F,N}	9	Winning
{A,T,N}	8	Winning
{F,T,N}	9	Winning
{A,F,T,N}	12	Winning

31.

Coalition	Weight		Critical Voters
{P,T}	9	Winning	P,T
{P,S}	7	Winning	P,S
{T,S}	6	Winning	T,S
{P,S,T}	11	Winning	None

33.

Coalition	Weight		Critical Voters
{A,F,T}	10	Winning	A,F,T
{A,F,N}	9	Winning	A,F,N
{A,T,N}	8	Winning	A,T,N
{F,T,N}	9	Winning	F,T,N
{A,F,T,N}	12	Winning	None

35. The only winning coalition in Exercise 7 is when all five members vote for the issue. Each voter would therefore have a Banzhaf power index of $\frac{1}{5}$.

37.

Coalition	Weight		Critical Voters
{A}	4		
{B}	4		
{C}	6		
{D}	7		
{E}	9		
{A,B}	8		
{A,C}	10		
{A,D}	11		
{A,E}	13		
{B,C}	10		
{B,D}	11		
{B,E}	13		
{C,D}	13		
{C,E}	15		
{D,E}	16		
{A,B,C}	14		
{A,B,D}	15		
{A,B,E}	17		
{A,C,D}	17		
{A,C,E}	19		
{A,D,E}	20		
{B,C,D}	17		
{B,C,E}	19		
{B,D,E}	20		
{C,D,E}	22		
{A,B,C,D}	21		
{A,B,C,E}	23		
{A,B,D,E}	24		
{A,C,D,E}	26	Winning	A,C,D,E
{B,C,D,E}	26	Winning	B,C,D,E
{A,B,C,D,E}	30	Winning	C,D,E

$$\frac{\text{The number of times A is critical in winning coalitions}}{\text{The total number of times voters are critical in winning coalitions}} = \frac{1}{11}$$

$$\frac{\text{The number of times B is critical in winning coalitions}}{\text{The total number of times voters are critical in winning coalitions}} = \frac{1}{11}$$

$$\frac{\text{The number of times C is critical in winning coalitions}}{\text{The total number of times voters are critical in winning coalitions}} = \frac{3}{11}$$

$$\frac{\text{The number of times D is critical in winning coalitions}}{\text{The total number of times voters are critical in winning coalitions}} = \frac{3}{11}$$

$$\frac{\text{The number of times E is critical in winning coalitions}}{\text{The total number of times voters are critical in winning coalitions}} = \frac{3}{11}$$

39.

Coalition	Weight		Critical Voters
{C,D}	12	Winning	C,D
{A,C,D}	13	Winning	C,D
{B,C,D}	15	Winning	C,D
{A,B,C,D}	16	Winning	C,D

$$\frac{\text{The number of times A is critical in winning coalitions}}{\text{The total number of times voters are critical in winning coalitions}} = \frac{0}{8} = 0$$

$$\frac{\text{The number of times B is critical in winning coalitions}}{\text{The total number of times voters are critical in winning coalitions}} = \frac{0}{8} = 0$$

$$\frac{\text{The number of times C is critical in winning coalitions}}{\text{The total number of times voters are critical in winning coalitions}} = \frac{4}{8} = \frac{1}{2}$$

$$\frac{\text{The number of times D is critical in winning coalitions}}{\text{The total number of times voters are critical in winning coalitions}} = \frac{4}{8} = \frac{1}{2}$$

41. a) No solution provided

b) Let the members of the firm be represented by {K,C,V,W,X,Y,Z}. The winning coalitions are given in the following table.

Coalition	Critical Voters
{K,C,V,W}	K,C,V,W
{K,C,V,X}	K,C,V,X
{K,C,V,Y}	K,C,V,Y
{K,C,V,Z}	K,C,V,Z
{K,C,W,X}	K,C,W,X
{K,C,W,Y}	K,C,W,Y
{K,C,W,Z}	K,C,W,Z
{K,C,X,Y}	K,C,X,Y
{K,C,X,Z}	K,C,X,Z
{K,C,Y,Z}	K,C,Y,Z
{K,C,V,W,X}	K,C
{K,C,V,W,Y}	K,C
{K,C,V,W,Z}	K,C
{K,C,V,X,Y}	K,C
{K,C,V,X,Z}	K,C
{K,C,V,Y,Z}	K,C
{K,C,W,X,Y}	K,C
{K,C,W,X,Z}	K,C
{K,C,W,Y,Z}	K,C
{K,C,X,Y,Z}	K,C
{K,C,V,W,X,Y}	K,C
{K,C,V,W,X,Z}	K,C
{K,C,V,W,Y,Z}	K,C
{K,C,V,X,Y,Z}	K,C
{K,C,W,X,Y,Z}	K,C
{K,C,V,W,X,Y,Z}	K,C

41. (continued)

$$\frac{\text{The number of times K is critical in winning coalitions}}{\text{The total number of times voters are critical in winning coalitions}} = \frac{26}{72} = \frac{13}{36}$$

$$\frac{\text{The number of times C is critical in winning coalitions}}{\text{The total number of times voters are critical in winning coalitions}} = \frac{26}{72} = \frac{13}{36}$$

$$\frac{\text{The number of times V is critical in winning coalitions}}{\text{The total number of times voters are critical in winning coalitions}} = \frac{4}{72} = \frac{1}{18}$$

$$\frac{\text{The number of times W is critical in winning coalitions}}{\text{The total number of times voters are critical in winning coalitions}} = \frac{4}{72} = \frac{1}{18}$$

$$\frac{\text{The number of times X is critical in winning coalitions}}{\text{The total number of times voters are critical in winning coalitions}} = \frac{4}{72} = \frac{1}{18}$$

$$\frac{\text{The number of times Y is critical in winning coalitions}}{\text{The total number of times voters are critical in winning coalitions}} = \frac{4}{72} = \frac{1}{18}$$

$$\frac{\text{The number of times Z is critical in winning coalitions}}{\text{The total number of times voters are critical in winning coalitions}} = \frac{4}{72} = \frac{1}{18}$$

43. a) No solution provided

   b) Let the members of the firm be represented by {K,C,H,X,Y,Z}. The winning coalitions are given in the following table.

Coalition	Critical Voters
{K,C,X,Y}	K,C,X,Y
{K,C,X,Z}	K,C,X,Z
{K,C,Y,Z}	K,C,Y,Z
{K,H,X,Y}	K,H,X,Y
{K,H,X,Z}	K,H,X,Z
{K,H,Y,Z}	K,H,Y,Z
{C,H,X,Y}	C,H,X,Y
{C,H,X,Z}	C,H,X,Z
{C,H,Y,Z}	C,H,Y,Z
{K,C,H,X,Y}	X,Y
{K,C,H,X,Z}	X,Z
{K,C,H,Y,Z}	Y,Z
{K,C,X,Y,Z}	K,C
{K,H,X,Y,Z}	K,H
{C,H,X,Y,Z}	C,H
{K,C,H,X,Y,Z}	None

288 Chapter 10: Voting

43. (continued)

$$\frac{\text{The number of times K is critical in winning coalitions}}{\text{The total number of times voters are critical in winning coalitions}} = \frac{8}{48} = \frac{1}{6}$$

$$\frac{\text{The number of times H is critical in winning coalitions}}{\text{The total number of times voters are critical in winning coalitions}} = \frac{8}{48} = \frac{1}{6}$$

$$\frac{\text{The number of times C is critical in winning coalitions}}{\text{The total number of times voters are critical in winning coalitions}} = \frac{8}{48} = \frac{1}{6}$$

$$\frac{\text{The number of times X is critical in winning coalitions}}{\text{The total number of times voters are critical in winning coalitions}} = \frac{8}{48} = \frac{1}{6}$$

$$\frac{\text{The number of times Y is critical in winning coalitions}}{\text{The total number of times voters are critical in winning coalitions}} = \frac{8}{48} = \frac{1}{6}$$

$$\frac{\text{The number of times Z is critical in winning coalitions}}{\text{The total number of times voters are critical in winning coalitions}} = \frac{8}{48} = \frac{1}{6}$$

45. Consider the votes to be labels A, B, C, D, E. There are $2^5 - 1 = 31$ different coalitions. Many of them have the same weight, however. The sixth line of Pascal's triangle (recall from earlier chapters) is

1 5 10 10 5 1.

This means there are 1 coalition with none (not counted), 5 with 2 (not a winning coalition because the quota is 3), 10 with 3, 10 with 4, 5 with 5, and 1 with 6. Of the 25 winning coalitions, only 10 of them will have critical voters.

Coalition	Critical Voters
{A,B,C}	A,B,C
{A,B,D}	A,B,D
{A,B,E}	A,B,E
{A,C,D}	A,C,D
{A,C,E}	A,C,E
{A,D,E}	A,D,E
{B,C,D}	B,C,D
{B,C,E}	B,C,E
{B,D,E}	B,D,E
{C,D,E}	C,D,E

Each voter would have the same Banzhaf power index of

$$\frac{\text{The number of times A is critical in winning coalitions}}{\text{The total number of times voters are critical in winning coalitions}} = \frac{6}{30} = \frac{1}{5}$$

47. a) The only winning coalitions are those that have the dictator included. Let voter A be the voter with 15 votes, voter B be the voter with 2 votes, voters C and D each with 3 votes, and voter E with 5 votes. The winning coalitions are

Coalition	Critical Voters
{A}	A
{A,B}	A
{A,C}	A
{A,D}	A
{A,E}	A
{A,B,C}	A
{A,B,D}	A
{A,B,E}	A
{A,C,D}	A
{A,C,E}	A
{A,D,E}	A
{A,B,C,D}	A
{A,B,C,E}	A
{A,B,D,E}	A
{A,C,D,E}	A
{A,B,C,D,E}	A

$$\frac{\text{The number of times A is critical in winning coalitions}}{\text{The total number of times voters are critical in winning coalitions}} = \frac{16}{16} = 1$$

The Banzhaf power index of A is 1, the index of the other voters is 0.

b) No solution provided

49. – 53. No solutions provided

Chapter Review Exercises

1. a) No. The total number of votes is $2{,}156 + 1{,}462 + 986 + 428 = 5{,}032$.
   $2156 \div 5032 \approx 0.428$, which is less than 50%.

   b) Myers

2. A with a total of 93 points

Preference	1st place votes × 4 (points)	2nd place votes × 3 (points)	3rd place votes × 2 (points)	4th place votes × 1 (points)	Total Points
A	15 × 4	6 × 3	3 × 2	9 × 1	93
B	7 × 4	13 × 3	6 × 2	7 × 1	86
C	6 × 4	10 × 3	17 × 2	0 × 1	88
D	5 × 4	4 × 3	7 × 2	17 × 1	63

3. Since education in the future (E) didn't have any first-place votes, it is removed. The new table would be

Preference	1,531	1,102	906	442	375
1st	G	R	S	S	G
2nd	R	S	G	R	S
3rd	S	G	R	G	R

Since role of government in a free society (R) has the fewest first-place votes, it is then eliminated

Preference	1,906	1,102	906	442
1st	G	S	S	S
2nd	S	G	G	G

By combining identical columns, we have

Preference	1,906	2,450
1st	G	S
2nd	S	G

Since social justice (S) now has the most first-place votes, it wins the election.

4. We must compare a) A with B, b) A with C, c) A with D, d) B with C, e) B with D, and f) C with D.

   a) In comparing A with B, we ignore C and D. Thus, 8 prefer A over B, and 4 + 5 + 6 = 15 prefer B over A. Thus, we award 1 point to option B.

   b) In comparing A with C, we ignore B and D. Thus, 8 prefer A over C, and 4 + 5 + 6 = 15 prefer C over A. Thus, we award 1 point to option C.

   c) In comparing A with D, we ignore B and C. Thus, 8 + 6 = 14 prefer A over D, and 4 + 5 = 9 prefer D over A. Thus, we award 1 point to option A.

   d) In comparing B with C, we ignore A and D. Thus, 8 + 4 + 5 = 17 prefer B over C, and 6 prefer C over B. Thus, we award 1 point to option B.

   e) In comparing B with D, we ignore A and C. Thus, 8 + 5 + 6 = 19 prefer B over D, and 4 prefer D over B. Thus, we award 1 point to option B.

   f) In comparing C with D, we ignore A and C. Thus, 8 + 6 = 14 prefer C over D, and 4 + 5 = 9 prefer D over C. Thus, we award 1 point to option C.

   Since B has 3 points, C has 2, and A has 1, option B is declared the winner.

5.

Preference	1st place votes × 4 (points)	2nd place votes × 3 (points)	3rd place votes × 2 (points)	4th place votes × 1 (points)	Total Points
R	2×4	0×3	0×2	1×1	9
D	1×4	2×3	0×2	0×1	10
P	0×4	1×3	1×2	1×1	6
Q	0×4	0×3	2×2	1×1	5

Option D is the winner by the Borda method. No, this election does not satisfy the majority criterion. Option R had the majority of the 1st place votes.

6.

Preference	1st place votes × 4 (points)	2nd place votes × 3 (points)	3rd place votes × 2 (points)	4th place votes × 1 (points)	Total Points
R	18×4	8×3	0×2	9×1	105
A	1×4	14×3	8×2	12×1	74
M	8×4	12×3	15×2	0×1	98
D	8×4	1×3	12×2	14×1	73

Option R is the winner by the Borda count. The Condorcet's criterion is satisfied in this example, because option R can beat all of the other options in a head-to-head competition and is the winner by Borda count. 26 out of 35 prefer R to A, 18 out of 35 prefer R to M, and 26 out of 35 prefer R to D.

7.

Preference	1st place votes × 4 (points)	2nd place votes × 3 (points)	3rd place votes × 2 (points)	4th place votes × 1 (points)	Total Points
A	3×4	10×3	23×2	0×1	88
B	13×4	14×3	3×2	6×1	106
C	17×4	3×3	4×2	12×1	97
D	3×4	9×3	6×2	18×1	69

Option B is the winner by the Borda count. The independence-of-irrelevant-alternatives criterion is not satisfied in this example because if option A were taken out, then the preference table would be

Preference	11	3	6	9	4	3
1st	C	D	C	B	B	C
2nd	B	B	D	D	C	B
3rd	D	C	B	C	D	D

According to the Borda count, C would now be declared the winner.

Preference	1st place votes × 3 (points)	2nd place votes × 2 (points)	3rd place votes × 1 (points)	Total Points
C	20×3	4×2	12×1	80
B	13×3	17×2	6×1	79
D	3×3	15×2	18×1	57

8. Remove option C, since it had the least number of first-place votes. The new table would be

Preference	10	7	2	4
1st	A	B	A	B
2nd	B	A	B	A

By combining identical columns, we have

Preference	12	11
1st	A	B
2nd	B	A

Since option A now has the most first-place votes, it wins the election.

Now if we remove option B, the new table would be

Preference	10	7	2	4
1st	A	A	C	C
2nd	C	C	A	A

By combining identical columns we have

Preference	17	6
1st	A	C
2nd	C	A

Since option A now has the most first-place votes, it still wins the election. The independence-of-irrelevant-alternatives criterion is satisfied.

9. The quota is 17; There are 4 voters. Voter A has 1 vote, voter B has 5, voter C has 7, and voter D has 8; There are no dictators; Voters B, C, and D each have veto power.

10. Let voter A be the voter with 2 votes, voter B with 3, voter C with 5, and voter D with 7.

Winning Coalition	Weight
{C,D}	12
{A,B,D}	12
{A,C,D}	14
{B,C,D}	15
{A,B,C,D}	17

11. Let voter A be the voter with 2 votes, voter B with 3, voter C with 5, and voter D with 7.

Coalition	Weight	Critical Voters
{C,D}	12	C,D
{A,B,D}	12	A,B,D
{A,C,D}	14	C,D
{B,C,D}	15	C,D
{A,B,C,D}	17	D

$$\frac{\text{The number of times A is critical in winning coalitions}}{\text{The total number of times voters are critical in winning coalitions}} = \frac{1}{10}$$

$$\frac{\text{The number of times B is critical in winning coalitions}}{\text{The total number of times voters are critical in winning coalitions}} = \frac{1}{10}$$

$$\frac{\text{The number of times C is critical in winning coalitions}}{\text{The total number of times voters are critical in winning coalitions}} = \frac{3}{10}$$

$$\frac{\text{The number of times D is critical in winning coalitions}}{\text{The total number of times voters are critical in winning coalitions}} = \frac{5}{10} = \frac{1}{2}$$

12. a) Let voter A be the voter with 1 vote, voter B with 2, voter C with 3, and voter D with 4.

Coalition	Weight	Critical Voters
{A,B,C,D}	10	A,B,C,D

Each voter would have the same Banzhaf power index as voter A

$$\frac{\text{The number of times A is critical in winning coalitions}}{\text{The total number of times voters are critical in winning coalitions}} = \frac{1}{4}$$

This is expected, because the sum of the voters' weights is the same as the quota. Only when they vote together will a winning coalition be formed. Each voter is a critical voter in this winning coalition.

12. (continued)

   b) Let voter A be the voter with 11 votes, voter B with 1, voters C and D with 3, and voter E with 2.

Coalition	Weight	Critical Voters
{A}	11	A
{A,B}	12	A
{A,C}	14	A
{A,D}	14	A
{A,E}	13	A
{A,B,C}	15	A
{A,B,D}	15	A
{A,B,E}	14	A
{A,C,D}	17	A
{A,C,E}	16	A
{A,D,E}	16	A
{A,B,C,D}	18	A
{A,B,C,E}	17	A
{A,B,D,E}	17	A
{A,C,D,E}	19	A
{A,B,C,D,E}	20	A

   Voter A has a Banzhaf power index of 1

   $$\frac{\text{The number of times A is critical in winning coalitions}}{\text{The total number of times voters are critical in winning coalitions}} = \frac{16}{16} = 1$$

   while the other four voter have a Banzhaf power index of 0

   $$\frac{\text{The number of times B,C,D,or E is critical in winning coalitions}}{\text{The total number of times voters are critical in winning coalitions}} = \frac{0}{16} = 0$$

   This result is expected, because the voter A is a dictator.

Chapter Test

1. a) No. The total number of votes is $2{,}543 + 1{,}532 + 892 + 473 = 5{,}440$. $2{,}543 \div 5{,}440 \approx 0.467$, which is less than 50%.

   b) Molina

2. B with a total of 77 points

Preference	1st place votes × 4 (points)	2nd place votes × 3 (points)	3rd place votes × 2 (points)	4th place votes × 1 (points)	Total Points
A	5×4	11×3	8×2	5×1	74
B	11×4	5×3	5×2	8×1	77
C	8×4	5×3	5×2	11×1	68
D	5×4	8×3	11×2	5×1	71

294   Chapter 10: Voting

3. We must compare a) A with B, b) A with C, c) A with D, d) B with C, e) B with D, and f) C with D.

   a) In comparing A with B, we ignore C and D. Thus, 23 + 83 = 106 prefer A over B, and 47 + 21 = 68 prefer B over A. Thus, we award 1 point to option A.
   b) In comparing A with C, we ignore B and D. Thus, 23 + 47 = 70 prefer A over C, and 83 + 21 = 104 prefer C over A. Thus, we award 1 point to option C.
   c) In comparing A with D, we ignore B and C. Thus, 23 + 47 + 21 = 91 prefer A over D, and 83 prefer D over A. Thus, we award 1 point to option A.
   d) In comparing B with C, we ignore A and D. Thus, 23 + 47 = 70 prefer B over C, and 83 + 21 = 104 prefer C over B. Thus, we award 1 point to option C.
   e) In comparing B with D, we ignore A and C. Thus, 23 + 47 + 21 = 91 prefer B over D, and 83 prefer D over B. Thus, we award 1 point to option B.
   f) In comparing C with D, we ignore A and C. Thus, 23 + 47 + 21 = 91 prefer C over D, and 83 prefer D over C. Thus, we award 1 point to option C.

   Since C has 3 points, A has 2, and B has 1, option C is declared the winner.

4. Since quality of life (Q) didn't have any first-place votes, it is removed. The new table would be

Preference	327	130	149	85	234
1st	E	R	A	E	R
2nd	R	E	E	R	E
3rd	A	A	R	A	A

   By combining identical columns, we have

Preference	412	365	149
1st	E	R	A
2nd	R	E	E
3rd	A	A	R

   Since attracting more women to science (A) has the fewest first-place votes, it is then eliminated

Preference	412	364	149
1st	E	R	E
2nd	R	E	R

   By combining identical columns, we have

Preference	561	364
1st	E	R
2nd	R	E

   Since equality in the workplace (E) now has the most first-place votes, it wins the election.

5.

Preference	1st place votes × 4 (points)	2nd place votes × 3 (points)	3rd place votes × 2 (points)	4th place votes × 1 (points)	Total Points
A	2×4	0×3	0×2	1×1	9
B	1×4	2×3	0×2	0×1	10
C	0×4	0×3	3×2	0×1	6
D	0×4	1×3	0×2	2×1	5

   Option B is the winner by the Borda method. No, this election does not satisfy the majority criterion. Option A had the majority of the 1st place votes.

6.

Preference	1st place votes × 4 (points)	2nd place votes × 3 (points)	3rd place votes × 2 (points)	4th place votes × 1 (points)	Total Points
A	19×4	21×3	6×2	8×1	159
B	21×4	7×3	20×2	6×1	151
C	6×4	8×3	23×2	17×1	111
D	8×4	18×3	5×2	23×1	119

Option A is the winner by the Borda count. The independence-of-irrelevant-alternatives criterion is not satisfied in this example because if option D were taken out, then the preference table would be

Preference	7	5	6	12	16	8
1st	A	B	C	A	B	C
2nd	B	A	A	B	A	B
3rd	C	C	B	C	C	A

According to the Borda count, B would now be declared the winner.

Preference	1st place votes × 3 (points)	2nd place votes × 2 (points)	3rd place votes × 1 (points)	Total Points
A	19×3	27×2	8×1	119
B	21×3	27×2	6×1	123
C	14×3	0×2	40×1	96

7.

Preference	1st place votes × 4 (points)	2nd place votes × 3 (points)	3rd place votes × 2 (points)	4th place votes × 1 (points)	Total Points
A	1,612×4	1,754×3	849×2	0×1	13,409
B	1,754×4	1,612×3	0×2	849×1	12,701
C	849×4	0×3	2,457×2	909×1	9,219
D	0×4	849×3	909×2	2,457×1	6,822

Option A is the winner by the Borda count. The Condorcet's criterion is satisfied in this example, because option A can beat all of the other options in a head-to-head competition and is the winner by Borda count. 2,461 out of 4,215 prefer A to B, 3,366 out of 4,215 prefer A to C, and 3,366 out of 4,215 prefer A to D.

8. Remove option D, since it had the least number of first-place votes. The new table would be

Preference	35	71	36	14
1st	A	B	C	A
2nd	B	A	A	C
3rd	C	C	B	B

Then remove option C, since it had the least number of first-place votes. The new table would be

Preference	35	71	36	14
1st	A	B	A	A
2nd	B	A	B	B

By combining identical columns, we have

Preference	85	71
1st	A	B
2nd	B	A

Since option A now has the most first-place votes, it wins the election.

8. (continued)

Now if we remove option C, the new table would be

Preference	35	71	36	14
1st	A	B	D	D
2nd	B	A	A	A
3rd	D	D	B	B

By combining identical columns we have

Preference	35	71	50
1st	A	B	D
2nd	B	A	A
3rd	D	D	B

Remove option A, since it had the least number of first-place votes. The new table would be

Preference	35	71	50
1st	B	B	D
2nd	D	D	B

By combining identical columns we have

Preference	104	50
1st	B	D
2nd	D	B

Since option B now has the most first-place votes, it wins the election instead of A. The independence-of-irrelevant-alternatives criterion is not satisfied.

9. Let voter A be the voter with 3 votes, voter B with 4, voter C with 6, and voter D with 8.

Winning Coalition	Weight
{A,B,D}	15
{A,C,D}	17
{B,C,D}	18
{A,B,C,D}	21

10. The quota is 15; Voter A has 5 vote, voter B has 3, voter C has 1, voter D has 3, voter E has 4, and voter F has 2; There are no dictators; Voters A and E each have veto power.

11. a) Let voter A be the voter with 2 votes, voter B with 8, voter C with 3, and voter D with 2.

Coalition	Weight	Critical Voters
{A,B,C,D}	15	A,B,C,D

Each voter would have the same Banzhaf power index as voter A

$$\frac{\text{The number of times A is critical in winning coalitions}}{\text{The total number of times voters are critical in winning coalitions}} = \frac{1}{4}$$

This is expected, because the sum of the voters' weights is the same as the quota. Only when they vote together will a winning coalition be formed. Each voter is a critical voter in this winning coalition.

11. (continued)

    b) Let voter A be the voter with 15 votes, voter B with 2, voter C with 4, voter D with 1, and voter E with 3.

Coalition	Weight	Critical Voters
{A}	15	A
{A,B}	17	A
{A,C}	19	A
{A,D}	16	A
{A,E}	18	A
{A,B,C}	21	A
{A,B,D}	18	A
{A,B,E}	20	A
{A,C,D}	20	A
{A,C,E}	22	A
{A,D,E}	19	A
{A,B,C,D}	22	A
{A,B,C,E}	24	A
{A,B,D,E}	21	A
{A,C,D,E}	23	A
{A,B,C,D,E}	25	A

    Voter A has a Banzhaf power index of 1

    $$\frac{\text{The number of times A is critical in winning coalitions}}{\text{The total number of times voters are critical in winning coalitions}} = \frac{16}{16} = 1$$

    while the other four voter have a Banzhaf power index of 0

    $$\frac{\text{The number of times B,C,D, or E is critical in winning coalitions}}{\text{The total number of times voters are critical in winning coalitions}} = \frac{0}{16} = 0$$

    This result is expected, because the voter A is a dictator.

Of Further Interest: The Shapley–Shubik Index

1. With the notation $\{A, B, C\}$, the order of the voters does not matter; with the notation (A, B, C), the order of the voters does matter.

2. We summed the weights of the voters until we reached the quota of 5.

3. We divided the number of times that A was pivotal in some permutation of the voters by total number of permutations of the voters.

4. $n! = n \times (n-1) \times (n-2) \ldots 3 \times 2 \times 1$

5. A pivotal voter for a coalition is the first player, as voters are added, who makes the coalition a winning coalition

6. A critical voter (used in calculating the Banzhaf index) is a voter in a winning coalition who, if removed, makes the coalition a non-winning one. A pivotal voter (used in calculating the Shapley–Shubik index) is the first voter in an ordered coalition who makes the sum of the weights greater than or equal to the quota.

7.

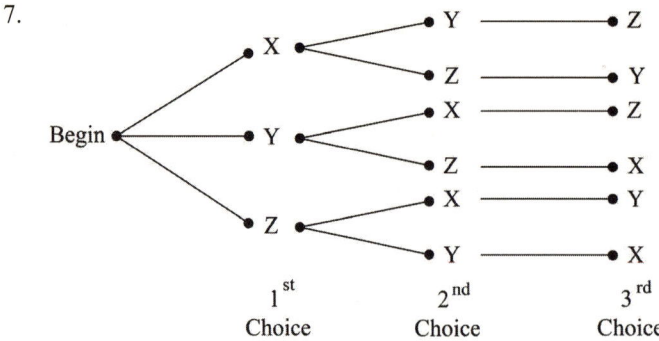

The permutations are (X,Y,Z), (X,Z,Y), (Y,X,Z), (Y,Z,X), (Z,X,Y), and (Z,Y,X).

8.

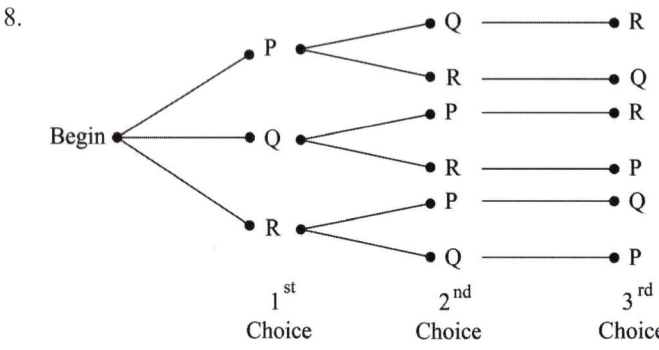

The permutations are (P,Q,R), (P,R,Q), (Q,P,R), (Q,R,P), (R,P,Q), and (R,Q,P).

9.

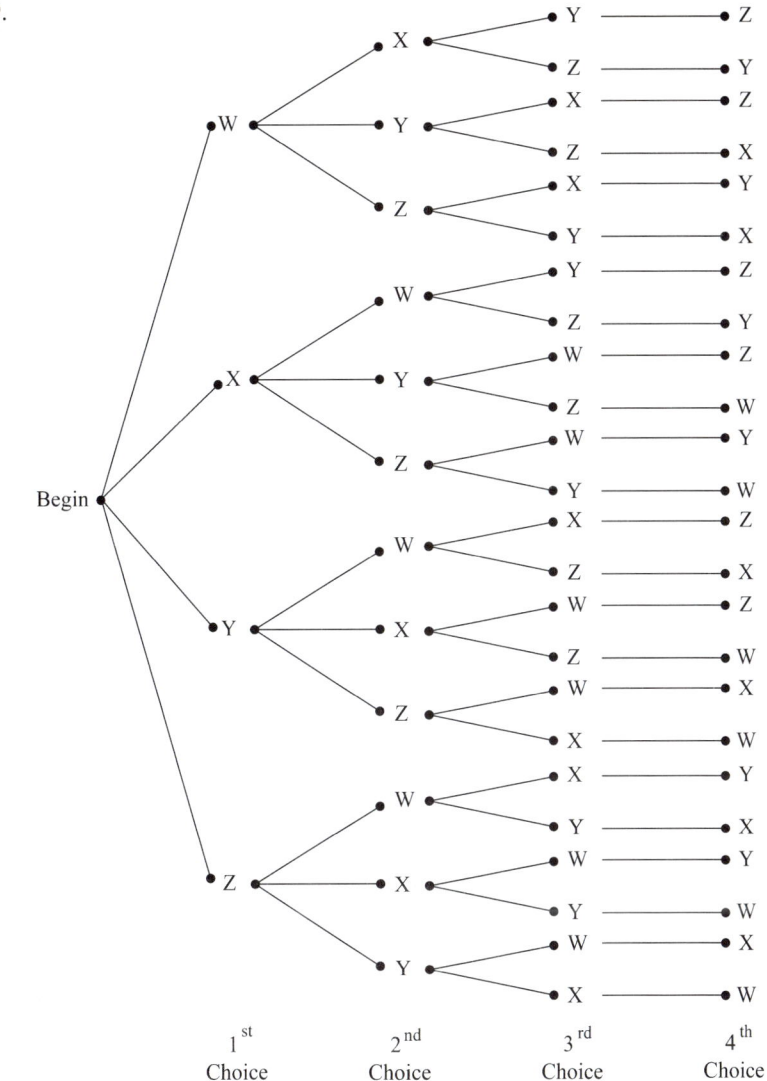

The permutations are (W,X,Y,Z), (W,X,Z,Y), (W,Y,X,Z), (W,Y,Z,X), (W,Z,X,Y), (W,Z,Y,X), (X,W,Y,Z), (X,W,Z,Y), (X,Y,W,Z), (X,Y,Z,W), (X,Z,W,Y), (X,Z,Y,W), (Y,W,X,Z), (Y,W,Z,X), (Y,X,W,Z), (Y,X,Z,W), (Y,Z,W,X), (Y,Z,X,W), (Z,W,X,Y), (Z,W,Y,X), (Z,X,W,Y), (Z,X,Y,W), (Z,Y,W,X), and (Z,Y,X,W).

10.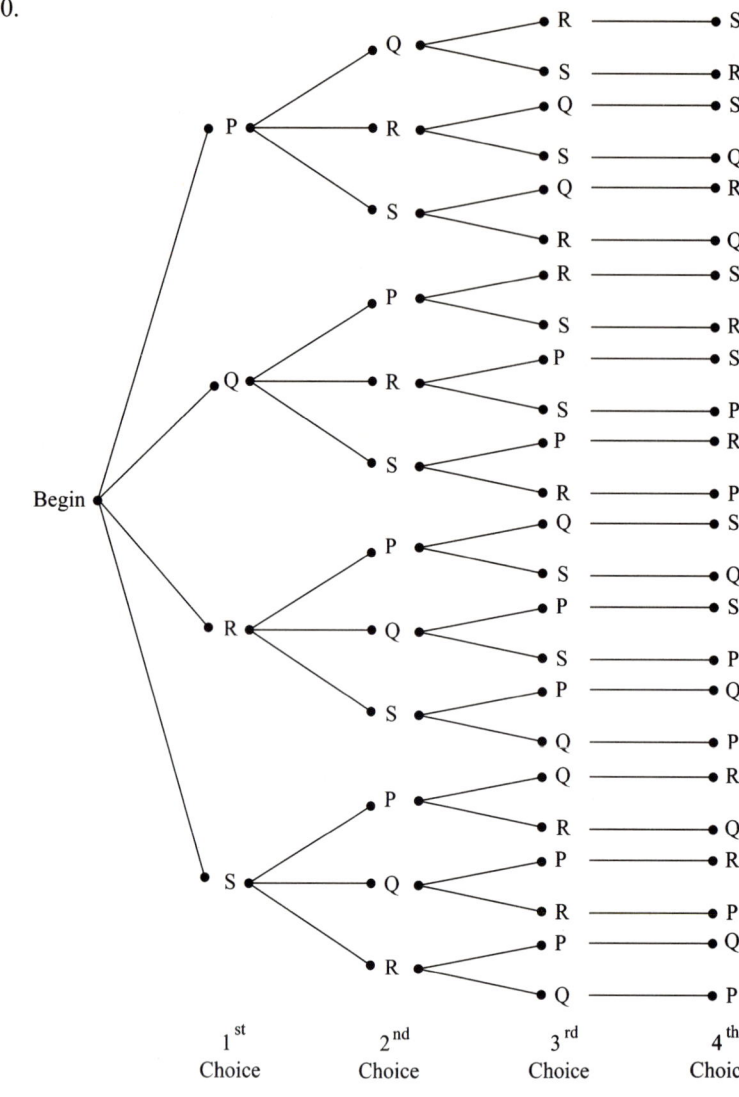

The permutations are (P,Q,R,S), (P,Q,S,R), (P,R,Q,S), (P,R,S,Q), (P,S,Q,R), (P,S,R,Q), (Q,P,R,S), (Q,P,S,R), (Q,R,P,S), (Q,R,S,P), (Q,S,P,R), (Q,S,R,P), (R,P,Q,S), (R,P,S,Q), (R,Q,P,S), (R,Q,S,P), (R,S,P,Q), (R,S,Q,P), (S,P,Q,R), (S,P,R,Q), (S,Q,P,R), (S,Q,R,P), (S,R,P,Q), and (S,R,Q,P).

11. $6! = 6 \times 5 \times 4 \times 3 \times 2 \times 1 = 720$

12. $8! = 8 \times 7 \times 6 \times 5 \times 4 \times 3 \times 2 \times 1 = 40,320$

13. $12! = 12 \times 11 \times 10 \times 9 \times 8 \times 7 \times 6 \times 5 \times 4 \times 3 \times 2 \times 1 = 479,001,600$

14. $13! = 13 \times 12 \times 11 \times 10 \times 9 \times 8 \times 7 \times 6 \times 5 \times 4 \times 3 \times 2 \times 1 = 6,227,020,800$

15.

Coalition	Weight after 1st, 2nd, and 3rd voter is added	Pivotal voter
a) (A,B,C)	3,5,7	B
b) (A,C,B)	3,5,7	C
c) (B,A,C)	2,5,7	A
d) (B,C,A)	2,4,7	A
e) (C,A,B)	2,5,7	A
f) (C,B,A)	2,4,7	A

16.

Coalition	Weight after 1st, 2nd, and 3rd voter is added	Pivotal voter
a) (A,B,C)	4,7,9	B
b) (A,C,B)	4,6,9	B
c) (B,A,C)	3,7,9	A
d) (B,C,A)	3,5,9	A
e) (C,A,B)	2,6,9	B
f) (C,B,A)	2,5,9	A

17. Let the voters be labeled A, B, and C.

Coalition	Weight after 1st, 2nd, and 3rd voter is added	Pivotal voter
(A,B,C)	3,6,8	B
(A,C,B)	3,5,8	B
(B,A,C)	3,6,8	A
(B,C,A)	3,5,8	A
(C,A,B)	2,5,8	B
(C,B,A)	2,5,8	A

$$\frac{\text{The number of times A is pivotal in some permutation of voters}}{\text{The total number of permutations of voters}} = \frac{3}{6} = \frac{1}{2}$$

$$\frac{\text{The number of times B is pivotal in some permutation of voters}}{\text{The total number of permutations of voters}} = \frac{3}{6} = \frac{1}{2}$$

$$\frac{\text{The number of times C is pivotal in some permutation of voters}}{\text{The total number of permutations of voters}} = \frac{0}{6} = 0$$

18. Let the voters be labeled A, B, and C.

Coalition	Weight after 1st, 2nd, and 3rd voter is added	Pivotal voter
(A,B,C)	4,7,10	C
(A,C,B)	4,7,10	B
(B,A,C)	3,7,10	C
(B,C,A)	3,6,10	A
(C,A,B)	3,7,10	B
(C,B,A)	3,6,10	A

$$\frac{\text{The number of times A is pivotal in some permutation of voters}}{\text{The total number of permutations of voters}} = \frac{2}{6} = \frac{1}{3}$$

$$\frac{\text{The number of times B is pivotal in some permutation of voters}}{\text{The total number of permutations of voters}} = \frac{2}{6} = \frac{1}{3}$$

$$\frac{\text{The number of times C is pivotal in some permutation of voters}}{\text{The total number of permutations of voters}} = \frac{2}{6} = \frac{1}{3}$$

19.

Coalition	Weight after 1st, 2nd, 3rd and 4th voter is added	Pivotal voter
(A,B,C,D)	3,6,8,10	C
(A,B,D,C)	3,6,8,10	D
(A,C,B,D)	3,5,8,10	B
(A,C,D,B)	3,5,7,10	B
(A,D,B,C)	3,5,8,10	B
(A,D,C,B)	3,5,7,10	B
(B,A,C,D)	3,6,8,10	C
(B,A,D,C)	3,6,8,10	D
(B,C,A,D)	3,5,8,10	A
(B,C,D,A)	3,5,7,10	A
(B,D,A,C)	3,5,8,10	A
(B,D,C,A)	3,5,7,10	A
(C,A,B,D)	2,5,8,10	B
(C,A,D,B)	2,5,7,10	B
(C,B,A,D)	2,5,8,10	A
(C,B,D,A)	2,5,7,10	A
(C,D,A,B)	2,4,7,10	B
(C,D,B,A)	2,4,7,10	A
(D,A,B,C)	2,5,8,10	B
(D,A,C,B)	2,5,7,10	B
(D,B,A,C)	2,5,8,10	A
(D,B,C,A)	2,5,7,10	A
(D,C,A,B)	2,4,7,10	B
(D,C,B,A)	2,4,7,10	A

$$\frac{\text{The number of times A is pivotal in some permutation of voters}}{\text{The total number of permutations of voters}} = \frac{10}{24} = \frac{5}{12}$$

$$\frac{\text{The number of times B is pivotal in some permutation of voters}}{\text{The total number of permutations of voters}} = \frac{10}{24} = \frac{5}{12}$$

$$\frac{\text{The number of times C is pivotal in some permutation of voters}}{\text{The total number of permutations of voters}} = \frac{2}{24} = \frac{1}{12}$$

$$\frac{\text{The number of times D is pivotal in some permutation of voters}}{\text{The total number of permutations of voters}} = \frac{2}{24} = \frac{1}{12}$$

20.

Coalition	Weight after 1st, 2nd, 3rd and 4th voter is added	Pivotal voter
(A,B,C,D)	4,7,10,11	C
(A,B,D,C)	4,7,8,11	C
(A,C,B,D)	4,7,10,11	B
(A,C,D,B)	4,7,8,11	B
(A,D,B,C)	4,5,8,11	C
(A,D,C,B)	4,5,8,11	B
(B,A,C,D)	3,7,10,11	C
(B,A,D,C)	3,7,8,11	C
(B,C,A,D)	3,6,10,11	A
(B,C,D,A)	3,6,7,11	A
(B,D,A,C)	3,4,8,11	C
(B,D,C,A)	3,4,7,11	A
(C,A,B,D)	3,7,10,11	B
(C,A,D,B)	3,7,8,11	B
(C,B,A,D)	3,6,10,11	A
(C,B,D,A)	3,6,7,11	A
(C,D,A,B)	3,4,8,11	B
(C,D,B,A)	3,4,7,11	A
(D,A,B,C)	1,5,8,11	C
(D,A,C,B)	1,5,8,11	B
(D,B,A,C)	1,4,8,11	C
(D,B,C,A)	1,4,7,11	A
(D,C,A,B)	1,4,8,11	B
(D,C,B,A)	1,4,7,11	A

$$\frac{\text{The number of times A is pivotal in some permutation of voters}}{\text{The total number of permutations of voters}} = \frac{8}{24} = \frac{1}{3}$$

$$\frac{\text{The number of times B is pivotal in some permutation of voters}}{\text{The total number of permutations of voters}} = \frac{8}{24} = \frac{1}{3}$$

$$\frac{\text{The number of times C is pivotal in some permutation of voters}}{\text{The total number of permutations of voters}} = \frac{8}{24} = \frac{1}{3}$$

$$\frac{\text{The number of times D is pivotal in some permutation of voters}}{\text{The total number of permutations of voters}} = \frac{0}{24} = 0$$

21.

Coalition	Weight after 1st, 2nd, 3rd and 4th voter is added	Pivotal voter
(A,B,C,D)	2,3,4,5	C
(A,B,D,C)	2,3,4,5	D
(A,C,B,D)	2,3,4,5	B
(A,C,D,B)	2,3,4,5	D
(A,D,B,C)	2,3,4,5	B
(A,D,C,B)	2,3,4,5	C
(B,A,C,D)	1,3,4,5	C
(B,A,D,C)	1,3,4,5	D
(B,C,A,D)	1,2,4,5	A
(B,C,D,A)	1,2,3,5	A
(B,D,A,C)	1,2,4,5	A
(B,D,C,A)	1,2,3,5	A
(C,A,B,D)	1,3,4,5	B
(C,A,D,B)	1,3,4,5	D
(C,B,A,D)	1,2,4,5	A
(C,B,D,A)	1,2,3,5	A
(C,D,A,B)	1,2,4,5	A
(C,D,B,A)	1,2,3,5	A
(D,A,B,C)	1,3,4,5	B
(D,A,C,B)	1,3,4,5	C
(D,B,A,C)	1,2,4,5	A
(D,B,C,A)	1,2,3,5	A
(D,C,A,B)	1,2,4,5	A
(D,C,B,A)	1,2,3,5	A

$$\frac{\text{The number of times A is pivotal in some permutation of voters}}{\text{The total number of permutations of voters}} = \frac{12}{24} = \frac{1}{2}$$

$$\frac{\text{The number of times B is pivotal in some permutation of voters}}{\text{The total number of permutations of voters}} = \frac{4}{24} = \frac{1}{6}$$

$$\frac{\text{The number of times C is pivotal in some permutation of voters}}{\text{The total number of permutations of voters}} = \frac{4}{24} = \frac{1}{6}$$

$$\frac{\text{The number of times D is pivotal in some permutation of voters}}{\text{The total number of permutations of voters}} = \frac{4}{24} = \frac{1}{6}$$

22.

Coalition	Weight after 1st, 2nd, 3rd and 4th voter is added	Pivotal voter
(A,B,C,D)	6,12,17,20	C
(A,B,D,C)	6,12,15,20	D
(A,C,B,D)	6,11,17,20	B
(A,C,D,B)	6,11,14,20	D
(A,D,B,C)	6,9,15,20	B
(A,D,C,B)	6,9,14,20	C
(B,A,C,D)	6,12,17,20	C
(B,A,D,C)	6,12,15,20	D
(B,C,A,D)	6,11,17,20	A
(B,C,D,A)	6,11,14,20	D
(B,D,A,C)	6,9,15,20	A
(B,D,C,A)	6,9,14,20	C
(C,A,B,D)	5,11,17,20	B
(C,A,D,B)	5,11,14,20	D
(C,B,A,D)	5,11,17,20	A
(C,B,D,A)	5,11,14,20	D
(C,D,A,B)	5,8,14,20	A
(C,D,B,A)	5,8,14,20	B
(D,A,B,C)	3,9,15,20	B
(D,A,C,B)	3,9,14,20	C
(D,B,A,C)	3,9,15,20	A
(D,B,C,A)	3,9,14,20	C
(D,C,A,B)	3,8,14,20	A
(D,C,B,A)	3,8,14,20	B

$$\frac{\text{The number of times A is pivotal in some permutation of voters}}{\text{The total number of permutations of voters}} = \frac{6}{24} = \frac{1}{4}$$

$$\frac{\text{The number of times B is pivotal in some permutation of voters}}{\text{The total number of permutations of voters}} = \frac{6}{24} = \frac{1}{4}$$

$$\frac{\text{The number of times C is pivotal in some permutation of voters}}{\text{The total number of permutations of voters}} = \frac{6}{24} = \frac{1}{4}$$

$$\frac{\text{The number of times D is pivotal in some permutation of voters}}{\text{The total number of permutations of voters}} = \frac{6}{24} = \frac{1}{4}$$

23. a) The Shapley–Shubik index for each person will be $\frac{1}{5}$.

   b) No solution provided

   c) No solution provided

24. a) The Shapley–Shubik index for each person will be $\frac{1}{12}$.

   b) No solution provided

   c) No solution provided

25. a) The Shapley–Shubik index for A is 1. It will be 0 for each of the others.
    b) No solution provided
    c) No solution provided

26. a) Since no voter is pivotal, it is not possible to find the Shapley–Shubik index for any person.
    b) No solution provided
    c) No solution provided

27.

Coalition	Pivotal voter
(A,B,C,D)	B
(A,B,D,C)	B
(A,C,B,D)	C
(A,C,D,B)	C
(A,D,B,C)	D
(A,D,C,B)	D
(B,A,C,D)	A
(B,A,D,C)	A
(B,C,A,D)	A
(B,C,D,A)	A
(B,D,A,C)	A
(B,D,C,A)	A
(C,A,B,D)	A
(C,A,D,B)	A
(C,B,A,D)	A
(C,B,D,A)	A
(C,D,A,B)	A
(C,D,B,A)	A
(D,A,B,C)	A
(D,A,C,B)	A
(D,B,A,C)	A
(D,B,C,A)	A
(D,C,A,B)	A
(D,C,B,A)	A

$$\frac{\text{The number of times A is pivotal in some permutation of voters}}{\text{The total number of permutations of voters}} = \frac{18}{24} = \frac{3}{4}$$

$$\frac{\text{The number of times B is pivotal in some permutation of voters}}{\text{The total number of permutations of voters}} = \frac{2}{24} = \frac{1}{12}$$

$$\frac{\text{The number of times C is pivotal in some permutation of voters}}{\text{The total number of permutations of voters}} = \frac{2}{24} = \frac{1}{12}$$

$$\frac{\text{The number of times D is pivotal in some permutation of voters}}{\text{The total number of permutations of voters}} = \frac{2}{24} = \frac{1}{12}$$

28. a) The permutations are (P,T,S), (P,S,T), (T,P,S), (T,S,P), (S,P,T), and (S,T,P)
    b) Since there are 11 voters, a simple majority is 6 or more.

Coalition	Weight after 1st, 2nd, and 3rd group is added	Pivotal group
(P,T,S)	5,9,11	T
(P,S,T)	5,7,11	S
(T,P,S)	4,9,11	P
(T,S,P)	4,6,11	S
(S,P,T)	2,7,11	P
(S,T,P)	2,6,11	T

c) $\dfrac{\text{The number of times P is pivotal in some permutation of voters}}{\text{The total number of permutations of voters}} = \dfrac{2}{6} = \dfrac{1}{3}$

$\dfrac{\text{The number of times T is pivotal in some permutation of voters}}{\text{The total number of permutations of voters}} = \dfrac{2}{6} = \dfrac{1}{3}$

$\dfrac{\text{The number of times S is pivotal in some permutation of voters}}{\text{The total number of permutations of voters}} = \dfrac{2}{6} = \dfrac{1}{3}$

29. a) (A,F,T,N),(A,F,N,T),(A,T,F,N),(A,T,N,F),(A,N,F,T),(A,N,T,F),(F,A,T,N),(F,A,N,T),(F,T,A,N),(F,T,N,A),(F,N,T,A),(F,N,A,T),(T,A,F,N),(T,A,N,F),(T,F,A,N),(T,F,N,A),(T,N,A,F),(T,N,F,A),(N,A,B,T),(N,A,C,F),(N,F,A,T),(N,F,T,A),(N,T,A,F), and (N,T,F,A)

b)

Coalition	Weight after 1st, 2nd, 3rd and 4th voter is added	Pivotal group
(A,F,T,N)	3,7,10,12	T
(A,F,N,T)	3,7,9,12	N
(A,T,F,N)	3,6,10,12	F
(A,T,N,F)	3,6,8,12	N
(A,N,F,T)	3,5,9,12	F
(A,N,T,F)	3,5,8,12	T
(F,A,T,N)	4,7,10,12	T
(F,A,N,T)	4,7,9,12	N
(F,T,A,N)	4,7,10,12	A
(F,T,N,A)	4,7,9,12	N
(F,N,A,T)	4,6,9,12	A
(F,N,T,A)	4,6,9,12	T
(T,A,F,N)	3,6,10,12	F
(T,A,N,F)	3,6,8,12	N
(T,F,A,N)	3,7,10,12	A
(T,F,N,A)	3,7,9,12	N
(T,N,A,F)	3,5,8,12	A
(T,N,F,A)	3,5,9,12	F
(N,A,F,T)	2,5,9,12	F
(N,A,T,F)	2,5,8,12	T
(N,F,A,T)	2,6,9,12	A
(N,F,T,A)	2,6,9,12	T
(N,T,A,F)	2,5,8,12	A
(N,T,F,A)	2,5,9,12	F

29. (continued)

c) $\dfrac{\text{The number of times A is pivotal in some permutation of voters}}{\text{The total number of permutations of voters}} = \dfrac{6}{24} = \dfrac{1}{4}$

$\dfrac{\text{The number of times F is pivotal in some permutation of voters}}{\text{The total number of permutations of voters}} = \dfrac{6}{24} = \dfrac{1}{4}$

$\dfrac{\text{The number of times T is pivotal in some permutation of voters}}{\text{The total number of permutations of voters}} = \dfrac{6}{24} = \dfrac{1}{4}$

$\dfrac{\text{The number of times N is pivotal in some permutation of voters}}{\text{The total number of permutations of voters}} = \dfrac{6}{24} = \dfrac{1}{4}$

30. Since there are $6! = 6 \times 5 \times 4 \times 3 \times 2 \times 1 = 720$ permutations, listing them all would be lengthy. In forming permutations of $\{K,C,W,X,Y,Z\}$ we will be filling slots.

$$\overline{1^{st}} \; \overline{2^{nd}} \; \overline{3^{rd}} \; \overline{4^{th}} \; \overline{5^{th}} \; \overline{6^{th}}$$

K cannot be pivotal unless he/she is in the 4th, 5th or 6th slot.

If K is in the 4th slot, then C must be in either the 1st or 2nd or 3rd slot. If C is in the 1st slot, then there are $4 \times 3 = 12$ ways to fill the other two slots. Also, there would $2 \times 1 = 2$ ways to fill slots 5 and 6. This would be similar for the case that C is in the 2nd or 3rd slot.

If K is in the 5th slot, then C must be in either the 1st or 2nd or 3rd or 4th slot. If C is in the 1st slot then there are $4 \times 3 \times 2 = 24$ ways to fill the other three slots. This would be similar for the case that C is in the 2nd or 3rd or 4th slot.

If K is in the 6th slot, then C must be in either the 1st or 2nd or 3rd or 4th or 5th slot. If C is in the 1st slot then there are $4 \times 3 \times 2 \times 1 = 24$ ways to fill the other four slots. This would be similar for the case that C is in the 2nd or 3rd or 4th or 5th slot.

So, K would be pivotal a total of $3 \times 12 \times 2 + 4 \times 24 + 5 \times 24 = 288$ times. The same would hold for C.

W can only be pivotal if he/she is in the 4th slot.

If W is in the 4th slot, then K and C must be in either the 1st or 2nd or 3rd slot along with either X, Y, or Z. If X is in either the 1st or 2nd or 3rd slot, then there are $3 \times 2 \times 1 = 6$ ways to arrange K, C, and X in the first 3 places. There are also $2 \times 1 = 2$ ways to fill the last two slots.

There are 2 other similar arrangements for Y and X. So, W is pivotal a total of $3 \times 6 \times 2 = 36$ times. The same would hold for X, Y, and Z.

$\dfrac{\text{The number of times K or C is pivotal in some permutation of voters}}{\text{The total number of permutations of voters}} = \dfrac{288}{720} = \dfrac{2}{5}$

$\dfrac{\text{The number of times W or X or Y or Z is pivotal in some permutation of voters}}{\text{The total number of permutations of voters}} = \dfrac{36}{720} = \dfrac{1}{20}$

Chapter 11
CONSUMER MATHEMATICS: The Mathematics Of Everyday Life

Section 11.1 Percent

1. We represented 19% as 19 hundredths, or 0.19, then placed the 32 in the two decimal places to the right of 0.19.

3. We found the difference between the selling price and the dealer's cost and then divided this value by the dealer's cost.

5. $\text{percent} \times \text{base} = \text{amount}$

7. 0.78

9. 0.08

11. 0.2735

13. 0.0035

15. 43%

17. 36.5%

19. 145%

21. 0.2%

23. $\dfrac{3}{4} = 0.75 = 75\%$

25. $\dfrac{5}{16} = 0.3125 = 31.25\%$

27. $\dfrac{5}{2} = 2.5 = 250\%$

29. $\dfrac{4}{250} = 0.016 = 1.6\%$

31. and 33. No solutions provided

35. $0.28 \times 350 = \text{amount}$
$98 = \text{amount}$

37. $\text{percent} \times 80 = 12$
$\text{percent} = \dfrac{12}{80} = 0.15 = 15\%$

39. $0.22 \times \text{base} = 77$
$\text{base} = \dfrac{77}{0.22} = 350$

41. $0.1225 \times 160 = \text{amount}$
$19.6 = \text{amount}$

43. $\text{percent} \times 48 = 8.4$
$\text{percent} = \dfrac{8.4}{48} = 0.175 = 17.5\%$

45. $0.2325 \times \text{base} = 29.76$
$\text{base} = \dfrac{29.76}{0.2325} = 128$

47. $\dfrac{606}{3{,}124} \approx 0.194 = 19.4\%$

49. The average price of a new home in the Midwest in 2003 represents 100% of the 2002 price plus the 4% increase or $100\% + 4\% = 104\%$.

$$1.04 \times base = 218,200 \Rightarrow base = \frac{218,200}{1.04} \approx \$210,000$$

The average price for a new home in the Midwest in 2002 was approximately $210,000.

51. $\dfrac{2,101}{2,101+2,192+1,119+984+760} = \dfrac{2,101}{7,156} \approx 0.2936 = 29.36\%$

53. $\dfrac{2,192-1,119}{1,119} = \dfrac{1,073}{1,119} \approx 0.95889 = 95.89\%$

55. $\dfrac{413}{413+398+300+300+269} = \dfrac{413}{1,680} \approx 0.2458 = 24.58\%$

57. $\dfrac{413-300}{300} = \dfrac{113}{300} \approx 0.3767 = 37.67\%$

59. The population in 2005 represents 100% of the 2000 population plus the 5.33% increase or $100\% + 5.33\% = 105.33\%$.

$$1.0533 \times base = 296 \Rightarrow base = \frac{296}{1.0533} \approx 281 \text{ million}$$

The population in 2000 was approximately 281 million.

61. The amount after the decline represents 100% of the amount before the decline minus the 78.2% decrease or $100\% - 78.2\% = 21.8\%$.

$$0.218 \times base = 23.6 \Rightarrow base = \frac{23.6}{0.218} = 108 \text{ million}$$

There were sales of approximately 108 million audio cassettes before the decline.

63. $\dfrac{16,065-14,875}{14,875} = \dfrac{1,190}{14,875} = 0.08 = 8\%$

65. She divided by 16,065 instead of 14,875.

$$\dfrac{16,065-14,875}{16,065} = \dfrac{1,190}{16,065} = 0.074 = 7.4\%$$

67. Amount of increase is $0.35 \times 28,000 = \$9,800$. Her new salary will be $28,000 + 9,800 = \$37,800$.

69. The sale price represents $100\% - 15\% = 85\%$ of the original price. $0.85 \times Base = \$578$. The original price was $Base = \dfrac{\$578}{0.85} = \$680$.

71. $\dfrac{155-124}{124} = \dfrac{31}{124} = 0.25 = 25\%$

73. $\dfrac{12,711-11,400}{11,400} = \dfrac{1,311}{11,400} = 0.115 = 11.5\%$

75. The amount that the fund is worth this quarter represents 100% of the amount the fund was worth last quarter minus the 12% decrease or $100\% - 12\% = 88\%$.

$$0.88 \times base = 11,264 \Rightarrow base = \frac{11,264}{0.88} = \$12,800$$

The fund was worth $12,800 last quarter.

77. The amount of gas mileage that the new car can get represents 100% of the amount that the old car could get plus the 26.3% increase or $100\% + 26.3\% = 126.3\%$.

$$1.263 \times base = 48 \Rightarrow base = \frac{48}{1.263} = 38 \text{ mpg}$$

Renaldo's old car got 24 mpg.

79. We must calculate the value for the 1st, 2nd, 3rd, then 4th year.

Value at the start of year		Depreciation	Value at the end of year
1	$18,000	$0.12 \times 18,000 = \$2,160$	$18,000 - 2,160 = \$15,840$
2	$15,840	$0.12 \times 15,840 = \$1,900.80$	$15,840 - 1,900.80 = \$13,939.20$
3	$13,939.20	$0.12 \times 13,939.20 \approx \$1,672.70$	$13,939.20 - 1,672.70 = \$12,266.50$
4	$12,266.50	$0.12 \times 12,266.50 = \$1,471.98$	$12,266.50 - 1,471.98 = \$10,794.52$

81. No; The price after the reduction will be less than the original price. Examples will vary.

Consider an original price of $1,000. In general, after an increase of $x\%$, we have a new price of
$$1,000 + 0.01 \cdot x \times 1,000 = 1,000 + 10x.$$

After a decrease of $x\%$, we have a new price of

$$(1,000 + 10x) - 0.01 \cdot x \times (1,000 + 10x) =$$
$$1,000 + 10x - 10x - 0.1 \cdot x^2 =$$
$$1,000 - 0.1x^2 =$$
$$1,000 - 0.0001 \cdot x^2 \times 1,000$$

or

The final price is $(0.01 \cdot x^2)\%$ less than the original price.

83. No; It is the same as an increase of 32%. Examples will vary.

One way to approach this problem is to start with an amount such as 10,000 and determine what this amount after the two increases.

After the 10% increase: $10,000 + 0.10 \times 10,000 = 10,000 + 1,000 = 11,000$

After the 20% increase: $11,000 + 0.20 \times 11,000 = 11,000 + 2,200 = 13,200$

A single increase of 32% would result in the amount.

$10,000 + 0.32 \times 10,000 = 10,000 + 3,200 = \$13,200$

Section 11.2 Interest

1. We substituted values for A, r, and t in the equation $A = P(1+rt)$, and then solved for P.

3. $\log(3^x) = x\log(3)$

5. Simple interest only pays interest on the principle. Compound interest pays interest on the principle and previously earned interest.

7. $I = P \times r \times t$
   $I = 1,000 \times 0.08 \times 3$
   $I = \$240$

9. $I = P \times r \times t$
   $700 = 3,500 \times r \times 4$
   $700 = 14,000r$
   $0.05 = r$
   $r = 5\%$

11. $A = P(1+rt)$
    $A = 2,500(1+(0.08)3)$
    $A = 2,500(1+0.24)$
    $A = 2,500(1.24)$
    $A = 3,100$
    $A = \$3,100$

13. $A = P(1+rt)$
    $1,770 = P(1+(0.06)3)$
    $1,770 = P(1+0.18)$
    $1,770 = P(1.18)$
    $1,500 = P$
    $P = \$1,500$

15. $A = P(1+rt)$
    $1,400 = 1,250(1+(r)2)$
    $1,400 = 1,250(1+2r)$
    $1,400 = 1,250 + 2500r$
    $150 = 2500r$
    $0.06 = r$
    $r = 6\%$

17.

	I (Interest)	A (Value of Account)
End of year 1	$160	$2,160
End of year 2	$172.80	$2,332.80
End of year 3	$186.62	$2,519.42

19. $\dfrac{r}{n} = \dfrac{0.18}{12} = 0.015 = 1.5\%$

21. $\dfrac{r}{n} = \dfrac{0.12}{365} = 12/365\%$

23. $A = P\left(1+\dfrac{r}{n}\right)^{nt}$
    $A = 5,000\left(1+\dfrac{0.05}{1}\right)^{1 \cdot 5}$
    $A = 5,000(1+0.05)^5$
    $A = 5,000(1.05)^5$
    $A \approx 6,381.41$
    $A \approx \$6,381.41$

25. $A = P\left(1+\dfrac{r}{n}\right)^{nt}$

$A = 4,000\left(1+\dfrac{0.08}{4}\right)^{4 \cdot 2}$

$A = 4,000(1+0.02)^8$

$A = 4,000(1.02)^8$

$A \approx 4,686.64$

$A \approx \$4,686.64$

27. $A = P\left(1+\dfrac{r}{n}\right)^{nt}$

$A = 20,000\left(1+\dfrac{0.08}{12}\right)^{12 \cdot 2}$

$A = 20,000\left(1+\dfrac{0.08}{12}\right)^{24}$

$A \approx 23,457.76$

$A \approx \$23,457.76$

29. $A = P\left(1+\dfrac{r}{n}\right)^{nt}$

$A = 4,000\left(1+\dfrac{0.10}{365}\right)^{365 \cdot 2}$

$A = 4,000\left(1+\dfrac{0.10}{365}\right)^{730}$

$A \approx 4,885.48$

$A \approx \$4,885.48$

31. You must compare $\left(1+\dfrac{r}{n}\right)^n$ for each case where n is the number of times the money compounds in a single year.

5% compounded yearly:

$\left(1+\dfrac{r}{n}\right)^n = \left(1+\dfrac{0.05}{1}\right)^1$

$= 1+0.05$

$= 1.05$

4.95% compounded quarterly:

$\left(1+\dfrac{r}{n}\right)^n = \left(1+\dfrac{0.0495}{4}\right)^4 \approx 1.05043$

4.95% compounded quarterly is better.

33. $A = P\left(1+\dfrac{r}{n}\right)^{nt}$

$30,000 = P\left(1+\dfrac{0.06}{4}\right)^{4 \cdot 15}$

$30,000 = P(1+0.015)^{60}$

$30,000 = P(1.015)^{60}$

$\dfrac{30,000}{1.015^{60}} = P$

$P \approx \$12,278.88$

35. $3^x = 10$

$\log 3^x = \log 10$

$x \log 3 = \log 10$

$x = \log 10 / \log 3$

$x \approx 2.096$

37. $(1.05)^x = 2$

$\log(1.05)^x = \log 2$

$x \log(1.05) = \log 2$

$x = \log 2 / \log(1.05)$

$x \approx 14.207$

39. $$x^3 = 10$$
$$\left(x^3\right)^{1/3} = (10)^{1/3}$$
$$x = 10^{1/3} \approx 2.1544$$

41. $$x^4 = 10$$
$$\left(x^4\right)^{1/4} = (10)^{1/4}$$
$$x = 10^{1/4} \approx 1.7783$$

43. $$A = P(1+r)^t$$
$$2500 = 2000(1+r)^5$$
$$1.25 = (1+r)^5$$
$$(1.25)^{1/5} = \left((1+r)^5\right)^{1/5}$$
$$(1.25)^{1/5} = 1+r$$
$$(1.25)^{1/5} - 1 = r$$
$$r \approx 0.0456 = 4.56\%$$

45. $$A = P(1+r)^t$$
$$1000 = 100(1+r)^{25}$$
$$10 = (1+r)^{25}$$
$$(10)^{1/25} = \left((1+r)^{25}\right)^{1/25}$$
$$(10)^{1/25} = 1+r$$
$$(10)^{1/25} - 1 = r$$
$$r \approx 0.0965 = 9.65\%$$

47. $$A = P(1+r)^t$$
$$1{,}500 = 1{,}000(1+0.04)^t$$
$$1.5 = (1.04)^t$$
$$\log(1.5) = \log(1.04)^t$$
$$\log(1.5) = t \cdot \log(1.04)$$
$$\log(1.5)/\log(1.04) = t$$
$$t \approx 10.34$$

49. $$A = P(1+r)^t$$
$$800 = 200(1+0.045)^t$$
$$4 = (1.045)^t$$
$$\log(4) = \log(1.045)^t$$
$$\log(4) = t \cdot \log(1.045)$$
$$\log(4)/\log(1.045) = t$$
$$t \approx 31.49$$

51. a) $36 \cdot 136 = \$4{,}896$
b) $4{,}896 - 3{,}600 = \$1{,}296$

53. $$I = P \times r \times t$$
$$I = 10{,}000 \times 0.08 \times \frac{1}{12}$$
$$I = \$66.67$$

55. $$I = P \times r \times t$$
$$800 = P \times 0.08 \times 2$$
$$800 = 0.16P$$
$$5000 = P$$
$$P = \$5{,}000$$

57. $$I = P \times r \times t$$
$$300 = 1{,}200 \times r \times 2$$
$$300 = 2{,}400r$$
$$0.125 = r$$
$$r = 12.5\%$$

59. $$I = P \times r \times t$$
$$I = 4{,}500 \times 0.15 \times \frac{4}{12}$$
$$I = 225$$
$$225 + 4{,}500 = \$4{,}725$$

61. $$I = P \times r \times t$$
$$I = 2{,}000 \times 0.12 \times \frac{3}{12}$$
$$I = 60$$
$$60 + 2{,}000 = \$2{,}060$$

63. $$I = 425 - 400 = 25$$
$$I = P \times r \times t$$
$$25 = 400 \times r \times \frac{1}{12}$$
$$25 = \frac{100}{3} r$$
$$0.75 = r$$
$$r = 75\%$$

65. a) She earned $2,450 - 2,375 = \$75$ in interest

$$I = P \times r \times t$$
$$75 = 2,375 \times r \times \frac{8}{12}$$
$$75 = \frac{4750}{3} \times r$$
$$\frac{9}{190} = r$$

$r \approx 0.04736842$ or approximately 4.74%

b) $$A = P\left(1+\frac{r}{n}\right)^{nt}$$
$$2,450 = 2,375\left(1+\frac{r}{12}\right)^{12 \cdot \frac{8}{12}}$$
$$\frac{98}{95} = \left(1+\frac{r}{12}\right)^8$$
$$\left(\frac{98}{95}\right)^{1/8} = \left(\left(1+\frac{r}{12}\right)^8\right)^{1/8}$$
$$\left(\frac{98}{95}\right)^{1/8} = 1+\frac{r}{12}$$
$$\left(\frac{98}{95}\right)^{1/8} - 1 = \frac{r}{12}$$
$$r \approx 4.67\%$$

67. $$A = P\left(1+\frac{r}{n}\right)^{n \cdot t}$$
$$20,000 = 9,420\left(1+\frac{0.075}{12}\right)^{12t}$$
$$20,000 = 9,420(1+0.00625)^{12t}$$
$$\frac{1,000}{471} = (1.00625)^{12t}$$
$$\log\left(\frac{1,000}{471}\right) = \log(1.00625)^{12t}$$
$$\log\left(\frac{1,000}{471}\right) = 12t \cdot \log(1.00625)$$
$$t \approx \frac{\log\left(\frac{1,000}{471}\right)/\log(1.00625)}{12} \approx 10.07 \text{ years}$$

69. $4.65(1.03)^{20} \approx \$8.40$

71. $35,000(1.04)^{20} \approx \$76,689.31$

73. $4.65(1+2.26)^5 = 4.65(3.26)^5 \approx \$1,712.15$

75. $\left(1+\frac{0.075}{12}\right)^{12} - 1 \approx 0.0776 = 7.76\%$

77. $\left(1+\frac{0.06}{4}\right)^4 - 1 \approx 0.0614 = 6.14\%$

79. $\left(1+\frac{0.10}{100,000}\right)^{100,000} - 1 \approx 0.105170863 = 10.5170863\%$

81. $e^r - 1 \approx 2.718281828^{0.10} - 1 \approx 0.105170918 = 10.5170918\%$

83. and 85. No solutions provided

Section 11.3 Consumer Loans

1. We added the loan amount and the interest, and then divided the sum by the number of payments.

3. It shows the balance for the loan for each day of September.

316   Chapter 11: Consumer Mathematics

5. Although the annual interest rate is 18%, for the last month of the loan, we pay $10.80 interest on a $30 loan, or $\frac{10.80}{30} = 0.36 = 36\%$ for one month, which is 432% annually.

7. $I = Prt = 900(0.12)(2) = 216$ Monthly Payment $= \frac{P+I}{n} = \frac{900+216}{2 \cdot 12} = \frac{1,116}{24} = \$46.50$

9. $I = Prt = 1,360(0.08)(4) = 435.20$
   Monthly Payment $= \frac{P+I}{n} = \frac{1,360+435.20}{4 \cdot 12} = \frac{1,795.20}{48} = \$37.40$

11. $I = Prt = 1,280(0.095)(2) = \$243.20$
    Monthly Payment $= \frac{P+I}{n} = \frac{1,280+243.20}{2 \cdot 12} = \frac{1,523.20}{24} \approx \$63.47$

13. $I = Prt = 6,480(0.1165)(4) = \$3,019.68$
    Monthly Payment $= \frac{P+I}{n} = \frac{6,480+3,019.68}{4 \cdot 12} = \frac{9,499.68}{48} = \$197.91$

15. $I = Prt = 900(0.12)(2) = 216$ Monthly Payment $= \frac{P+I}{n} = \frac{900+216}{2 \cdot 12} = \frac{1,116}{24} = \$46.50$

    $\frac{900}{24} = \$37.50$ of the loan amount is paid each month   46.50-37.50 = \$9 in interest each month

    $\frac{9}{37.50} = 0.24 = 24\%$ per month or $12 \cdot 24\% = 288\%$ annual interest rate

17. $I = Prt = 1360(0.08)(4) = 435.20$ Monthly Payment $= \frac{P+I}{n} = \frac{1,360+435.20}{4 \cdot 12} = \frac{1,795.20}{48} = \$37.40$

    $\frac{1,360}{48} \approx \$28.33$ of the loan amount is paid each month   37.40-28.33 = \$9.07 in interest each month

    $\frac{9.07}{28.33} \approx 0.32 = 32\%$ per month or $12 \cdot 32\% = 384\%$ annual interest rate

19. a) $I = Prt = 11,000(0.092)(4) = 4,048$

    Monthly Payment $= \frac{P+I}{n} = \frac{11,000+4048}{4 \cdot 12} = \frac{15,048}{48} = \$313.50$

    b) $I = Prt = 9,000(0.092)(4) = 3,312$
    Monthly Payment $= \frac{P+I}{n} = \frac{9,000+3,312}{4 \cdot 12} = \frac{12,312}{48} = \$256.50$
    His monthly payments will be reduced by 313.50-256.50=\$57.00

    c) Monthly Payment $= 200 = \frac{P+I}{48} \Rightarrow P+I = 9,600 = P+P(0.092)(4) = P+0.368P = 1.368P \Rightarrow$
    $9,600 = 1.368P \Rightarrow P \approx \$7,017.54$ This represents the amount financed, so
    $11,000 - 7,017.54 = \$3,982.46$ should be the down payment.

21. a) $I = Prt = 15,000(0.096)(3) = 4,320$

$$\text{Monthly Payment} = \frac{P+I}{n} = \frac{15,000+4,320}{3 \cdot 12} = \frac{19,320}{36} \approx \$536.67$$

b) $I = Prt = 12,000(0.096)(3) = 3,456$

$$\text{Monthly Payment} = \frac{P+I}{n} = \frac{12,000+3,456}{3 \cdot 12} = \frac{15,456}{36} \approx \$429.34$$

Her monthly payments will be reduced by 536.67 − 429.34=$107.33

c) Monthly Payment $= 300 = \dfrac{P+I}{36} \Rightarrow P+I = 10,800 = P + P(0.096)(3) = P + 0.288P = 1.288P \Rightarrow$

$10,800 = 1.288P \Rightarrow P \approx \$8,385.09$ This represents the amount financed, so

$15,000 - 8,385.09 = \$6,614.91$ should be the down payment.

23.

Method	Finance Charge = (Last Month's Balance)rt	P	r	t	Finance Charge = $I=Prt$
Unpaid Balance	$475 \times (0.18/12) \approx 7.13$	Last Month's Balance + Finance − Payment + New Charges − Returned Charges = 475 + 7.13 − 225 + 180 − 145 = 292.13	18%	$\dfrac{1}{12}$	$292.13 \times (0.18/12)$ $\approx \$4.38$

25.

Method	Finance Charge = (Last Month's Balance)rt	P	r	t	Finance Charge = $I=Prt$
Unpaid Balance	$640 \times (0.165/12) = 8.80$	Last Month's Balance + Finance − Payment + New Charges = 640 + 8.80−320 + 140 + 35 + 75 = 578.80	16.5%	$\dfrac{1}{12}$	$578.80 \times (0.165/12)$ $\approx \$7.96$

27.

Method	Finance Charge = (Last Month's Balance)rt	P	r	t	Finance Charge = $I=Prt$
Unpaid Balance	$460 \times (0.188/12) \approx 7.21$	Last Month's Balance + Finance − Payment + New Charges = 460 + 7.21 − 300 + 140 + 135 + 175 = 617.21	18.8%	$\dfrac{1}{12}$	$617.21 \times (0.188/12)$ $\approx \$9.67$

29.

Day	Balance	Number of days × Balance
1,2,3,4	$280	4 × 280 = 1,120
5,6,7,8,9,10,11,12,13,14	$205	10 × 205 = 2,050
15,16,17,18,19,20	$340	6 × 340 = 2,040
21,22,23	$356	3 × 356 = 1,068
24,25,26,27,28,29,30,31	$382	8 × 382 = 3,056

Average daily balance $\frac{1,120+2,050+2,040+1,068+3,056}{31} = \frac{9,334}{31} \approx 301.10$

Finance charge $I = Prt = 301.10(.21)(31/365) \approx \$5.37$

31.

Day	Balance	Number of days × Balance
1,2	$240	2 × 240 = 480
3,4,5,6,7,8,9,10,11,12	$375	10 × 375 = 3,750
13,14,15,16,17,18,19,20,21,22	$225	10 × 225 = 2,250
23,24,25,26,27	$255	5 × 255 = 1,275
28,29,30	$283	3 × 283 = 849

Average daily balance $\frac{480+3,750+2,250+1,275+849}{30} = \frac{8,604}{30} = 286.80$

Finance charge $I = Prt = 286.80(.21)(30/365) \approx \$4.95$

33. Finance charge on unpaid balance $I = Prt = 280 \cdot 0.21 \cdot (1/12) = \$4.90$

Method	P	r	t	Finance Charge = $I = Prt$
Unpaid Balance	Last month's balance + finance charge + purchases − payment = 280 + 4.90 + 135 + 16 + 26 − 75 = 386.90	21%	$\frac{1}{12}$	$386.90 \times (0.21/12)$ $\approx \$6.77$

35. Finance charge on unpaid balance $I = Prt = 240 \cdot 0.21 \cdot (1/12) = \$4.20$

Method	P	r	t	Finance Charge = $I = Prt$
Unpaid Balance	Last month's balance + finance charge + purchases − payment = 240 + 4.20 + 135 + 30 + 28 − 150 = 287.20	21%	$\frac{1}{12}$	$287.20 \times (0.21/12)$ $\approx \$5.03$

37. Add-on Interest: $I = Prt = 1{,}000(0.105)\left(\dfrac{10}{12}\right) = 87.50$

Unpaid Balance Method

Beginning of Month	Finance Charge = (Last Month's Balance)rt	Last Month's Balance + Finance Charge – Payment
2	1,000 × (0.18/12) = 15.00	1,000 + 15.00 – (100 + 15.00) = 900
3	900 × (0.18/12) = 13.50	900 + 13.50 – (100 + 13.50) = 800
4	800 × (0.18/12) = 12.00	800 + 12.00 – (100 + 12.00) = 700
5	700 × (0.18/12) = 10.50	700 + 10.50 – (100 + 10.50) = 600
6	600 × (0.18/12) = 9.00	600 + 9.00 – (100 + 9.00) = 500
7	500 × (0.18/12) = 7.50	500 + 7.50 – (100 + 7.50) = 400
8	400 × (0.18/12) = 6.00	400 + 6.00 – (100 + 6.00) = 300
9	300 × (0.18/12) = 4.50	300 + 4.50 – (100 + 4.50) = 200
10	200 × (0.18/12) = 3.00	200 + 3.00 – (100 + 3.00) = 100
11	100 × (0.18/12) = 1.50	100 + 1.50 – (100 + 1.50) = 0

The total interest paid for the unpaid balance method is $82.50, so the unpaid balance method would accumulate less finance charges.

39. Unpaid Balance Method

Beginning of Month	Finance Charge = (Last Month's Balance)r	Last Month's Balance + Finance Charge
2	1,150 × (0.0175) ≈ 20.13	1,150 + 20.13 = 1,170.13
3	1,170.13 × (0.0175) ≈ 20.48	1,170.13 + 20.48 = 1,190.61
4	1,190.61 × (0.0175) ≈ 20.84	1,190.61 + 20.84 = 1,211.45

The total interest accumulated with the unpaid balance method is $61.45 –or – By using the formula $A = P(1+r)^n$ where $P=1{,}150$, $r=0.0175$, and $n=3$ we get

$$A = 1{,}150(1+0.0175)^3 = 1{,}150 \cdot 1.0175^3 \approx 1{,}211.44$$
$$1{,}211.44 - 1{,}150 = \$61.44$$

41. – 45. No solutions provided

Section 11.4 Annuities

1. The deposits are earning interest for different periods of time.

3. The exponent property; $\log a^x = x \log a$.

5. $\dfrac{x^5 - 1}{x - 1}$

7. $\dfrac{x^8 - 1}{x - 1}$

Your calculations may differ slightly from ours due to rounding at intermediate steps.

9. $\dfrac{r}{n} = \dfrac{0.12}{12} = 0.01$; $nt = 12 \cdot \dfrac{6}{12} = 6$; $A = 150\left(\dfrac{(1+0.01)^6 - 1}{0.01}\right) \approx \$922.80$

11. $\dfrac{r}{n} = \dfrac{0.03}{12} = 0.0025$; $nt = 12 \cdot \dfrac{4}{12} = 4$; $A = 200\left(\dfrac{(1+0.0025)^4 - 1}{0.0025}\right) \approx \$803.01$

13. $\dfrac{r}{n} = \dfrac{0.06}{12}$; $nt = 8$; $A = 150\left(\dfrac{\left(1+\dfrac{0.06}{12}\right)^{8}-1}{0.06/12}\right) \approx \$1{,}221.21$

15. $\dfrac{r}{n} = \dfrac{0.095}{12}$; $nt = 48$; $A = 400\left(\dfrac{\left(1+\dfrac{0.095}{12}\right)^{48}-1}{0.095/12}\right) \approx \$23{,}247.07$

17. $\dfrac{r}{n} = \dfrac{0.08}{4}$ and $nt = 4 \cdot 5 = 20$; $A = 500\left(\dfrac{\left(1+\dfrac{0.08}{12}\right)^{20}-1}{0.08/12}\right) \approx \$12{,}148.68$

19. $\dfrac{r}{n} = \dfrac{0.09}{12}$ and $nt = 12 \cdot 4 = 48$; $A = 400\left(\dfrac{\left(1+\dfrac{0.09}{12}\right)^{48}-1}{0.09/12}\right) \approx \$23{,}008.28$

21. $\dfrac{r}{n} = \dfrac{0.095}{12}$ and $nt = 12 \cdot 8 = 96$; $A = 600\left(\dfrac{\left(1+\dfrac{0.095}{12}\right)^{96}-1}{0.095/12}\right) \approx \$85{,}785.11$

23. $\dfrac{r}{n} = \dfrac{0.06}{12} = 0.005$ and $nt = 12 = 12 \cdot 1$

$2{,}000 = R\left(\dfrac{(1+0.005)^{12}-1}{0.005}\right)$

$2{,}000 \approx R(12.3355623729)$

$R \approx \dfrac{2{,}000}{12.3355623729} \approx \$162.14$

25. $\dfrac{r}{n} = \dfrac{0.075}{12} = 0.00625$ and $nt = 24 = 12 \cdot 2$

$5{,}000 = R\left(\dfrac{(1+0.00625)^{24}-1}{0.00625}\right)$

$5{,}000 \approx R(25.8067228988)$

$R \approx \dfrac{5{,}000}{25.8067228988} \approx \$193.75$

27. $3^x = 20$

$\log 3^x = \log 20$

$x \cdot \log 3 = \log 20$

$x = \dfrac{\log 20}{\log 3} \approx 2.7268$

29. $8^x = 10$

$\log 8^x = \log 10$

$x \cdot \log 8 = \log 10$

$x = \dfrac{\log 10}{\log 8} \approx 1.1073$

31. $\dfrac{8^x - 5}{6} = 20$

$8^x - 5 = 120$

$8^x = 125$

$\log 8^x = \log 125$

$x \cdot \log 8 = \log 125$

$x = \dfrac{\log 125}{\log 8} \approx 2.3219$

33. $\dfrac{8^x + 2}{5} = 12$

$8^x + 2 = 60$

$8^x = 58$

$\log 8^x = \log 58$

$x \cdot \log 8 = \log 58$

$x = \dfrac{\log 58}{\log 8} \approx 1.9527$

35. $10{,}000 = 200 \left(\dfrac{\left(1 + \dfrac{0.09}{12}\right)^{nt} - 1}{0.09/12} \right)$

$50 = \left(\dfrac{(1.0075)^{nt} - 1}{0.0075} \right)$

$0.375 = 1.0075^{nt} - 1$

$1.375 = 1.0075^{nt}$

$\log 1.375 = \log 1.0075^{nt}$

$\log 1.375 = nt \cdot \log 1.0075$

$\dfrac{\log 1.375}{\log 1.0075} = nt$

$nt \approx 42.62$

37. $5{,}000 = 150 \left(\dfrac{\left(1 + \dfrac{0.06}{12}\right)^{nt} - 1}{0.06/12} \right)$

$\dfrac{100}{3} = \left(\dfrac{(1.005)^{nt} - 1}{0.005} \right)$

$0.1666667 \approx 1.005^{nt} - 1$

$1.1666667 \approx 1.005^{nt}$

$\log 1.1666667 \approx \log 1.005^{nt}$

$\log 1.1666667 \approx nt \cdot \log 1.005$

$\dfrac{\log 1.1666667}{\log 1.005} \approx nt$

$nt \approx 30.91$

39. $6{,}000 = 250 \left(\dfrac{\left(1 + \dfrac{0.075}{12}\right)^{nt} - 1}{0.075/12} \right)$

$24 = \left(\dfrac{(1.00625)^{nt} - 1}{0.00625} \right)$

$0.15 = 1.00625^{nt} - 1$

$1.15 = 1.00625^{nt}$

$\log 1.15 = \log 1.00625^{nt}$

$\log 1.15 = nt \cdot \log 1.00625$

$\dfrac{\log 1.15}{\log 1.00625} = nt$

$nt \approx 22.43$

41. $\dfrac{r}{n} = \dfrac{0.065}{12}$ and $nt = 30 = 12 \cdot \dfrac{30}{12}$; $A = 75 \left(\dfrac{\left(1 + \dfrac{0.065}{12}\right)^{30} - 1}{0.065/12} \right) \approx \$2{,}435.99$

322   Chapter 11: Consumer Mathematics

43. $\dfrac{r}{n} = \dfrac{0.102}{12}$ and $nt = 12 \cdot 3 = 36$; $A = 150\left(\dfrac{\left(1+\dfrac{0.102}{12}\right)^{36} - 1}{0.102/12}\right) \approx \$6,286.36$

45. $\dfrac{r}{n} = \dfrac{0.065}{12}$ and $nt = 12 \cdot 15 = 180$; $A = 400\left(\dfrac{\left(1+\dfrac{0.065}{12}\right)^{180} - 1}{0.065/12}\right) \approx \$121,417.91$

47. $nt = 12 \cdot 8 = 96$

$14{,}000 = R\left(\dfrac{(1+0.007)^{96} - 1}{0.007}\right)$

$14{,}000 \approx R(136.224411498)$

$R \approx \dfrac{14{,}000}{136.2244114988} \approx \$102.78$

49. $\dfrac{r}{n} = \dfrac{0.082}{12}$ and $nt = 6 = 12 \cdot \dfrac{6}{12}$

$600 = R\left(\dfrac{\left(1+\dfrac{0.082}{12}\right)^{6} - 1}{0.082/12}\right)$

$600 \approx R(6.10343868814)$

$R \approx \dfrac{600}{6.10343868814} \approx \$98.31$

51. and 53. No solutions provided

55. $400{,}000 = 5{,}000\left(\dfrac{\left(1+\dfrac{0.108}{12}\right)^{nt} - 1}{0.108/12}\right)$

$80 = \left(\dfrac{(1+0.009)^{nt} - 1}{0.009}\right)$

$0.72 = 1.009^{nt} - 1$

$1.72 = 1.009^{nt}$

$\log 1.72 = \log 1.009^{nt}$

$\log 1.72 = nt \cdot \log 1.009$

$\dfrac{\log 1.72}{\log 1.009} = nt$

$nt \approx 61$ months

57. $30{,}000 = 550\left(\dfrac{\left(1+\dfrac{0.078}{12}\right)^{nt} - 1}{0.078/12}\right)$

$\dfrac{600}{11} = \left(\dfrac{(1+0.0065)^{nt} - 1}{0.0065}\right)$

$0.3545455 \approx 1.0065^{nt} - 1$

$1.3545455 \approx 1.0065^{nt}$

$\log 1.3545455 \approx \log 1.0065^{nt}$

$\log 1.3545455 \approx nt \cdot \log 1.0065$

$\dfrac{\log 1.3545455}{\log 1.0065} \approx nt$

$nt \approx 47$ months

59. $\dfrac{r}{n} = \dfrac{0.0935}{12}$ and $nt = 12 \cdot 20 = 240$

$$98{,}695.16 = R\left(\dfrac{\left(1+\dfrac{0.0935}{12}\right)^{240} - 1}{0.0935/12}\right)$$

$98{,}695.16 \approx R(698.366876471)$

$$R \approx \dfrac{98{,}695.16}{698.366876471} \approx \$141.33$$

61. $\dfrac{r}{n} = \dfrac{0.06}{12} = 0.005$ and $nt = 12 \cdot 10 = 120$

$$A = 200\left(\dfrac{(1+0.005)^{120} - 1}{0.005}\right) \approx \$32{,}775.87$$

$$A = P\left(1 + \dfrac{r}{n}\right)^{nt}$$

$A = 32{,}775.87\left(1 + \dfrac{0.06}{12}\right)^{12 \cdot 30}$

$A = 32{,}775.87(1 + 0.005)^{360}$

$A = 32{,}775.87(1.005)^{360}$

$A \approx 197{,}395.14$

$A \approx \$197{,}395.14$

Section 11.5 Amortization

1. The left side represents the amount owed on the loan after 48 months of compounding. The right side represents the amount that must be in the annuity to pay off the loan.

3. We saw the left expression in calculating compound interest and the right expression in computing the future value of annuities.

5. $\dfrac{r}{n} = \dfrac{0.10}{12}$ and $nt = 12 \cdot 4 = 48$

$$5{,}000\left(1 + \dfrac{0.10}{12}\right)^{48} = R\left(\dfrac{\left(1+\dfrac{0.10}{12}\right)^{48} - 1}{0.10/12}\right)$$

$7{,}446.77049303 \approx R(58.7224918326)$

$$R \approx \dfrac{7{,}446.77049303}{58.722491826} \approx \$126.82$$

7. $\dfrac{r}{n} = \dfrac{0.075}{12} = 0.00625$ and $n = 12 \cdot 4 = 48$

$$8{,}000(1+0.00625)^{48} = R\left(\dfrac{(1+0.00625)^{48} - 1}{0.00625}\right)$$

$10{,}788.7932107 \approx R(55.7758642148)$

$$R \approx \dfrac{10{,}788.7932107}{55.7758642148} \approx \$193.44$$

9. $\dfrac{r}{n} = \dfrac{0.0825}{12} = 0.006875$ and $n = 12 \cdot 4 = 48$

$$12{,}500(1+0.006875)^{48} = R\left(\dfrac{(1+0.006875)^{48} - 1}{0.006875}\right)$$

$17{,}367.4791291 \approx R(56.6397571391)$

$$R \approx \dfrac{17{,}367.4791291}{56.6397571391} \approx \$306.64$$

11. $\dfrac{r}{n} = \dfrac{0.08}{12}$

$$1{,}900\left(1+\dfrac{0.08}{12}\right)^{18} = R\left(\dfrac{\left(1+\dfrac{0.08}{12}\right)^{18}-1}{0.08/12}\right)$$

$2{,}141.39107973 \approx R(19.0571905052)$

$R \approx \dfrac{2{,}141.39107973}{19.0571905052} \approx \$112.37$

13. $\dfrac{r}{n} = \dfrac{0.108}{12} = 0.009$

$$1{,}250(1+0.009)^{20} = R\left(\dfrac{(1+0.009)^{20}-1}{0.009}\right)$$

$1{,}495.31723064 \approx R(21.8059760573)$

$R \approx \dfrac{1{,}495.31723064}{21.8059760573} \approx \$68.58$

15.

Payment Number	Amount of Payment	Interest Payment	Applied to Principal	Balance
				$5,000.00
1	$126.82	$41.67	$85.15	$4,914.85
2	$126.82	$40.96	$85.86	$4,828.99
3	$126.82	$40.24	$86.58	$4,742.41

17.

Payment Number	Amount of Payment	Interest Payment	Applied to Principal	Balance
				$12,500.00
1	$306.64	$85.94	$220.70	$12,279.30
2	$306.64	$84.42	$222.22	$12,057.08
3	$306.64	$82.89	$223.75	$11,833.33

19. a) $\dfrac{r}{n} = \dfrac{0.07}{12}$ and $nt = 12 \cdot 30 = 360$

$$100{,}000\left(1+\dfrac{0.07}{12}\right)^{360} = R\left(\dfrac{\left(1+\dfrac{0.07}{12}\right)^{360}-1}{0.07/12}\right)$$

$R \approx \dfrac{811{,}649.747526}{1{,}219.97099576} \approx \$665.31$

$811{,}649.747526 \approx R(1{,}219.97099576)$

Payment Number	Amount of Payment	Interest Payment	Applied to Principal	Balance
				$100,000.00
1	$665.31	$583.33	$81.98	$99,918.02
2	$665.31	$582.86	$82.45	$99,835.57
3	$665.31	$582.37	$82.94	$99,752.63

19. (continued)

    b)

Payment Number	Amount of Payment	Interest Payment	Applied to Principal	Balance
				$100,000.00
1	$765.31	$583.33	$181.98	$99,818.02
2	$765.31	$582.27	$183.04	$99,634.98
3	$765.31	$581.20	$184.11	$99,450.87

    c) The 4th payment line for a payment of $665.31 is

4	$665.31	$581.89	$83.42	$99,669.21

    The 4th payment line for a payment of $765.31 is

4	$765.31	$580.13	$185.18	$99,265.69

    The difference is $581.89 - 580.13 = \$1.76$.

21. a) $\dfrac{r}{n} = \dfrac{0.072}{12} = 0.006$ and $nt = 12 \cdot 4 = 48$

    $$11{,}500(1+0.006)^{48} = R\left[\dfrac{(1+0.006)^{48} - 1}{0.006}\right]$$

    $15{,}325.0152341 \approx R(55.4350033931)$

    $R \approx \dfrac{15{,}325.0152341}{55.4350033931} \approx \$276.46$

    b) $48 \cdot 276.46 = 13{,}270.08$ and $13{,}270.08 - 11{,}500.00 = \$1{,}770.08$

    This represents the approximate amount of interest paid. The actual amount is slightly lower due to the fact that we round our payments up to the next penny and our final payment is slightly less than the other 47 payments.

23. a) $\dfrac{r}{n} = \dfrac{0.15}{12} = 0.0125$ and $nt = 18$

    $$1900(1+0.0125)^{18} = R\left[\dfrac{(1+0.0125)^{18} - 1}{0.0125}\right]$$

    $2{,}376.09704886 \approx R(20.046191531)$

    $R \approx \dfrac{2{,}376.09704886}{20.046191531} \approx \$118.54$

    b) $18 \cdot 118.54 = 2{,}133.72$ and $2{,}133.72 - 1{,}900.00 = \$233.72$

    This represents the approximate amount of interest paid. The actual amount is slightly lower due to the fact that we round our payments up to the next penny and our final payment is slightly less than the other 17 payments.

25. a) $\dfrac{r}{n} = \dfrac{0.18}{12} = 0.015$ and $nt = 5$

$$275(1+0.015)^5 = R\left[\dfrac{(1+0.015)^5 - 1}{0.015}\right]$$

$296.253101068 \approx R(5.15226692563)$

$R \approx \dfrac{296.253101068}{5.15226692563} \approx \$57.50$

b) $5 \cdot 57.50 = 287.50$ and $287.50 - 275.00 = \$12.50$

This represents the approximate amount of interest paid. The actual amount is slightly lower due to the fact that we round our payments up to the next penny and our final payment is slightly less than the other 4 payments.

27. a) $\dfrac{r}{n} = \dfrac{0.14}{12}$ and $nt = 15$

$$1575\left(1+\dfrac{0.14}{12}\right)^{15} = R\left[\dfrac{\left(1+\dfrac{0.14}{12}\right)^{15} - 1}{0.14/12}\right]$$

$1{,}874.31322051 \approx R(16.2891548575)$

$R \approx \dfrac{1{,}874.31322051}{16.2891548575} \approx \$115.07$

b) $15 \cdot 115.07 = 1{,}726.05$ and $1{,}726.05 - 1{,}575.00 = \$151.05$

This represents the approximate amount of interest paid. The actual amount is slightly lower due to the fact that we round our payments up to the next penny and our final payment is slightly less than the other 14 payments.

29. $\dfrac{r}{n} = \dfrac{0.06}{12} = 0.005$ and $nt = 12 \cdot 3 = 36$

$$A = 30\left[\dfrac{(1+0.005)^{36} - 1}{0.005}\right] \approx \$1{,}180.08$$

$$A = P\left(1 + \dfrac{r}{n}\right)^{nt}$$

$1{,}180.08 = P(1+0.005)^{36}$

$1{,}180.08 = P(1.005)^{36}$

$1{,}180.08 = P(1.005)^{36}$

$\dfrac{1{,}180.08}{1.005^{36}} = P$

$P \approx \$986.13$

Section 11.5: Amortization   327

31. $A = 50{,}000\left(\dfrac{(1+0.10)^{20}-1}{0.10}\right) \approx \$2{,}863{,}749.97$

$$A = P\left(1+\dfrac{r}{n}\right)^{nt}$$

$2{,}863{,}749.97 = P(1+0.10)^{20}$

$2{,}863{,}749.97 = P(1.10)^{20}$

$\dfrac{2{,}863{,}749.97}{1.10^{20}} = P$

$P \approx \$425{,}678.19$

The present value of the annual payments is slightly better than the lump sum of $425,000.

33. $\dfrac{r}{n} = \dfrac{0.108}{12} = 0.009$ and $nt = 12 \cdot 4 = 48$

$P(1+0.009)^{48} = 350\left(\dfrac{(1+0.009)^{48}-1}{0.009}\right)$

$P(1.53736142389) \approx 20{,}897.3887067$

$P \approx \dfrac{20{,}897.3887067}{1.53736142389} \approx \$13{,}593.02$

35. $\dfrac{r}{n} = \dfrac{0.09}{12} = 0.0075$ and $n = 12 \cdot 20 = 240$

$A = 350\left(\dfrac{(1+0.0075)^{240}-1}{0.0075}\right) \approx \$233{,}760.40$

$A = P\left(1+\dfrac{r}{n}\right)^{nt}$

$233{,}760.40 = P(1+0.0075)^{240}$

$233{,}760.40 = P(1.0075)^{240}$

$\dfrac{233{,}760.40}{1.0075^{240}} = P$

$P \approx \$38{,}900.73$

The present value of the annual payments is slightly worse than the lump sum of $40,000.

37. We need to find the monthly payments first, where $\dfrac{r}{n} = \dfrac{0.085}{12}$ and $nt = 12 \cdot 5 = 60$.

$12{,}000\left(1+\dfrac{0.085}{12}\right)^{60} = R\left(\dfrac{\left(1+\dfrac{0.085}{12}\right)^{60}-1}{0.085/12}\right)$

$118{,}327.6071744 \approx R(74.4424373453)$

$R \approx \dfrac{118{,}327.6071744}{74.4424373453} \approx \$246.20$

Now find the unpaid balance where $nt = 12 \cdot 3 = 36$.

$U = 12{,}000\left(1+\dfrac{0.085}{12}\right)^{36} - 246.20\left(\dfrac{\left(1+\dfrac{0.085}{12}\right)^{36}-1}{0.085/12}\right) \approx 15{,}471.63 - 10{,}055.46 = \$5{,}416.17$

328   Chapter 11: Consumer Mathematics

39. We need to find the monthly payments first where $\dfrac{r}{n} = \dfrac{0.07}{12}$ and $nt = 12 \cdot 30 = 360$.

$$120{,}000\left(1+\dfrac{0.07}{12}\right)^{360} = R\left(\dfrac{\left(1+\dfrac{0.07}{12}\right)^{360}-1}{0.07/12}\right) \qquad \begin{aligned}973{,}979.697032 &\approx R(1{,}219.97099576)\\ R &\approx \dfrac{973{,}979.697032}{1{,}219.97099576} \approx \$798.37\end{aligned}$$

Now find the unpaid balance where $nt = 12 \cdot 8 = 96$.

$$U = 120{,}000\left(1+\dfrac{0.07}{12}\right)^{96} - 798.37\left(\dfrac{\left(1+\dfrac{0.07}{12}\right)^{96}-1}{0.07/12}\right) \approx 209{,}739.17 - 102{,}350.09 = \$107{,}389.08$$

41. We need to find the monthly payments first where $\dfrac{r}{n} = \dfrac{0.08}{12}$ and $nt = 12 \cdot 2 = 24$.

$$8{,}000\left(1+\dfrac{0.08}{12}\right)^{24} = R\left(\dfrac{\left(1+\dfrac{0.08}{12}\right)^{24}-1}{0.08/12}\right) \qquad \begin{aligned}9{,}383.10345397 &\approx R(25.9331897619)\\ R &\approx \dfrac{9{,}383.10345397}{25.9331897619} \approx \$361.82\end{aligned}$$

Now find the unpaid balance where $nt - 12$.

$$U = 8{,}000\left(1+\dfrac{0.08}{12}\right)^{12} - 361.82\left(\dfrac{\left(1+\dfrac{0.08}{12}\right)^{12}-1}{0.08/12}\right) \approx 8{,}664.00 - 4{,}504.63 = \$4{,}159.37$$

43. a) We need to find the unpaid balance where $\dfrac{r}{n} = \dfrac{0.08}{12}$.

$$U = 10{,}000\left(1+\dfrac{0.08}{12}\right)^{24} - 244.13\left(\dfrac{\left(1+\dfrac{0.08}{12}\right)^{24}-1}{0.08/12}\right) \approx 11{,}728.88 - 6{,}331.07 = \$5{,}397.81$$

Next we need to determine the new monthly payments where $\dfrac{r}{n} = \dfrac{0.065}{12}$.

$$5{,}397.81\left(1+\dfrac{0.065}{12}\right)^{24} = R\left(\dfrac{\left(1+\dfrac{0.065}{12}\right)^{24}-1}{0.065/12}\right)$$

$$6{,}145.02307866 \approx R(25.5561107016)$$
$$R \approx \dfrac{6{,}145.02307866}{25.5561107016} \approx \$240.46$$

b) $(244.13 - 240.46) \cdot 24 = \$88.08$

45. a) We need to find the unpaid balance where $\dfrac{r}{n} = \dfrac{0.08}{12}$.

$$U = 100{,}000\left(1+\dfrac{0.08}{12}\right)^{60} - 836.45\left(\dfrac{\left(1+\dfrac{0.08}{12}\right)^{60}-1}{0.08/12}\right) \approx 148{,}984.57 - 61{,}459.72 = \$87{,}524.85$$

Next we need to determine the new monthly payments where $\dfrac{r}{n} = \dfrac{0.07}{12}$.

$$87{,}524.85\left(1+\dfrac{0.07}{12}\right)^{180} = R\left(\dfrac{\left(1+\dfrac{0.07}{12}\right)^{180}-1}{0.07/12}\right)$$

$$249{,}353.635276 \approx R(316.962296718) \Rightarrow R \approx \dfrac{249{,}353.635276}{316.962296718} \approx \$786.70$$

b) $(836.45 - 786.70) \cdot 180 = \$8{,}955.00$

47. a) We need to find the unpaid balance where $\dfrac{r}{n} = \dfrac{0.08}{12}$.

$$U = 40{,}000\left(1+\dfrac{0.08}{12}\right)^{24} - 976.52\left(\dfrac{\left(1+\dfrac{0.08}{12}\right)^{24}-1}{0.08/12}\right) \approx 46{,}915.52 - 25{,}324.28 = \$21{,}591.24$$

Next we need to determine the new monthly payments where $\dfrac{r}{n} = \dfrac{0.075}{12} = 0.00625$.

$$21{,}591.24(1+0.00625)^{24} = R\left(\dfrac{(1+0.00625)^{24}-1}{0.00625}\right)$$

$$25{,}073.7346733 \approx R(25.8067228988) \Rightarrow R \approx \dfrac{25{,}073.7346733}{25.8067228988} \approx \$971.60$$

b) $(976.52 - 971.60) \cdot 24 = \$118.08$

49. and 51. No solutions provided

Chapter Review Exercises

1. 12.45%

2. 0.01365

3. $\dfrac{11}{16} = 0.6875 = 68.75\%$

4. $percent \times 3{,}400 = 2{,}890$

$$percent = \dfrac{2{,}890}{3{,}400} = 0.85 = 85\%$$

5. Amount of increase is $0.12 \times 1{,}380 = \$165.60$. The selling price will be $1{,}380.00 + 165.60 = \$1{,}545.60$.

6. $I = P \times r \times t$
   $I = 1{,}500 \times 0.09 \times 2$
   $I = 270$
   $1{,}500 + 270 = \$1{,}770$

7. $I = 400 \cdot 24 - 8{,}000 = 1{,}600$
   $I = P \times r \times t$
   $1{,}600 = 8{,}000 \times r \times 2$
   $1{,}600 = 16{,}000r$
   $0.10 = r$
   $r = 10\%$

8. $A = P\left(1 + \dfrac{r}{n}\right)^{nt}$
   $A = 5{,}000\left(1 + \dfrac{0.06}{12}\right)^{12 \cdot 5}$
   $A = 5{,}000(1 + 0.005)^{60}$
   $A \approx 6{,}744.25$
   $A \approx \$6{,}744.25$

9. $A = P\left(1 + \dfrac{r}{n}\right)^{nt}$
   $10{,}000 = P\left(1 + \dfrac{0.06}{12}\right)^{12 \cdot 10}$
   $10{,}000 = P(1 + 0.005)^{120}$
   $10{,}000 = P(1.005)^{120}$
   $\dfrac{10{,}000}{1.005^{120}} = P$
   $P \approx \$5{,}496.33$

10. $A = P\left(1 + \dfrac{r}{n}\right)^{nt}$
    $2{,}000 = 1{,}000\left(1 + \dfrac{0.064}{12}\right)^{nt}$
    $2 \approx (1.00533333333)^{nt}$
    $\log(2) \approx \log(1.00533333333)^{nt}$
    $\log(2) \approx nt \cdot \log(1.00533333333)$
    $nt \approx \dfrac{\log(2)}{\log(1.00533333333)} \approx 130.311$
    $nt \approx 131$ months

11. $A = P(1 + r)^t$
    $1{,}400 = 1{,}200(1 + r)^5$
    $\dfrac{7}{6} = (1 + r)^5$
    $\left(\dfrac{7}{6}\right)^{1/5} = \left((1+r)^5\right)^{1/5}$
    $\left(\dfrac{7}{6}\right)^{1/5} = 1 + r$
    $\left(\dfrac{7}{6}\right)^{1/5} - 1 = r$
    $r \approx 0.0313 = 3.13\%$

12. $I = Prt = 1{,}320(0.0825)(3) = \$326.70$

    Monthly Payment $= \dfrac{P + I}{n} = \dfrac{1{,}320 + 326.70}{3 \cdot 12} = \dfrac{1{,}646.70}{36} \approx \$45.75$

13.

Day	Balance	Number of days × Balance
1,2,3,4,5	$275	5 × 275 = 1,375
6,7,8,9,10,11	$200	6 × 200 = 1,200
12,13,14,15,16,17,18	$315	7 × 315 = 2,205
19,20,21,22,23	$335	5 × 335 = 1,675
24,25,26,27,28,29,30,31	$351	8 × 351 = 2,808

Average daily balance $\dfrac{1,375+1,200+2,205+1,675+2,808}{31} = \dfrac{9,263}{31} \approx 298.81$

Finance charge $I = Prt = 298.81(0.18)(31/365) \approx \$4.57$

14. $r = 0.0935/12$ and $n = 10 \cdot 12 = 120$

$$A = 175\left(\dfrac{\left(1+\dfrac{0.0935}{12}\right)^{120}-1}{0.0935/12}\right) \approx \$34,543.31$$

15. $r = 0.06/12 = 0.005$

$$2,000 = R\left(\dfrac{(1+0.005)^{36}-1}{0.005}\right)$$

$2,000 \approx R(39.3361049646)$

$R \approx \dfrac{2,000}{39.3361049646} \approx \$50.85$

16. $\dfrac{3^x - 4}{2} = 10$

$3^x - 4 = 20$

$3^x = 24$

$\log 3^x = \log 24$

$x \cdot \log 3 = \log 24$

$x = \dfrac{\log 24}{\log 3} \approx 2.8928$

17. $\dfrac{r}{n} = \dfrac{0.09}{12} = 0.0075$

$$10,000 = 300\left(\dfrac{(1+0.0075)^{nt}-1}{0.0075}\right)$$

$$\dfrac{100}{3} = \left(\dfrac{(1.0075)^{nt}-1}{0.0075}\right)$$

$0.25 = 1.0075^{nt} - 1$

$1.25 = 1.0075^{nt}$

$\log 1.25 = \log 1.0075^{nt}$

$\log 1.25 = nt \cdot \log 1.0075$

$\dfrac{\log 1.25}{\log 1.0075} = nt$

$nt \approx 30$ months

18. $\dfrac{r}{n} = \dfrac{0.10}{12}$ and $nt = 12 \cdot 4 = 48$

$$5,000\left(1+\dfrac{0.10}{12}\right)^{48} = R\left(\dfrac{\left(1+\dfrac{0.10}{12}\right)^{48}-1}{0.10/12}\right)$$

$7,446.77049303 \approx R(58.7224918326)$

$R \approx \dfrac{7,446.77049303}{58.722491826} \approx \$126.82$

332 Chapter 11: Consumer Mathematics

19. We need to determine the monthly payments where $\dfrac{r}{n} = \dfrac{0.08}{12}$ and $nt = 12 \cdot 20 = 240$.

$$100,000\left(1+\dfrac{0.08}{12}\right)^{240} = R\left(\dfrac{\left(1+\dfrac{0.08}{12}\right)^{240}-1}{0.08/12}\right)$$

$$492,680.277085 \approx R(589.020415627)$$

$$R \approx \dfrac{492,680.277085}{589.020415627} \approx \$836.45$$

Payment Number	Amount of Payment	Interest Payment	Applied to Principal	Balance
				$100,000.00
1	$836.45	$666.67	$169.78	$99,830.22
2	$836.45	$665.53	$170.92	$99,659.30

20. $A = 50,000\left(\dfrac{(1+0.08)^{20}-1}{0.08}\right) \approx \$2,288,098.21$

$$A = P\left(1+\dfrac{r}{n}\right)^{nt}$$

$$2,288,098.21 = P(1+0.08)^{20}$$

$$2,288,098.21 \approx P(4.66095714385)$$

$$\dfrac{2,288,098.21}{4.66095714385} \approx P$$

$$P \approx \$490,907.37$$

The present value is $490,907.37; the lump sum of $500,000 is a better deal.

21. We need to find the monthly payments first.

$$\dfrac{r}{n} = \dfrac{0.075}{12} = 0.00625 \text{ and } nt = 12 \cdot 5 = 60$$

$$13,000(1+0.00625)^{60} = R\left(\dfrac{(1+0.00625)^{60}-1}{0.00625}\right)$$

$$18,892.8273076 \approx R(72.5271053242)$$

$$R \approx \dfrac{18,892.82730764}{72.5271053242} \approx \$260.50$$

Now find the unpaid balance where $nt = 12 \cdot 2 = 24$.

$$U = 13,000(1+0.00625)^{24} - 260.50\left(\dfrac{(1+0.00625)^{24}-1}{0.00625}\right)$$

$$U \approx 15,096.80 - 6,722.65 = \$8,374.15$$

Chapter Test

1. 36.24%

2. 0.2345

3. $\dfrac{7}{16} = 0.4375 = 43.75\%$

4. $I = P \times r \times t$
   $I = 3,400 \times 0.025 \times 3$
   $I = 255$
   $3,400 + 255 = \$3,655$

5. $\text{percent} \times 2,840 = 994$
   $\text{percent} = \dfrac{994}{2,840} = 0.35 = 35\%$

6. The amount of decrease in price is $169.99 - $149.99 = $20.
   The percent reduction in price is $\dfrac{\$20}{\$169.99} \approx 0.1177 = 11.77\%$.

7. $I = 162.50 \cdot 24 - 3,000 = 900$

   $I = P \times r \times t$
   $900 = 3,000 \times r \times 2$
   $900 = 6,000r$
   $0.15 = r$
   $r = 15\%$

8. $A = P\left(1 + \dfrac{r}{n}\right)^{nt}$

   $A = 4,000\left(1 + \dfrac{0.036}{12}\right)^{12 \cdot 4}$

   $A = 4,000(1 + 0.003)^{48}$

   $A \approx 4,618.54$

   $A \approx \$4,618.54$

9. $A = P\left(1 + \dfrac{r}{n}\right)^{nt}$

   $2,000 = 1,000\left(1 + \dfrac{0.048}{12}\right)^{nt}$

   $2 \approx (1.004)^{nt}$

   $\log(2) \approx \log(1.004)^{nt}$

   $\log(2) \approx nt \cdot \log(1.004)$

   $nt \approx \dfrac{\log(2)}{\log(1.004)} \approx 173.633$

   $nt \approx 174$ months

10. $A = P\left(1 + \dfrac{r}{n}\right)^{nt}$

    $15,000 = P\left(1 + \dfrac{0.042}{12}\right)^{12 \cdot 18}$

    $15,000 = P(1 + 0.0035)^{216}$

    $15,000 = P(1.0035)^{216}$

    $\dfrac{15,000}{1.0035^{216}} = P$

    $P \approx \$7,052.42$

11. $I = Prt = 1,560(0.105)(2) = \$327.60$

    Monthly Payment $= \dfrac{P + I}{n} = \dfrac{1,560 + 327.60}{2 \cdot 12} = \dfrac{1,887.60}{24} = \$78.65$

12.
$$A = P(1+r)^t$$
$$2{,}400 = 2{,}100(1+r)^3$$
$$\frac{8}{7} = (1+r)^3$$
$$\left(\frac{8}{7}\right)^{1/3} = \left((1+r)^3\right)^{1/3}$$
$$\left(\frac{8}{7}\right)^{1/3} = 1+r$$
$$\left(\frac{8}{7}\right)^{1/3} - 1 = r$$

$r \approx 0.0455 = 4.55\%$

13.

Day	Balance	Number of days × Balance
1,2,3	$425	3 × 425 = 1,275
4,5,6,7,8,9	$340	6 × 340 = 2,040
10,11,12,13,14	$365	5 × 365 = 1,825
15,16,17,18,19,20,21,22,23,24	$380	10 × 380 = 3,800
25,26,27,28,29,30	$460	6 × 460 = 2,760

Average daily balance $\dfrac{1{,}275 + 2{,}040 + 1{,}825 + 3{,}800 + 2{,}760}{30} = \dfrac{11{,}700}{30} = 390.00$

Finance charge $I = Prt = 390.00(0.21)(30/365) \approx \$6.74$

14. $r = 0.0515/12$ and $n = 8 \cdot 12 = 96$

$$A = 200 \left[\frac{\left(1 + \frac{0.0515}{12}\right)^{96} - 1}{0.0515/12} \right] \approx \$23{,}697.27$$

15.
$$\frac{5^x - 4}{3} = 10$$
$$5^x - 4 = 30$$
$$5^x = 34$$
$$\log 5^x = \log 34$$
$$x \cdot \log 5 = \log 34$$
$$x = \frac{\log 34}{\log 5} \approx 2.1911$$

16. $r = 0.04/12 = 0.0033333333$

$$1{,}800 = R \left(\frac{(1+0.0033333333)^{36} - 1}{0.0033333333} \right)$$
$$1{,}800 \approx R(38.181562)$$
$$R \approx \frac{1{,}800}{38.181562} \approx \$47.15$$

17. $\dfrac{r}{n} = \dfrac{0.09}{12} = 0.0075$ and $nt = 12 \cdot 8 = 96$

$$20{,}000(1+0.0075)^{96} = R \left(\frac{(1+0.0075)^{96} - 1}{0.0075} \right)$$
$$40{,}978.4245646 \approx R(139.85613674)$$
$$R \approx \frac{40{,}978.4245646}{139.85613674} \approx \$293.01$$

18. $\dfrac{r}{n} = \dfrac{.0375}{12} = 0.003125$

$$9{,}000 = 450\left(\dfrac{(1+0.003125)^{nt} - 1}{0.003125}\right)$$

$$20 = \left(\dfrac{(1.003125)^{nt} - 1}{0.003125}\right)$$

$$0.0625 = 1.003125^{nt} - 1$$

$$1.0625 = 1.003125^{nt}$$

$$\log 1.0625 = \log 1.003125^{nt}$$

$$\log 1.0625 = nt \cdot \log 1.003125$$

$$\dfrac{\log 1.0625}{\log 1.003125} = nt$$

$$nt \approx 20 \text{ months}$$

19. $A = 100{,}000\left(\dfrac{(1+0.04)^{10} - 1}{0.04}\right) \approx \$1{,}200{,}610.71$

$$A = P\left(1 + \dfrac{r}{n}\right)^{nt}$$

$$1{,}200{,}610.71 = P(1 + 0.04)^{10}$$

$$1{,}200{,}610.71 \approx P(1.480244285)$$

$$\dfrac{1{,}200{,}610.71}{1.480244285} \approx P$$

$$P \approx \$811{,}089.58$$

The annuity is the better deal because its present value is $811,089.58.

20. We need to determine the monthly payments where $\dfrac{r}{n} = \dfrac{0.075}{12}$ and $nt = 12 \cdot 20 = 240$.

$$140{,}000\left(1 + \dfrac{0.075}{12}\right)^{240} = R\left(\dfrac{\left(1 + \dfrac{0.075}{12}\right)^{240} - 1}{0.075/12}\right)$$

$$624{,}514.384397 \approx R(553.730725)$$

$$R \approx \dfrac{624{,}514.384397}{553.730725} \approx \$1127.84$$

Payment Number	Amount of Payment	Interest Payment	Applied to Principal	Balance
				$140,000.00
1	$1,127.84	$875.00	$252.84	$139,747.16
2	$1,127.84	$873.42	$254.42	$139,492.74

21. We need to find the monthly payments first.

$$\frac{r}{n} = \frac{0.096}{12} = 0.008 \text{ and } nt = 12 \cdot 5 = 60$$

$$16{,}500(1+0.008)^{60} = R\left(\frac{(1+0.008)^{60}-1}{0.008}\right)$$

$26{,}614.350421808 \approx R(76.6238668)$ Now find the unpaid balance where $nt = 12 \cdot 2 = 24$.

$$R \approx \frac{26{,}614.350421808}{76.6238668} \approx \$347.34$$

$$U = 16{,}500(1+0.008)^{24} - 347.34\left(\frac{(1+0.008)^{24}-1}{0.008}\right)$$

$U \approx 19{,}977.30 - 9{,}150.03 = \$10{,}827.27$

Of Further Interest: The Annual Percentage Rate

1. You had paid \$300 on an outstanding balance of \$1,000, so the interest rate was $\dfrac{\$1{,}000}{\$300} = 0.30$ or 30%.

2. We calculated $\dfrac{\$2{,}150 \text{ (finance charge)}}{\$9{,}850 \text{ (amount borrowed)}} = 0.2183$ and then multiplied by 100.

3. First calculate the finance charge per \$100 of the loan. Next, use the line corresponding to the number of loan payment to locate the amount that is closest to the finance charge per \$100. The percent shown at the top of that column is the approximate APR for the loan.

4. With the rent-to-own agreement, the consumer can terminate the agreement without making all payments as would be the case with the add-on interest loan.

5. $I = P \times r \times t$
   $I = 6{,}000 \times 0.08 \times 3$
   $I = \$1{,}440.00$

   $6{,}000 \times a \times 1 + 4{,}000 \times a \times 1 + 2{,}000 \times a \times 1 = 1{,}440$
   $12{,}000a = 1{,}440$
   $a = \dfrac{1{,}440}{12{,}000} = 0.12 = 12\%$

6. $I = P \times r \times t$
   $I = 8{,}000 \times 0.12 \times 4$
   $I = \$3{,}840.00$

   $8{,}000 \times a \times 1 + 6{,}000 \times a \times 1 + 4{,}000 \times a \times 1 + 2{,}000 \times a \times 1 = 3{,}840$
   $20{,}000a = 3{,}840$
   $a = \dfrac{3{,}840}{20{,}000} = 0.192 = 19.2\%$

7. $\dfrac{\text{finance charge}}{\text{amount borrowed}} \times 100 = \dfrac{270}{1,800} \times 100 = \$15.00$

8. $\dfrac{\text{finance charge}}{\text{amount borrowed}} \times 100 = \dfrac{840}{3,000} \times 100 = \$28.00$

9. $\dfrac{\text{finance charge}}{\text{amount borrowed}} \times 100 = \dfrac{260}{2,000} \times 100 = \$13.00$

10. $\dfrac{\text{finance charge}}{\text{amount borrowed}} \times 100 = \dfrac{1,125}{5,000} \times 100 = \$22.50$

11. $\dfrac{\text{finance charge}}{\text{amount borrowed}} \times 100 = \dfrac{420}{3,000} \times 100 = \$14.00$

    From Table 11.4, we see that for a 24-payment loan, this corresponds to an APR of roughly 13%.

12. $\dfrac{\text{finance charge}}{\text{amount borrowed}} \times 100 = \dfrac{600}{4,500} \times 100 \approx \$13.33$

    From Table 11.4, we see that for a 24-payment loan, this corresponds to an APR of roughly 12%.

13. $\dfrac{\text{finance charge}}{\text{amount borrowed}} \times 100 = \dfrac{165}{4,000} \times 100 \approx \$4.13$

    From Table 11.4, we see that for a 6-payment loan, this corresponds to an APR of roughly 14%.

14. $\dfrac{\text{finance charge}}{\text{amount borrowed}} \times 100 = \dfrac{310}{5,000} \times 100 = \$6.20$

    From Table 11.4, we see that for a 12-payment loan, this corresponds to an APR of roughly 11%.

15. Finance charge $= 24 \cdot 485 - 10,000 = 11,640 - 10,000 = 1,640$

    $\dfrac{\text{finance charge}}{\text{amount borrowed}} \times 100 = \dfrac{1,640}{10,000} \times 100 = \$16.40$

    From Table 11.4, we see that for a 24-payment loan, this corresponds to an APR of roughly 15%.

16. Finance charge $= 36 \cdot 270 - 8,000 = 9,720 - 8,000 = 1,720$

    $\dfrac{\text{finance charge}}{\text{amount borrowed}} \times 100 = \dfrac{1,720}{8,000} \times 100 = \$21.50$

    From Table 11.4, we see that for a 36-payment loan, this corresponds to an APR of roughly 13%.

17. Finance charge $= 48 \cdot 116.50 - 4,500 = 5,592 - 4,500 = 1,092$

    $\dfrac{\text{finance charge}}{\text{amount borrowed}} \times 100 = \dfrac{1,092}{4,500} \times 100 \approx \$24.27$

    From Table 11.4, we see that for a 48-payment loan, this corresponds to an APR of roughly 11%.

338   Chapter 11: Consumer Mathematics

18. Finance charge $= 24 \cdot 71.25 - 1,500 = 1,710 - 1,500 = 210$

    $\dfrac{\text{finance charge}}{\text{amount borrowed}} \times 100 = \dfrac{210}{1,500} \times 100 = \$14.00$

    From Table 11.4 we see that for a 24-payment loan, this corresponds to an APR of roughly 13%.

19. $I = Prt = 2,000 \times 0.08 \times 2 = 320$

    $\dfrac{\text{finance charge}}{\text{amount borrowed}} \times 100 = \dfrac{320}{2,000} \times 100 = \$16.00$

    From Table 11.4, we see that for a 24-payment loan, this corresponds to an APR of roughly 15%.

20. $I = Prt = 26,000 \times 0.079 \times 4 = 8,216$

    $\dfrac{\text{finance charge}}{\text{amount borrowed}} \times 100 = \dfrac{8,216}{26,000} \times 100 = \$31.60$

    From Table 11.4, we see that for a 48-payment loan, this corresponds to an APR of roughly 14%.

21. $I = Prt = P \times 0.082 \times 3 = 0.246P$

    $\dfrac{\text{finance charge}}{\text{amount borrowed}} \times 100 = \dfrac{0.246P}{P} \times 100 = \$24.60$

    From Table 11.4, we see that for a 36-payment loan, this corresponds to an APR of roughly 15%.

22. $I = Prt = P \times 0.0875 \times 4 = 0.35P$

    $\dfrac{\text{finance charge}}{\text{amount borrowed}} \times 100 = \dfrac{0.35P}{P} \times 100 = \$35.00$

    From Table 11.4, we see that for a 48-payment loan, this corresponds to an APR of roughly 16%.

23. b) is better

    An add-on interest loan at 8.4% for three years:
    $I = Prt = 5,000 \times 0.084 \times 3 = 1,260$

    $\dfrac{\text{finance charge}}{\text{amount borrowed}} \times 100 = \dfrac{1,260}{5,000} \times 100 = \$25.20$

    From Table 11.4, we see that for a 36-payment loan, this corresponds to an APR of roughly 15%.

    24 payments of $230:
    Finance charge $= 24 \cdot 230 - 5,000 = 5,520 - 5,000 = 520$

    $\dfrac{\text{finance charge}}{\text{amount borrowed}} \times 100 = \dfrac{520}{5,000} \times 100 = \$10.40$

    From Table 11.4, we see that for a 24-payment loan, this corresponds to an APR of roughly 10%.

24. b) is better

    An add-on interest loan at 7.2% for two years:
    $I = Prt = 5{,}000 \times 0.072 \times 2 = 720$

    $\dfrac{\text{finance charge}}{\text{amount borrowed}} \times 100 = \dfrac{720}{5{,}000} \times 100 = \$14.40$

    From Table 11.4, we see that for a 24-payment loan, this corresponds to an APR of roughly 13%.

    36 payments of $165:
    Finance charge $= 36 \cdot 165 - 5{,}000 = 5{,}940 - 5{,}000 = 940$

    $\dfrac{\text{finance charge}}{\text{amount borrowed}} \times 100 = \dfrac{940}{5{,}000} \times 100 = \$18.80$

    From Table 11.4, we see that for a 36-payment loan, this corresponds to an APR of roughly 12%.

25. b) is better

    An add-on interest loan at 8.9% for four years:
    $I = Prt = 5{,}000 \times 0.089 \times 4 = 1{,}780$

    $\dfrac{\text{finance charge}}{\text{amount borrowed}} \times 100 = \dfrac{1{,}780}{5{,}000} \times 100 = \$35.60$

    From Table 11.4, we see that for a 48-payment loan, this corresponds to an APR of roughly 16%.

    36 payments of $165:
    Finance charge $= 36 \cdot 165 - 5{,}000 = 5{,}940 - 5{,}000 = 940$

    $\dfrac{\text{finance charge}}{\text{amount borrowed}} \times 100 = \dfrac{940}{5{,}000} \times 100 = \$18.80$

    From Table 11.4, we see that for a 36-payment loan, this corresponds to an APR of roughly 12%.

26. b) is better

    An add-on interest loan at 8.4% for one year:
    $I = Prt = 5{,}000 \times 0.084 \times 1 = 420$

    $\dfrac{\text{finance charge}}{\text{amount borrowed}} \times 100 = \dfrac{420}{5{,}000} \times 100 = \$8.40$

    From Table 11.4, we see that for a 12-payment loan, this corresponds to an APR of roughly 15%.

    24 payments of $240:
    Finance charge $= 24 \cdot 240 - 5{,}000 = 5{,}760 - 5{,}000 = 760$

    $\dfrac{\text{finance charge}}{\text{amount borrowed}} \times 100 = \dfrac{760}{5{,}000} \times 100 = \$15.20$

    From Table 11.4, we see that for a 24-payment loan, this corresponds to an APR of roughly 14%.

27. 24 payments of $18.75:
    Finance charge $= 24 \cdot 18.75 - 375 = 450 - 375 = 75$

    $\dfrac{\text{finance charge}}{\text{amount borrowed}} \times 100 = \dfrac{75}{375} \times 100 = \$20.00$

    From Table 11.4, we see that for a 24-payment loan, this corresponds to an APR of over 16%.

28. 36 payments of $49:

Finance charge = $36 \cdot 49 - 1,375 = 1,764 - 1,375 = 389$

$$\frac{\text{finance charge}}{\text{amount borrowed}} \times 100 = \frac{389}{1,375} \times 100 \approx \$28.29$$

From Table 11.4, we see that for a 36-payment loan, this corresponds to an APR of over 16%.

29. – 31. No solutions provided

Chapter 12
COUNTING: Just How Many Are There?

Section 12.1 Introduction To Counting Methods

1. We first listed pairs that began with C, then all pairs that began with O, then all pairs that began with F, and finally all pairs that began with T.

3. We thought of flipping the coins one at a time. We also thought of rolling the dice one at a time.

5. AB, AC, AD, AE, BC, BD, BE, CD, CE, DE

7. AB, AC, AD, AE, BA, BC, BD, BE, CA, CB, CD, CE, DA, DB, DC, DE, EA, EB, EC, ED

9. No solution provided

Use the following diagram for the solutions to Exercises 11 and 13.

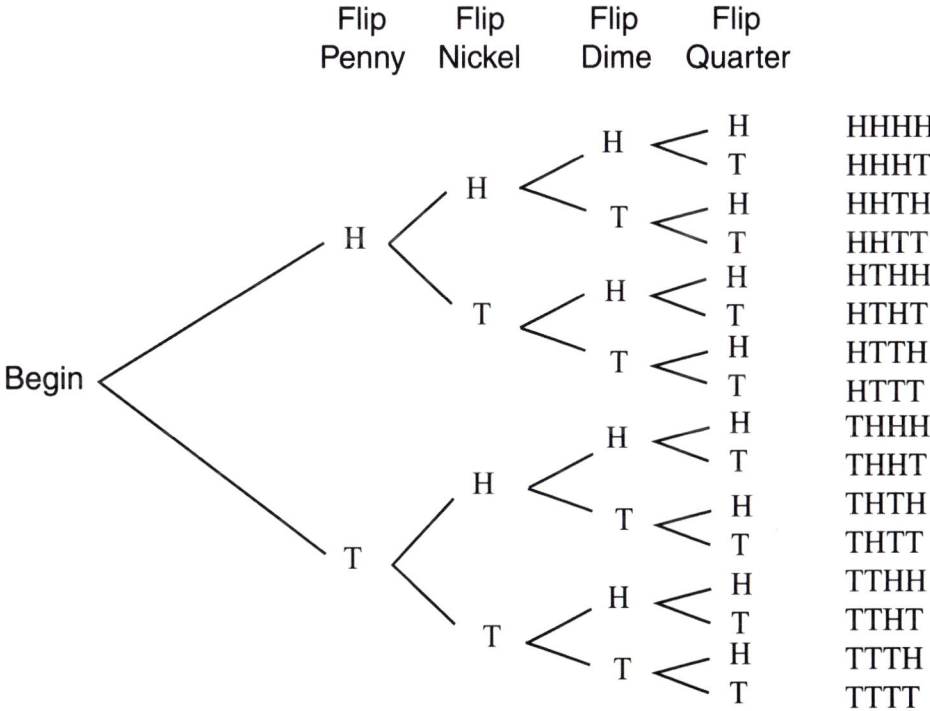

11. 4, the ways are HTTT, THTT, TTHT, and TTTH

13. 6, the ways are HHTT, HTHT, HTTH, THHT, THTH, and TTHH

15. Without drawing the tree, you can imagine that the tree begins and there are six branches on the first number choice. Each of those branches split into five branches for the second number choice for a total of thirty branches. There would therefore be **thirty** two-digit numbers that could be formed.

341

342   Chapter 12: Counting

17. Without drawing the tree, you can imagine that the tree begins and there are six branches on the first number choice. Each of those branches split into five branches for the second number choice for a total of thirty branches. Each of those branches split into four branches for the third number choice for a total of 120 branches. There would therefore be **120** three-digit numbers that could be formed.

Use the diagram on the next page for the solutions to Exercises 19 – 23.

19. 4 ways; (1,4), (2,3), (3,2), (4,1)

21. 6 ways; (1,1), (2,2), (3,3), (4,4), (5,5), (6,6)

23. 10 ways; (1,1), (1,2), (1,3), (1,4), (2,1), (2,2), (2,3), (3,1), (3,2), (4,1)

25. No solution provided

27. Without drawing the tree, you can imagine that the tree begins and there are six branches on the first top choice. Each of those branches split into five branches for the second choice of pants for a total of thirty branches. Each of those branches split into four branches for the choice of a jacket for a total of 120 branches. There would therefore be **120** outfits that could be formed.

29. Assume that the ordering is that the first person buys beverages, the second arranges for food and the third sends out invitations. Let the first letter of the name indicate the name of Susan's friend. The possible arrangements are:
    VLM, VLT, VMT, VTM, VTL, VML
    LVM, LVT, LMT, LTM, LTV, LMV
    MTV, MTL, MVL, MLV, MLT, MVT
    TVL, TVM, TLM, TML, TMV, TLV   for a total of 24 ways.

31. Without drawing the tree, you can imagine that the tree begins and there are two branches on the first flip. Each of those branches split into two branches on the second flip for a total of four branches. Each of those branches split into two branches on the third flip for a total of eight branches. Each of those branches split into two branches on the fourth flip for a total of sixteen branches. Finally, each of those branches split into two branches on the fifth flip for a total of thirty-two branches. There would therefore be **thirty-two** ways five coins could be flipped.

33. Without drawing the tree, you can imagine that the tree begins and there are four branches on the first roll. Each of those branches split into four branches on the second roll for a total of sixteen branches. There would therefore be **sixteen** ways two tetrahedral dice could be rolled.

35. 16

37. 9

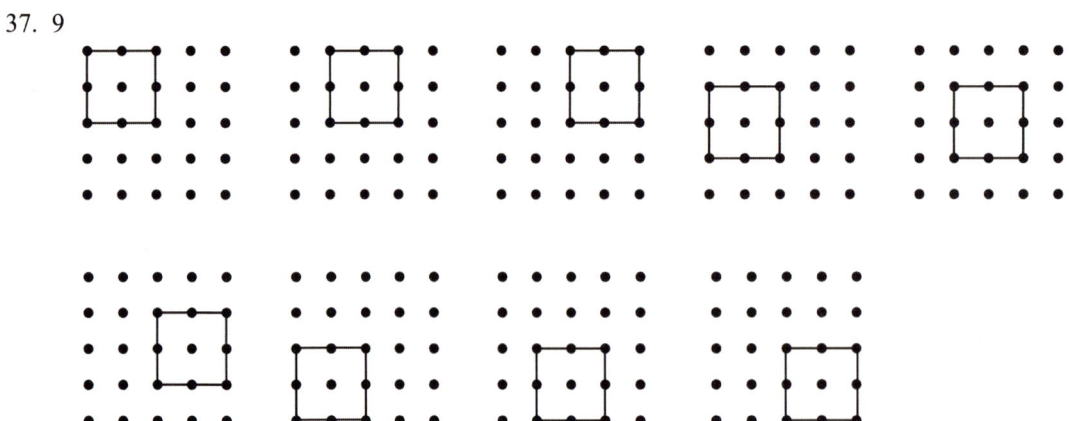

Use this diagram for the solutions to Exercises 19 – 23.

Red	Green	Combination	Sum
1	1	(1,1)	2
	2	(1,2)	3
	3	(1,3)	4
	4	(1,4)	5
	5	(1,5)	6
	6	(1,6)	7
2	1	(2,1)	3
	2	(2,2)	4
	3	(2,3)	5
	4	(2,4)	6
	5	(2,5)	7
	6	(2,6)	8
3	1	(3,1)	4
	2	(3,2)	5
	3	(3,3)	6
	4	(3,4)	7
	5	(3,5)	8
	6	(3,6)	9
4	1	(4,1)	5
	2	(4,2)	6
	3	(4,3)	7
	4	(4,4)	8
	5	(4,5)	9
	6	(4,6)	10
5	1	(5,1)	6
	2	(5,2)	7
	3	(5,3)	8
	4	(5,4)	9
	5	(5,5)	10
	6	(5,6)	11
6	1	(6,1)	7
	2	(6,2)	8
	3	(6,3)	9
	4	(6,4)	10
	5	(6,5)	11
	6	(6,6)	12

344 Chapter 12: Counting

39. 16

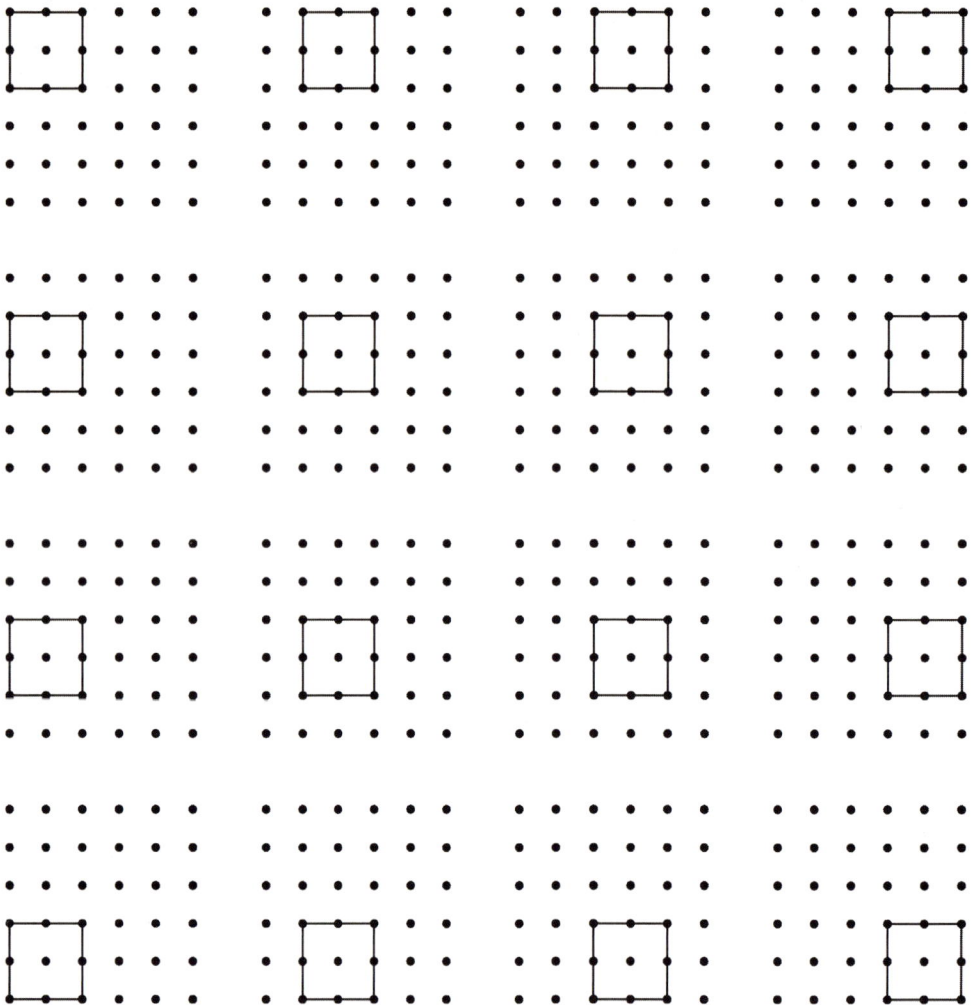

41. Without drawing the tree, you can imagine that the tree begins and there are three branches for the first choice of letters. Each of those branches split into two branches for the second choice of the letter not chosen on the first choice. Each of those branches extends to the third and last letter choice. Each of these six branches split into three branches for the first number choice. Each of these 18 branches splits into two branches for the second number choice and each of these 36 branches extends for the third and final number choice. There would therefore be **thirty-six** license plates to be investigated.

43. 24

Seat 1	Seat 2	Seat 3	Seat 4
Al	Monica	Ed	Vicki
Al	Monica	Vicki	Ed
Al	Ed	Monica	Vicki
Al	Ed	Vicki	Monica
Al	Vicki	Monica	Ed
Al	Vicki	Ed	Monica
Monica	Al	Ed	Vicki
Monica	Al	Vicki	Ed
Monica	Ed	Al	Vicki
Monica	Ed	Vicki	Al
Monica	Vicki	Al	Ed
Monica	Vicki	Ed	Al

Seat 1	Seat 2	Seat 3	Seat 4
Ed	Al	Monica	Vicki
Ed	Al	Vicki	Monica
Ed	Monica	Al	Vicki
Ed	Monica	Vicki	Al
Ed	Vicki	Al	Monica
Ed	Vicki	Monica	Al
Vicki	Al	Ed	Monica
Vicki	Al	Monica	Ed
Vicki	Monica	Al	Ed
Vicki	Monica	Ed	Al
Vicki	Ed	Al	Monica
Vicki	Ed	Monica	Al

45. 12

Seat 1	Seat 2	Seat 3	Seat 4
Monica	Al	Vicki	Ed
Vicki	Al	Monica	Ed
Monica	Ed	Vicki	Al
Vicki	Ed	Monica	Al
Monica	Al	Ed	Vicki
Vicki	Al	Ed	Monica
Monica	Ed	Al	Vicki
Vicki	Ed	Al	Monica
Al	Monica	Ed	Vicki
Al	Vicki	Ed	Monica
Ed	Monica	Al	Vicki
Ed	Vicki	Al	Monica

47. In the bottom row, there are $5 \times 5 = 25$ oranges. In the next row there will be $4 \times 4 = 16$ oranges. This is followed by oranges, then $2 \times 2 = 4$ oranges, and finally one orange on top. There is a total of $25 + 16 + 9 + 4 + 1 = 55$ oranges.

Use the diagram on the next page for the solutions to Exercises 49 and 51.

49. 27

51. 18

Vanilla – Vanilla – Strawberry, Vanilla – Vanilla – Chocolate, Vanilla – Strawberry – Vanilla, Vanilla – Strawberry – Strawberry, Vanilla – Chocolate – Vanilla, Vanilla-Chocolate – Chocolate, Strawberry – Vanilla – Vanilla, Strawberry – Vanilla – Strawberry, Strawberry – Strawberry – Vanilla, Strawberry – Strawberry – Chocolate, Strawberry – Chocolate – Strawberry, Strawberry – Chocolate – Chocolate, Chocolate – Vanilla – Vanilla, Chocolate – Vanilla – Chocolate, Chocolate – Strawberry – Strawberry, Chocolate – Strawberry – Chocolate, Chocolate – Chocolate – Vanilla, and Chocolate -Chocolate – Strawberry

Use the diagram below for the solutions to Exercises 49 and 51.

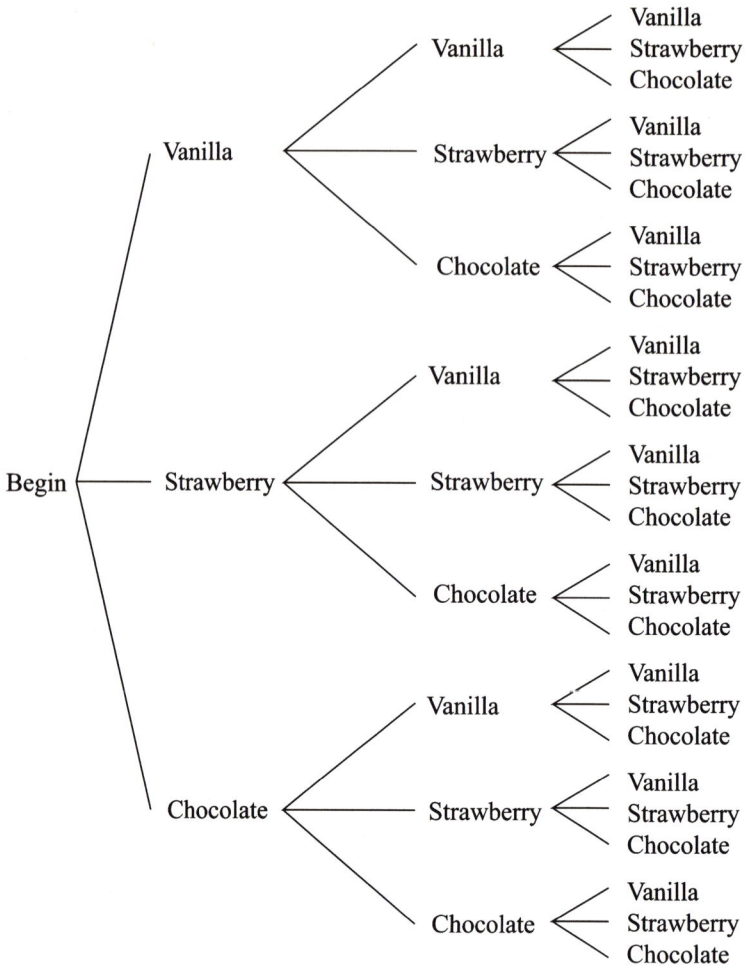

53. Follow the branches in the tree below to make the schedules.

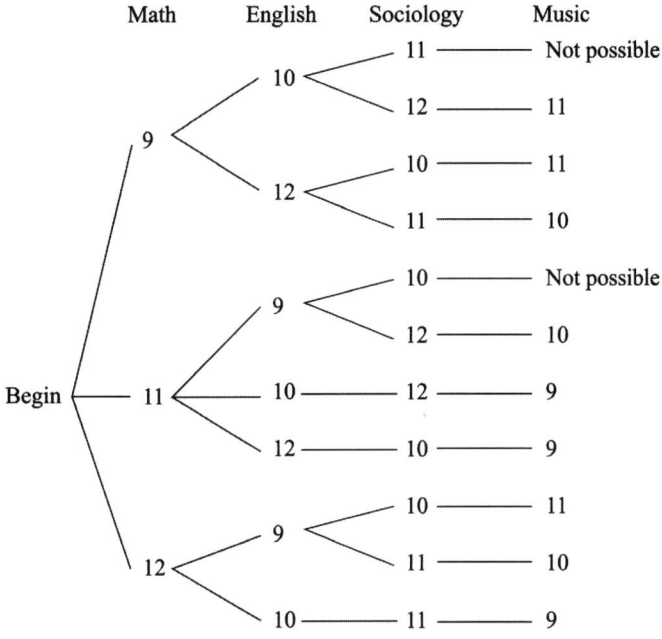

His nine possible schedules are:

9:00	10:00	11:00	12:00
Math	English	Music	Sociology
Math	Sociology	Music	English
Math	Music	Sociology	English
English	Music	Math	Sociology
Music	English	Math	Sociology
Music	Sociology	Math	English
English	Sociology	Music	Math
English	Music	Sociology	Math
Music	English	Sociology	Math

Use the following diagram for the solutions to Exercises 55 and 57.

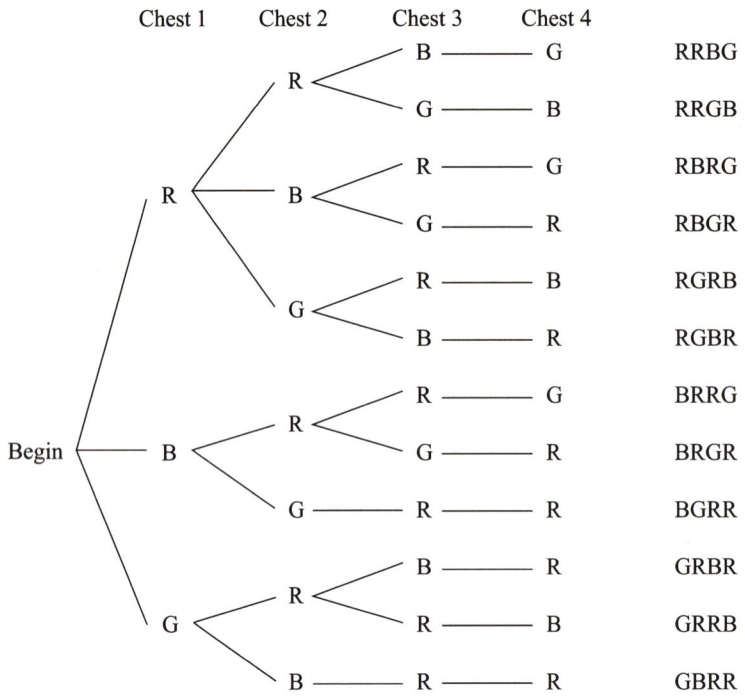

55. RRBG, RBRG, BRRG

57. RBGR, RGBR, BRGR, BGRR, GRBR, GBRR

59. No solution provided

Section 12.2 The Fundamental Counting Principle

1. In (a), we thought of flipping the four coins one at a time instead of flipping them all at once.
   In (b), we thought of rolling the red die, then the blue die, and then the green die.

3. In example 5, we considered the condition that Louise and her tutor must sit next to each other first.

5. $7 \times 6 = 42$

7. $8 \times 7 \times 6 = 336$

9. $4 \times 8 \times 3 \times 5 = 480$

11. $5 \times 6 \times 13 \times 4 = 1{,}560$

13. $8 \times 8 = 64$

15. $8 \times 12 = 96$

17. $9 \times 10 \times 10 \times 10 = 9{,}000$

19. $5 \times 8 \times 7 \times 4 = 1{,}120$

21. If the number must be greater than 5,000 and odd, the first digit must be 5 or greater and the last digit must be odd, so $5 \times 10 \times 10 \times 5 = 2{,}500$.

23. $1 \times 26 \times 26 \times 26 = 17{,}576$

25. $12 \times 11 \times 10 \times 9 \times 8 \times 7 \times 6 \times 5 \times 4 \times 3 \times 2 \times 1 = 479{,}001{,}600$

27. $3 \times 3 \times 3 \times 3 \times 3 = 243$

29. a) $2 \times 2 \times 2 \times 2 \times 2 \times 2 \times 2 \times 2 = 256$

   b) $256 \cdot 2 = 512$ minutes which is $\dfrac{512}{60} = 8$ hours and 32 minutes or over $8\frac{1}{2}$ hours

31. 6; Explanations may vary.
    Let n be the number of combinations of toppings.
    Consider that there are two main choices, crust and toppings. There are three ways to choose the crust and n ways to choose the toppings. By the fundamental counting principle we have

    $$3 \times n < 200$$
    $$n < 66\tfrac{2}{3}$$

    From Chapter 1, we have that there will be 2^t combinations, where t is the number of toppings. Since $2^6 = 64 = n$, there must be 6 toppings for $3 \cdot 64 = 192$ combinations.

33. Answers may vary.
    We can use a tree diagram to visualize the total number of possibilities that we count when we use the Fundamental Counting Principle.

35. Louise and her tutor can sit in the front in 7 different ways. Since the tutor must sit to the right of Louise, we would have $7 \times 1 \times 10 \times 9 \times 8 \times 7 \times 6 \times 5 = 1,058,400$ ways to arrange the front row.

37. $6 \times 5 \times 4 \times 3 \times 2 \times 1 = 720$ 39. $6 \times 2 \times 1 \times 3 \times 2 \times 1 = 72$

41. – 47. No solutions provided

Section 12.3 Permutations And Combinations

1. Objects cannot be repeated in a permutation.

3. One combination of three people corresponds to $3! = 6$ permutations.

5. Not all choices of four people were acceptable; for example, four women could not be selected. First, the number of ways to select two women was determined and then the number of ways to select two men was determined. These values were then multiplied together.

7. $4! = 4 \cdot 3 \cdot 2 \cdot 1 = 24$ 9. $(8-5)! = 3! = 3 \cdot 2 \cdot 1 = 6$

11. $\dfrac{10!}{7!} = \dfrac{10 \cdot 9 \cdot 8 \cdot 7 \cdot 6 \cdot 5 \cdot 4 \cdot 3 \cdot 2 \cdot 1}{7 \cdot 6 \cdot 5 \cdot 4 \cdot 3 \cdot 2 \cdot 1} = 10 \cdot 9 \cdot 8 = 720$

13. $\dfrac{10!}{7!3!} = \dfrac{10 \cdot 9 \cdot 8 \cdot 7 \cdot 6 \cdot 5 \cdot 4 \cdot 3 \cdot 2 \cdot 1}{7 \cdot 6 \cdot 5 \cdot 4 \cdot 3 \cdot 2 \cdot 1 \cdot 3 \cdot 2 \cdot 1} = \dfrac{10 \cdot 9 \cdot 8}{3 \cdot 2 \cdot 1} = 10 \cdot 3 \cdot 4 = 120$

15. $P(6,2) = \dfrac{6!}{(6-2)!} = \dfrac{6!}{4!} = \dfrac{6 \cdot 5 \cdot 4 \cdot 3 \cdot 2 \cdot 1}{4 \cdot 3 \cdot 2 \cdot 1} = 6 \cdot 5 = 30$

17. $C(10,3) = \dfrac{10!}{3!(10-3)!} = \dfrac{10!}{3!7!} = \dfrac{10 \cdot 9 \cdot 8 \cdot 7 \cdot 6 \cdot 5 \cdot 4 \cdot 3 \cdot 2 \cdot 1}{3 \cdot 2 \cdot 1 \cdot 7 \cdot 6 \cdot 5 \cdot 4 \cdot 3 \cdot 2 \cdot 1} = \dfrac{10 \cdot 9 \cdot 8}{3 \cdot 2 \cdot 1} = 10 \cdot 3 \cdot 4 = 120$

19. $P(8,3) = \dfrac{8!}{(8-3)!} = \dfrac{8!}{5!} = \dfrac{8\cdot 7\cdot 6\cdot 5\cdot 4\cdot 3\cdot 2\cdot 1}{5\cdot 4\cdot 3\cdot 2\cdot 1} = 8\cdot 7\cdot 6 = 336$

21. $P(10,8) = \dfrac{10!}{(10-8)!} = \dfrac{10!}{2!} = \dfrac{10\cdot 9\cdot 8\cdot 7\cdot 6\cdot 5\cdot 4\cdot 3\cdot 2\cdot 1}{2\cdot 1} = 10\cdot 9\cdot 8\cdot 7\cdot 6\cdot 5\cdot 4\cdot 3 = 1{,}814{,}400$

23. $C(8,3) = \dfrac{8!}{3!(8-3)!} = \dfrac{8!}{3!5!} = \dfrac{8\cdot 7\cdot 6\cdot 5\cdot 4\cdot 3\cdot 2\cdot 1}{3\cdot 2\cdot 1\cdot 5\cdot 4\cdot 3\cdot 2\cdot 1} = \dfrac{8\cdot 7\cdot 6}{3\cdot 2\cdot 1} = 8\cdot 7 = 56$

25. $C(10,8) = \dfrac{10!}{8!(10-8)!} = \dfrac{10!}{8!2!} = \dfrac{10\cdot 9\cdot 8\cdot 7\cdot 6\cdot 5\cdot 4\cdot 3\cdot 2\cdot 1}{8\cdot 7\cdot 6\cdot 5\cdot 4\cdot 3\cdot 2\cdot 1\cdot 2\cdot 1} = \dfrac{10\cdot 9}{2\cdot 1} = 5\cdot 9 = 45$

27. The number of permutations of two selections that can be made from 10 objects.

29.
```
              1
            1   1
          1   2   1
        1   3   3   1
      1   4   6   4   1
    1   5  10  10   5   1
  1   6  15  20  15   6   1
1   7  21  35  35  21   7   1
1  8  28  56  70  56  28  8  1
```

31. Seventh Row: 1   7   **21**   35   35   21   7   1

33. The second entry of the 18th row

35. The sixth entry of the 20th row

37. $P(8,8)$

39. $C(17,3)$

41. $P(5,5)$

43. $C(10,3)$

45. $C(9,5)$

47. $C(17,8)$

49. $C(25,6) = \dfrac{25!}{6!(25-6)!} = \dfrac{25!}{6!19!}$

$= \dfrac{25\cdot 24\cdot 23\cdot 22\cdot 21\cdot 20\cdot 19\cdot 18\cdot 17\cdot 16\cdot 15\cdot 14\cdot 13\cdot 12\cdot 11\cdot 10\cdot 9\cdot 8\cdot 7\cdot 6\cdot 5\cdot 4\cdot 3\cdot 2\cdot 1}{6\cdot 5\cdot 4\cdot 3\cdot 2\cdot 1\cdot 19\cdot 18\cdot 17\cdot 16\cdot 15\cdot 14\cdot 13\cdot 12\cdot 11\cdot 10\cdot 9\cdot 8\cdot 7\cdot 6\cdot 5\cdot 4\cdot 3\cdot 2\cdot 1}$

$= \dfrac{25\cdot 24\cdot 23\cdot 22\cdot 21\cdot 20}{6\cdot 5\cdot 4\cdot 3\cdot 2\cdot 1} = 5\cdot 23\cdot 11\cdot 7\cdot 20 = 177{,}100$

51. $P(17,8) = \dfrac{17!}{(17-8)!} = \dfrac{17!}{9!} = \dfrac{17 \cdot 16 \cdot 15 \cdot 14 \cdot 13 \cdot 12 \cdot 11 \cdot 10 \cdot 9 \cdot 8 \cdot 7 \cdot 6 \cdot 5 \cdot 4 \cdot 3 \cdot 2 \cdot 1}{9 \cdot 8 \cdot 7 \cdot 6 \cdot 5 \cdot 4 \cdot 3 \cdot 2 \cdot 1}$

$= 17 \cdot 16 \cdot 15 \cdot 14 \cdot 13 \cdot 12 \cdot 11 \cdot 10 = 980,179,200$ minutes

There are $60 \cdot 24 \cdot 365 = 525,600$ minutes per year, so it would take $\dfrac{980,179,200}{525,600} \approx 1,864.9$ years or about 1,865 years.

53. $P(15,5) = \dfrac{15!}{(15-5)!} = \dfrac{15!}{10!} = \dfrac{15 \cdot 14 \cdot 13 \cdot 12 \cdot 11 \cdot 10 \cdot 9 \cdot 8 \cdot 7 \cdot 6 \cdot 5 \cdot 4 \cdot 3 \cdot 2 \cdot 1}{10 \cdot 9 \cdot 8 \cdot 7 \cdot 6 \cdot 5 \cdot 4 \cdot 3 \cdot 2 \cdot 1} = 15 \cdot 14 \cdot 13 \cdot 12 \cdot 11 = 360,360$

55. $P(15,5) \times P(15,5) \times P(15,4) \times P(15,5) \times P(15,5) = 552,446,474,061,128,648,601,600,000$ (a VERY large number)

57. $C(6,2) \cdot C(8,3) = \dfrac{6!}{2!(6-2)!} \cdot \dfrac{8!}{3!(8-3)!} = \dfrac{6!}{2!4!} \cdot \dfrac{8!}{3!5!} = \dfrac{6 \cdot 5 \cdot 4 \cdot 3 \cdot 2 \cdot 1}{2 \cdot 1 \cdot 4 \cdot 3 \cdot 2 \cdot 1} \cdot \dfrac{8 \cdot 7 \cdot 6 \cdot 5 \cdot 4 \cdot 3 \cdot 2 \cdot 1}{3 \cdot 2 \cdot 1 \cdot 5 \cdot 4 \cdot 3 \cdot 2 \cdot 1}$

$= \dfrac{6 \cdot 5}{2 \cdot 1} \cdot \dfrac{8 \cdot 7 \cdot 6}{3 \cdot 2 \cdot 1} = (3 \cdot 5) \cdot (8 \cdot 7) = 15 \cdot 56 = 840$

59. $C(7,3) \cdot C(5,2) \cdot C(11,3) = \dfrac{7!}{3!(7-3)!} \cdot \dfrac{5!}{2!(5-2)!} \cdot \dfrac{11!}{3!(11-3)!} = \dfrac{7!}{3!4!} \cdot \dfrac{5!}{2!3!} \cdot \dfrac{11!}{3!8!} =$

$\dfrac{7 \cdot 6 \cdot 5 \cdot 4 \cdot 3 \cdot 2 \cdot 1}{3 \cdot 2 \cdot 1 \cdot 4 \cdot 3 \cdot 2 \cdot 1} \cdot \dfrac{5 \cdot 4 \cdot 3 \cdot 2 \cdot 1}{2 \cdot 1 \cdot 3 \cdot 2 \cdot 1} \cdot \dfrac{11 \cdot 10 \cdot 9 \cdot 8 \cdot 7 \cdot 6 \cdot 5 \cdot 4 \cdot 3 \cdot 2 \cdot 1}{3 \cdot 2 \cdot 1 \cdot 8 \cdot 7 \cdot 6 \cdot 5 \cdot 4 \cdot 3 \cdot 2 \cdot 1} = \dfrac{7 \cdot 6 \cdot 5}{3 \cdot 2 \cdot 1} \cdot \dfrac{5 \cdot 4}{2 \cdot 1} \cdot \dfrac{11 \cdot 10 \cdot 9}{3 \cdot 2 \cdot 1} =$

$(7 \cdot 5) \cdot (5 \cdot 2) \cdot (11 \cdot 5 \cdot 3) = 35 \cdot 10 \cdot 165 = 57,750$

61. $C(5,2) \cdot C(4,2) = \dfrac{5!}{2!(5-2)!} \cdot \dfrac{4!}{2!(4-2)!} = \dfrac{5!}{2!3!} \cdot \dfrac{4!}{2!2!} = \dfrac{5 \cdot 4 \cdot 3 \cdot 2 \cdot 1}{2 \cdot 1 \cdot 3 \cdot 2 \cdot 1} \cdot \dfrac{4 \cdot 3 \cdot 2 \cdot 1}{2 \cdot 1 \cdot 2 \cdot 1} = \dfrac{5 \cdot 4}{2 \cdot 1} \cdot \dfrac{4 \cdot 3}{2 \cdot 1}$

$= (5 \cdot 2) \cdot (2 \cdot 3) = 10 \cdot 6 = 60$

63. $P(12,2) \cdot C(10,2) = \dfrac{12!}{(12-2)!} \cdot \dfrac{10!}{2!(10-2)!} = \dfrac{12!}{10!} \cdot \dfrac{10!}{2!8!}$

$= \dfrac{12 \cdot 11 \cdot 10 \cdot 9 \cdot 8 \cdot 7 \cdot 6 \cdot 5 \cdot 4 \cdot 3 \cdot 2 \cdot 1}{10 \cdot 9 \cdot 8 \cdot 7 \cdot 6 \cdot 5 \cdot 4 \cdot 3 \cdot 2 \cdot 1} \cdot \dfrac{10 \cdot 9 \cdot 8 \cdot 7 \cdot 6 \cdot 5 \cdot 4 \cdot 3 \cdot 2 \cdot 1}{2 \cdot 1 \cdot 8 \cdot 7 \cdot 6 \cdot 5 \cdot 4 \cdot 3 \cdot 2 \cdot 1} = 12 \cdot 11 \cdot \dfrac{10 \cdot 9}{2 \cdot 1}$

$= 132 \cdot (5 \cdot 9) = 132 \cdot 45 = 5,940$

65. $P(5,2) \cdot C(11,3) = \dfrac{5!}{(5-2)!} \cdot \dfrac{11!}{3!(11-3)!} = \dfrac{5!}{3!} \cdot \dfrac{11!}{3!8!}$

$= \dfrac{5 \cdot 4 \cdot 3 \cdot 2 \cdot 1}{3 \cdot 2 \cdot 1} \cdot \dfrac{11 \cdot 10 \cdot 9 \cdot 8 \cdot 7 \cdot 6 \cdot 5 \cdot 4 \cdot 3 \cdot 2 \cdot 1}{3 \cdot 2 \cdot 1 \cdot 8 \cdot 7 \cdot 6 \cdot 5 \cdot 4 \cdot 3 \cdot 2 \cdot 1} = 5 \cdot 4 \cdot \dfrac{11 \cdot 10 \cdot 9}{3 \cdot 2 \cdot 1}$

$= 20 \cdot (11 \cdot 5 \cdot 3) = 20 \cdot 165 = 3,300$

67. As you look at the intersections, you'll see that there is a pattern as to how many routes can be created as you leave the mall on the way to the bank. To choose direct routes, you must always be traveling down and/or to the right. The numbers indicate how many ways there are to get to the intersection below and to the right of the number. The total length of the route is 8 blocks. There are 70 possible routes.

	Mall	1	1	1	1
	1	2	3	4	5
	1	3	6	10	15
	1	4	10	20	35
	1	5	15	35	70

Bank

Also, since there are 8 blocks to be traveled and you could turn right at any of these intersections four times. So there are $C(8,4) = \dfrac{8!}{4!(8-4)!} = \dfrac{8!}{4!4!} = \dfrac{8 \cdot 7 \cdot 6 \cdot 5 \cdot 4 \cdot 3 \cdot 2 \cdot 1}{4 \cdot 3 \cdot 2 \cdot 1 \cdot 4 \cdot 3 \cdot 2 \cdot 1} = \dfrac{8 \cdot 7 \cdot 6 \cdot 5}{4 \cdot 3 \cdot 2 \cdot 1} = 7 \cdot 2 \cdot 5 = 70$ paths.

Also, since there are 8 blocks to be traveled and you could go down at any of these intersections four times. So there are $C(8,4) = \dfrac{8!}{4!(8-4)!} = \dfrac{8!}{4!4!} = \dfrac{8 \cdot 7 \cdot 6 \cdot 5 \cdot 4 \cdot 3 \cdot 2 \cdot 1}{4 \cdot 3 \cdot 2 \cdot 1 \cdot 4 \cdot 3 \cdot 2 \cdot 1} = \dfrac{8 \cdot 7 \cdot 6 \cdot 5}{4 \cdot 3 \cdot 2 \cdot 1} = 7 \cdot 2 \cdot 5 = 70$ paths.

69. $C(13,2) \cdot C(13,3) = \dfrac{13!}{2!(13-2)!} \cdot \dfrac{13!}{3!(13-3)!} = \dfrac{13!}{2!11!} \cdot \dfrac{13!}{3!10!} =$

$\dfrac{13 \cdot 12 \cdot 11 \cdot 10 \cdot 9 \cdot 8 \cdot 7 \cdot 6 \cdot 5 \cdot 4 \cdot 3 \cdot 2 \cdot 1}{2 \cdot 1 \cdot 11 \cdot 10 \cdot 9 \cdot 8 \cdot 7 \cdot 6 \cdot 5 \cdot 4 \cdot 3 \cdot 2 \cdot 1} \cdot \dfrac{13 \cdot 12 \cdot 11 \cdot 10 \cdot 9 \cdot 8 \cdot 7 \cdot 6 \cdot 5 \cdot 4 \cdot 3 \cdot 2 \cdot 1}{3 \cdot 2 \cdot 1 \cdot 10 \cdot 9 \cdot 8 \cdot 7 \cdot 6 \cdot 5 \cdot 4 \cdot 3 \cdot 2 \cdot 1} = \dfrac{13 \cdot 12}{2 \cdot 1} \cdot \dfrac{13 \cdot 12 \cdot 11}{3 \cdot 2 \cdot 1} =$

$(13 \cdot 6) \cdot (13 \cdot 2 \cdot 11) = 78 \cdot 286 = 22{,}308$

71. Because of the special conditions on the numbers, we cannot use all digits at each position in the number.

73. a) This is the number of ways to choose no elements from a 5-element set.

   b) The set we are choosing is the empty set, which can be chosen in only one way.

   c) $C(5,0) = \dfrac{5!}{0!(5-0)!} = \dfrac{5!}{0!5!} = \dfrac{5 \cdot 4 \cdot 3 \cdot 2 \cdot 1}{1 \cdot 5 \cdot 4 \cdot 3 \cdot 2 \cdot 1} = 1$

75. a) This is the number of ways to choose a four-element set from a 5-element set.

   b) Since there are five elements, this can be done is five ways, since we need to eliminate one element.

   c) $C(5,4) = \dfrac{5!}{4!(5-4)!} = \dfrac{5!}{4!1!} = \dfrac{5 \cdot 4 \cdot 3 \cdot 2 \cdot 1}{4 \cdot 3 \cdot 2 \cdot 1 \cdot 1} = 5$

77. The number of ways to do something must be an integer.

79. No solution provided

81.

83. 6; for six lines

85. $C(n-1, r-1)$

87. $C(n-1, r-1) + C(n-1, r) = C(n, r)$

89. The number of ways that we can choose an r-element set from n elements is the same as the number of ways that we can choose $n - r$ elements from n elements; for example, the number of ways that we can choose two elements from five elements is the same as the number of ways that we can choose three elements from five elements.

Chapter Review Exercises

1. PQ, PR, PS, QP, QR, QS, RP, RQ, RS, SP, SQ, SR

2. Without drawing the tree, you can imagine that the tree begins and there are four branches on the first roll. Each of those branches split into four branches on the second roll for a total sixteen branches. Each of those branches split into four branches on the third roll for a total of sixty-four branches. There would therefore be **sixty-four** ways three four-sided dice could be rolled.

3. Without drawing the tree, you can imagine that the tree begins and there are five branches on the first shirt choice. Each of those branches split into four branches for the second choice of pants for a total of twenty branches. Each of those branches split into six branches for the choice of a tie for a total of 120 branches. Each of those branches split into two branches for the choice of a jacket for a total of 240 branches. There would therefore be **240** outfits that could be formed.

4. $14 \times 13 \times 12 = 2{,}184$

5. $4 \times 12 \times 6 = 288$

6. $5 \times 3 \times 2 \times 3 = 90$

7.

Seat 1	Seat 2	Seat 3	Seat 4	Seat 5	Seat 6
Alex	Bonnie	X	X	X	X
X	Alex	Bonnie	X	X	X
X	X	Alex	Bonnie	X	X
X	X	X	Alex	Bonnie	X
X	X	X	X	Alex	Bonnie

Alex and Bonnie can occupy five different groupings of two seats. There are two different arrangements for each of these five possibilities. The number of ways that the students could be arranged in these six seats is $5 \times 2 \times 4 \times 3 \times 2 \times 1 = 240$.

8. In a permutation, order is important; in a combination, it is not.

9. $P(8, 8) = \dfrac{8!}{(8-0)!} = \dfrac{8!}{0!} = \dfrac{8 \cdot 7 \cdot 6 \cdot 5 \cdot 4 \cdot 3 \cdot 2 \cdot 1}{1} = 40{,}320$

10. $C(17, 3) = \dfrac{17!}{3!(17-3)!} = \dfrac{17!}{3! 14!} = \dfrac{17 \cdot 16 \cdot 15 \cdot 14 \cdot 13 \cdot 12 \cdot 11 \cdot 10 \cdot 9 \cdot 8 \cdot 7 \cdot 6 \cdot 5 \cdot 4 \cdot 3 \cdot 2 \cdot 1}{3 \cdot 2 \cdot 1 \cdot 14 \cdot 13 \cdot 12 \cdot 11 \cdot 10 \cdot 9 \cdot 8 \cdot 7 \cdot 6 \cdot 5 \cdot 4 \cdot 3 \cdot 2 \cdot 1}$

$= \dfrac{17 \cdot 16 \cdot 15}{3 \cdot 2 \cdot 1} = 17 \cdot 8 \cdot 5 = 680$

11. $P(26,3) \cdot P(10,2) = \dfrac{26!}{(26-3)!} \cdot \dfrac{10!}{(10-2)!} = \dfrac{26!}{23!} \cdot \dfrac{10!}{8!} =$

$\dfrac{26 \cdot 25 \cdot 24 \cdot 23 \cdot 22 \cdot 21 \cdot 20 \cdot 19 \cdot 18 \cdot 17 \cdot 16 \cdot 15 \cdot 14 \cdot 13 \cdot 12 \cdot 11 \cdot 10 \cdot 9 \cdot 8 \cdot 7 \cdot 6 \cdot 5 \cdot 4 \cdot 3 \cdot 2 \cdot 1}{23 \cdot 22 \cdot 21 \cdot 20 \cdot 19 \cdot 18 \cdot 17 \cdot 16 \cdot 15 \cdot 14 \cdot 13 \cdot 12 \cdot 11 \cdot 10 \cdot 9 \cdot 8 \cdot 7 \cdot 6 \cdot 5 \cdot 4 \cdot 3 \cdot 2 \cdot 1} \cdot$

$\dfrac{10 \cdot 9 \cdot 8 \cdot 7 \cdot 6 \cdot 5 \cdot 4 \cdot 3 \cdot 2 \cdot 1}{8 \cdot 7 \cdot 6 \cdot 5 \cdot 4 \cdot 3 \cdot 2 \cdot 1} = (26 \cdot 25 \cdot 24) \cdot (10 \cdot 9) = 15,600 \cdot 90 = 1,404,000$

12. $C(n,r) = \dfrac{P(n,r)}{r!}$

13. 21

```
            1
           1 1
          1 2 1
         1 3  3 1
        1 4  6  4 1
       1 5 10 10  5 1
      1 6 15 20 15  6 1
     1 7 (21) 35 35 21  7 1
```

14. Beginning with the zeroth entry, $C(18,2)$ represents the 2nd entry of the 18th row.

Chapter Test

1. AB, AC, AD, BA, BC, BD, CA, CB, CD, DA, DB, DC

2. Without drawing the tree, you can imagine that the tree begins and there are three branches for the biology class choice. Each of those branches split into four branches for the diversity class choice for a total of twelve branches. Each of those branches split into six branches for the choice of a writing class for a total of seventy-two branches. There would therefore be **72** class schedules that could be formed.

3. Without drawing the tree, you can imagine that the tree begins and there are eight branches on the first roll. Each of those branches split into eight branches on the second roll for a total of sixty-four branches. There would therefore be **64** ways two octahedral dice could be rolled.

4. $P(15,4) = 15 \times 14 \times 13 \times 12 = 32,760$

5. $5 \times 4 \times 11 \times 6 = 1,320$

6.

Seat 1	Seat 2	Seat 3	Seat 4	Seat 5
Devaun	Emily	X	X	X

Seat 1	Seat 2	Seat 3	Seat 4	Seat 5
X	Devaun	Emily	X	X

Seat 1	Seat 2	Seat 3	Seat 4	Seat 5
X	X	Devaun	Emily	X

Seat 1	Seat 2	Seat 3	Seat 4	Seat 5
X	X	X	Devaun	Emily

Devaun and Emily can occupy four different groupings of two seats. There are two different arrangements for each of these four possibilities. The number of ways that the students could be arranged in these six seats is $4 \times 2 \times 3 \times 2 \times 1 = 48$.

7. $3 \times 9 \times 2 \times 2 = 108$

8. $P(6,6) = \dfrac{6!}{(6-0)!} = \dfrac{6!}{0!} = \dfrac{6 \cdot 5 \cdot 4 \cdot 3 \cdot 2 \cdot 1}{1} = 720$

9. $C(12,3) = \dfrac{12!}{3!(12-3)!} = \dfrac{12!}{3!(9!)} = \dfrac{12 \cdot 11 \cdot 10 \cdot 9 \cdot 8 \cdot 7 \cdot 6 \cdot 5 \cdot 4 \cdot 3 \cdot 2 \cdot 1}{3 \cdot 2 \cdot 1 \cdot 9 \cdot 8 \cdot 7 \cdot 6 \cdot 5 \cdot 4 \cdot 3 \cdot 2 \cdot 1} = \dfrac{12 \cdot 11 \cdot 10}{3 \cdot 2 \cdot 1} = 4 \cdot 11 \cdot 5 = 220$

10. $P(26,3) \cdot P(10,4) = \dfrac{26!}{(26-3)!} \cdot \dfrac{10!}{(10-4)!} = \dfrac{26!}{23!} \cdot \dfrac{10!}{6!} =$

    $\dfrac{26 \cdot 25 \cdot 24 \cdot 23 \cdot 22 \cdot 21 \cdot 20 \cdot 19 \cdot 18 \cdot 17 \cdot 16 \cdot 15 \cdot 14 \cdot 13 \cdot 12 \cdot 11 \cdot 10 \cdot 9 \cdot 8 \cdot 7 \cdot 6 \cdot 5 \cdot 4 \cdot 3 \cdot 2 \cdot 1}{23 \cdot 22 \cdot 21 \cdot 20 \cdot 19 \cdot 18 \cdot 17 \cdot 16 \cdot 15 \cdot 14 \cdot 13 \cdot 12 \cdot 11 \cdot 10 \cdot 9 \cdot 8 \cdot 7 \cdot 6 \cdot 5 \cdot 4 \cdot 3 \cdot 2 \cdot 1} \cdot$

    $\dfrac{10 \cdot 9 \cdot 8 \cdot 7 \cdot 6 \cdot 5 \cdot 4 \cdot 3 \cdot 2 \cdot 1}{6 \cdot 5 \cdot 4 \cdot 3 \cdot 2 \cdot 1} = (26 \cdot 25 \cdot 24) \cdot (10 \cdot 9 \cdot 8 \cdot 7) = 15,600 \cdot 5040 = 78,624,000$

11. 20

    ```
                1
              1   1
            1   2   1
          1   3   3   1
        1   4   6   4   1
      1   5   10  10  5   1
    1   6   15 (20) 15  6   1
    ```

12. In a permutation, order is important; in a combination, it is not.

13. $C(n,r) = \dfrac{P(n,r)}{r!}$

14. $C(12,5)$ represents the 5th entry of the 12th row.

Of Further Interest: Counting And Gambling

1. The number of ways each of the three wheels could stop were multiplied together.

2. The number of ways to choose three cards of the same rank was multiplied by the number of ways to choose the remaining two cards of the same rank.

3. We are not simply choosing three cards from a possible 52.

4. In example 6, you were encouraged to imagine the tree without actually drawing it.

5. $2 \times 5 \times 12 = 120$

6. $2 \times 5 \times 8 = 80$

7. $5 \times 2 \times 2 = 20$

8. $4 \times 3 \times 7 = 84$

9. $1 \times 5 \times 2 = 10$

10. $2 \times 3 \times 1 = 6$

11. The number of ways we can obtain the payoff for three cherries is $2 \times 5 \times 8 = 80$.
    The number of ways we can obtain the payoff for three oranges is $5 \times 2 \times 2 = 20$.
    Since there are fewer ways to obtain three oranges, we would expect the payoff to be higher.

12. The number of ways we can obtain the payoff for three plums is $4 \times 3 \times 7 = 84$
    The number of ways we can obtain the payoff for three bells is $1 \times 5 \times 2 = 10$
    Since there are fewer ways to obtain three bells, we would expect the payoff to be higher.

13. 4, since there are four suits

14. a) 4
    b) Since the ace can be considered either as a one or higher than a king, there are 10 ways to choose a sequence of five cards within a suit.

A	2	3	4	5	6	7	8	9	10	J	Q	K	A

    c) Although there are $4 \cdot 10 = 40$ ways to choose a sequence of five cards, four of these ways are also royal flushes. So there are $40 - 4 = 36$ ways to construct a straight flush.

15. a) 4
    b) $C(13,5) = \dfrac{13!}{5!(13-5)!} = \dfrac{13!}{5!8!} = \dfrac{13 \cdot 12 \cdot 11 \cdot 10 \cdot 9 \cdot 8 \cdot 7 \cdot 6 \cdot 5 \cdot 4 \cdot 3 \cdot 2 \cdot 1}{5 \cdot 4 \cdot 3 \cdot 2 \cdot 1 \cdot 8 \cdot 7 \cdot 6 \cdot 5 \cdot 4 \cdot 3 \cdot 2 \cdot 1}$
    $= \dfrac{13 \cdot 12 \cdot 11 \cdot 10 \cdot 9}{5 \cdot 4 \cdot 3 \cdot 2 \cdot 1} = 13 \cdot 11 \cdot 9 = 1,287$

    c) Although there are $4 \cdot 1,287 = 5,148$ ways to choose five cards from the same suit, 40 of these ways are either straight flushes or royal flushes. So, there are $5,148 - 40 = 5,108$ ways to construct a flush.

16. a) Since the ace can be considered either as a one or higher than a king, there are 10 ways to choose a sequence of five cards within thirteen different valued cards.
    b) $4 \times 4 \times 4 \times 4 \times 4 = 1,024$
    c) $10 \cdot 1,024 = 10,240$
    d) $10,240 - 40 = 10,200$

17. We can construct a two-pair hand in stages:
    Stage 1: Pick the rank of the card of which we will have a pair. This decision can be made in 13 ways.
    Stage 2: Now choose these two cards from the four cards of a rank. This can be done in $C(4,2) = 6$ ways
    Stage 3: Now that you have chosen two cards, we select the next two for a pair. This decision can be made in 12 ways, because we don't want four-of-a-kind.
    Stage 4: Now choose these two cards from the four cards of a rank. This can be done in $C(4,2) = 6$ ways
    Stage 5: Now the fifth card can be chosen in 44 ways since eight of the cards must be excluded.

    Since there are two ways to choose the two ranks, we must divide by two.

    So there are $\dfrac{13 \times 6 \times 12 \times 6}{2} \times 44 = \dfrac{5,616}{2} \times 44 = 2,808 \times 44 = 123,552$ ways to choose a two-pair hand.

18. We can construct the one-pair hand in stages:
    Stage 1: Pick the rank of the card of which we will have a pair. This decision can be made in 13 ways.
    Stage 2: Now choose these two cards from the four cards of a rank. This can be done in $C(4,2) = 6$ ways
    Stage 3: Now the third card can be chosen in 48 ways, since four of the cards must be excluded.
    Stage 4: Now the forth card can be chosen in 44 ways, since eight of the cards must be excluded.
    Stage 5: Now the fifth card can be chosen in 40 ways, since twelve of the cards must be excluded.

    In the selection of the last three cards we must consider that since order in not important, we need to divide the ways of selecting these last three cards by the number of ways of selecting three cards. Thus the total number of ways to construct a poker hand that contains one pair and no other cards of any value is $13 \times 6 \times \dfrac{48 \times 44 \times 40}{3 \times 2 \times 1} = 1,098,240$.

19. a) There are 4 ways to make a Royal Flush (Exercise 13). There are 36 ways to make a Straight Flush (Exercise 14). There are 624 ways to make a Four of a Kind (Example 4). There are 3,744 ways to make a Full House (Example 6). There are 5,108 ways to make a Flush (Exercise 15). There are 10,200 ways to make a Straight (Exercise 16). There are 54,912 ways to make a Three of a Kind (Example 7). There are 123,552 ways to make Two Pair (Exercise 17). There are 1,098,240 ways to make One Pair. This totals to be 1,296,420 winning hands.

    b) There are 2,598,960 ways to make a poker hand (Example 3). 2,598,960 − 1,296,420 = 1,302,540.

Chapter 13
PROBABILITY: What Are The Chances?

Section 13.1 The Basics Of Probability Theory

1. We drew a tree diagram.

3. We know from Chapter 12 that there are $C(52,5)$ ways to choose five cards from 52.

5. $P(E) = \dfrac{\text{number of times } E \text{ occurs}}{\text{number of times experiment is performed}}$

7. All plants had one yellow and one green gene, and yellow was dominant.

9. If we let "H" represent a heads and "T" represent tails then the sample space would have eight outcomes.
   The sample space would be {HHH, HHT, HTH, HTT, THH, THT, TTH, TTT}.

11. The sample space would have sixteen pairs. The sample space would be {(1,1), (1,2), (1,3), (1,4), (2,1), (2,2), (2,3), (2,4), (3,1), (3,2), (3,3), (3,4), (4,1), (4,2), (4,3), (4,4)}.

13. The sample space would be {(b,f), (b,r), (i,f), (i,r), (h,f), (h,r)}.

15. The sample space would be {(i,i), (i,d), (i,s), (d,i), (d,d), (d,s), (s,i), (s,d), (s,s)}.

17. {(1,6), (2,5), (3,4), (4,3), (5,2), (6,1)}

19. If we let "h" represent a heads and "t" represent tails, then the outcomes would be {hhh, hht, hth, thh}.

21. {(b,f), (i,f), (h,f)}   23. {(i,d), (i,s), (d,i), (s,i)}

25. If we let "r" represent red, "b" represent blue, and "y" represent yellow, then the outcomes would be {(r,b), (r,y), (b,r), (y,r)}.

27. If we let "r" represent red, "b" represent blue, and "y" represent yellow, then the outcomes would be {(r,r,b), (r,r,y), (r,b,r), (r,y,r), (b,r,r), (y,r,r)}.

29. A total of nine can occur in four ways. They are: {(3,6), (4,5), (5,4), (6,3)}. Since there are 36 pairs that can occur, the probability of rolling a total of nine on two fair dice is $\dfrac{4}{36} = \dfrac{1}{9} \approx 0.11$

31. Since there are 13 hearts out of the 52 cards, the probability of drawing a heart is $\dfrac{13}{52} = \dfrac{1}{4} = 0.25$.

33. The sixteen possible outcomes are {HHHH, HHHT, HHTH, **HHTT**, HTHH, **HTHT**, **HTTH**, HTTT, THHH, **THHT**, **THTH**, THTT, **TTHH**, TTHT, TTTH, TTTT}. The bolded outcomes indicate the six ways we can toss four fair coins and obtain exactly two heads. The probability of this is therefore $\dfrac{6}{16} = \dfrac{3}{8} \approx 0.38$.

359

35. Since $P(E) = \frac{1}{9}$, $P(E') = 1 - \frac{1}{9} = \frac{9}{9} - \frac{1}{9} = \frac{8}{9}$. Thus, the odds against the event are

$\frac{P(E')}{P(E)} = \frac{8/9}{1/9} = \frac{8}{1}$, or 8 to 1.

37. Since $P(E) = \frac{1}{4}$, $P(E') = 1 - \frac{1}{4} = \frac{4}{4} - \frac{1}{4} = \frac{3}{4}$. Thus, the odds against the event are

$\frac{P(E')}{P(E)} = \frac{3/4}{1/4} = \frac{3}{1}$, or 17 to 1.

39. Since $P(E) = \frac{3}{8}$, $P(E') = 1 - \frac{3}{8} = \frac{8}{8} - \frac{3}{8} = \frac{5}{8}$. Thus, the odds against the event are

$\frac{P(E')}{P(E)} = \frac{5/8}{3/8} = \frac{5}{3}$, or 5 to 3.

41. $\frac{C(13,5)}{C(52,5)} = \frac{1,287}{2,598,960} \approx 0.000495$

43. $4 \times \frac{C(13,5)}{C(52,5)} = 4 \times \frac{1,287}{2,598,960} \approx 0.002$

45. Since the sample space is {(s,c), (s,w), (s,d), (s,h), (c,s), (c,w), (c,d), (c,h), (w,s), (w,c), (w,d), (w,h), (d,s), (d,c), (d,w), (d,h), (h,s), (h,c), (h,w), (h,d)}, the bolded elements satisfy the condition of exactly one star appearing on the two cards. Thus, the probability is $\frac{8}{20} = \frac{2}{5}$.

47. Since there are 100 balls in the jar and 30 of them are not red, the probability of choosing a non-red ball is $\frac{30}{100} = \frac{3}{10}$.

49. Since $P(E) = \frac{2}{5}$, $P(E') = 1 - \frac{2}{5} = \frac{5}{5} - \frac{2}{5} = \frac{3}{5}$. Thus, the odds against the event are

$\frac{P(E')}{P(E)} = \frac{3/5}{2/5} = \frac{3}{2}$, or 3 to 2.

51. Since $P(E) = \frac{3}{10}$, $P(E') = 1 - \frac{3}{10} = \frac{10}{10} - \frac{3}{10} = \frac{7}{10}$. Thus the odds against the event are

$\frac{P(E')}{P(E)} = \frac{7/10}{3/10} = \frac{7}{3}$, or 7 to 3.

53. Answers may vary.
An outcome is an element in a sample space; an event is a subset of a sample space.

55. Answers may vary.
Rolling a total less than thirteen when rolling two dice.

57. No solution provided

59. Since there are 1,951,379 females out of the 1,951,379 + 2,048,861 = 4,000,240 children, the probability of selecting a female is $\frac{1,951,379}{4,000,240} \approx 0.49$

61. Since there are 277 short plants out of the 787 + 277 = 1,064 plants, the probability that the plant will be short is $\frac{277}{1,064} \approx 0.26$. This is consistent with the theoretical results that there should be a 3 to 1 ratio in the second generation of dominant trait to recessive trait; that is, the theoretical probability of that a plant selected from the second generation will be short is 0.25.

63. The probability that the child is a carrier is $\frac{2}{4} = \frac{1}{2}$.

		Second	Parent
		s	n
First	s	ss	sn
Parent	s	ss	sn

67. a)

		Second	Parent
		N	c
First	N	NN	Nc
Parent	c	cN	cc

b) $P(\text{disease}) = \frac{1}{4}$

65. a)

		First	Gen.
		r	r
First	w	wr	wr
Gen.	w	wr	wr

b) All flowers will be pink. $P(\text{pink})=1$; all other probabilities are 0.

69. There are 109 + 29 = 138 students out of the 320 surveyed that has a GPA of at least 2.5. The probability of selecting a student with a GPA of at least 2.5 is $\frac{138}{320} = \frac{69}{160} \approx 0.43$.

71. Since you are rolling two dice, there are 36 possible combinations. St. James Place, Tennessee Avenue, and New York Avenue are within 12 spaces from the Electric Company. St. James Place is 4 spaces from the Electric Company. You can land on St. James Place if you roll (1,3), (2,2), or (3,1). Tennessee Avenue is 6 spaces from the Electric Company. You can land on Tennessee Avenue if you roll (1,5), (2,4), (3,3), (4,2), or (5,1). New York Avenue is 7 spaces from the Electric Company. You can land on New York Avenue if you roll (1,6), (2,5), (3,4), (4,3), (5,2), or (6,1). Since there are 14 ways to go bankrupt, there are $36 - 14 = 22$ ways to avoid going bankrupt. Thus, the probability of avoiding bankruptcy is $\frac{22}{36} = \frac{11}{18} \approx 0.61$.

73. Since you are rolling two dice, there are 36 possible combinations. There are two railroads within 12 spaces from Virginia Avenue. The closest one is one space away. You cannot land on it, since the minimum roll would be a total of 2. The second railroad is 11 spaces from Virginia Avenue. You can land on it if you roll (5,6) or (6,5). Thus the probability of landing on a railroad is $\frac{2}{36} = \frac{1}{18} \approx 0.06$.

75. The sample space, S, consists of $1,015 + 20,117 + 35,860 = 56,992$ thousand people in this age group. The event, call it A, is the set of 1,015 thousand people who earn less than $5.16 an hour.

    The desired probability is $P(A) = \dfrac{n(A)}{n(S)} = \dfrac{1,015,000}{56,992,000} \approx 0.0178$.

77. The sample space, S, consists of $640 + 5,743 + 618 + 660 + 6,729 + 3,387 = 17,777$ thousand people in this age group. The event, call it A is the set of $640 + 5,743 + 660 + 6,729 = 13,772$ thousand people who earn less than $10.00 an hour.

    The desired probability is $P(A) = \dfrac{n(A)}{n(S)} = \dfrac{13,772,000}{17,777,000} \approx 0.7747$.

79. Since order is important, the number of ways of choosing the horses would be a permutation. There are $P(8,3) = 336$ ways to choose the horses. Since only one way is a winner, the probability of winning is $\dfrac{1}{336} \approx 0.002976$.

81. Since the hacker can enter a password every 10 seconds, he will have 6 opportunities each minute. Since the hacker is allowed three minutes, he has a total of 18 opportunities. Since the hacker is randomly trying passwords, it is conceivable that the hacker might try the same password more than once. Since there are $26 \times 26 \times 10 \times 10 \times 10 = 676,000$ possible passwords, the probability that the hacker would be successful in discovering a valid password is $18 \times \dfrac{1}{676,000} \approx 0.000027$.

83. The odds against E are $\dfrac{5}{2} = \dfrac{5/7}{2/7} = \dfrac{P(E')}{P(E)}$. Thus, the probability of E is $\dfrac{2}{7}$.

85. The odds against W are $\dfrac{7}{5} = \dfrac{7/12}{5/12} = \dfrac{P(W')}{P(W)}$. Thus, the probability of winning is $\dfrac{5}{12}$.

87. a) The odds in favor of E (winning) are $\dfrac{P(E)}{P(E')} = \dfrac{0.30}{1-0.30} = \dfrac{0.30}{0.70} = \dfrac{3/10}{7/10} = \dfrac{3}{7}$, or 3 to 7.

    b) The odds against E (winning) are $\dfrac{P(E')}{P(E)} = \dfrac{1-0.30}{0.30} = \dfrac{0.70}{0.30} = \dfrac{7/10}{3/10} = \dfrac{7}{3}$, or 7 to 3.

89. – 93. No solutions provided

Section 13.2 Complements And Unions Of Events

1. It was easier to compute $P(A')$ and then subtract it from one instead of computing $P(A)$ directly.

3. Because $E \cup E' = S$, and $P(S) = 1$.

5. $1 - 0.015 = 0.985$

7. $1 - \dfrac{1}{1,000} = \dfrac{1,000}{1,000} - \dfrac{1}{1,000} = \dfrac{999}{1,000}$

9. Let E be the event that either die shows a five on it. This can happen in 11 ways: {(1,5), (2,5), (3,5), (4,5), (5,5), (6,5), (5,1), (5,2), (5,3), (5,4), (5,6)}. Since there are 36 ways to roll two dice, the probability that neither die shows a five on it is $P(E') = 1 - P(E) = 1 - \dfrac{11}{36} = \dfrac{25}{36}$.

11. Let E be the event that no coin shows a head. This can happen in only one way: {TTTTT}. Since there are $2 \times 2 \times 2 \times 2 \times 2 = 32$ ways to flip five coins, the probability of obtaining at least one head is
$P(E') = 1 - P(E) = 1 - \dfrac{1}{32} = \dfrac{31}{32}$.

13. Let F be the event that we choose a five, and let R be the event that we draw a red card. There are 4 fives in a deck, 26 red cards, and 2 red fives.
$$P(F \cup R) = P(F) + P(R) - P(F \cap R) = \dfrac{4}{52} + \dfrac{26}{52} - \dfrac{2}{52} = \dfrac{28}{52} = \dfrac{7}{13}$$

15. Let O be the event of rolling an odd total and E be the event of rolling a total greater than 8. There are 18 ways of rolling an odd total: {(1,2), (1,4), (1,6), (2,1), (2,3), (2,5), (3,2), (3,4), **(3,6)**, (4,1), (4,3), **(4,5)**, (5,2), **(5,4)**, **(5,6)**, (6,1), **(6,3)**, **(6,5)**}. There are 10 ways to roll a total greater than 8: {**(3,6)**, **(4,5)**, (4,6), **(5,4)**, (5,5), **(5,6)**, **(6,3)**, (6,4), **(6,5)**, (6,6)}. There are 6 ways to roll an odd total greater than 8 (bolded).
$$P(O \cup E) = P(O) + P(E) - P(O \cap E) = \dfrac{18}{36} + \dfrac{10}{36} - \dfrac{6}{36} = \dfrac{22}{36} = \dfrac{11}{18}$$

17. $P(A \cup B) = P(A) + P(B) - P(A \cap B)$
$0.85 = 0.55 + 0.40 - P(A \cap B)$
$0.85 = 0.95 - P(A \cap B)$
$-0.10 = -P(A \cap B)$
$P(A \cap B) = 0.10$

19. $P(A \cup B) = P(A) + P(B) - P(A \cap B)$
$0.70 = 0.40 + P(B) - 0.25$
$0.70 = 0.15 + P(B)$
$P(B) = 0.55$

21. Let H be the event that we select a heart and F be the event that we select a face card.

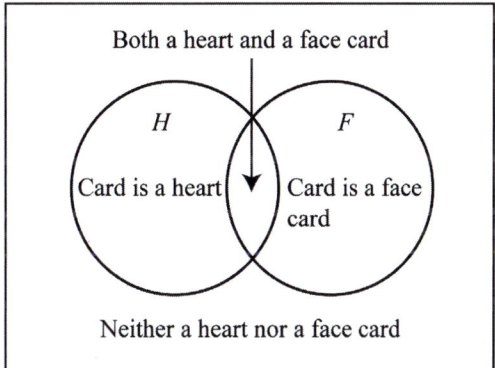

$P\left((H \cup F)'\right) = 1 - P(H \cup F) = 1 - [P(H) + P(F) - P(H \cap F)]$
$= 1 - \left[\dfrac{13}{52} + \dfrac{12}{52} - \dfrac{3}{52}\right] = 1 - \dfrac{22}{52} = \dfrac{30}{52} = \dfrac{15}{26}$

23. There are 1,483 (in thousands) surveyed. The total (in thousands) of those that are 55 or older is $61 + 53 = 114$. The probability of selecting a person less than 55 years old is
$P(E) = 1 - P(E') = 1 - \dfrac{114}{1,483} = \dfrac{1,369}{1,483} \approx 0.923$.

25. Let T be the event that we select a consumer that spends 0-2 hours per month shopping on the Internet, and let A be the event that we select a consumer that has an annual income below \$40,000.

$$P(T \cup A) = P(T) + P(A) - P(T \cap A) = \frac{544}{1,600} + \frac{592}{1,600} - \frac{272}{1,600} = \frac{864}{1,600} = \frac{27}{50} = 0.54$$

27. Let T be the event that we select a consumer that spends more than 2 hours per month shopping on the Internet, and let A be the event that we select a consumer that has an annual income of \$60,000 or less.

$$P((T \cup A)') = 1 - P(T \cup A) = 1 - [P(T) + P(A) - P(T \cap A)]$$

$$= 1 - \left[\frac{480 + 576}{1,600} + \frac{512 + 592}{1,600} - \frac{160 + 128 + 208 + 192}{1,600} \right]$$

$$= 1 - \left[\frac{1,056}{1,600} + \frac{1,104}{1,600} - \frac{688}{1,600} \right] = 1 - \frac{1,472}{1,600} = 1 - \frac{23}{25} = \frac{2}{25} = 0.08$$

29. Let E be the event that Joanna will earn less than \$1,000 in commissions.

$$P(E') = 1 - P(E) = 1 - 0.08 = 0.92$$

31. Let E be the event that Joanna will earn more than \$1,999 in commissions.

$$P(E') = 1 - P(E) = 1 - (0.05 + 0.08 + 0.03) = 1 - 0.16 = 0.84$$

Use the following table for Exercises 33 and 35. The bolded information needed to be determined.

	Probation	Not on Probation	Totals
Satisfied	**38**	110	148
Not Satisfied	32	20	52
Totals	70	**130**	200

33. Let E be the event that the student chosen is not on academic probation.

$$P(E) = \frac{130}{200} = 0.65$$

35. Let E be the event that the student chosen is on academic probation and is satisfied with advisement.

$$P(E) = \frac{38}{200} = 0.19$$

37. False; this would be true only if the intersection of events A and B was empty.

39. True; since the Rule for Computing the Probability of a Union of Two Events is $P(A \cup B) = P(A) + P(B) - P(A \cap B)$, we can solve for $P(A \cap B)$ to obtain $P(A \cap B) = P(A) + P(B) - P(A \cup B)$.

41. If E and F are disjoint sets then $P(E \cap F) = 0$.

43. No solution provided

45. Explanations will vary; $P(A) + P(B) + P(C) - P(A \cap B) - P(B \cap C)$.

Section 13.3 Conditional Probability

1. The probability that the total was odd, given that the total was greater than 9.

3. We found that $P(G\,|\,F) \neq P(G)$, which meant that knowing the event F occurred had an effect on the probability of G.

5. Multiply both sides of the equation by $P(E)$ to get $P(E \cap F) = P(F\,|\,E) \cdot P(E)$.

For Exercises 7 and 9, refer to the tree diagram on the next page.

7. $P(F) = \dfrac{6}{36} = \dfrac{1}{6}$

   There are 18 times out of the 36 outcomes that yield an odd sum. Of these 18 occurrences, 6 of them yield a sum of 7.
   $P(F\,|\,E) = \dfrac{6}{18} = \dfrac{1}{3}$

9. $P(F) = \dfrac{6}{36} = \dfrac{1}{6}$

   There are 11 times out of the 36 outcomes that a three shows on at least one of the dice. Of these 11 occurrences, 2 of them yield a sum of less than 5.
   $P(F\,|\,E) = \dfrac{2}{11}$

11. $P(F\,|\,E) = \dfrac{50}{61}$

13. $P(F\,|\,E) = \dfrac{2}{61}$

15. $P(\text{heart}|\text{red}) = \dfrac{P(\text{red} \cap \text{heart})}{P(\text{red})} = \dfrac{13/52}{26/52} = \dfrac{13}{26} = \dfrac{1}{2}$

17. $P(\text{seven}|\text{non-face card}) = \dfrac{P(\text{non-face card} \cap \text{seven})}{P(\text{non-face card})} = \dfrac{4/52}{40/52} = \dfrac{4}{40} = \dfrac{1}{10}$

For Exercises 19 and 21, refer to the tree diagram on the next page.

19. $P(E\,|\,F) = \dfrac{P(E \cap F)}{P(F)} = \dfrac{2/36}{18/36} = \dfrac{2}{18} = \dfrac{1}{9}$; $P(F\,|\,E) = \dfrac{P(F \cap E)}{P(E)} = \dfrac{2/36}{6/36} = \dfrac{2}{6} = \dfrac{1}{3}$

21. $P(E\,|\,F) = \dfrac{P(E \cap F)}{P(F)} = \dfrac{6/36}{18/36} = \dfrac{6}{18} = \dfrac{1}{3}$; $P(F\,|\,E) = \dfrac{P(F \cap E)}{P(E)} = \dfrac{6/36}{11/36} = \dfrac{6}{11}$

Use the following tree diagram for Exercises 7 and 9 and 19 and 21.

First	Second	Outcome	Sum
1	1	(1,1)	2
	2	(1,2)	3
	3	(1,3)	4
	4	(1,4)	5
	5	(1,5)	6
	6	(1,6)	7
2	1	(2,1)	3
	2	(2,2)	4
	3	(2,3)	5
	4	(2,4)	6
	5	(2,5)	7
	6	(2,6)	8
3	1	(3,1)	4
	2	(3,2)	5
	3	(3,3)	6
	4	(3,4)	7
	5	(3,5)	8
	6	(3,6)	9
4	1	(4,1)	5
	2	(4,2)	6
	3	(4,3)	7
	4	(4,4)	8
	5	(4,5)	9
	6	(4,6)	10
5	1	(5,1)	6
	2	(5,2)	7
	3	(5,3)	8
	4	(5,4)	9
	5	(5,5)	10
	6	(5,6)	11
6	1	(6,1)	7
	2	(6,2)	8
	3	(6,3)	9
	4	(6,4)	10
	5	(6,5)	11
	6	(6,6)	12

Section 13.3: Conditional Probability   367

For Exercises 23 and 25, let M = the student has Mono and T = the test was positive.

23. $P(M \mid T) = \dfrac{P(M \cap T)}{P(T)} = \dfrac{72/140}{76/140} = \dfrac{18}{19} \approx 0.947$

25. $P(T \mid M) = \dfrac{P(T \cap M)}{P(M)} = \dfrac{72/140}{80/140} = \dfrac{9}{10} \approx 0.90$

27. a) $P(J \cap J) = P(J) \cdot P(J \mid J) = \dfrac{4}{52} \cdot \dfrac{3}{51} = \dfrac{1}{13} \cdot \dfrac{1}{17} = \dfrac{1}{221}$

b) $P(J \cap J) = P(J) \cdot P(J) = \dfrac{4}{52} \cdot \dfrac{4}{52} = \dfrac{1}{13} \cdot \dfrac{1}{13} = \dfrac{1}{169}$

29. a) $P(F \cap N) = P(F) \cdot P(N \mid F) = \dfrac{12}{52} \cdot \dfrac{40}{51} = \dfrac{3}{13} \cdot \dfrac{40}{51} = \dfrac{120}{663} = \dfrac{40}{221}$

b) $P(F \cap N) = P(F) \cdot P(N) = \dfrac{12}{52} \cdot \dfrac{40}{52} = \dfrac{3}{13} \cdot \dfrac{10}{13} = \dfrac{30}{169}$

31. a) $P(J \cap K) = P(J) \cdot P(K \mid J) = \dfrac{4}{52} \cdot \dfrac{4}{51} = \dfrac{1}{13} \cdot \dfrac{4}{51} = \dfrac{4}{663}$

$P(K \cap J) = P(K) \cdot P(J \mid K) = \dfrac{4}{52} \cdot \dfrac{4}{51} = \dfrac{1}{13} \cdot \dfrac{4}{51} = \dfrac{4}{663}$

$\dfrac{4}{663} + \dfrac{4}{663} = \dfrac{8}{663}$

b) $P(J \cap K) = P(J) \cdot P(K) = \dfrac{4}{52} \cdot \dfrac{4}{52} = \dfrac{1}{13} \cdot \dfrac{1}{13} = \dfrac{1}{169}$

$P(K \cap J) = P(K) \cdot P(J) = \dfrac{4}{52} \cdot \dfrac{4}{52} = \dfrac{1}{13} \cdot \dfrac{1}{13} = \dfrac{1}{169}$

$\dfrac{1}{169} + \dfrac{1}{169} = \dfrac{2}{169}$

33. Let E be the event that we draw at least one face card. $\overline{E}$ is therefore the event that we draw no face cards. If N denotes a non-face card, then

$P(E) = 1 - P(\overline{E}) = 1 - P(N \cap N) = 1 - P(N) \cdot P(N \mid N) = 1 - \dfrac{40}{52} \cdot \dfrac{39}{51} = 1 - \dfrac{1{,}560}{2{,}652} = 1 - \dfrac{10}{17} = \dfrac{7}{17}$

35. $P(F \cap F \cap F) = \dfrac{12}{52} \cdot \dfrac{11}{51} \cdot \dfrac{10}{50} = \dfrac{3}{13} \cdot \dfrac{11}{51} \cdot \dfrac{1}{5} = \dfrac{33}{3{,}315} = \dfrac{11}{1{,}105}$

37. $P(H \cap H \cap N) + P(H \cap N \cap H) + P(N \cap H \cap H) =$

$\dfrac{13}{52} \cdot \dfrac{12}{51} \cdot \dfrac{39}{50} + \dfrac{13}{52} \cdot \dfrac{39}{51} \cdot \dfrac{12}{50} + \dfrac{39}{52} \cdot \dfrac{13}{51} \cdot \dfrac{12}{50} = \dfrac{6{,}084}{132{,}600} + \dfrac{6{,}084}{132{,}600} + \dfrac{6{,}084}{132{,}600} = \dfrac{18{,}252}{132{,}600} = \dfrac{117}{850}$

368   Chapter 13: Probability

39. A total of five can occur in four ways: (1,4), (2,3), (3,2), and (4,1). So, out of the 36 possible ways to throw a pair a dice, the probability of throwing a five is $\frac{4}{36} = \frac{1}{9}$. Therefore, the probability of a total of five on each of the three rolls is: $P(F \cap F \cap F) = \frac{1}{9} \cdot \frac{1}{9} \cdot \frac{1}{9} = \frac{1}{729}$

For Exercises 41 and 43, let B = the student voted for Bush, K = the student voted for Kerry, R = the student is a Republican, D = the student is a Democrat, and I = the student is an Independent.

41. $P(D \mid K) = \dfrac{P(D \cap K)}{P(K)} = \dfrac{105/240}{142/240} = \dfrac{105}{142} \approx 0.739$

43. $P(K \mid I) = \dfrac{P(K \cap I)}{P(I)} = \dfrac{25/240}{40/240} = \dfrac{5}{8} \approx 0.625$

For Exercises 45 and 47, refer to the tree diagram for Exercises 7 and 9.

45. dependent

$P(F \mid E) = \dfrac{P(F \cap E)}{P(E)} = \dfrac{4/36}{6/36} = \dfrac{4}{6} = \dfrac{2}{3} \neq P(F) = \dfrac{18}{36} = \dfrac{1}{2}$

47. independent

$P(F \mid E) = \dfrac{P(F \cap E)}{P(E)} = \dfrac{3/36}{6/36} = \dfrac{3}{6} = \dfrac{1}{2} = P(F) = \dfrac{18}{36} = \dfrac{1}{2}$

49. $P(F \mid E) = \dfrac{P(F \cap E)}{P(E)} = \dfrac{6/100}{18/100} = \dfrac{6}{18} = \dfrac{1}{3}$ 51. $P(F \mid E) = \dfrac{P(F \cap E)}{P(E)} = \dfrac{8/100}{15/100} = \dfrac{8}{15}$

53. Forty percent of the available spaces in dorm Y are rooms.

55. 0.50

57.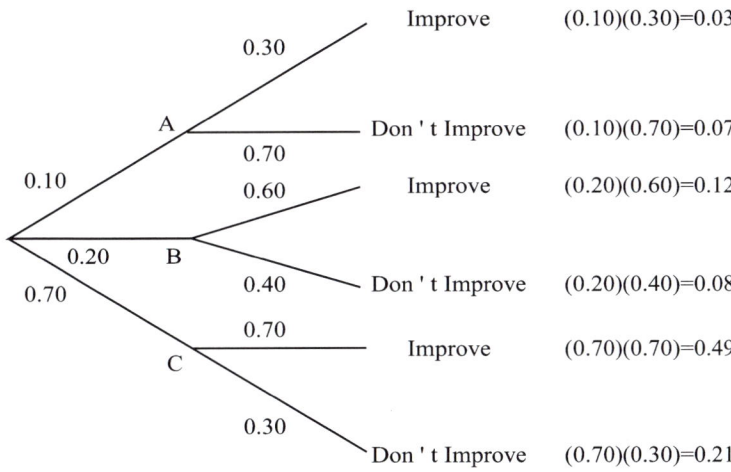

59. $0.03 + 0.12 + 0.49 = 0.64$

61. No solution provided

63. Answers will vary.
Generally $P(B\mid A)$ is larger. If you consider the equation $P(B\mid A)P(A) = P(B\cap A)$, you see that we have to multiply $P(B\mid A)$ by $P(A)$, which is usually less than 1, to get $P(B\cap A)$.

65. $\dfrac{0.18}{0.18+0.30} = \dfrac{0.18}{0.48} = 0.375$

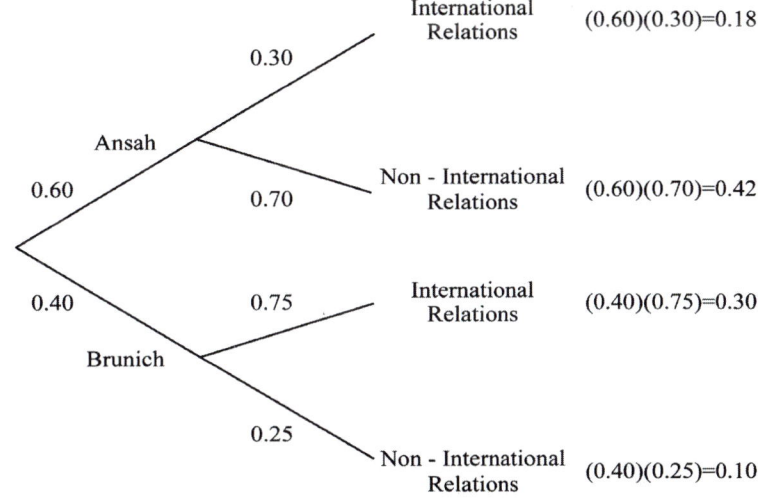

370   Chapter 13: Probability

Use the following tree diagram for Exercises 67 and 69.

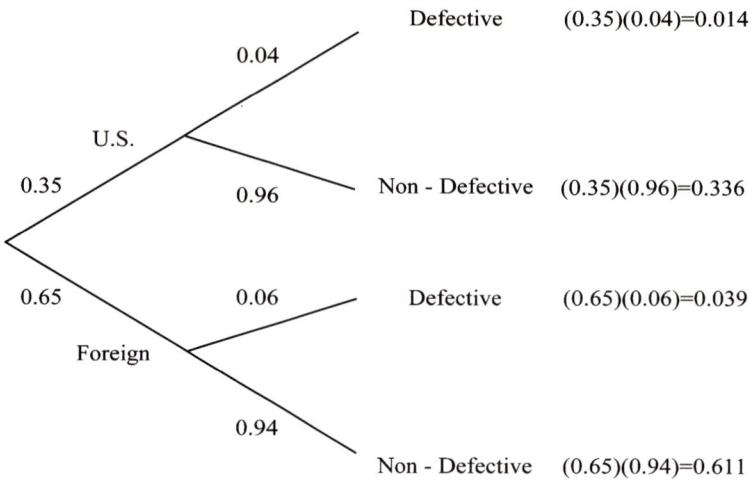

67. 0.336

69. $\dfrac{0.039}{0.014+0.039} = \dfrac{0.039}{0.053} \approx 0.736$

Use the following tree diagram for Exercise 71.

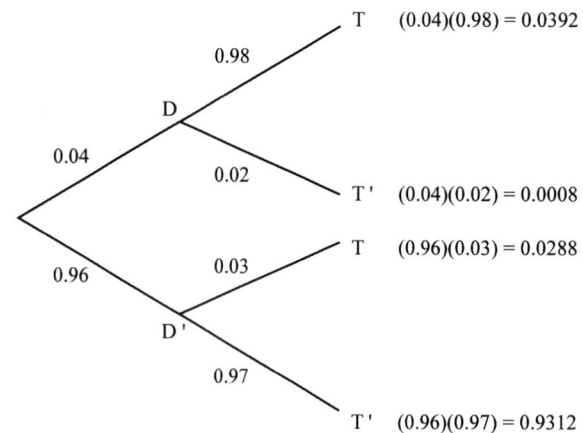

71. $\dfrac{0.0288}{0.0288+0.0392} = \dfrac{0.0288}{0.068} \approx 0.424$

73. and 75. No solutions provided

77. $P(\text{duplication of birthdays}) = 1 - P(\text{no duplication of birthdays}) =$

$1 - \dfrac{364}{365} \cdot \dfrac{363}{365} \cdot \dfrac{362}{365} \cdot \dfrac{361}{365} \cdot \dfrac{360}{365} \cdot \dfrac{359}{365} \cdot \dfrac{358}{365} \cdot \dfrac{357}{365} \cdot \dfrac{356}{365} = 1 - \dfrac{P(364,9)}{365^9} \approx 0.117$

79. For 21 people: $1 - \dfrac{364}{365} \cdot \dfrac{363}{365} \cdot \ldots \cdot \dfrac{346}{365} \cdot \dfrac{345}{365} = 1 - \dfrac{P(364, 20)}{365^{20}} \approx 0.444$

For 22 people: $1 - \dfrac{364}{365} \cdot \dfrac{363}{365} \cdot \ldots \cdot \dfrac{345}{365} \cdot \dfrac{344}{365} = 1 - \dfrac{P(364, 21)}{365^{21}} \approx 0.476$

For **23 people**: $1 - \dfrac{364}{365} \cdot \dfrac{363}{365} \cdot \ldots \cdot \dfrac{344}{365} \cdot \dfrac{343}{365} = 1 - \dfrac{P(364, 22)}{365^{22}} \approx 0.507$

Section 13.4 Expected Value

1. Julio worked 30 hours for three of the thirty-six weeks, so the probability of this working 30 hours was $\dfrac{3}{36}$. We multiplied the probability $\dfrac{3}{36}$ times the value of 30 in computing the expected value.

3. Answers may vary

5. The expected value of the game is:

$$\left(\dfrac{1}{6}\right)\cdot(+1) + \left(\dfrac{1}{6}\right)\cdot(-2) + \left(\dfrac{1}{6}\right)\cdot(+3) + \left(\dfrac{1}{6}\right)\cdot(-4) + \left(\dfrac{1}{6}\right)\cdot(+5) + \left(\dfrac{1}{6}\right)\cdot(-6)$$
$$= \dfrac{1-2+3-4+5-6}{6} = \dfrac{-3}{6} = -0.50$$

This is not a fair game.

For Exercise 7, refer to the tree diagram for Exercises 7 and 9 in Section 13.3.

7. A 6, 7 or 8 will occur in 16 out of the 36 ways to roll two dice. A 2 or a 12 will occur in 2 out of the 36 ways to roll two dice.

The expected value of the game is:

$$\left(\dfrac{16}{36}\right)\cdot(+4) + \left(\dfrac{2}{36}\right)\cdot(+2) + \left(\dfrac{18}{36}\right)\cdot(-1) = \dfrac{16\cdot 4 + 2\cdot 2 - 18\cdot 1}{36} = \dfrac{50}{36} \approx \$1.39$$

This is not a fair game.

9. $\left(\dfrac{13}{52}\right)\cdot(+5) + \left(\dfrac{39}{52}\right)\cdot(-5) = \dfrac{13\cdot 5 - 39\cdot 5}{52} = \dfrac{-130}{52} = -\$2.50$

Use the following table for Exercise 11.

Possible Outcomes	
(h,h,h)	(t,h,h)
(h,h,t)	(t,h,t)
(h,t,h)	(t,t,h)
(h,t,t)	(t,t,t)

11. $\left(\dfrac{2}{8}\right)\cdot(+2) + \left(\dfrac{6}{8}\right)\cdot(-1) = \dfrac{2\cdot 2 - 6\cdot 1}{8} = \dfrac{-2}{8} = -\$0.25$

13. $1 + 1.39 = \$2.39$

15. $-0.25 + 1.00 = \$0.75$

17. $\left(\dfrac{1}{1,000}\right)\cdot(+599)+\left(\dfrac{999}{1,000}\right)\cdot(-1)=\dfrac{1\cdot 599-999\cdot 1}{1,000}=\dfrac{-400}{1,000}=-\$0.40$

This is not a fair game. To make it a fair game, one should charge $-0.40+1.00=\$0.60$

19. $\left(\dfrac{1}{500}\right)\cdot(+495)+\left(\dfrac{2}{500}\right)\cdot(+245)+\left(\dfrac{5}{500}\right)\cdot(+95)+\left(\dfrac{492}{500}\right)\cdot(-5)=$

$\dfrac{1\cdot 495+2\cdot 245+5\cdot 95-492\cdot 5}{500}=\dfrac{-1,000}{500}=-\$2.00$

This is not a fair game. To make it a fair game, one should charge $-2.00+5.00=\$3.00$

21. $\left(\dfrac{5}{38}\right)\cdot(+6)+\left(\dfrac{33}{38}\right)\cdot(-1)=\dfrac{5\cdot 6-33\cdot 1}{38}=\dfrac{-3}{38}\approx -\$0.08$

23. $\left(\dfrac{3}{38}\right)\cdot(+8)+\left(\dfrac{35}{38}\right)\cdot(-1)=\dfrac{3\cdot 8-35\cdot 1}{38}=\dfrac{-11}{38}\approx -\$0.29$

25. The expected value is $\left(\dfrac{1}{4}\right)\cdot(+1)+\left(\dfrac{3}{4}\right)\cdot\left(-\dfrac{1}{4}\right)=\dfrac{4}{16}+\dfrac{-3}{16}=\dfrac{1}{16}$. Since this is positive, the student should guess.

27. The expected value when ruling out no option is: $\left(\dfrac{1}{5}\right)\cdot(+1)+\left(\dfrac{4}{5}\right)\cdot\left(-\dfrac{1}{2}\right)=\dfrac{2}{10}+\dfrac{-4}{10}=-\dfrac{2}{10}=-\dfrac{1}{5}$.

The expected value when ruling out one option is: $\left(\dfrac{1}{4}\right)\cdot(+1)+\left(\dfrac{3}{4}\right)\cdot\left(-\dfrac{1}{2}\right)=\dfrac{2}{8}+\dfrac{-3}{8}=-\dfrac{1}{8}$.

The expected value when ruling out **two options** is: $\left(\dfrac{1}{3}\right)\cdot(+1)+\left(\dfrac{2}{3}\right)\cdot\left(-\dfrac{1}{2}\right)=\dfrac{2}{6}+\dfrac{-2}{6}=0$.

29. $(0.02)\cdot(+972.50)+(0.98)\cdot(-27.50)=-\$7.50$

31. 44.00

$(0.02)\cdot(+2,200-x)+(0.98)\cdot(-x)=0$

$44-.02x-0.98x=0$

$44-x=0$

$44=x$

33. $(0.60)\cdot(+45,000)+(0.40)\cdot(-5,000)=\$25,000$

35. The total number who earned a profit is $4+8+13+21+3+1=50$.

$\left(\dfrac{4}{50}\right)\cdot(100)+\left(\dfrac{8}{50}\right)\cdot(200)+\left(\dfrac{13}{50}\right)\cdot(300)+\left(\dfrac{21}{50}\right)\cdot(400)+\left(\dfrac{3}{50}\right)\cdot(500)+\left(\dfrac{1}{50}\right)\cdot(600)$

$=\dfrac{400+1,600+3,900+8,400+1,500+600}{50}=\dfrac{16,400}{50}=\$328$

37. a) $\left(\dfrac{3}{20}\right)\cdot(+0.25)\cdot 130 + \left(\dfrac{6}{20}\right)(+0.25)\cdot 130 + \left(\dfrac{5}{20}\right)(+0.25)\cdot 130 + \left(\dfrac{6}{20}\right)(+0.25\cdot 120 + (-0.65)\cdot 10) =$
$29.80

b) $\left(\dfrac{3}{20}\right)\cdot(+0.25)\cdot 140 + \left(\dfrac{6}{20}\right)(+0.25)\cdot 140 + \left(\dfrac{5}{20}\right)(+0.25\cdot 130 + (-0.65)\cdot 10) +$
$\left(\dfrac{6}{20}\right)(+0.25\cdot 120 + (-0.65)\cdot 20) = \$27.35$

39. and 41. No solutions provided

Chapter Review Exercises

1. a) If we let "H" represent a heads and "T" represent tails then we would have three outcomes. The set of outcomes would be {HHT, HTH, THH}.

    b) We would have five outcomes. The set of outcomes would be {(2,6), (3,5), (4,4), (5,3), (6,2)}.

2. Since there are 6 red face cards in a deck, the probability of choosing one is $\dfrac{6}{52} = \dfrac{3}{26}$

3. No solution provided

4. The Punnett square is

	First Gen.	
	w	r
First w	ww	wr
Gen. r	rw	rr

    $P(\text{white}) = \dfrac{1}{4}$

5. a) The odds against E are $\dfrac{7}{3} = \dfrac{7/10}{3/10} = \dfrac{P(E')}{P(E)}$. Thus, the probability of E is $\dfrac{3}{10}$.

    b) The odds against E are $\dfrac{P(E')}{P(E)} = \dfrac{1-0.55}{0.55} = \dfrac{0.45}{0.55} = \dfrac{45/100}{55/100} = \dfrac{45}{55} = \dfrac{9}{11}$.

6. a) $P(E') = 1 - P(E)$

    b) No solution provided

    c) No solution provided

374 Chapter 13: Probability

7. Let F be the event that we select a face card and R be the event that we select a red card.

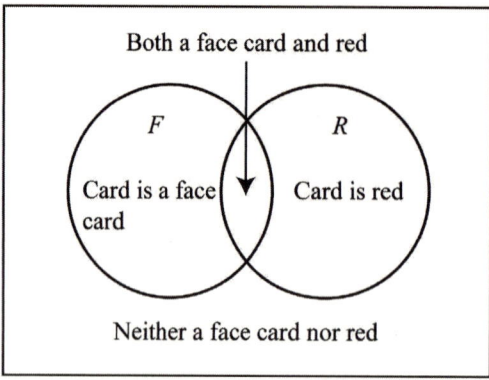

$$P(F \cup R) = P(F) + P(R) - P(F \cap R) = \frac{12}{52} + \frac{26}{52} - \frac{6}{52} = \frac{32}{52} = \frac{8}{13}$$

8. No solution provided

9. There are 26 ways to roll a total less than nine; they are {(1,1), (1,2), (1,3), **(1,4)**, (1,5), (1,6), (2,1), (2,2), **(2,3)**, (2,4), (2,5), (2,6), (3,1), **(3,2)**, (3,3), (3,4), (3,5), **(4,1)**, (4,2), (4,3), (4,4), (5,1), (5,2), (5,3), (6,1), (6,2)}. Of these, there are 4 ways to roll a total of 5; thus, the probability of rolling a total of five given that the total is less than nine is $\frac{4}{26} = \frac{2}{13}$.

10. a) $P(H \cap H) = P(H) \cdot P(H \mid H) = \frac{13}{52} \cdot \frac{12}{51} = \frac{1}{4} \cdot \frac{12}{51} = \frac{12}{204} = \frac{1}{17}$

   b) $P(Q \cap A) = P(Q) \cdot P(A \mid Q) = \frac{4}{52} \cdot \frac{4}{51} = \frac{1}{13} \cdot \frac{4}{51} = \frac{4}{663}$

11. There are 18 out of the 36 ways to roll an odd total. There are 10 ways to roll a total less than six, and they are {(1,1), (1,2), (1,3), (1,4), (2,1), (2,2), (2,3), (3,1), (3,2), (4,1)}.

$$P(F \mid E) = \frac{P(F \cap E)}{P(E)} = \frac{6/36}{18/36} = \frac{6}{18} = \frac{1}{3} \neq P(F) = \frac{10}{36} = \frac{5}{18}$$

These two events are therefore dependent.

12. $\dfrac{0.036}{0.036+0.0576} = \dfrac{0.036}{0.0936} \approx 0.385$

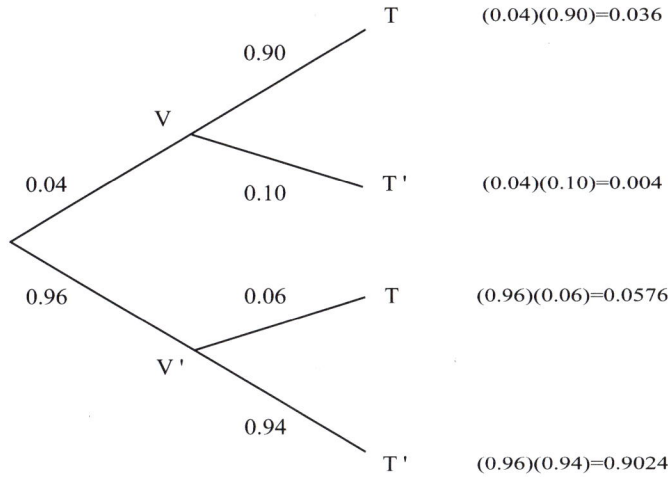

13. $\left(\dfrac{12}{52}\right)\cdot(+15) + \left(\dfrac{40}{52}\right)\cdot(-4) = \dfrac{12\cdot 15 - 40\cdot 4}{52} = \dfrac{20}{52} \approx \$0.38$

14. There are 16 ways to flip 4 coins. Of these 16 ways, two of them will yield all heads or all tails. The expected value of this game is $\left(\dfrac{2}{16}\right)\cdot(+5) + \left(\dfrac{14}{16}\right)\cdot(0) = \dfrac{2\cdot 5 + 14\cdot 0}{16} = \dfrac{10}{16} \approx \$0.63$

    The price to play this game should be $0.63 to make this game fair.

Chapter Test

1. a) There would be six outcomes. The set of outcomes would be {(4,6), (6,4), (5,5), (5,6), (6,5), (6,6)}.
   b) If "H" represents heads and "T" represent tails, there would be four outcomes. The set of outcomes would be {HHHT, HHTH, HTHH, THHH}.

2. Since there are 2 black kings in a deck, the probability of choosing one is $\dfrac{2}{52} = \dfrac{1}{26}$.

3. Empirical: The probability of the vaccine in Example 3 having serious side effects was 0.03.
   Theoretical: The probability of a total of four showing when rolling two dice in Example 6.

4. The Punnett square is

	First Gen.	
	w	r
First Gen. w	ww	wr
First Gen. r	rw	rr

   $P(\text{pink}) = \dfrac{1}{2}$

5. a) $P(E)+P(E')=1$

   b) No solution provided

6. a) The odds against E are $\dfrac{P(E')}{P(E)}=\dfrac{1-0.15}{0.15}=\dfrac{0.85}{0.15}=\dfrac{85/100}{15/100}=\dfrac{85}{15}=\dfrac{17}{3}$.

   b) The odds against E are $\dfrac{11}{5}=\dfrac{11/16}{5/16}=\dfrac{P(F')}{P(F)}$. Thus, the probability of F is $\dfrac{5}{16}$.

7. Let H be the event that we select a heart and K be the event that we select a king.

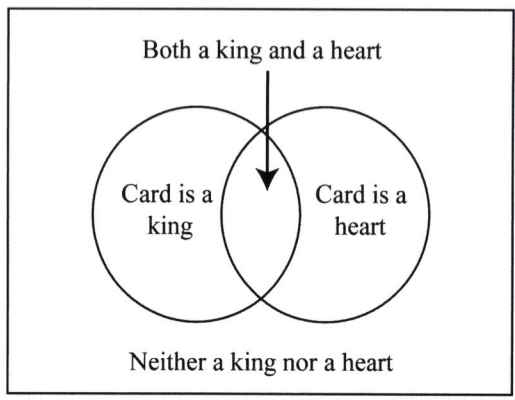

$P(H\cup K)=P(H)+P(K)-P(H\cap K)=\dfrac{13}{52}+\dfrac{4}{52}-\dfrac{1}{52}=\dfrac{16}{52}=\dfrac{4}{13}$.

8. $P(B\,|\,A)$ is the probability of B assuming that A has occurred. $P(A\,|\,B)$ is the probability of A assuming that B has occurred.

9. There are 6 ways to roll a total less than five; they are $\{(1,1), (1,2), (1,3), (3,1), (2,1), (2,2)\}$. Of these, there are 4 ways to roll an even total; thus, the probability of rolling an even total given that the total is less than five is $\dfrac{4}{6}=\dfrac{2}{3}$.

10. There are 6 out of the 36 ways to roll the same number. There are 8 ways to roll a total greater than 8, and they are $\{(4,5), (5,4), (5,5), (4,6), (6,4), (5,6), (6,5), (6,6)\}$.

$$P(F\,|\,E)=\dfrac{P(F\cap E)}{P(E)}=\dfrac{2/36}{8/36}=\dfrac{2}{8}=\dfrac{1}{4}\neq P(F)=\dfrac{8}{36}=\dfrac{4}{9}$$

These two events are therefore dependent.

11. a) $P(F\cap F)=P(F)\cdot P(F\,|\,F)=\dfrac{12}{52}\cdot\dfrac{11}{51}=\dfrac{3}{13}\cdot\dfrac{11}{51}=\dfrac{1}{13}\cdot\dfrac{11}{17}=\dfrac{11}{221}$.

    b) $P(K\,\&\,A)=P(K\cap A)+P(A\cap K)=P(K)\cdot P(A\,|\,K)+P(A)\cdot P(K\,|\,A)=$
    $\dfrac{4}{52}\cdot\dfrac{4}{51}+\dfrac{4}{52}\cdot\dfrac{4}{51}=\dfrac{1}{13}\cdot\dfrac{4}{51}+\dfrac{1}{13}\cdot\dfrac{4}{51}=\dfrac{8}{663}$

12. Since it costs $1 to play, the expected value of this game is $\left(\frac{6}{36}\right)\cdot(+4)+\left(\frac{30}{36}\right)\cdot(-1)=\frac{6\cdot 4-30\cdot 1}{36}$

    $=-\frac{30}{36}=-\frac{1}{6}\approx -\$0.17.$

13. $\dfrac{0.019}{0.019+0.049}=\dfrac{0.019}{0.068}\approx 0.279$

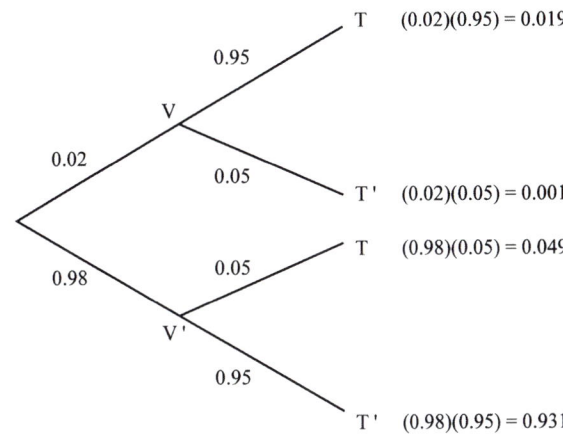

14. Since is costs $2 to play, the expected value of the raffle is
    $\left(\frac{1}{500}\right)\cdot(+248)+\left(\frac{3}{500}\right)\cdot(98)+\left(\frac{5}{500}\right)\cdot(48)+\left(\frac{491}{500}\right)\cdot(-2)=\frac{1\cdot 248+3\cdot 98+5\cdot 48-2\cdot 491}{500}=$
    $-\dfrac{200}{500}=-\$0.40.$

Of Further Interest: Binomial Experiments

1. The dice were rolled a fixed number of times; the experiment had two outcomes (seven and nonseven); the probability of rolling a seven was constant; the rolls were independent.

2. B means the experiment was binomial; 20 is the number of trials; 15 is the number of successes; 0.75 is the probability of success.

3. The experiment has a fixed number of trials. The experiment has two outcomes: "success" and "failure". The probability of success is the same from trial to trial. The trials are independent of each other.

4. The expected number of trials before a success in a binomial experiment with $p = \frac{1}{5}$ is $\dfrac{1}{\frac{1}{5}} = 5$.

5. yes   8. yes

6. yes   9. yes

7. No; the number of trials is not fixed.   10. yes

378 Chapter 13: Probability

11. A total of five can be made in 4 ways, (1,4), (2,3), (3,2), (4,1). The probability of obtaining a total of five on a single roll is $\frac{4}{36} = \frac{1}{9}$. The probability of rolling a total of five exactly once is

$$B\left(3,1,\frac{1}{9}\right) = C(3,1) \cdot \left(\frac{1}{9}\right)^1 \cdot \left(1-\frac{1}{9}\right)^{3-1} = \frac{3!}{1!2!} \cdot \frac{1}{9} \cdot \left(\frac{8}{9}\right)^2 = 3 \cdot \frac{1}{9} \cdot \frac{64}{81} = \frac{1}{3} \cdot \frac{64}{81} = \frac{64}{243} \approx 0.2634.$$

12. A total of eight can be made in 5 ways, (2,6), (3,5), (4,4), (5,3), (6,2). The probability of obtaining a total of eight on a single roll is $\frac{5}{36}$. The probability of rolling a total of eight exactly twice is

$$B\left(3,2,\frac{5}{36}\right) = C(3,2) \cdot \left(\frac{5}{36}\right)^2 \cdot \left(1-\frac{5}{36}\right)^{3-2} = \frac{3!}{2!1!} \cdot \frac{25}{1296} \cdot \left(\frac{31}{36}\right) = 3 \cdot \frac{25}{1296} \cdot \frac{31}{36} = \frac{25}{432} \cdot \frac{31}{36} = \frac{775}{15552} \approx 0.0498.$$

13. An experiment is performed 5 times. $B\left(5,3,\frac{1}{4}\right)$ represents the probability of obtaining a certain outcome 3 times, where the probability that an individual trial results in that certain outcome is $\frac{1}{4}$.

$$B\left(5,3,\frac{1}{4}\right) = C(5,3) \cdot \left(\frac{1}{4}\right)^3 \cdot \left(1-\frac{1}{4}\right)^{5-3} = \frac{5!}{3!2!} \cdot \frac{1}{64} \cdot \left(\frac{3}{4}\right)^2 = \frac{5 \cdot 4}{2} \cdot \frac{1}{64} \cdot \frac{9}{16} = \frac{180}{2048} = \frac{45}{512} \approx 0.0879$$

14. An experiment is performed 6 times. $B\left(6,2,\frac{1}{2}\right)$ represents the probability of obtaining a certain outcome 2 times, where the probability that an individual trial results in that certain outcome is $\frac{1}{2}$.

$$B\left(6,2,\frac{1}{2}\right) = C(6,2) \cdot \left(\frac{1}{2}\right)^2 \cdot \left(1-\frac{1}{2}\right)^{6-2} = \frac{6!}{2!4!} \cdot \frac{1}{4} \cdot \left(\frac{1}{2}\right)^4 = \frac{6 \cdot 5}{2} \cdot \frac{1}{4} \cdot \frac{1}{16} = \frac{30}{128} = \frac{15}{64} \approx 0.2344$$

15. k is larger than n

16. p is greater than 1

17. $B\left(12,9,\frac{1}{2}\right) = C(12,9) \cdot \left(\frac{1}{2}\right)^9 \cdot \left(1-\frac{1}{2}\right)^{12-9} = \frac{12!}{9!3!} \cdot \frac{1}{512} \cdot \left(\frac{1}{2}\right)^3 = \frac{12 \cdot 11 \cdot 10}{3 \cdot 2} \cdot \frac{1}{512} \cdot \frac{1}{8} = \frac{1,320}{24,576}$

$= \frac{55}{1,024} \approx 0.0537$

18. $B\left(8,6,\frac{1}{4}\right) = C(8,6) \cdot \left(\frac{1}{4}\right)^6 \cdot \left(1-\frac{1}{4}\right)^{8-6} = \frac{8!}{6!2!} \cdot \frac{1}{4096} \cdot \left(\frac{3}{4}\right)^2 = \frac{8 \cdot 7}{2} \cdot \frac{1}{4096} \cdot \frac{9}{16} = \frac{504}{131,072}$

$= \frac{63}{16,384} \approx 0.0038$

19. $B\left(12,9,\frac{1}{2}\right) = C(12,9) \cdot \left(\frac{1}{2}\right)^9 \cdot \left(1-\frac{1}{2}\right)^{12-9} = \frac{12!}{9!3!} \cdot \frac{1}{512} \cdot \left(\frac{1}{2}\right)^3 = \frac{12 \cdot 11 \cdot 10}{3 \cdot 2} \cdot \frac{1}{512} \cdot \frac{1}{8} = \frac{1,320}{24,576}$

$= \frac{55}{1,024} = 0.0537109375$

19. (continued)

$$B\left(12,10,\frac{1}{2}\right) = C(12,10) \cdot \left(\frac{1}{2}\right)^{10} \cdot \left(1-\frac{1}{2}\right)^{12-10} = \frac{12!}{10!2!} \cdot \frac{1}{1{,}024} \cdot \left(\frac{1}{2}\right)^2 = \frac{12 \cdot 11}{2} \cdot \frac{1}{1{,}024} \cdot \frac{1}{4}$$

$$= \frac{132}{8{,}192} \approx 0.01611328125$$

$$B\left(12,11,\frac{1}{2}\right) = C(12,11) \cdot \left(\frac{1}{2}\right)^{11} \cdot \left(1-\frac{1}{2}\right)^{12-11} = \frac{12!}{11!1!} \cdot \frac{1}{2{,}048} \cdot \left(\frac{1}{2}\right)^1 = 12 \cdot \frac{1}{2{,}048} \cdot \frac{1}{2} = \frac{12}{4{,}096}$$

$$= \frac{3}{1{,}024} = 0.0029296875$$

$$B\left(12,12,\frac{1}{2}\right) = C(12,12) \cdot \left(\frac{1}{2}\right)^{12} \cdot \left(1-\frac{1}{2}\right)^{12-12} = 1 \cdot \frac{1}{4{,}096} \cdot \left(\frac{1}{2}\right)^0 = \frac{1}{4{,}096} \approx 0.0002441406$$

$$0.0537109375 + 0.01611328125 + 0.0029296875 + 0.0002441406 = 0.07299804685 \approx 0.0730$$

20. $$B\left(6,4,\frac{1}{5}\right) = C(6,4) \cdot \left(\frac{1}{5}\right)^4 \cdot \left(1-\frac{1}{5}\right)^{6-4} = \frac{6!}{4!2!} \cdot \frac{1}{625} \cdot \left(\frac{4}{5}\right)^2 = \frac{6 \cdot 5}{2} \cdot \frac{1}{625} \cdot \frac{16}{25} = \frac{480}{31{,}250} = \frac{48}{3{,}125} \approx 0.01536$$

$$B\left(6,5,\frac{1}{5}\right) = C(6,5) \cdot \left(\frac{1}{5}\right)^5 \cdot \left(1-\frac{1}{5}\right)^{6-5} = \frac{6!}{5!1!} \cdot \frac{1}{3{,}125} \cdot \left(\frac{4}{5}\right)^1 = 6 \cdot \frac{1}{3{,}125} \cdot \frac{4}{5} = \frac{24}{15{,}625} = 0.001536$$

$$B\left(6,6,\frac{1}{5}\right) = C(6,6) \cdot \left(\frac{1}{5}\right)^6 \cdot \left(1-\frac{1}{5}\right)^{6-6} = \frac{6!}{6!0!} \cdot \frac{1}{15{,}625} \cdot \left(\frac{4}{5}\right)^0 = 1 \cdot \frac{1}{15{,}625} \cdot 1 = \frac{1}{15{,}625} = 0.000064$$

$$0.01536 + 0.001536 + 0.000064 = 0.01696 \approx 0.0170$$

21. $$B(10,8,0.80) = C(10,8) \cdot (0.80)^8 \cdot (1-0.80)^{10-8} = \frac{10!}{8!2!} \cdot 0.16777216 \cdot (0.20)^2$$

$$= \frac{10 \cdot 9}{2} \cdot 0.16777216 \cdot 0.04 = 0.301989888$$

$$B(10,9,0.80) = C(10,9) \cdot (0.80)^9 \cdot (1-0.80)^{10-9} = \frac{10!}{9!1!} \cdot 0.1342217728 \cdot (0.20)^1$$

$$= 10 \cdot 0.134217728 \cdot 0.20 = 0.268435456$$

$$B(10,10,0.80) = C(10,10) \cdot (0.80)^{10} \cdot (1-0.80)^{10-10} = \frac{10!}{10!0!} \cdot 0.1073741824 \cdot (0.20)^0$$

$$= 1 \cdot 0.1073741824 \cdot 1 = 0.1073741824$$

$$0.301989888 + 0.268435456 + 0.1073741824 = 0.6777995264 \approx 0.6778$$

22. $$B(3,1,0.30) = C(3,1) \cdot (0.30)^1 \cdot (1-0.30)^{3-1} = \frac{3!}{1!2!} \cdot 0.30 \cdot (0.70)^2 = 3 \cdot 0.30 \cdot 0.49 = 0.4410$$

$$B(3,0,0.30) = C(3,0) \cdot (0.30)^0 \cdot (1-0.30)^{3-0} = \frac{3!}{0!3!} \cdot 1 \cdot (0.70)^3 = 1 \cdot 1 \cdot 0.343 = 0.3430$$

$$0.4410 + 0.3430 = 0.7840$$

23. $B(5,2,0.250) = C(5,2) \cdot (0.250)^2 \cdot (1-0.250)^{5-2} = \dfrac{5!}{2!3!} \cdot 0.0625 \cdot (0.750)^3$

$= \dfrac{5 \cdot 4}{2} \cdot 0.0625 \cdot 0.421875 = 0.263671875 \approx 0.2637$

24. $B(8,4,0.50) = C(8,4) \cdot (0.50)^4 \cdot (1-0.50)^{8-4} = \dfrac{8!}{4!4!} \cdot 0.0625 \cdot (0.5)^4$

$= \dfrac{8 \cdot 7 \cdot 6 \cdot 5}{4 \cdot 3 \cdot 2} \cdot 0.0625 \cdot 0.0625 = 0.2734375 \approx 0.2734$

25. We need to determine the probability that the hospital does not meet the need so we need to determine the probability that the hospital has 0, 1, or 2 donors with type A+ blood waiting.

$B(12,0,0.30) = C(12,0) \cdot (0.30)^0 \cdot (1-0.30)^{12-0} = \dfrac{12!}{0!12!} \cdot 1 \cdot (0.70)^{12}$

$\approx 1 \cdot 1 \cdot 0.013841287201 = 0.013841287201$

$B(12,1,0.30) = C(12,1) \cdot (0.30)^1 \cdot (1-0.30)^{12-1} = \dfrac{12!}{1!11!} \cdot 0.30 \cdot (0.70)^{11}$

$= 12 \cdot 0.30 \cdot 0.01977326743 = 0.071183762748$

$B(12,2,0.30) = C(12,2) \cdot (0.30)^2 \cdot (1-0.30)^{12-2} = \dfrac{12!}{2!10!} \cdot 0.09 \cdot (0.70)^{10}$

$\approx \dfrac{12 \cdot 11}{2} \cdot 0.09 \cdot 0.0282475249 \approx 0.167790297906$

$0.013841287201 + 0.071183762748 + 0.167790297906 = 0.252815347855$

$1 - 0.252815347855 = 0.747184652145 \approx 0.7472$

26. We need to first determine the probability that the network will not fail.

$B(15,12,0.95) = C(15,12) \cdot (0.95)^{12} \cdot (1-0.95)^{15-12} \approx \dfrac{15!}{12!3!} \cdot 0.540360087663 \cdot (0.05)^3$

$\approx \dfrac{15 \cdot 14 \cdot 13}{3 \cdot 2} \cdot 0.540360087663 \cdot 0.000125 \approx 0.030732979986$

$B(15,13,0.95) = C(15,13) \cdot (0.95)^{13} \cdot (1-0.95)^{15-13} \approx \dfrac{15!}{13!2!} \cdot 0.51334208328 \cdot (0.05)^2$

$\approx \dfrac{15 \cdot 14}{2} \cdot 0.51334208328 \cdot 0.0025 \approx 0.134752296861$

$B(15,14,0.95) = C(15,14) \cdot (0.95)^{14} \cdot (1-0.95)^{15-14} \approx \dfrac{15!}{14!1!} \cdot 0.487674979116 \cdot (0.05)^1$

$\approx 15 \cdot 0.487674979116 \cdot 0.05 \approx 0.365756234337$

$B(15,15,0.95) = C(15,15) \cdot (0.95)^{15} \cdot (1-0.95)^{15-15} \approx \dfrac{15!}{15!0!} \cdot 0.46329123016 \cdot 1$

$\approx 1 \cdot 0.46329123016 \cdot 1 = 0.46329123016$

$0.030732979986 + 0.134752296861 + 0.365756234337 + 0.46329123016 = 0.994532741343$

$1 - 0.994532741343 = 0.005467258657 \approx 0.0055$

27. $\frac{3}{4}$

28. 7

29. The total number of purchases to obtain a complete set T is: the number of purchases to obtain the first new figure + the number of purchases to obtain the second new figure + the number of purchases to obtain the third new figure + the number of purchases to obtain the fourth new figure + the number of purchases to obtain the fifth new figure + the number of purchases to obtain the sixth new figure =

$$1 + \frac{1}{5/6} + \frac{1}{4/6} + \frac{1}{3/6} + \frac{1}{2/6} + \frac{1}{1/6} = 1 + \frac{6}{5} + \frac{6}{4} + \frac{6}{3} + \frac{6}{2} + \frac{6}{1} = \frac{147}{10} = 14.7$$

The child should expect to purchase 15 boxes.

30. The total number of purchases to obtain a complete set T is: the number of purchases to obtain the first new card + the number of purchases to obtain the second new card + the number of purchases to obtain the third new card + the number of purchases to obtain the fourth new card + the number of purchases to obtain the fifth new card + the number of purchases to obtain the sixth new card + the number of purchases to obtain the seventh new card + the number of purchases to obtain the eighth new card + the number of purchases to obtain the ninth new card + the number of purchases to obtain the tenth new card =

$$1 + \frac{1}{9/10} + \frac{1}{8/10} + \frac{1}{7/10} + \frac{1}{6/10} + \frac{1}{5/10} + \frac{1}{4/10} + \frac{1}{3/10} + \frac{1}{2/10} + \frac{1}{1/10} =$$

$$1 + \frac{10}{9} + \frac{10}{8} + \frac{10}{7} + \frac{10}{6} + \frac{10}{5} + \frac{10}{4} + \frac{10}{3} + \frac{10}{2} + \frac{10}{1} = \frac{7{,}381}{252} \approx 29.29$$

The child should expect to purchase 30 packages.

31. We need to determine the probability that 4 or fewer people will come down with the cold without the vaccine.

$$B(10,4,0.50) = C(10,4) \cdot (0.50)^4 \cdot (1-0.50)^{10-4} = \frac{10!}{4!6!} \cdot 0.0625 \cdot (0.50)^6 = \frac{10 \cdot 9 \cdot 8 \cdot 7}{4 \cdot 3 \cdot 2} \cdot 0.0625 \cdot 0.015625$$
$$= 0.205078125$$

$$B(10,3,0.50) = C(10,3) \cdot (0.50)^3 \cdot (1-0.50)^{10-3} = \frac{10!}{3!7!} \cdot 0.125 \cdot (0.50)^7 = \frac{10 \cdot 9 \cdot 8}{3 \cdot 2} \cdot 0.125 \cdot 0.0078125$$
$$= 0.1171875$$

$$B(10,2,0.50) = C(10,2) \cdot (0.50)^2 \cdot (1-0.50)^{10-2} = \frac{10!}{2!8!} \cdot 0.25 \cdot (0.50)^8 = \frac{10 \cdot 9}{2} \cdot 0.25 \cdot 0.00390625$$
$$= 0.0439453125$$

$$B(10,1,0.50) = C(10,1) \cdot (0.50)^1 \cdot (1-0.50)^{10-1} = \frac{10!}{1!9!} \cdot 0.50 \cdot (0.50)^9 = 10 \cdot 0.50 \cdot 0.001953125$$
$$= 0.009765625$$

$$B(10,0,0.50) = C(10,0) \cdot (0.50)^0 \cdot (1-0.50)^{10-0} = \frac{10!}{0!10!} \cdot 1 \cdot (0.50)^{10} = 1 \cdot 1 \cdot 0.0009765625$$
$$= 0.0009765625$$

$0.205078125 + 0.1171875 + 0.0439453125 + 0.009765625 + 0.0009765625 = 0.376953125 \approx 0.3770$

32. $B(10,2,0.25) = C(10,2) \cdot (0.25)^2 \cdot (1-0.25)^{10-2} = \dfrac{10!}{2!8!} \cdot 0.0625 \cdot (0.75)^8$

$\approx \dfrac{10 \cdot 9}{2} \cdot 0.0625 \cdot 0.100112915039 \approx 0.281597573547$

$B(10,1,0.25) = C(10,1) \cdot (0.25)^1 \cdot (1-0.25)^{10-1} = \dfrac{10!}{1!9!} \cdot 0.25 \cdot (0.75)^9$

$\approx 10 \cdot 0.25 \cdot 0.075084686279 \approx 0.187711715698$

$B(10,0,0.25) = C(10,0) \cdot (0.25)^0 \cdot (1-0.25)^{10-0} = \dfrac{10!}{0!10!} \cdot 1 \cdot (0.75)^{10}$

$= 1 \cdot 1 \cdot 0.056313514709 \approx 0.056313514709$

$0.281567573547 + 0.187711715698 + 0.056313514709 = 0.525592803955$

$1 - 0.525592803955 = 0.474407196045 \approx 0.4744$

Chapter 14
DESCRIPTIVE STATISTICS: What A Data Set Tells Us

Section 14.1 Organizing And Visualizing Data

1. selection bias and leading-question bias

3. Example 4 dealt with weight, which is a continuous variable, so a bar graph would not be appropriate.

5. The population is the set of all objects being studied; a sample is a subset of the population.

7. A continuous variable can take on arbitrary values; a discrete variable cannot.

9.

x	Frequency
2	1
3	0
4	0
5	3
6	2
7	4
8	4
9	3
10	3

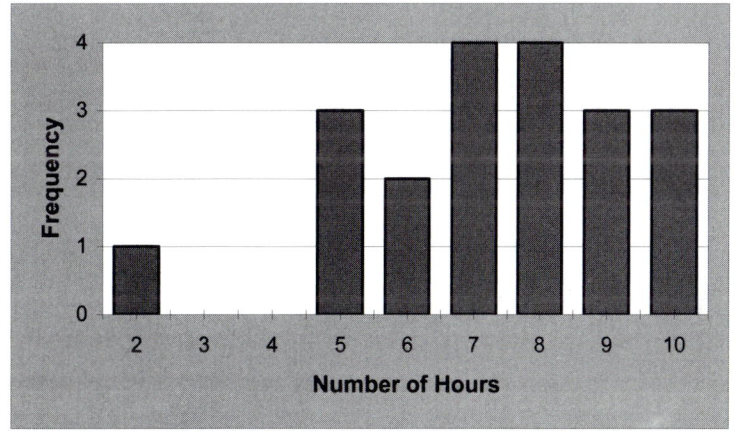

x	Relative Frequency
2	0.05
3	0.00
4	0.00
5	0.15
6	0.10
7	0.20
8	0.20
9	0.15
10	0.15

11.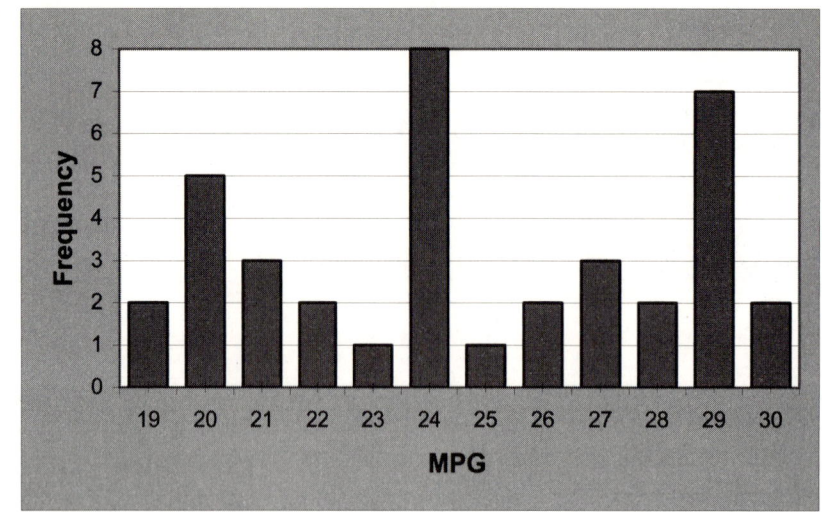

13.

x	Frequency
21	4
22	6
23	10
24	9
25	5
26	8
27	10
28	5
29	3

x	Relative Frequency	
21	4/60	0.067=6.7%
22	6/60=0.10=10%	
23	10/60	0.167=16.7%
24	9/60=0.15=15%	
25	5/60	0.083=8.3%
26	8/60	0.133=13.3%
27	10/60	0.167=16.7%
28	5/60	0.083=8.3%
29	3/60=0.05=5%	

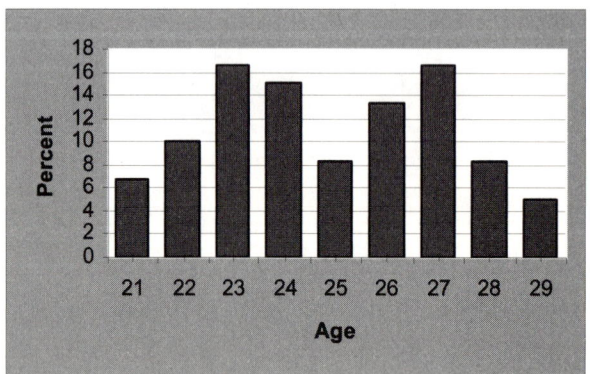

15.

x	Frequency
65.5+ to 67.5	1
67.5+ to 69.5	6
69.5+ to 71.5	8
71.5+ to 73.5	10
73.5+ to 75.5	8
75.5+ to 77.5	10
77.5+ to 79.5	0
79.5+ to 81.5	1

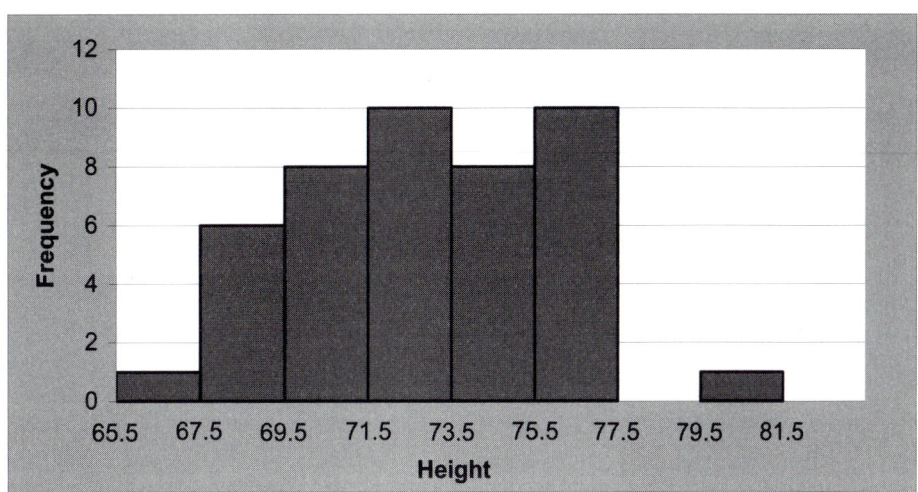

17.

```
       A            |   |       B
              8 8   | 1 | 3 6 8
          9 6 2 2 1 | 2 | 0 1 1 2 2 2 3 4 6 8 9
      9 9 8 7 4 3 2 | 3 | 2 3 3 8 9
          7 3 3 3 2 2 | 4 | 7
```

19.

x	Frequency
1.1+ to 1.2	2
1.2+ to 1.3	1
1.3+ to 1.4	1
1.4+ to 1.5	10
1.5+ to 1.6	5
1.6+ to 1.7	5
1.7+ to 1.8	3
1.8+ to 1.9	4
1.9+ to 1.2	1
2.0+ to 2.1	4

19. (continued)

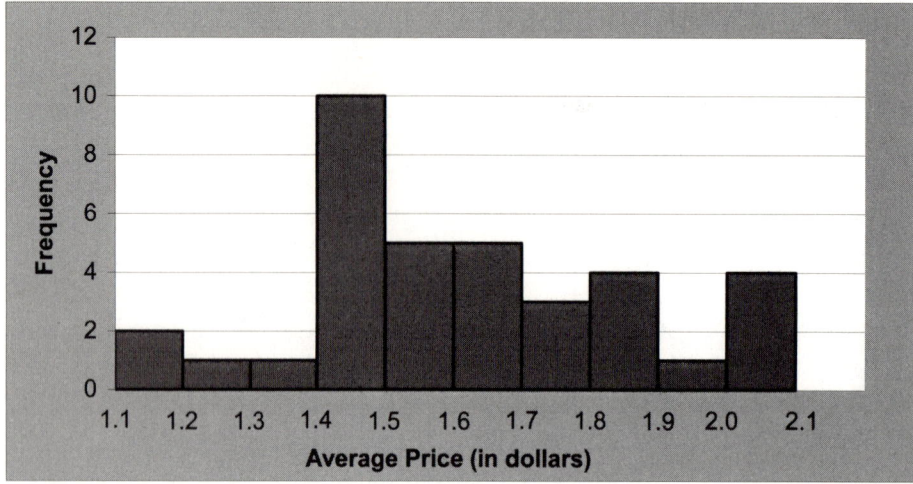

21. No solution provided

23. a) 0 occurred thirteen times
    b) 12 occurred two times
    c) 6
    d) $13 + 11 + 5 + 8 + 12 + 9 + 6 + 5 + 2 = 71$
    e) $\dfrac{9+6+5+2}{71} = \dfrac{22}{71}$

25. a) 5+3=8
    b) 3
    c) 5+3+4=12
    d) $3 + 2 + 3 + 5 + 3 = 16$

27. Answers may vary.
    This exercise was based on real–world data. An approximation of 10% would be reasonable. The exact value is 10.2%.

29. 25 and older, earning at least $10.00 per hour

31. 16 to 19, earning $5.15 or less per hour

33. The distribution is shifted to higher numbers, so it appears that the orientation program is succeeding.

    ```
           No Training   |   | With Training
                      9  | 1 |
            9 8 8 7 6 6 3 2 | 2 | 1 8 8 9 9
                      4 2 | 3 | 2 2 3 6 6 7 9
                  3 3 2 1 0 | 4 | 1 3 4 5
    ```

35. – 39. No solutions provided

41. 234, 245, 249, 257, 268, 277, 282, 323, 345, 362, 373, 378, 412, 434, 482, 489, 493

Section 14.2 Measures Of Central Tendency

1. They ignore the frequencies and divided by 6, which is the number of different scores.

3. A median is the middle strip between a divided highway. This helps us remember that the median is the middle score after the scores have been arranged in order.

5. the min, the first quartile, the median, the third quartile, and the max

7. Place the data in order.
   3, 4, 4, 5, 6, 7, 8, 9, 11
   Mean: $\bar{x} = \dfrac{\sum x}{n} = \dfrac{57}{9} \approx 6.3$; Median: 6; Mode: 4

9. Place the data in order.
   3, 4, 4, 6, 6, 7, 9, 9, 10, 11
   Mean: $\bar{x} = \dfrac{\sum x}{n} = \dfrac{69}{10} = 6.9$; Median: $\dfrac{6+7}{2} = \dfrac{13}{2} = 6.5$; Mode: no mode

11. Place the data in order.
    1, 2, 3, 4, 5, 5, 6, 7, 8, 9
    Mean: $\bar{x} = \dfrac{\sum x}{n} = \dfrac{50}{10} = 5$; Median: $\dfrac{5+5}{2} = \dfrac{10}{2} = 5$; Mode: 5

13. Place the data in order.
    2, 5, 5, 5, 6, 6, 7, 7, 7, 7, 8, 8, 8, 8, 9, 9, 9, 10, 10, 10
    Mean: $\bar{x} = \dfrac{\sum x}{n} = \dfrac{146}{20} = 7.3$; Median: $\dfrac{7+8}{2} = \dfrac{15}{2} = 7.5$; Mode: 7, 8

15. Place the data in order.
    21, 21, 22, 23, 24, 24, 24, 25, 25, 25, 26, 26, 26
    Mean: $\bar{x} = \dfrac{\sum x}{n} = \dfrac{312}{13} = 24$; Median: 24; Mode: no mode
    No explanation provided as to which measure is typical

17. Place this data in order.
    6.5, 7, 7, 7, 8, 18, 21, 23, 35, 40
    Mean: $\bar{x} = \dfrac{\sum x}{n} = \dfrac{172.5}{10} = 17.25$; Median: $\dfrac{8+18}{2} = 13$; Mode: 7
    No explanation provided as to which measure is typical

19. $\sum(x \cdot f) = 3 \cdot 2 + 4 \cdot 5 + 6 \cdot 3 + 7 \cdot 1 + 8 \cdot 4 + 10 \cdot 3 + 11 \cdot 2 = 6 + 20 + 18 + 7 + 32 + 30 + 22 = 135$
    $n = \sum f = 2 + 5 + 3 + 1 + 4 + 3 + 2 = 20$; Mean: $\bar{x} = \dfrac{\sum(x \cdot f)}{\sum f} = \dfrac{135}{20} = 6.75$
    10th piece of data: 6; 11th piece of data: 7; Median: $\dfrac{6+7}{2} = \dfrac{13}{2} = 6.5$; Mode: 4

21. $\sum(x \cdot f) = 5 \cdot 1 + 8 \cdot 3 + 11 \cdot 6 + 12 \cdot 2 + 14 \cdot 5 + 15 \cdot 4 + 18 \cdot 3 = 5 + 24 + 66 + 24 + 70 + 60 + 54 = 303$
    $n = \sum f = 1 + 3 + 6 + 2 + 5 + 4 + 3 = 24$; Mean: $\bar{x} = \dfrac{\sum(x \cdot f)}{\sum f} = \dfrac{303}{24} \approx 12.6$
    12th piece of data: 12; 13th piece of data: 14; Median: $\dfrac{12+14}{2} = \dfrac{26}{2} = 13$; Mode: 11

388 Chapter 14: Descriptive Statistics

23. a) Place the data in order.
    11, 17, 18, 23, 25, 26, 26, 31, 31, 33, 41, 44, 45, 48, 53
    lower half: 11, 17, 18, 23, 25, 26, 26
    upper half: 31, 33, 41, 44, 45, 48, 53
    minimum: 11
    first quartile is the median of the lower half: $Q_1 = 23$

    median: 31

    third quartile is the median of the upper half: $Q_3 = 44$

    maximum: 53

    b)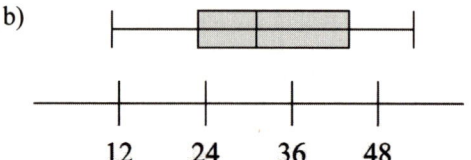

25. a) Place the data in order.
    25, 28, 29, 30, 31, 32, 33, 33, 34, 37, 38, 41, 41, 45, 49
    lower half: 25, 28, 29, 30, 31, 32, 33
    upper half: 34, 37, 38, 41, 41, 45, 49
    minimum: 25
    first quartile is the median of the lower half: $Q_1 = 30$

    median: 33

    third quartile is the median of the upper half: $Q_3 = 41$

    maximum: 49

    b)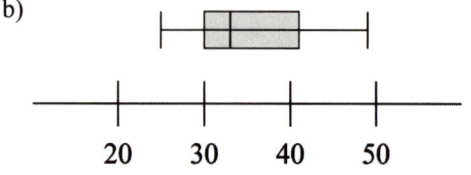

27. Place the data in order.
    0, 21, 30, 31, 37, 43, 44, 66, 78, 181, 250, 635

    Mean: $\bar{x} = \dfrac{\sum x}{n} = \dfrac{1{,}416}{12} = 118$; Median: $\dfrac{43+44}{2} = 43.5$; Mode: no mode

    No explanation provided as to which measure is typical

29. Let x be the incorrect final exam score. Assuming that the three exams are equally weighted we have
    $$\dfrac{84+86+x}{3} = 69$$
    $$\dfrac{170+x}{3} = 69$$
    $$170+x = 207$$
    $$x = 37.$$
    The correct grade was 73.

31. Mean: $\bar{x} = \dfrac{\sum x \cdot f}{\sum f} = \dfrac{1{,}435}{58} \approx 24.74$

Median: 25; You would need to add together the 29th and the 30th values and divide by two. Both of these numbers are 25.

Mode: 24, 25

No explanation provided as to which measure is typical

33. mean in both cases

35. a) $\dfrac{78+82+56+72+x}{5} = 70$

$\dfrac{288+x}{5} = 70$

$288 + x = 350$

$x = 62$

b) $\dfrac{78+82+56+72+x}{5} = 80$

$\dfrac{288+x}{5} = 80$

$288 + x = 400$

$x = 112$

37. and 39. No solutions provided

41. $\dfrac{9 \cdot 50 + 15 \cdot 125 + 7 \cdot 245}{31} = \dfrac{4{,}040}{31} \approx \$130.32$

43. 1975–1989: Place the data in order.

31, 36, 37, 37, 37, 38, 38, 39, 40, 40, 47, 48, 48, 49, 52

lower half: 31, 36, 37, 37, 37, 38, 38

upper half: 40, 40, 47, 48, 48, 49, 52

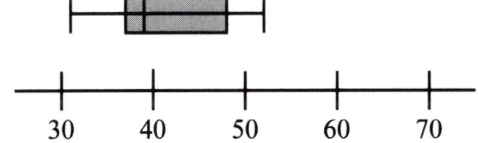

minimum: 31

first quartile is the median of the lower half: $Q_1 = 37$

median: 39

third quartile is the median of the upper half: $Q_3 = 48$

maximum: 52

1990–2004: Place the data in order.

35, 38, 40, 40, 43, 46, 47, 47, 48, 49, 49, 50, 65, 70, 73

lower half: 35, 38, 40, 40, 43, 46, 47

upper half: 48, 49, 49, 50, 65, 70, 73

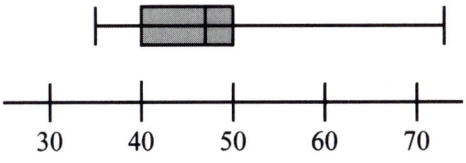

minimum: 35

first quartile is the median of the lower half: $Q_1 = 40$

median: 47

third quartile is the median of the upper half: $Q_3 = 50$

maximum: 73

45. liberal arts

47. education; no; No discussion provided

49. The maximum and minimum salaries are not too far from the middle 50% of the salaries.

51. – 59. No solutions provided

Section 14.3 Measures Of Dispersion

1. One outlier can make the range meaningless.

3. In calculating a sample's standard deviation, we divide by $n-1$; for a population, we divide by n.

5. Place data in order.
   17, 18, 18, 18, 19, 19, 20, 20, 21, 22
   Range: largest – smallest = 22 – 17 = 5
   Mean: $\bar{x} = \dfrac{\sum x}{n} = \dfrac{192}{10} = 19.2$

   Standard deviation:

Data Value	$x - \bar{x}$	$(x - \bar{x})^2$
17	17 – 19.2 = – 2.2	4.84
18	18 – 19.2 = – 1.2	1.44
18	18 – 19.2 = – 1.2	1.44
18	18 – 19.2 = – 1.2	1.44
19	19 – 19.2 = – 0.2	0.04
19	19 – 19.2 = – 0.2	0.04
20	20 – 19.2 = 0.8	0.64
20	20 – 19.2 = 0.8	0.64
21	21 – 19.2 = 1.8	3.24
22	22 – 19.2 = 2.8	7.84
Total	0	21.60

   $s = \sqrt{\dfrac{\sum(x-\bar{x})^2}{n-1}} = \sqrt{\dfrac{21.60}{10-1}} = \sqrt{\dfrac{21.60}{9}} \approx 1.55$

7. Place data in order.
   4, 5, 6, 7, 7, 8, 9, 10
   Range: largest – smallest = 10 – 4 = 6
   Mean: $\bar{x} = \dfrac{\sum x}{n} = \dfrac{56}{8} = 7$

   Standard deviation:

Data Value	$x - \bar{x}$	$(x - \bar{x})^2$
4	4 – 7 = – 3	9
5	5 – 7 = – 2	4
6	6 – 7 = – 1	1
7	7 – 7 = 0	0
7	7 – 7 = 0	0
8	8 – 7 = 1	1
9	9 – 7 = 2	4
10	10 – 7 = 3	9
Total	0	28

   $s = \sqrt{\dfrac{\sum(x-\bar{x})^2}{n-1}} = \sqrt{\dfrac{28}{8-1}} = \sqrt{\dfrac{28}{7}} = 2$

9. Place data in order.

   2, 3, 5, 8, 9, 11, 12, 14

   Range: largest − smallest = 14 − 2 = 12

   Mean: $\bar{x} = \dfrac{\sum x}{n} = \dfrac{64}{8} = 8$

   Standard deviation:

Data Value	$x - \bar{x}$	$(x - \bar{x})^2$
2	2 − 8 = −6	36
3	3 − 8 = −5	25
5	5 − 8 = −3	9
8	8 − 8 = 0	0
9	9 − 8 = 1	1
11	11 − 8 = 3	9
12	12 − 8 = 4	16
14	14 − 8 = 6	36
Total	0	132

   $s = \sqrt{\dfrac{\sum(x-\bar{x})^2}{n-1}} = \sqrt{\dfrac{132}{8-1}} = \sqrt{\dfrac{132}{7}} \approx 4.34$

11. Place data in order.

    3, 7, 7, 7, 8, 9, 13, 18

    Range: largest − smallest = 18 − 3 = 15

    Mean: $\bar{x} = \dfrac{\sum x}{n} = \dfrac{72}{8} = 9$

    Standard deviation:

Data Value	$x - \bar{x}$	$(x - \bar{x})^2$
3	3 − 9 = −6	36
7	7 − 9 = −2	4
7	7 − 9 = −2	4
7	7 − 9 = −2	4
8	8 − 9 = −1	1
9	9 − 9 = 0	0
13	13 − 9 = 4	16
18	18 − 9 = 9	81
Total	0	146

    $s = \sqrt{\dfrac{\sum(x-\bar{x})^2}{n-1}} = \sqrt{\dfrac{146}{8-1}} = \sqrt{\dfrac{146}{7}} \approx 4.57$

13. Place data in order.

    3, 3, 3, 3, 3, 3, 3

    Range: largest − smallest = 3 − 3 = 0

    Mean: $\bar{x} = \dfrac{\sum x}{n} = \dfrac{21}{7} = 3$

    Standard deviation:

Data Value	$x - \bar{x}$	$(x - \bar{x})^2$
3	3 − 3 = 0	0
3	3 − 3 = 0	0
3	3 − 3 = 0	0
3	3 − 3 = 0	0
3	3 − 3 = 0	0
3	3 − 3 = 0	0
3	3 − 3 = 0	0
Total	0	0

    $s = \sqrt{\dfrac{\sum(x-\bar{x})^2}{n-1}} = \sqrt{\dfrac{0}{7-1}} = \sqrt{\dfrac{0}{6}} = 0$

15. $\bar{x} = \dfrac{\sum x \cdot f}{\sum f} = \dfrac{140}{20} = 7$ and $s = \sqrt{\dfrac{\sum (x-\bar{x})^2 \cdot f}{n-1}} = \sqrt{\dfrac{86}{20-1}} = \sqrt{\dfrac{86}{19}} \approx 2.13$

Number x	Frequency f	Product $x \cdot f$	Deviation $x - \bar{x}$	Deviation^2 $(x-\bar{x})^2$	Product $(x-\bar{x})^2 \cdot f$
2	1	2	2 − 7 = − 5	25	25·1 = 25
3	1	3	3 − 7 = − 4	16	16·1 = 16
4	0	0	4 − 7 = − 3	9	9·0 = 0
5	2	10	5 − 7 = − 2	4	4·2 = 8
6	3	18	6 − 7 = − 1	1	1·3 = 3
7	4	28	7 − 7 = 0	0	0·4 = 0
8	4	32	8 − 7 = 1	1	1·4 = 4
9	3	27	9 − 7 = 2	4	4·3 = 12
10	2	20	10 − 7 = 3	9	9·2 = 18
$\sum f = 20$		$\sum x \cdot f = 140$			$\sum (x-\bar{x})^2 \cdot f = 86$

17. $\bar{x} = \dfrac{\sum x \cdot f}{\sum f} = \dfrac{48}{12} = 4$ and $s = \sqrt{\dfrac{\sum (x-\bar{x})^2 \cdot f}{n-1}} = \sqrt{\dfrac{28}{12-1}} = \sqrt{\dfrac{28}{11}} \approx 1.60$

Number x	Frequency f	Product $x \cdot f$	Deviation $x - \bar{x}$	Deviation^2 $(x-\bar{x})^2$	Product $(x-\bar{x})^2 \cdot f$
2	2	4	2 − 4 = − 2	4	4·2 = 8
3	4	12	3 − 4 = − 1	1	1·4 = 4
4	2	8	4 − 4 = 0	0	0·2 = 0
5	0	0	5 − 4 = 1	1	1·0 = 0
6	4	24	6 − 4 = 2	4	4·4 = 16
$\sum f = 12$		$\sum x \cdot f = 48$			$\sum (x-\bar{x})^2 \cdot f = 28$

19. $\bar{x} = \dfrac{\sum x \cdot f}{\sum f} = \dfrac{105}{15} = 7$ and $s = \sqrt{\dfrac{\sum (x-\bar{x})^2 \cdot f}{n-1}} = \sqrt{\dfrac{42}{15-1}} = \sqrt{\dfrac{42}{14}} \approx 1.73$

Number x	Frequency f	Product $x \cdot f$	Deviation $x - \bar{x}$	Deviation^2 $(x-\bar{x})^2$	Product $(x-\bar{x})^2 \cdot f$
3	1	3	3 − 7 = − 4	16	16·1 = 16
4	1	4	4 − 7 = − 3	9	9·1 = 9
5	0	0	5 − 7 = − 2	4	4·0 = 0
6	2	12	6 − 7 = − 1	1	1·2 = 2
7	5	35	7 − 7 = 0	0	0·5 = 0
8	3	24	8 − 7 = 1	1	1·3 = 3
9	3	27	9 − 7 = 2	4	4·3 = 12
$\sum f = 15$		$\sum x \cdot f = 105$			$\sum (x-\bar{x})^2 \cdot f = 42$

21. $\bar{x} = \dfrac{\sum x \cdot f}{\sum f} = \dfrac{1{,}580}{20} = 79$ and $s = \sqrt{\dfrac{\sum (x-\bar{x})^2 \cdot f}{n-1}} = \sqrt{\dfrac{826}{20-1}} = \sqrt{\dfrac{826}{19}} \approx 6.59$

Number x	Frequency f	Product $x \cdot f$	Deviation $x - \bar{x}$	Deviation^2 $(x-\bar{x})^2$	Product $(x-\bar{x})^2 \cdot f$
72	7	504	$72 - 79 = -7$	49	$49 \cdot 7 = 343$
73	1	73	$73 - 79 = -6$	36	$36 \cdot 1 = 36$
78	3	234	$78 - 79 = -1$	1	$1 \cdot 3 = 3$
84	6	504	$84 - 79 = 5$	25	$25 \cdot 6 = 150$
86	2	172	$86 - 79 = 7$	49	$49 \cdot 2 = 98$
93	1	93	$93 - 79 = 14$	196	$196 \cdot 1 = 196$
$\sum f = 20$		$\sum x \cdot f = 1{,}580$			$\sum (x-\bar{x})^2 \cdot f = 826$

23. $\bar{x} = \dfrac{\sum x}{n} = \dfrac{1{,}877}{11} \approx 170.64$ and $s = \sqrt{\dfrac{\sum (x-\bar{x})^2}{n-1}} = \sqrt{\dfrac{1{,}336.56}{11-1}} = \sqrt{\dfrac{1{,}336.56}{10}} \approx 11.56$

Number x	Deviation $x - \bar{x}$	Deviation^2 $(x-\bar{x})^2$
198	$148 - 170.6 = -22.6$	510.76
206	$160 - 170.6 = -10.6$	112.36
190	$172 - 170.6 = 1.4$	1.96
148	$170 - 170.6 = -0.6$	0.36
160	$165 - 170.6 = -5.6$	31.36
170	$169 - 170.6 = -1.6$	2.56
165	$169 - 170.6 = -1.6$	2.56
125	$180 - 170.6 = 9.4$	88.36
172	$169 - 170.6 = -1.6$	2.56
160	$185 - 170.6 = 14.4$	207.36
165	$190 - 170.6 = 19.4$	376.36
$\sum x = 1{,}877$		$\sum (x-\bar{x})^2 = 1{,}336.56$

25. $\bar{x} = \dfrac{\sum x \cdot f}{\sum f} = \dfrac{313}{71} \approx 4.41$ and $s = \sqrt{\dfrac{\sum(x-\bar{x})^2 \cdot f}{n-1}} = \sqrt{\dfrac{801.1551}{71-1}} = \sqrt{\dfrac{801.1551}{70}} \approx 3.38$

Number x	Frequency f	Product $x \cdot f$	Deviation $x - \bar{x}$	Deviation^2 $(x-\bar{x})^2$	Product $(x-\bar{x})^2 \cdot f$
0	13	0	0 − 4.41 = − 4.41	19.4481	252.8253
1	11	11	1 − 4.41 = − 3.41	11.6281	127.9091
2	5	10	2 − 4.41 = − 2.41	5.8081	29.0405
3	0	0	3 − 4.41 = − 1.41	1.9881	0
4	0	0	4 − 4.41 = − 0.41	0.1681	0
5	8	40	5 − 4.41 = 0.59	0.3481	2.7848
6	12	72	6 − 4.41 = 1.59	2.5281	30.3372
7	9	63	7 − 4.41 = 2.59	6.7081	60.3729
8	6	48	8 − 4.41 = 3.59	12.8881	77.3286
9	5	45	9 − 4.41 = 4.59	21.0681	105.3405
10	0	0	10 − 4.41 = 5.59	31.2481	0
11	0	0	11 − 4.41 = 6.59	43.4281	0
12	2	24	12 − 4.41 = 7.59	57.6081	115.2162
	$\sum f = 71$	$\sum x \cdot f = 313$			$\sum(x-\bar{x})^2 \cdot f = 801.1551$

27. False; The mean is 0, but the standard deviation is not. Additional explanations may vary.

$\bar{x} = \dfrac{\sum x}{n} = \dfrac{0}{8} = 0$ and $s = \sqrt{\dfrac{\sum(x-\bar{x})^2}{n-1}} = \sqrt{\dfrac{32}{8-1}} = \sqrt{\dfrac{32}{7}} \approx 2.14$

Number x	Deviation $x - \bar{x}$	Deviation^2 $(x-\bar{x})^2$
−2	−2 − 0 = − 2	4
2	2 − 0 = 2	4
−2	−2 − 0 = − 2	4
2	2 − 0 = 2	4
−2	−2 − 0 = − 2	4
2	2 − 0 = 2	4
−2	−2 − 0 = − 2	4
2	2 − 0 = 2	4
$\sum x = 0$		$\sum(x-\bar{x})^2 = 32$

29. True; They both do have the same standard deviation. Additional explanations may vary.

$$\bar{x} = \frac{\sum x}{n} = \frac{35}{5} = 7 \text{ and } s = \sqrt{\frac{\sum(x-\bar{x})^2}{n-1}} = \sqrt{\frac{10}{5-1}} = \sqrt{\frac{10}{4}} \approx 1.58$$

Number x	Deviation $x - \bar{x}$	Deviation² $(x-\bar{x})^2$
5	$5 - 7 = -2$	4
6	$6 - 7 = -1$	1
7	$7 - 7 = 0$	0
8	$8 - 7 = 1$	1
9	$9 - 7 = 2$	4
$\sum x = 35$		$\sum(x-\bar{x})^2 = 10$

$$\bar{x} = \frac{\sum x}{n} = \frac{135}{5} = 27 \text{ and } s = \sqrt{\frac{\sum(x-\bar{x})^2}{n-1}} = \sqrt{\frac{10}{5-1}} = \sqrt{\frac{10}{4}} \approx 1.58$$

Number x	Deviation $x - \bar{x}$	Deviation² $(x-\bar{x})^2$
25	$25 - 27 = -2$	4
26	$26 - 27 = -1$	1
27	$27 - 27 = 0$	0
28	$28 - 27 = 1$	1
29	$29 - 27 = 2$	4
$\sum x = 135$		$\sum(x-\bar{x})^2 = 10$

31. $\bar{x} = \dfrac{\sum x}{n} = \dfrac{392}{8} = 49$ and $s = \sqrt{\dfrac{\sum(x-\bar{x})^2}{n-1}} = \sqrt{\dfrac{18}{8-1}} = \sqrt{\dfrac{18}{7}} \approx 1.60$

Number x	Deviation $x - \bar{x}$	Deviation² $(x-\bar{x})^2$
47	$47 - 49 = -2$	4
48	$48 - 49 = -1$	1
50	$50 - 49 = 1$	1
49	$49 - 49 = 0$	0
51	$51 - 49 = 2$	4
47	$47 - 49 = -2$	4
49	$49 - 49 = 0$	0
51	$51 - 49 = 2$	4
$\sum x = 392$		$\sum(x-\bar{x})^2 = 18$

$\dfrac{51 - 49}{1.6} = 1.25$; Family H's income is 1.25 standard deviations above the mean.

33. $\bar{x} = \dfrac{\sum x}{n} = \dfrac{800}{10} = 80$ and $s = \sqrt{\dfrac{\sum(x-\bar{x})^2}{n-1}} = \sqrt{\dfrac{184}{10-1}} = \sqrt{\dfrac{184}{9}} \approx 4.52$

Number x	Deviation $x - \bar{x}$	Deviation^2 $(x-\bar{x})^2$
80	80 − 80 = 0	0
76	76 − 80 = −4	16
81	81 − 80 = 1	1
84	84 − 80 = 4	16
79	79 − 80 = −1	1
80	80 − 80 = 0	0
90	90 − 80 = 10	100
75	75 − 80 = −5	25
75	75 − 80 = −5	25
80	80 − 80 = 0	0
$\sum x = 800$		$\sum(x-\bar{x})^2 = 184$

$\dfrac{76-80}{4.52} \approx -0.88$; A grade of 76 is 0.88 standard deviations below the mean. This would be a grade of "D".

35. Magnum Industrial is more volatile.

Magnum Industrial: $CV = \dfrac{s}{\bar{x}} \cdot 100\% = \dfrac{12.3}{123.76} \cdot 100\% \approx 9.9\%$

GrandCore Incorporated: $CV = \dfrac{s}{\bar{x}} \cdot 100\% = \dfrac{7.2}{78.6} \cdot 100\% \approx 9.2\%$

37. WebMaster is more volatile.

DJIA: $CV = \dfrac{s}{\bar{x}} \cdot 100\% = \dfrac{72.17}{11261.12} \cdot 100\% \approx 0.6\%$;

WebMaster: $CV = \dfrac{s}{\bar{x}} \cdot 100\% = \dfrac{1.7}{37.6} \cdot 100\% \approx 4.5\%$

39. Coffee

$\bar{x} = \dfrac{\sum x}{n} = \dfrac{34.17}{12} \approx 2.85$ and $s = \sqrt{\dfrac{\sum(x-\bar{x})^2}{n-1}} = \sqrt{\dfrac{0.0333}{12-1}} = \sqrt{\dfrac{0.0333}{11}} \approx 0.055$

Price x	Deviation $x - \bar{x}$	Deviation^2 $(x-\bar{x})^2$
2.89	2.89 − 2.85 = 0.04	0.0016
2.86	2.86 − 2.85 = 0.01	0.0001
2.93	2.93 − 2.85 = 0.08	0.0064
2.91	2.91 − 2.85 = 0.06	0.0036
2.83	2.83 − 2.85 = −0.02	0.0004
2.75	2.75 − 2.85 = −0.10	0.0100
2.88	2.88 − 2.85 = 0.03	0.0009
2.87	2.87 − 2.85 = 0.02	0.0004
2.84	2.84 − 2.85 = −0.01	0.0001
2.78	2.78 − 2.85 = −0.07	0.0049
2.78	2.78 − 2.85 = −0.07	0.0049
2.85	2.85 − 2.85 = 0.00	0.0000
$\sum x = 34.17$		$\sum(x-\bar{x})^2 = 0.0333$

39. (continued)

Gasoline

$$\bar{x} = \frac{\sum x}{n} = \frac{22.56}{12} = 1.88 \text{ and } s = \sqrt{\frac{\sum(x-\bar{x})^2}{n-1}} = \sqrt{\frac{0.2288}{12-1}} = \sqrt{\frac{0.2288}{11}} \approx 0.144$$

Price x	Deviation $x - \bar{x}$	Deviation^2 $(x-\bar{x})^2$
1.59	1.59 − 1.88 = −0.29	0.0841
1.67	1.67 − 1.88 = −0.21	0.0441
1.77	1.77 − 1.88 = −0.11	0.0121
1.83	1.83 − 1.88 = −0.05	0.0025
2.01	2.01 − 1.88 = 0.13	0.0169
2.04	2.04 − 1.88 = 0.16	0.0256
1.94	1.94 − 1.88 = 0.06	0.0036
1.90	1.90 − 1.88 = 0.02	0.0004
1.89	1.89 − 1.88 = 0.01	0.0001
2.03	2.03 − 1.88 = 0.15	0.0225
2.01	2.01 − 1.88 = 0.13	0.0169
1.88	1.88 − 1.88 = 0.00	0.0000

$\sum x = 22.56$ $\sum(x-\bar{x})^2 = 0.2288$

41. G went from the mean income to 0.935 (0.94 rounded) standard deviations above the mean.

Family	Year One $\dfrac{x-\bar{x}}{s}$	Year Two $\dfrac{x-\bar{x}}{s}$	Change in Standard Deviations
A	$\dfrac{47-49}{1.60} = -1.25$	$\dfrac{49-51}{1.07} \approx -1.87$	Down 0.62
B	$\dfrac{48-49}{1.60} = -0.625$	$\dfrac{51-51}{1.07} = 0$	Up 0.625
C	$\dfrac{50-49}{1.60} = 0.625$	$\dfrac{52-51}{1.07} \approx 0.935$	Up 0.31
D	$\dfrac{49-49}{1.60} = 0$	$\dfrac{51-51}{1.07} = 0$	No Change
E	$\dfrac{51-49}{1.60} = 1.25$	$\dfrac{51-51}{1.07} = 0$	Down 1.25
F	$\dfrac{47-49}{1.60} = -1.25$	$\dfrac{50-51}{1.07} \approx -0.935$	Up 0.315
G	$\dfrac{49-49}{1.60} = 0$	$\dfrac{52-51}{1.07} \approx 0.935$	**Up 0.935**
H	$\dfrac{51-49}{1.60} = 1.25$	$\dfrac{52-51}{1.07} \approx 0.935$	Down 0.315

43. − 47. No solutions provided

Section 14.4 The Normal Distribution

1. See box on page 821.

3. The normal curve is symmetric about the mean.

5. Since 12 is one standard deviation above the mean, we would expect $\frac{68\%}{2} = 34\%$ of the values to lie between 10 and 12.

7. Since 14 is two standard deviations above the mean, we would expect $\frac{95\%}{2} = 47.5\%$ of the values to lie between 10 and 14. Therefore, 50% – 47.5%=2.5% of the values should fall above 14.

9. Since 12 is one standard deviation above the mean, we would expect $\frac{68\%}{2} = 34\%$ of the values to lie between 10 and 12. Therefore, 50% – 34%=16% of the values should fall above 12.

11. Since 9 is one standard deviation below the mean, we would expect $\frac{68\%}{2} = 34\%$ of the values to lie between 9 and 12. Therefore, 50% – 34%=16% of the values should fall below 9.

13. Since 6 is two standard deviations below the mean, we would expect $\frac{95\%}{2} = 47.5\%$ of the values to lie between 6 and 12. Therefore, 50% + 47.5%=97.5% of the values should fall above 6.

15. Since 18 is two standard deviations above the mean, we would expect $\frac{95\%}{2} = 47.5\%$ of the values to lie between 12 and 18. Similarly, since 15 is one standard deviation above the mean, we would expect $\frac{68\%}{2} = 34\%$ of the values to lie between 12 and 15. Therefore, we would expect 47.5% – 34%=13.5% of the values to lie between 15 and 18.

17. 39.1%

19. 42.4% – 31.3% = 11.1%

21. 27.3%

23. 47.4% – 39.4% = 8%

25. 27.6% –14.8% = 12.8%

27. 50% – 42.7% = 7.3%

29. 50% – 41.9% = 8.1%

31. 50% + 40.8% = 90.8%

33. 50% + 30.0% = 80%

35. We need to find a z–score that corresponds to 0.5 – 0.1 = 0.4. The closest z–score is 1.28.

37. We need to find a z–score that corresponds to 0.5 – 0.12 = 0.38. Since 0.38 is halfway between 0.379 and 0.381, the z–score would be 1.175. This z–score needs to be –1.175, however, since it indicates an area less than 50% to the left of the mean.

39. We need to find a z–score that corresponds to 0.6 – 0.5 = 0.1. Since 0.1 is one-fourth of the way between 0.099 and 0.103, the z–score would be 0.2525.

41. We need to find a z–score that corresponds to 0.75 – 0.5 = 0.25. Since 0.25 is one-third of the way between 0.249 and 0.252, the z–score would be about 0.6733. This z–score needs to be – 0.6733 however, since it indicates an area more than 50%.

43. $z = \dfrac{87-80}{5} = 1.40$

45. $z = \dfrac{14-21}{4} = -1.75$

47. $z = \dfrac{48-38}{10.3} \approx 0.97$

49. $0.84 = \dfrac{x-60}{5}$
    $4.2 = x - 60$
    $64.2 = x$

51. $-0.45 = \dfrac{x-35}{3}$
    $-1.35 = x - 35$
    $33.65 = x$

53. $1.64 = \dfrac{x-28}{2.25}$
    $3.69 = x - 28$
    $31.69 = x$

55. a) 50%

    b) Since 7.5 is one standard deviation below the mean, we would expect $\dfrac{68\%}{2} = 34\%$ of the values to lie between 7.5 and 8. Therefore, 50% – 34% = 16% of the cups should have less that 7.5 ounces.

57. a) Since 202 is one standard deviation above the mean, we would expect $\dfrac{68\%}{2} = 34\%$ of the bags to have between 200 and 202 pieces in them.

    b) 50% – 34% = 16%

59. The score is almost two standard deviations above the mean.

61. By looking at the graph of the normal curve, there is clearly more area under the curve between 0.0 and 0.7 than there is between 1.3 and 2.0.

63. $z = \dfrac{140-120}{12} \approx 1.67$

65. 6 feet 8 inches corresponds to 80 inches and 7 feet corresponds to 84 inches. $z = \dfrac{84-80}{3} \approx 1.33$, and the percent of area that corresponds to players between 80 inches and 84 inches is 40.8%. Therefore, 50% – 40.8% = 9.2% have a height over 7 feet. Since $0.092 \cdot 324 = 29.808$, we would expect about 29 or 30 players out of the 324 to be over 7 feet.

67. $z = \dfrac{70-68}{4} = 0.50$, and this corresponds to an area of 19.2% + 50% = 69.2% for women with heart rates of less than 70 beats per minute. Since $0.692 \cdot 200 = 138.4$, we would expect 138 or 139 out of the 200 women to have heart rates of less that 70 beats per minute.

69. Since roughly 95% of data values are within 2 standard deviations of the mean, the standard deviation should be 9 since 250 is 18 below 268 and 286 is 18 above 268. We now need to find the z–score associated with 275 days. $z = \dfrac{275-268}{9} \approx 0.78$ and therefore, the percentage of pregnancies that would be expected to last at least 275 days would be 50% – 28.2% = 21.8%.

71. $1.5 = \dfrac{x-37}{11}$

$16.5 = x - 37$

$x = 53.5$

73. Carew was more dominant.

Jackie Robinson: $z = \dfrac{0.342 - 0.267}{0.0326} \approx 2.30$ above the mean

Rod Carew: $z = \dfrac{0.350 - 0.261}{0.0317} \approx 2.81$ above the mean

75. The z–score that corresponds to 80% – 50% = 30% = 0.300 is 0.84.

77. The z–score that approximately corresponds to 75% – 50% = 25% = 0.250 is 0.673.

$0.673 = \dfrac{x - 40}{4} \Rightarrow 2.692 = x - 40 \Rightarrow x = 42.692$

79. $z = \dfrac{75 - 68}{4} = 1.75$ corresponds to an accumulated area of 50% + 46% = 96%. This implies the 96th percentile.

81. The z–score that corresponds to 50% – 4% = 46% = 0.46 is 1.75. Moreover, this z–score should be –1.75, since the accumulated area is less than 50%.

$-1.75 = \dfrac{x - 2,000}{800} \Rightarrow -1,400 = x - 2,000 \Rightarrow x = 600$ hours of play. Since a game is played for about 2 hours a day, the number of days the game is played is $\dfrac{600}{2} = 300$. This implies that the warrantee should be for about 10 months.

83. a) $z = \dfrac{9 - 7.8}{1.3} \approx 0.92$ and this corresponds to an area of 50% – 32.1% = 17.9%. So of the past 15 years, you would expect to receive at least 9% on the investments $0.179 \cdot 15 = 2.685$ or 2 or 3 years.

b) $z = \dfrac{6 - 7.8}{1.3} \approx -1.38$, and this corresponds to an area of 50% – 41.6% = 8.4%. So of the past 15 years, you would expect to receive less than 6% on the investments $0.084 \cdot 15 = 1.26$ or 1 or 2 years.

85. $z = \dfrac{480 - 500}{100} = -0.20$, and this corresponds to 50% – 7.9% = 42.1%.

87. Answers may vary.
The mean is 20 in both distributions; standard deviation is about 2 or 3 in distribution 1; standard deviation is about 4 or 5 in distribution 2. The estimation for the means comes from determining for what value the curves are symmetric about. The estimation for the standard deviations comes from applying the properties that govern how much data will fall within 1, 2, or 3 standard deviations from the mean.

89. and 91. No solutions provided

Chapter Review Exercises

1.
Number of Accidents	Frequency
4	1
5	3
6	6
7	2
8	10
9	4
10	5

2.
Number of Accidents	Relative Frequency
4	0.03
5	0.10
6	0.19
7	0.06
8	0.32
9	0.13
10	0.16

3.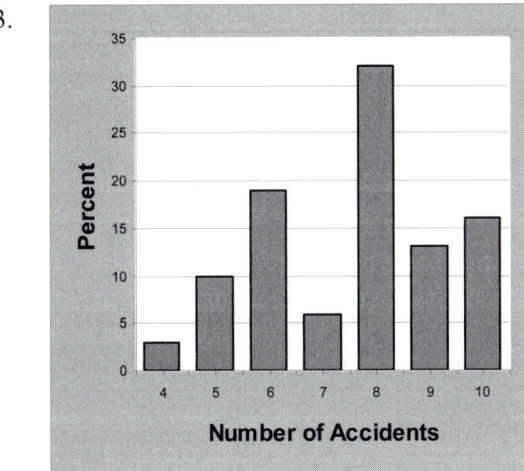

4. a) Four riders occurred three times.
   b) five riders
   c) $3 + 8 + 6 + 7 + 3 + 5 + 3 + 2 = 37$

5. The programs are much the same in their effectiveness. B might be a little better.

```
       A       |   |      B
               9 | 1 |
         98766 5 | 2 | 1 6 6 8 8 9
       7 6 4 3 2 2 1 | 3 | 2 3 6 6 7 9 9
           7 3 0 | 4 | 1 3 4 5
               2 | 5 | 3
```

6. Place the data in order.
   4, 6, 6, 6, 7, 7, 7, 7, 7, 8, 9, 10, 10, 10, 12, 13, 13

   Mean: $\bar{x} = \dfrac{\sum x}{n} = \dfrac{142}{17} \approx 8.35$

   Median: 7
   Mode: 7

402 Chapter 14: Descriptive Statistics

7. The mean is the arithmetic average; the median is the middle score; the mode is the most frequent score.

8. Place the data in order.
   3, 4, 6, 6, 6, 7, 7, 8, 9, 10, 11, 12, 13, 17, 20
   lower half: 3, 4, 6, 6, 6, 7, 7
   upper half: 9, 10, 11, 12, 13, 17, 20
   minimum: 3
   first quartile is the median of the lower half: $Q_1 = 6$
   median: 8
   third quartile is the median of the upper half: $Q_3 = 12$
   maximum: 20

9.

x	Frequency
4	3
5	8
6	6
7	7
8	3
9	5
10	0
11	3
12	2

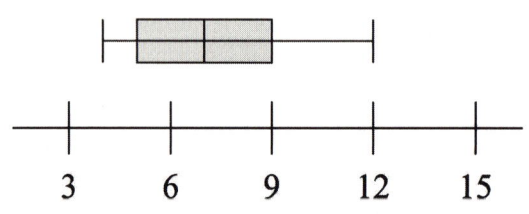

lower half: The first 18 values are in the lower half; 4, 4, 4, 5, 5, 5, 5, 5, 5, 5, 5, 6, 6, 6, 6, 6, 6, 7
upper half: The last 18 values are in the upper half; 7, 7, 7, 7, 7, 8, 8, 8, 9, 9, 9, 9, 9, 11, 11, 11, 12, 12
minimum: 4
first quartile is the median of the lower half: $Q_1 = \dfrac{5+5}{2} = \dfrac{10}{2} = 5$
median: 7
third quartile is the median of the upper half: $Q_3 = \dfrac{9+9}{2} = \dfrac{18}{2} = 9$
maximum: 12

10. $\bar{x} = \dfrac{\sum x}{n} = \dfrac{40}{8} = 5$ and $s = \sqrt{\dfrac{\sum(x-\bar{x})^2}{n-1}} = \sqrt{\dfrac{12}{8-1}} = \sqrt{\dfrac{12}{7}} \approx 1.31$

Number x	Deviation $x - \bar{x}$	Deviation^2 $(x-\bar{x})^2$
4	$4-5=-1$	1
6	$6-5=1$	1
7	$7-5=2$	4
3	$3-5=-2$	4
5	$5-5=0$	0
6	$6-5=1$	1
4	$4-5=-1$	1
5	$5-5=0$	0
$\sum x = 40$		$\sum(x-\bar{x})^2 = 12$

11. the spread of the distribution

12. See box on page 821.

13. $z = \dfrac{85-80}{7} \approx 0.71$

14. $1.35 = \dfrac{x-60}{5}$
    $6.75 = x - 60$
    $x = 66.75$

15. The history exam score is relatively better, since it has a higher z-score.

    History: $z = \dfrac{82-78}{3} \approx 1.33$

    Anthropology: $z = \dfrac{84-79}{4} = 1.25$

16. No solution provided

Chapter Test

1.

Number of Visits	Frequency
4	1
5	4
6	8
7	9
8	7
9	6
10	2
11	2
12	1

2.

Number of Visits	Relative Frequency
4	0.025
5	0.100
6	0.200
7	0.225
8	0.175
9	0.150
10	0.050
11	0.050
12	0.025

3.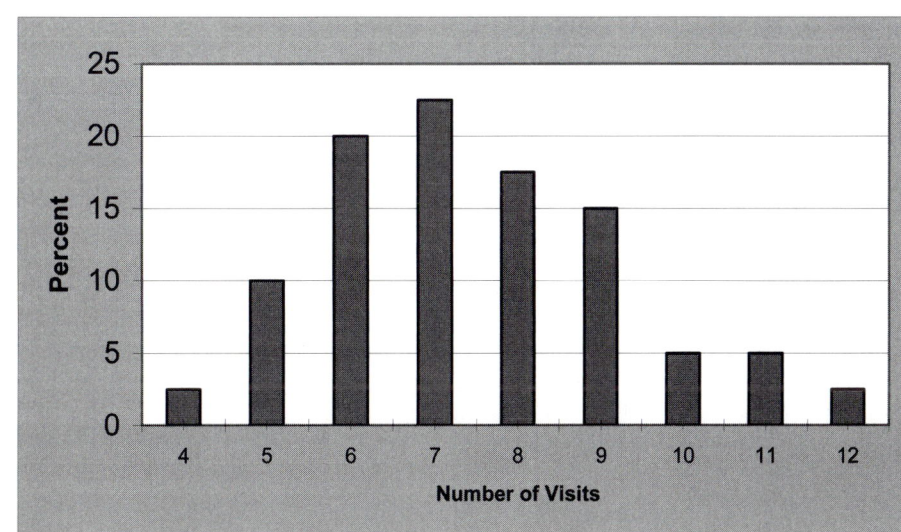

404   Chapter 14: Descriptive Statistics

4.
```
     Ruth           Aaron
        5 2 | 2 |
          5 4 | 3 | 0 2 4 4 8 9 9
    9 7 6 6 6 1 1 | 4 | 0 0 4 4 4 4 5 7
          9 4 4 | 5 |
              0 | 6 |
```

5. a) Three hits per minute occurred seven times.
    b) four and six hits per minute
    c) $7 + 8 + 7 + 8 + 5 + 3 = 38$

6. The mean is the arithmetic average; the median is the middle score; the mode is the most frequent score.

7. Place the data in order.
    3, 5, 5, 6, 6, 6, 7, 7, 8, 8, 9, 9, 9, 9, 9, 9, 10, 10, 11, 11

    Mean: $\bar{x} = \dfrac{\sum x}{n} = \dfrac{157}{20} = 7.85$

    Median: $\dfrac{8+9}{2} = \dfrac{17}{2} = 8.5$

    Mode: 9

8. Place the data in order.
    2, 3, 3, 4, 5, 6, 6, 7, 8, 9, 9, 9, 10, 11, 12, 14, 17, 20
    lower half: 2, 3, 3, 4, 5, 6, 6, 7, 8
    upper half: 9, 9, 9, 10, 11, 12, 14, 17, 20
    minimum: 2
    first quartile is the median of the lower half: $Q_1 = 5$

    median: $\dfrac{8+9}{2} = \dfrac{17}{2} = 8.5$

    third quartile is the median of the upper half: $Q_3 = 11$
    maximum: 20

9.
    minimum: 2
    first quartile: $Q_1 = 5$
    median: 8.5
    third quartile: $Q_3 = 11$
    maximum: 20

    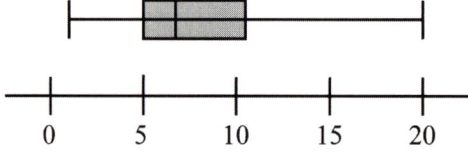

10. the spread of the distribution

11. $\bar{x} = \dfrac{\sum x}{n} = \dfrac{54}{9} = 6$ and $s = \sqrt{\dfrac{\sum(x-\bar{x})^2}{n-1}} = \sqrt{\dfrac{32}{9-1}} = \sqrt{\dfrac{32}{8}} = 2$

Number x	Deviation $x - \bar{x}$	Deviation^2 $(x-\bar{x})^2$
3	3 − 2 = −3	9
4	4 − 2 = −2	4
5	5 − 2 = −1	1
6	6 − 2 = 0	0
5	5 − 2 = −1	1
6	6 − 2 = 0	0
8	8 − 2 = 2	4
9	9 − 2 = 3	9
8	8 − 2 = 2	4
$\sum x = 54$		$\sum(x-\bar{x})^2 = 32$

12. See box on page 821.

13. The statistics exam score is relatively better, since it has a higher z-score.

    Statistics: $z = \dfrac{85 - 78}{4} = 1.75$

    Sociology: $z = \dfrac{88 - 80}{5} = 1.60$

14. $1.83 = \dfrac{x - 50}{6}$

    $10.98 = x - 50$

    $x = 60.98 \approx 61$

15. $z = \dfrac{82 - 75}{5} = 1.4$

Of Further Interest: Linear Correlation

1. Because the grades generally increased as the number of tutoring sessions increased, there was some relationship between the two variables.

2. There was a strong positive linear correlation.

3. We can be 99% confident that there is a significant linear correlation between the variables of car weight and mileage.

4. It gives us a linear approximation of the set of data points.

5. positive correlation

6. significant negative correlation

7. a)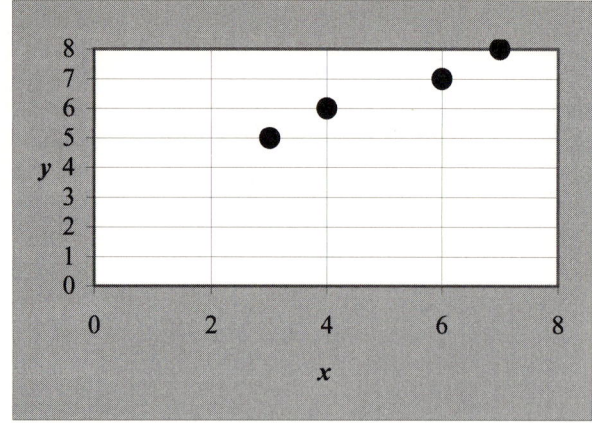

b) No solution provided

c)

x	y	x^2	y^2	xy
3	5	9	25	15
7	8	49	64	56
4	6	16	36	24
6	7	36	49	42
$\sum x = 20$	$\sum y = 26$	$\sum x^2 = 110$	$\sum y^2 = 174$	$\sum xy = 137$

$$r = \frac{n\sum xy - (\sum x)(\sum y)}{\sqrt{n(\sum x^2) - (\sum x)^2}\sqrt{n(\sum y^2) - (\sum y)^2}} = \frac{4 \cdot 137 - 20 \cdot 26}{\sqrt{4 \cdot 110 - 20^2}\sqrt{4 \cdot 174 - 26^2}}$$

$$= \frac{548 - 520}{\sqrt{440 - 400}\sqrt{696 - 676}} = \frac{28}{\sqrt{40}\sqrt{20}} = \frac{28}{\sqrt{800}} \approx 0.99$$

8. a)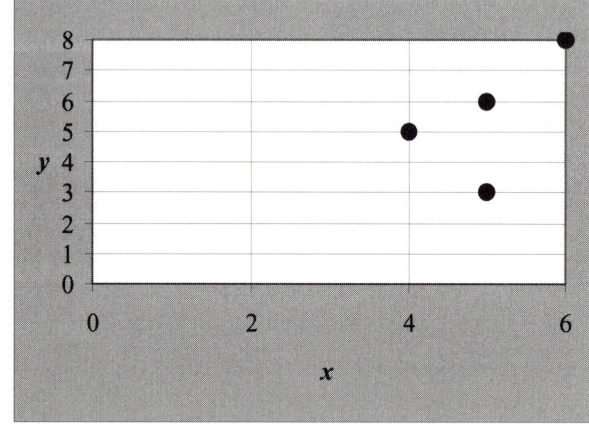

b) No solution provided

8. (continued)

c)

x	y	x^2	y^2	xy
4	5	16	25	20
6	8	36	64	48
5	6	25	36	30
5	3	25	9	15
$\sum x = 20$	$\sum y = 22$	$\sum x^2 = 102$	$\sum y^2 = 134$	$\sum xy = 113$

$$r = \frac{n\sum xy - (\sum x)(\sum y)}{\sqrt{n(\sum x^2) - (\sum x)^2}\sqrt{n(\sum y^2) - (\sum y)^2}} = \frac{4 \cdot 113 - 20 \cdot 22}{\sqrt{4 \cdot 102 - 20^2}\sqrt{4 \cdot 134 - 22^2}}$$

$$= \frac{452 - 440}{\sqrt{408 - 400}\sqrt{536 - 484}} = \frac{12}{\sqrt{8}\sqrt{52}} = \frac{12}{\sqrt{416}} \approx 0.59$$

9. a)

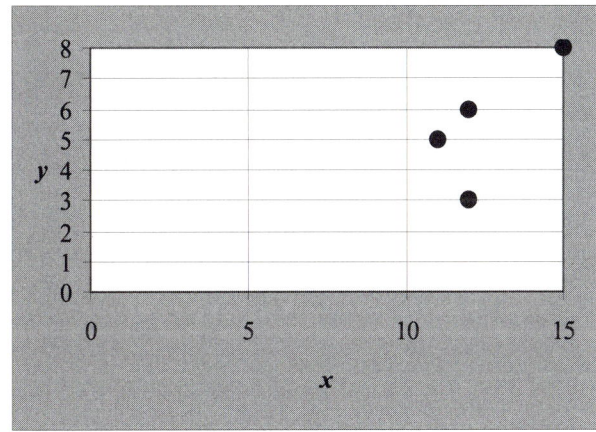

b) No solution provided

c)

x	y	x^2	y^2	xy
11	5	121	25	55
15	8	225	64	120
12	3	144	9	36
12	6	144	36	72
$\sum x = 50$	$\sum y = 22$	$\sum x^2 = 634$	$\sum y^2 = 134$	$\sum xy = 283$

$$r = \frac{n\sum xy - (\sum x)(\sum y)}{\sqrt{n(\sum x^2) - (\sum x)^2}\sqrt{n(\sum y^2) - (\sum y)^2}} = \frac{4 \cdot 283 - 50 \cdot 22}{\sqrt{4 \cdot 634 - 50^2}\sqrt{4 \cdot 134 - 22^2}}$$

$$= \frac{1{,}132 - 1{,}100}{\sqrt{2{,}536 - 2{,}500}\sqrt{536 - 484}} = \frac{32}{\sqrt{36}\sqrt{52}} = \frac{32}{\sqrt{1{,}872}} \approx 0.74$$

10. a)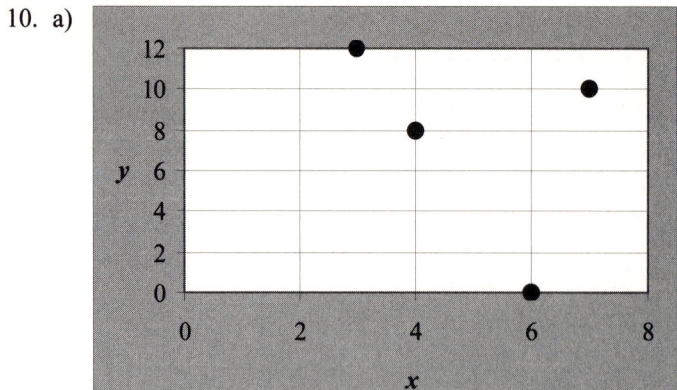

b) No solution provided

c)

x	y	x^2	y^2	xy
3	12	9	144	36
7	10	49	100	70
4	8	16	64	32
6	0	36	0	0
$\sum x = 20$	$\sum y = 30$	$\sum x^2 = 110$	$\sum y^2 = 308$	$\sum xy = 138$

$$r = \frac{n\sum xy - (\sum x)(\sum y)}{\sqrt{n(\sum x^2)-(\sum x)^2}\sqrt{n(\sum y^2)-(\sum y)^2}} = \frac{4 \cdot 138 - 20 \cdot 30}{\sqrt{4 \cdot 110 - 20^2}\sqrt{4 \cdot 308 - 30^2}}$$

$$= \frac{552-600}{\sqrt{440-400}\sqrt{1,232-900}} = \frac{-48}{\sqrt{40}\sqrt{332}} = \frac{-48}{\sqrt{13,280}} \approx -0.42$$

11. We can be 95% confident, but not 99% confident.

12. neither   13. neither

14. neither

15. $y = 0.7x + 3$

x	y	x^2	xy
3	5	9	15
7	8	49	56
4	6	16	24
6	7	36	42
$\sum x = 20$	$\sum y = 26$	$\sum x^2 = 110$	$\sum xy = 137$

$$m = \frac{n\sum xy - (\sum x)(\sum y)}{n(\sum x^2)-(\sum x)^2} = \frac{4 \cdot 137 - 20 \cdot 26}{4 \cdot 110 - 20^2} = \frac{548-520}{440-400} = \frac{28}{40} = 0.7$$

$$b = \frac{\sum y - m(\sum x)}{n} = \frac{26 - 0.7 \cdot 20}{4} = \frac{26-14}{4} = \frac{12}{4} = 3$$

16. $y = 1.5x - 2$

x	y	x^2	xy
4	5	16	20
6	8	36	48
5	3	25	15
5	6	25	30
$\sum x = 20$	$\sum y = 22$	$\sum x^2 = 102$	$\sum xy = 113$

$$m = \frac{n\sum xy - (\sum x)(\sum y)}{n(\sum x^2) - (\sum x)^2} = \frac{4 \cdot 113 - 20 \cdot 22}{4 \cdot 102 - 20^2} = \frac{452 - 440}{408 - 400} = \frac{12}{8} = 1.5$$

$$b = \frac{\sum y - m(\sum x)}{n} = \frac{22 - 1.5 \cdot 20}{4} = \frac{22 - 30}{4} = \frac{-8}{4} = -2$$

17. $y = 0.89x - 5.61$

x	y	x^2	xy
11	5	121	55
15	8	225	120
12	3	144	36
12	6	144	72
$\sum x = 50$	$\sum y = 22$	$\sum x^2 = 634$	$\sum xy = 283$

$$m = \frac{n\sum xy - (\sum x)(\sum y)}{n(\sum x^2) - (\sum x)^2} = \frac{4 \cdot 283 - 50 \cdot 22}{4 \cdot 634 - 50^2} = \frac{1{,}132 - 1{,}100}{2{,}536 - 2{,}500} = \frac{32}{36} = \frac{8}{9} \approx 0.89$$

$$b = \frac{\sum y - m(\sum x)}{n} = \frac{22 - \frac{8}{9} \cdot 50}{4} = \frac{198 - 8 \cdot 50}{36} = \frac{198 - 400}{36} = \frac{-202}{36} = -\frac{101}{18} \approx -5.61$$

18. $y = -1.2x + 13.5$

x	y	x^2	xy
3	12	9	36
7	10	49	70
4	8	16	32
6	0	36	0
$\sum x = 20$	$\sum y = 30$	$\sum x^2 = 110$	$\sum xy = 138$

$$m = \frac{n\sum xy - (\sum x)(\sum y)}{n(\sum x^2) - (\sum x)^2} = \frac{4 \cdot 138 - 20 \cdot 30}{4 \cdot 110 - 20^2} = \frac{552 - 600}{440 - 400} = \frac{-48}{40} = -1.2$$

$$b = \frac{\sum y - m(\sum x)}{n} = \frac{30 - (-1.2) \cdot 20}{4} = \frac{30 + 24}{4} = \frac{54}{4} = 13.5$$

19. We can be 99% confident that there is positive linear correlation.

x	y	x^2	y^2	xy
0	23	0	529	0
1	22	1	484	22
2	27	4	729	54
2	28	4	784	56
5	35	25	1,225	175
$\sum x = 10$	$\sum y = 135$	$\sum x^2 = 34$	$\sum y^2 = 3,751$	$\sum xy = 307$

$$r = \frac{n\sum xy - (\sum x)(\sum y)}{\sqrt{n(\sum x^2)-(\sum x)^2}\sqrt{n(\sum y^2)-(\sum y)^2}} = \frac{5 \cdot 307 - 10 \cdot 135}{\sqrt{5 \cdot 34 - 10^2}\sqrt{5 \cdot 3,751 - 135^2}}$$

$$= \frac{1,535 - 1,350}{\sqrt{170 - 100}\sqrt{18,755 - 18,225}} = \frac{185}{\sqrt{70}\sqrt{530}} = \frac{185}{\sqrt{37,100}} \approx 0.96$$

20. neither

x	y	x^2	y^2	xy
901	44	811,801	1,936	39,644
903	35	815,409	1,225	31,605
1,040	30	1,081,600	900	31,200
994	29	988,036	841	28,826
993	26	986,049	676	25,818
$\sum x = 4,831$	$\sum y = 164$	$\sum x^2 = 4,682,895$	$\sum y^2 = 5,578$	$\sum xy = 157,093$

$$r = \frac{n\sum xy - (\sum x)(\sum y)}{\sqrt{n(\sum x^2)-(\sum x)^2}\sqrt{n(\sum y^2)-(\sum y)^2}} = \frac{5 \cdot 157,093 - 4,831 \cdot 164}{\sqrt{5 \cdot 4,682,895 - 4,831^2}\sqrt{5 \cdot 5,578 - 164^2}}$$

$$= \frac{785,465 - 792,284}{\sqrt{23,414,475 - 23,338,561}\sqrt{27,890 - 26,896}} = \frac{-6,819}{\sqrt{75,914}\sqrt{994}} = \frac{-6,819}{\sqrt{75,458,516}} \approx -0.78$$

21. neither

x	y	x^2	y^2	xy
30	26	900	676	780
40	31	1,600	961	1,240
50	33	2,500	1,089	1,650
60	31	3,600	961	1,860
70	26	4,900	676	1,820
$\sum x = 250$	$\sum y = 147$	$\sum x^2 = 13,500$	$\sum y^2 = 4,363$	$\sum xy = 7,350$

$$r = \frac{n\sum xy - (\sum x)(\sum y)}{\sqrt{n(\sum x^2)-(\sum x)^2}\sqrt{n(\sum y^2)-(\sum y)^2}} = \frac{5 \cdot 7,350 - 250 \cdot 147}{\sqrt{5 \cdot 13,500 - 250^2}\sqrt{5 \cdot 4,363 - 147^2}}$$

$$= \frac{36,750 - 36,750}{\sqrt{67,500 - 62,500}\sqrt{21,815 - 21,609}} = \frac{0}{\sqrt{5,000}\sqrt{206}} = \frac{0}{\sqrt{1,030,000}} = 0.00$$

22. neither

x	y	x^2	y^2	xy
8	62.7	64	3,931.29	501.6
10	61.1	100	3,733.21	611.0
12	65.5	144	4,290.25	786.0
16	57.6	256	3,317.76	921.6
18	53.7	324	2,883.69	966.6
$\sum x = 64$	$\sum y = 300.6$	$\sum x^2 = 888$	$\sum y^2 = 18,156.2$	$\sum xy = 3,786.8$

$$r = \frac{n\sum xy - (\sum x)(\sum y)}{\sqrt{n(\sum x^2) - (\sum x)^2}\sqrt{n(\sum y^2) - (\sum y)^2}} = \frac{5 \cdot 3,786.8 - 64 \cdot 300.6}{\sqrt{5 \cdot 888 - 64^2}\sqrt{5 \cdot 18,156.2 - 300.6^2}}$$

$$= \frac{18,934 - 19,238.4}{\sqrt{4,440 - 4,096}\sqrt{90,781 - 90,360.36}} = \frac{-304.4}{\sqrt{344}\sqrt{420.64}} = \frac{-304.4}{\sqrt{144,700.16}} \approx -0.80$$

23. $y = 2.64x + 21.71$

x	y	x^2	xy
0	23	0	0
1	22	1	22
2	27	4	54
2	28	4	56
5	35	25	175
$\sum x = 10$	$\sum y = 135$	$\sum x^2 = 34$	$\sum xy = 307$

$$m = \frac{n\sum xy - (\sum x)(\sum y)}{n(\sum x^2) - (\sum x)^2} = \frac{5 \cdot 307 - 10 \cdot 135}{5 \cdot 34 - 10^2} = \frac{1,535 - 1,350}{170 - 100} = \frac{185}{70} = \frac{37}{14} \approx 2.64$$

$$b = \frac{\sum y - m(\sum x)}{n} = \frac{135 - \frac{37}{14} \cdot 10}{5} = \frac{1,890 - 370}{70} = \frac{1,520}{70} = \frac{152}{7} \approx 21.71$$

24. $y = -0.09x + 119.59$ *

x	y	x^2	xy
901	44	811,801	39,644
903	35	815,409	31,605
1,040	30	1,081,600	31,200
994	29	988,036	28,826
993	26	986,049	25,818
$\sum x = 4,831$	$\sum y = 164$	$\sum x^2 = 4,682,895$	$\sum xy = 157,093$

$$m = \frac{n\sum xy - (\sum x)(\sum y)}{n(\sum x^2) - (\sum x)^2} = \frac{5 \cdot 157,093 - 4,831 \cdot 164}{5 \cdot 4,682,895 - 4,831^2} = \frac{785,465 - 792,284}{23,414,475 - 23,338,561} = \frac{-6819}{75,914} \approx -0.09$$

$$b = \frac{\sum y - m(\sum x)}{n} = \frac{164 - \left(-\frac{6819}{75,914}\right) \cdot 4,831}{5} = \frac{12,449,896 + 32,942,589}{379,570} = \frac{45,392,485}{379,570} \approx 119.59$$

*If you use a four decimal approximation for m (-0.0898), the b value will be 119.56 (rounded).

412 Chapter 14: Descriptive Statistics

25. $y = 0 \cdot x + 29.4$

x	y	x^2	xy
30	26	900	780
40	31	1600	1240
50	33	2500	1650
60	31	3600	1860
70	26	4900	1820
$\sum x = 250$	$\sum y = 147$	$\sum x^2 = 13{,}500$	$\sum xy = 7{,}350$

$$m = \frac{n\sum xy - (\sum x)(\sum y)}{n(\sum x^2) - (\sum x)^2} = \frac{5 \cdot 7{,}350 - 250 \cdot 147}{5 \cdot 13{,}500 - 250^2} = \frac{36{,}750 - 36{,}750}{67{,}500 - 62{,}500} = \frac{0}{5{,}000} = 0$$

$$b = \frac{\sum y - m(\sum x)}{n} = \frac{147 - 0 \cdot 250}{5} = \frac{147 - 0}{5} = \frac{147}{5} = 29.4$$

26. $y = -0.88x + 71.45$

x	y	x^2	xy
8	62.7	64	501.6
10	61.1	100	611
12	65.5	144	786
16	57.6	256	921.6
18	53.7	324	966.6
$\sum x = 64$	$\sum y = 300.6$	$\sum x^2 = 888$	$\sum xy = 3{,}786.8$

$$m = \frac{n\sum xy - (\sum x)(\sum y)}{n(\sum x^2) - (\sum x)^2} = \frac{5 \cdot 3{,}786.8 - 64 \cdot 300.6}{5 \cdot 888 - 64^2} = \frac{18{,}934 - 19{,}238.4}{4{,}440 - 4{,}096} = \frac{-304.4}{344} = -\frac{761}{860} \approx -0.88$$

$$b = \frac{\sum y - m(\sum x)}{n} = \frac{300.6 - \left(-\frac{761}{860}\right) \cdot 64}{5} = \frac{258{,}516 + 48{,}704}{4{,}300} = \frac{307{,}220}{4{,}300} \approx 71.45$$

27. No solution provided

28. 1

29. 1